THEORETICAL AND COMPUTATIONAL AERODYNAMICS

Aerospace Series List

Title	Author	Date
Theoretical and Computational Aerodynamics	Sengupta	September 2014
Aerospace Propulsion	Lee	October 2013
Aircraft Flight Dynamics and Control	Durham	August 2013
Civil Avionics Systems, 2nd Edition	Moir, Seabridge and Jukes	August 2013
Modelling and Managing Airport Performance	Zografos, Andreatta and Odoni	July 2013
Advanced Aircraft Design: Conceptual Design, Analysis and Optimization of Subsonic Civil Airplanes	Torenbeek	June 2013
Design and Analysis of Composite Structures: With Applications to Aerospace Structures, 2nd Edition	Kassapoglou	April 2013
Aircraft Systems Integration of Air-Launched Weapons	Rigby	April 2013
Design and Development of Aircraft Systems, 2nd Edition	Moir and Seabridge	November 2012
Understanding Aerodynamics: Arguing from the Real Physics	McLean	November 2012
Aircraft Design: A Systems Engineering Approach	Sadraey	October 2012
Introduction to UAV Systems 4e	Fahlstrom and Gleason	August 2012
Theory of Lift: Introductory Computational Aerodynamics with MATLAB and Octave	McBain	August 2012
Sense and Avoid in UAS: Research and Applications	Angelov	April 2012
Morphing Aerospace Vehicles and Structures	Valasek	April 2012
Gas Turbine Propulsion Systems	MacIsaac and Langton	July 2011
Basic Helicopter Aerodynamics, 3rd Edition	Seddon and Newman	July 2011
Advanced Control of Aircraft, Spacecraft and Rockets	Tewari	July 2011
Cooperative Path Planning of Unmanned Aerial Vehicles	Tsourdos et al	November 2010
Principles of Flight for Pilots	Swatton	October 2010
Air Travel and Health: A Systems Perspective	Seabridge et al	September 2010
Design and Analysis of Composite Structures: With applications to aerospace Structures	Kassapoglou	September 2010
Unmanned Aircraft Systems: UAVS Design, Development and Deployment	Austin	April 2010
Introduction to Antenna Placement & Installations	Macnamara	April 2010
Principles of Flight Simulation	Allerton	October 2009
Aircraft Fuel Systems	Langton et al	May 2009
The Global Airline Industry	Belobaba	April 2009
Computational Modelling and Simulation of Aircraft and the Environment: Volume 1 – Platform Kinematics and Synthetic Environment	Diston	April 2009
Handbook of Space Technology	Ley, Wittmann Hallmann	April 2009
Aircraft Performance Theory and Practice for Pilots	Swatton	August 2008
Aircraft Systems, 3rd Edition	Moir & Seabridge	March 2008
Introduction to Aircraft Aeroelasticity And Loads	Wright & Cooper	December 2007
Stability and Control of Aircraft Systems	Langton	September 2006
Military Avionics Systems	Moir & Seabridge	February 2006
Design and Development of Aircraft Systems	Moir & Seabridge	June 2004
Aircraft Loading and Structural Layout	Howe	May 2004
Aircraft Display Systems	Jukes	December 2003
Civil Avionics Systems	Moir & Seabridge	December 2002

THEORETICAL AND COMPUTATIONAL AERODYNAMICS

Tapan K. Sengupta
IIT Kanpur, India

This edition first published 2015

Registered office
John Wiley & Sons Ltd, The Atrium, Southern Gate, Chichester, West Sussex, PO19 8SQ, United Kingdom

For details of our global editorial offices, for customer services and for information about how to apply for permission to reuse the copyright material in this book please see our website at www.wiley.com.

Library of Congress Cataloging-in-Publication Data

Sengupta, Tapan Kumar, 1955-
Theoretical and computational aerodynamics / Tapan K. Sengupta.
pages cm
Includes bibliographical references and index.
ISBN 978-1-118-78759-5 (cloth)
1. Aerodynamics. I. Title.
TL570.S4426 2015
629.132′3001–dc23

2014015240

A catalogue record for this book is available from the British Library.

ISBN: 978-1-118-78759-5

Set in 10/12 TimesLTStd-Roman by Thomson Digital, Noida, India

1 2015

I would like to dedicate this book to all my family members since my childhood, whose many sacrifices helped me reach here; without their support and care this would not have been possible.

Contents

Series Preface

The field of aerospace is multidisciplinary and wide ranging, covering a large variety of products, disciplines and domains, not merely in engineering but in many related supporting activities. These combine to enable the aerospace industry to produce exciting and technologically advanced vehicles. The wealth of knowledge and experience that has been gained by expert practitioners in the various aerospace fields needs to be passed onto others working in the industry, including those just entering from university.

The *Aerospace Series* aims to be a practical, topical and relevant series of books aimed at people working in the aerospace industry, including engineering professionals and operators, allied professions such commercial and legal executives, and also engineers in academia. The range of topics is intended to be broad, covering design and development, manufacture, operation and support of aircraft, as well as topics such as infrastructure operations and developments in research and technology.

Aerodynamics is the fundamental enabling science that underpins the worldwide aerospace industry–without the ability to generate lift from airflow passing over aircraft wings, helicopter rotor blades and jet engine turbine blades, it would not be possible to fly the sophisticated heavier-than-air vehicles that we take for granted nowadays. Much of the development of current highly efficient aircraft has been due to the ability to model aerodynamic flows accurately, and thus to design high performance wings.

This book, *Theoretical and Computational Aerodynamics*, provides a detailed description of the underlying mathematical aerodynamic models and their computation with application to fixed wing aircraft. It is a very welcome addition to the Wiley Aerospace Series. Starting with the basic principles, the book takes the reader through aerodynamic theories in increasing complexity, from potential flow theory through to boundary layers and the Navier-Stokes equations. Of particular note are the sections relating to computational aerodynamics for incompressible and compressible flow around airfoils across the entire speed range, and a number of specific applications including drag reduction, low Reynolds number aerodynamics, high lift devices and flow control.

Peter Belobaba, Jonathan Cooper and Allan Seabridge

Preface

In writing the preface of a book, the author gets another opportunity to share with readers the thinking that motivates one to write the book. This is more so in the area of aerodynamics, where many books have been written and the reader is spoilt for choice, if not bewildered, by the multiplicity of textbooks. As the author, I would like to take the opportunity to explain what I have done here, hoping that readers understand this. I note that most books on aerodynamics emphasize classical approximate methods and less effort is spent on topics of current interests, such as transonic flows, low Reynolds number flows, natural laminar flow (NLF) airfoils and drag reductions.

Aerodynamics as a subject has been developed over more than a century and has seen extraordinary growth during this period. The Aeronautical Society of Great Britain, formed in 1866, stated as one of its initial goals the development of methods to estimate the lift and drag experienced on a flat plate! Although George Cayley laid the foundation of flight at the beginning of the nineteenth century, indicating the plausible shapes of wing, concepts of flight stability etc., it is also equally true that the correct theoretical models of fluid flow were established by Euler, Navier, Saint Venant and Stokes by 1850 (see the excellent book on the history of aerodynamics by J. D. Anderson for further details). However, these could not be used due to theoretical intractability and the search was on for further simplified models of fluid flows. This originated with the concept of velocity potential by Lagrange and its practical utility shown by Helmholtz through his vortex theorems. This route was followed over the decades with insights from conformal mapping, thin airfoil theory, finite wing theory for potential flows, which still form the bedrock of understanding the subject, as discussed here. The concurrent development of boundary layer theory to embed viscous effects helps bridge the chasm between ideal and real fluid flow as an aid in the analysis and design of aircraft for its steady state operation by focusing more on lifting surface behaviour.

Spectacular progress has been made in aerodynamics due to the growth of computing which significantly reduced the design cycle time of aircraft. Today it constitutes the preliminary step in a new design, relegating costly, time-consuming wind tunnel testing to the final stages. But a major stumbling block in understanding viscous flow is the slower progress in understanding transition and turbulence, without which the correct estimate of drag is impossible. Methods used for many decades have been semi-empirical, yet the linear viscous instability has helped in developing NLF airfoils currently used in civil aviation. Now this last barrier has been crossed and one can solve canonical flow transition problems by solving the Navier-Stokes equation from first principles, without any models.

Since the 1960s, computers have aided aerodynamics in studying viscous-inviscid interaction, a very useful analysis tool of the complete vehicle. One cannot overstate the role of panel methods to analyse complete aircraft at low speeds. Panel methods can also be used in the supersonic flight regime, which allows linearization. However, the transonic flight (the most efficient flight speed) regime evaded proper analysis until the early 1970s. Since then, the development of transonic small disturbance theory and full potential equation have made transonic flow analyses by computers amenable for aircraft and rotor craft applications. This was followed by computing Euler equation for full aircraft in the 1980s. Concurrently, computers also enabled early developments of studying fully developed time-averaged turbulent flows by solving the Reynolds averaged Navier-Stokes (RANS) equation with various turbulence models. The results thus obtained still do not allow one to study off-design performances or vehicles that depend upon unsteady aerodynamics. We also note that historically incompressible and compressible flows have been solved as belonging to distinct branches of fluid dynamics in analytical and experimental framework. This also prompted computing such flows as distinct activities. The present book shows this as unnecessary and the same computing methods work for compressible and incompressible flows, even for transonic flows.

Spectacular progress in scientific computing have enabled the study of unsteady flows using large eddy simulation (LES) and direct numerical simulation (DNS) as a unified activity. It is now feasible to study flow past aerofoils for unsteady flight conditions or even to study the flow transition from laminar to turbulent states by DNS. Such studies can refine NLF airfoils, as we understand transition phenomena better than before. Today we can simulate very high Reynolds number flows past NLF aerofoils and know how to delay transition. In the present context of high fuel cost, aerodynamics is still the pacing item for efficient aerodynamic design.

Thus, we are seeing progress in aerodynamics studies which was impossible for practical parameter ranges. For example, earlier developed theoretical analysis tools were for steady/time-averaged flows at high Reynolds numbers. Now, advances have been made for low Reynolds numbers for truly unsteady flight, such as in the flights of insects, birds and micro air vehicles. Further progress in these areas is possible by DNS/LES of such flow fields. These are all dealt with in this book.

This book can be used as a single text for two modules of courses on aerodynamics. In the first module, all chapters dealing with theoretical approaches for incompressible flows given in Chapters 1 to 7 can be covered (omitting the sections on compressible flows in Chapter 2). This can be supplemented with Chapter 13, discussing only the high lift devices part. Time permitting, one can include flow instability, as in Sections 9.1 to 9.4, and additional materials in Chapter 10 on NLF airfoils. The second module can contain Chapter 2 (as review and preparatory material for compressible flows) and from Chapters 8 to 13, for an advanced level course on the subject. This same module can also be taught as a course on computational aerodynamics by including Chapters 3 to 5.

Some of the chapters from this book can also be adapted for a special course on aerodynamics, emphasizing scientific computing as used here for transonic and low Reynolds number flows. While students can use commercial codes available to obtain time-averaged flow fields for conventional aircrafts (with limited accuracy and usage) using models for transition and turbulence, such tools are practically useless for low Reynolds number flows, which may seem easier as compared to high Reynolds number flows. Readers should note that the same numerical method has been used for transonic flow, which has been used for incompressible flow and this allows a very easy introduction to transonic flows, without going through the substantial efforts which have been invested in developing approximate theories involving transonic small perturbation theory and full potential equation.

As indicated above, there are new elements in this present book, which are not covered elsewhere. However, an aerospace professional should know these as essential minimum, which the author felt while teaching basic courses on aerodynamics at IIT Kanpur. This has motivated the author to write this book and hopefully this will be a welcome addition to the existing literature on the subject.

Any such undertaking requires the help and generosity of many peers, and also from various students and colleagues at IIT Kanpur. I am particularly indebted to Prof. T. T. Lim in providing various figures on vortex bursting in Chapter 6. Prof. Huu Duc Vo and Prof. Eric Laurendeau are gratefully acknowledged for various discussions and help with plasma models and transonic flows. I acknowledge the NPMICAV program of DRDO by supporting research on low Reynolds number aerodynamics and also acknowledge the help of Dr Mudkavi, the head of CTFD division of NAL Bangalore in providing many figures providing panel method results for Dornier DO-228 aircraft. All the results in Chapter 11 on transonic flows are due to the combined efforts of N. A. Sreejith and Ashish Bhole, two very competent graduate students. The results in Chapter 12 have been provided by Satish B. Krishnan and Pramod M. Bagade. Also acknowledged are Dr Y. G. Bhumkar and Dr Swagata Bhaumik, who provided many figures continuously. Secretarial assistance in typing using LaTeX has been done by Mrs Shashi Shukla and Baby Gaur.

I also take the opportunity to express my sincere appreciation for the unstinting support provided by my family, without which this would have been impossible. I would also like to express my gratitude to all other past and present students who have been generous with many suggestions. I accepts all responsibility for any errors or omissions and invites all readers and users of this book to point out inadvertent mistakes and provide critiques to improve the book further. Finally, I have only words of appreciation for the professional yet so accommodative team from Wiley. It is a pleasant duty to acknowledge the help provided by Tom Carter and Anne Hunt of Wiley and their team members.

Tapan K. Sengupta
April 2014

Acknowledgements

I am deeply indebted for the help and generosity of many peers, and also from various students and colleagues. Particular mention must be made for the help from Prof. T. T. Lim, Prof. Huu Duc Vo and Prof. Eric Laurendeau in the form of materials, suggestions and various discussions. I also acknowledge, ARDB and NPMICAV program of DRDO, Govt. of India, who supported researches on NLF airfoils for bypass transition, plasma actuation for flow control and low Reynolds number aerodynamics relevant for micro air vehicles. I would fail in my duty if I do not especially acknowledge the great help received from Dr. Mudkavi, the head of CTFD division, NAL Bangalore in providing many figures for panel method results obtained for Dornier DO-228 aircraft and discussions on various topics covered in the book. It is also my pleasant duty to acknowledge help provided by Tom Carter and Anne Hunt of Wiley and their team members in preparing the manuscript.

1

Introduction to Aerodynamics and Atmosphere

1.1 Motivation and Scope of Aerodynamics

Study of aerodynamics involves the ability to predict aerodynamic forces and moments acting on an airborne vehicle. However, it all began with the search for the quintessential shape that will make anything airborne in a sustained manner. Historically, the search for human flight began with lighter than air vehicles, now known variously as aerostats. While airships, blimps and/or dirigibles are still in use, the original search for vehicles heavier than air was the main attraction for human flight. In our quest for flight, we always wanted to emulate the birds, but even today this appears unattainable with present-day technologies. A bird flies in which the flapping wing performs the dual role of propulsive and aerodynamic devices. In man-made devices the propulsive device produces power or thrust for the vehicle to overcome resistance, while the aerodynamic device creates the necessary force to keep the body aloft in a dynamical equilibrium. Any device that imitates the flight of birds is known as an ornithopter and it is to the genius of Sir George Cayley (1773–1857) mankind owes a debt for the conventional aircraft shape and design. In a marked departure, he suggested that such propulsive devices did not exist and that it was more important to understand the analysis and design of aerodynamic aspect of the vehicle first. To do so, he advocated the study of powerless flight of aeronautical shapes of interest in a stable manner. This way of compartmentalizing the different aspects of flight into aerodynamics, propulsion, structures, performance, stability and control is now one essential component in the study of the discipline and was started by the need to understand the basics of flight as pioneered by Cayley. For an historic account of the development of flight and aerodynamics in particular, readers are advised to study it in Anderson (1997).

In this book, the sole motivation would be to study the aerodynamic forces and moments acting on an aircraft. Before we discuss the motivation of studying aerodynamics further, we should familiarize ourselves with different parts of an aircraft and their roles in flight. In Figure 1.1, we show the different main parts of a small aircraft as viewed externally.

Theoretical and Computational Aerodynamics, First Edition. Tapan K. Sengupta.

Companion Website: www.wiley.com/go/sengupta

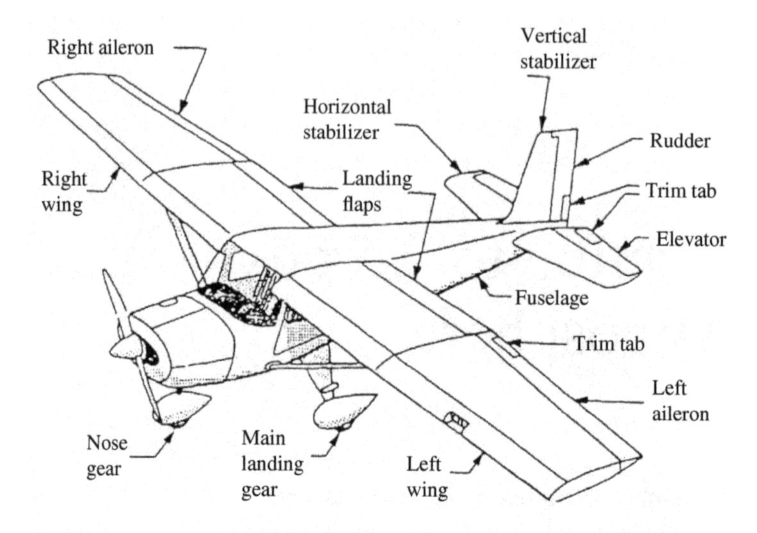

Figure 1.1 An external view of a small propeller aircraft with tricycle landing gear, identifying different parts of aerodynamic and control surfaces

The external shape of an aircraft is central to the study of aerodynamics, flight stability and controls. As an aircraft is heavier than the ambient displaced air, various surface stresses acting on different parts of the aircraft should be such that there must be a resultant force acting which will sustain the weight of the aircraft, when the aircraft is flying level and steady. This resultant force acting over the aircraft sustaining the weight is called the lift force. In a conventional aircraft, lift force is created by the flow over the wing. This is often modelled by the normal stress or the static pressure acting on the wing surface and is obtained by considering an ideal flow over the wing. When and how this is possible, will be the recurring theme of this book. Such normal stress or pressure distribution acting over the wing also creates a moment acting about a general point along the chord of the wing section and is a concomitant liability of producing the lift force. This is counteracted upon by the smaller amount of lift created on the horizontal stabilizer. One realizes that the main purpose of an aerospace vehicle is to carry payload and this is the reason for having a fuselage. Such a tubular shape of the fuselage, along with different aerodynamic and control surfaces, also creates resistance or drag for the aircraft. The required thrust to overcome drag is provided by the propeller engine located at the nose of the aircraft. The horizontal and vertical stabilizers are required for maintaining the stability of the aircraft for different flight regime.

One of the rudimentary aspects of flight operation is the cruise configuration, in which the aircraft flies level and steady at constant altitude. In Figure 1.2, we show a conventional aircraft flying level and steady in the longitudinal plane. External forces are shown to be in equilibrium, i.e. the weight of the aircraft is balanced by the vertical component of the aerodynamic forces acting on the aircraft called the lift and denoted by L. Most of it is created by the aircraft wing. The horizontal component of the aerodynamic force is termed the drag and denoted by D. Almost half the drag of an aircraft is created by the fuselage and a large portion of it is caused by the creation of lift itself. The total drag is overcome by the thrust

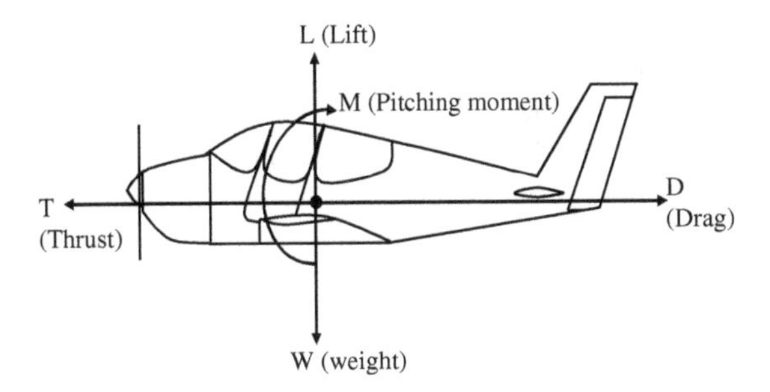

Figure 1.2 A conventional aircraft flying level and steady in the longitudinal plane. Forces and moments acting on the aircraft are shown in equilibrium; the pitching moment is balanced by the lift acting on the tailplane

(indicated by T), created by the power plant. Thrust is also a form of aerodynamic force created by internal flows through the engine where energy is added by burning fuel and ejecting a well-directed stream of fluid which by reaction creates the thrust. But this is outside the scope of this book and will not be discussed further.

Note that the drawn free body diagram in Figure 1.2, is an idealization as the forces are not collinear and also the forces and moment created by the stabilizer are not shown in this diagram. At the outset, we want to state the directions of the aerodynamic force as consisting of: the drag force, which is always along the oncoming flow direction, and lift, which is perpendicular to it. For a general motion of the aircraft, additionally another component of aerodynamic force will be experienced, which is perpendicular to both lift and drag and is termed as the side force. The moment experienced by the aircraft in the longitudinal plane has been indicated by the pitching moment. For general motion of the rigid aircraft will give rise to two more components of moment termed as the rolling (about the fuselage axis) and yawing (about the vertical axis of the aircraft) moments. In Figure 1.3, we show a sketch of all three forces and three moments relevant for a rigid aircraft. The reader is made aware of the fact that the aircraft is hardly ever a rigid body and in general the problem constitutes an

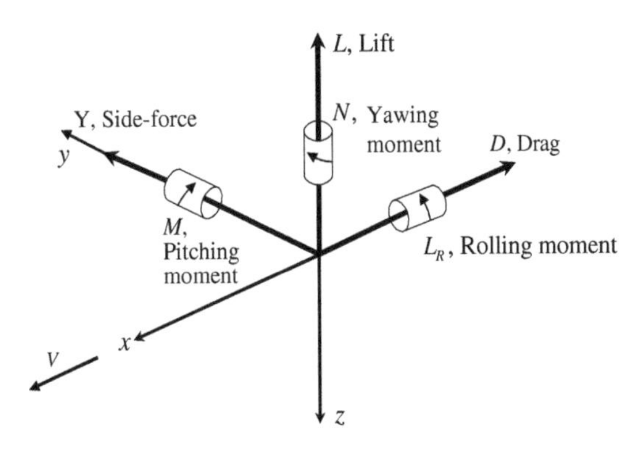

Figure 1.3 Axes and coordinate system used for any rigid aircraft. Forces and moments acting on the aircraft are shown with conventional usage

analysis for very large degrees of freedom. (The static tip deflection of Boeing 747 in cruise from the static position in ground is more than 12 feet!) This constitutes the interesting field of study called aero-elasticity where simultaneous consideration of fluid dynamics and structural dynamics are taken into account. When the design of aircraft is complicated by the fact that the stability and control of the aircraft depends and directly interacts with fluid and structural dynamics of the aircraft, then one is forced to consider aero-servo-elasticity. However creating additional complications by various forces interacting is important for many applications, but the primary goal of this book is to study the aerodynamics of simple lifting surfaces. The moot point is: How have such aerodynamic surfaces evolved to their present state of development?

1.2 Conservation Principles

We start this discussion with the innocuous question: How does an aerodynamic surface creates lift? Answer to this is related to the simple observation that if a bound vortex induces a circulatory motion in a uniform steady flow by an aerodynamic surface, the resultant superposition of the two yields the flow pattern seen around a lifting airfoil with the flow leaving the trailing edge with a small downward component of velocity. But will any bound vortex do, as in the case of a rotating cylinder in a uniform flow experiencing the Robins-Magnus effect, as explained in White (2008)? It will simply not work due to the fact that, apart from creating lift via rotation, such a body (also referred to as a bluff body) will experience large value of drag. Any aerodynamic surface, to perform efficiently, must provide not only lift but must also yield a very high value of lift to drag ratio. This ratio is also called the aerodynamic efficiency. Instead of getting into technical details of aerofoil aerodynamics, let us first understand what is expected from an aerodynamic surface from first principle of conservation laws in fluid dynamics. For low-speed applications, these would be nothing more than conservation of mass and momentum.

1.2.1 Conservation Laws and Reynolds Transport Theorem (RTT)

All conservation laws used in mechanics are nothing but interaction between system and surrounding, which are separated by real or imaginary boundaries. If the system mass is m, then for a control mass system the conservation of mass implies

$$\frac{dm}{dt} = 0. \tag{1.1}$$

If the system is moving with a translational velocity, $\vec{V}$, then the conservation of linear momentum is given by

$$\vec{F} = \frac{d}{dt}(m\vec{V}), \tag{1.2}$$

where $\vec{F}$ is the vector sum of all the applied forces acting on the system. In fluid mechanics or aerodynamics, above control mass analysis is often of limited value, as we do not track individual fluid particles. Instead we focus upon a fixed space in the flow domain and this is the rationale for control volume analysis. We will make use of the Reynolds transport theorem as given in White (2008) for an arbitrary moving and deformable control volume by noting the following. Let P be the property of the fluid (which could be mass, momentum or energy) and the corresponding intensive property be β, i.e. the property per unit mass. Considering incompressible flow with density, ρ of a control volume (CV) defined by $\mathcal{V}$, the property in the control volume will be denoted as $P_{CV} = \int_{CV} \beta\rho \, d\mathcal{V}$. The rate of change of this property of a control mass system $\left[\frac{dP_{sys}}{dt}\right]$ will be determined by, (i) a change within the control volume $\left[\frac{dP_{CV}}{dt}\right]$; (ii) outflow of β from the CV through the control surface (CS): $\int_{CS} \beta\rho\vec{V}\cos\theta \, dA_{out}$, where θ is the angle the elementary surface area dA_{out} makes with the flow velocity at the outflow and (iii) inflow of β to the CV: $\int_{CS} \beta\rho\vec{V}\cos\theta \, dA_{in}$. Written as a mathematical equation, this turns out to be

$$\frac{dP_{sys}}{dt} = \frac{dP_{CV}}{dt} + \int_{CS} \beta\rho\vec{V}\cos\theta \, dA_{out} - \int_{CS} \beta\rho\vec{V}\cos\theta \, dA_{in}. \tag{1.3}$$

The last two terms taken together constitute the net flux in the CV, which could also be written out as $\int \beta \, d\dot{m}_{out} - \int \beta \, d\dot{m}_{in}$, where $d\dot{m} = \rho(\vec{V}\cdot\hat{n})\, dA$, with the unit vector $\hat{n}$ changes sign from inflow to outflow ports. This is the Reynolds transport theorem written in compact form as

$$\frac{dP_{sys}}{dt} = \frac{dP_{CV}}{dt} + \int_{CS} \beta\rho(\vec{V}\cdot\hat{n})\, dA. \tag{1.4}$$

If the CV of fixed shape move at a uniform velocity $\vec{V}_s(t)$, then any observer fixed to the CV will note the velocity of fluid crossing CS as $\vec{V}_r = \vec{V} - \vec{V}_s(t)$ which will be used in calculating the flux term, so that the Reynolds transport theorem in this case will be written as

$$\frac{dP_{sys}}{dt} = \frac{dP_{CV}}{dt} + \int_{CS} \beta\rho(\vec{V}_r\cdot\hat{n})\, dA. \tag{1.5}$$

Finally, if the CV is arbitrary deformable and move with space-time varying speed, $\vec{V}_s(r,t)$, so that $\vec{V}_r = \vec{V} - \vec{V}_s(r,t)$, then Reynolds transport equation is stated as in Equation (1.5).

Example 1.1: To derive the conservation of mass, we note that $P = m$ and then $\beta = 1$. Hence, $\frac{dP_{sys}}{dt} = 0$ and Reynolds transport theorem provides

$$0 = \frac{d}{dt}\left(\int_{CV} \rho \, d\mathcal{V}\right) + \int_{CS} \rho\,(\vec{V}_r\cdot\hat{n})\, dA. \tag{1.6}$$

For a fixed CV, this further simplifies to

$$0 = \frac{d}{dt}\left(\int_{CV} \rho \, d\mathcal{V}\right) + \int_{CS} \rho\,(\vec{V}\cdot\hat{n})\, dA. \tag{1.7}$$

For a fixed CV, if the flow is steady, $\frac{\partial\rho}{\partial t} = 0$ and Equation (1.7) further simplifies to

$$\int_{CS} \rho(\vec{V}\cdot\hat{n})\, dA = 0. \tag{1.8}$$

If one uses Gauss's divergence theorem, then the above further simplifies to $\int_{CV} \text{Div}(\rho \vec{V})\, d\mathcal{V} = 0$, i.e.

$$\nabla \cdot (\vec{V}) = 0. \tag{1.9}$$

1.2.2 Application of RTT: Conservation of Linear Momentum

In this case, $P = m\vec{V}$ and $\beta = \vec{V}$. Hence, Reynolds transport theorem for conservation of linear momentum is given by

$$\frac{d(m\vec{V})_{sys}}{dt} = \sum \vec{F} = \frac{d}{dt}\left(\int_{CV} \rho \vec{V}\, d\mathcal{V}\right) + \int_{CS} \rho \vec{V}(\vec{V}_r \cdot \hat{n})\, dA, \tag{1.10}$$

where $\vec{F}$ represent all body and surface forces. As $\vec{V}$ is referred to inertial frame, the x-component of Equation (1.10) can be written as

$$\sum F_x = \frac{d}{dt}\left(\int_{CV} \rho\, u\, d\mathcal{V}\right) + \int_{CS} \rho u\, (\vec{V}_r \cdot \hat{n})\, dA. \tag{1.11}$$

Now for a fixed CV, $\vec{V}_r \equiv \vec{V}$, the conservation of linear momentum equation can be written in vector notation as

$$\sum \vec{F} = \frac{d}{dt}\left(\int_{CV} \rho \vec{V}\, d\mathcal{V}\right) + \int_{CS} \rho \vec{V}(\vec{V} \cdot \hat{n})\, dA. \tag{1.12}$$

1.3 Origin of Aerodynamic Forces

To begin with, we lay down the principles by which an efficient aerodynamic surface generates lift, while producing significantly lower drag. Let us consider the aerodynamic device as shown in Figure 1.4 in the block diagram, which shows the input of the device indicated by a mass flow rate, $\dot{m}$ and a uniform flow, V with the direction aligned with the x-axis. The output of the device is also characterized by the same mass flow rate and speed. However, the velocity vector is indicated by a small turning angle, θ at the outflow. The aerodynamic device is characterized by the same cross-sectional area at the inflow and outflow as A.

The free body diagram, also shown in Figure 1.4, indicates that the momentum passing through the inflow to the interior of the control volume (identified by the *black-box*, termed as the aerodynamic device) is given by $\dot{m}V$, aligned along the x-axis. The momentum outflow through the control volume is also $\dot{m}V$, but the vector subtends the angle θ with the x-axis, as indicated in the free body diagram. The force acting on the aerodynamic device is indicated by the vector, $\vec{F}$, having components F_x and F_y, in x- and y-directions, respectively. Thus, the force acting on the device is given by

$$\vec{F} = \dot{m}_{out} V_{out} - \dot{m}_{in} V_{in}. \tag{1.13}$$

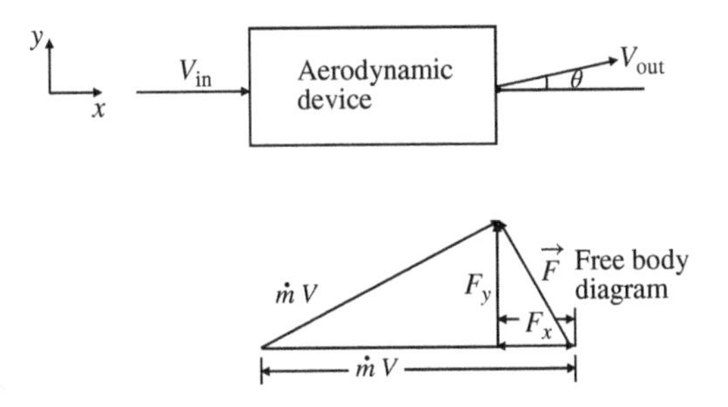

Figure 1.4 A systems approach to the analysis of aerodynamic devices. Free body diagram helps explain the force components active in flight, for $|V_{in}| = |V_{out}|$

For this case, $V_{out} = V_{in} = V$ and $\dot{m}_{out} = \dot{m}_{in} = \dot{m} = \rho AV$. Thus, the force components in the figure are given by

$$F_x = \dot{m}V(\cos\theta - 1) \tag{1.14}$$

and

$$F_y = \dot{m}V\sin\theta. \tag{1.15}$$

Thus, the net force magnitude is given by

$$|\vec{F}| = \sqrt{F_x^2 + F_y^2} = 2\dot{m}V\sin\theta/2. \tag{1.16}$$

And its orientation is given by

$$\phi = \pi - \tan^{-1}\frac{F_y}{F_x} = \pi/2 + \theta/2. \tag{1.17}$$

The interesting aspect of efficiency of this device is understood when the turning angle approaches a small value, i.e. $\theta \to 0$, then the magnitude and orientation of the resultant force is given by

$$|\vec{F}| = \dot{m}V\theta \quad \text{and} \quad \phi = \pi/2, \tag{1.18}$$

This will be an ideal device, for which the resultant force is normal to the oncoming flow, i.e. only lift is produced, while the drag is vanishingly small. The efficiency of aerodynamic device is often quoted in terms of L/D which approaches a very large value in the ideal situation. It is also clearly evident that if the outflow is deflected more, then F_x also increases and it helps us identify that large flow turning indicates that the drag is lift-dependent, even when the inflow and outflow velocities are the same. In an actual flow, there will be viscous losses that will make $V_{out} < V_{in}$. This aspect of viscous losses is dealt with next, again using Reynolds transport theorem.

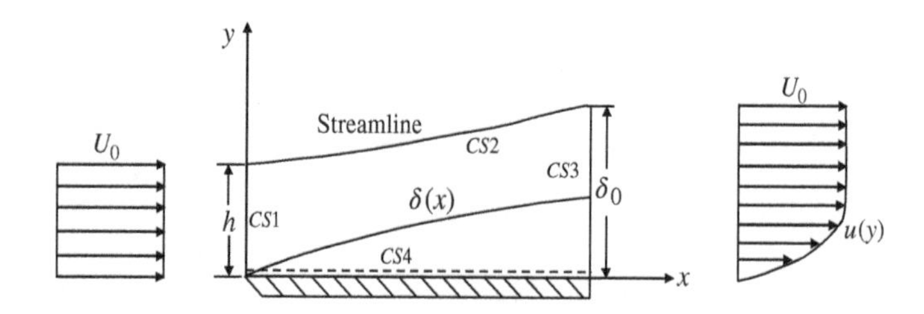

Figure 1.5 Control volume analysis of boundary layer forming over a flat plate

1.3.1 Momentum Integral Theory: Real Fluid Flow

This is the integral form of boundary layer theory developed by Von Kármán (1921). Consider the uniform flow over a flat plate with a sharp leading edge as shown in Figure 1.5, with the axis system fixed to the leading edge of the plate. This flow is over a flat plate in the absence of any pressure gradient in the free stream. Later on in Chapter 7, we will prove by boundary layer equations in differential form that the shear layer transmits the pressure through it unaltered, specifically if the flow is laminar. Hence the pressure is uniform over the depicted control volume in the figure and there is no force created by the pressure.

There are certain other features of the depicted control volume and the enclosing control surface segments. It is easy to identify the segments CS1 and CS3 through which the flow enters and leaves the control volume. While the incoming flow through CS1 is uniform, the boundary layer velocity profile is noted at CS3. In the figure, the boundary layer thickness is denoted by $\delta(x)$ and indicates the locus of the points at which the local velocity reaches a certain fraction (say, 99.9 %) of the free stream speed, indicated here by U_0. The segment CS4 of the control surface just grazes over the flat plate and a surface force caused by shear at the solid surface is transmitted through this segment on to the control volume. This contribution is denoted $-D\hat{i}$ where $\hat{i}$ is the unit vector in x-direction. This is indicative of total drag experienced by such a flat plate at zero incidence angle. Along CS4, we also note that $\vec{V} \cdot \hat{n} = 0$, as this is a no-slip wall with $\vec{V} = 0$.

The most interesting aspect of the choice of control surface is the segment CS2, which is identified to be above the boundary layer thickness at each and every x location and is considered to be along a streamline. From the definition of streamline, there cannot be any flow across it and thus, $\vec{V} \cdot \hat{n} = 0$ along this segment. This would be true, even if the flow is turbulent and then the segment CS2 would represent instantaneous streamline. We note that this is not the only choice of control surface segments, but a very convenient one. For example, if we consider a straight line as segment CS2 parallel to the flat plate, then $\vec{V} \cdot \hat{n} \neq 0$ and it would lead to additional calculations. Note specifically that there would be a net mass flow through such straight segment.

Let us consider that the flow is incompressible and steady. Hence, conservation of linear momentum equation (Equation (1.12)) becomes

$$\sum \vec{F} = \frac{d}{dt}\left(\int_{CV} \rho \vec{V} d\mathcal{V}\right) + \int_{CS} \rho \vec{V}(\vec{V} \cdot \hat{n})\, dA. \tag{1.19}$$

with the first term on the right-hand side equal to zero due to steady state assumption. Considering the x-component of this equation we get

$$\sum F_x = -D = \int_{CS1} \rho u\,(\vec{V}\cdot\hat{n})\,dA + \int_{CS3} \rho u\,(\vec{V}\cdot\hat{n})\,dA$$
$$= \rho \int_0^h U_0(-U_0)\,b\,dy + \rho \int_0^{\delta_0} u^2 b\,dy, \tag{1.20}$$

where b is the breadth of the plate. Thus, the drag is estimated as

$$D = \rho U_0^2 bh - \rho b \int_0^{\delta_0} u^2 dy. \tag{1.21}$$

We also note that h and δ_0 are related to each other, as these points belong to the same streamline and the mass conservation requires

$$0 = \rho \int_0^h (-U_0)\,b\,dy + \rho \int_0^{\delta_0} u\,b\,dy,$$

which upon simplification yields

$$U_0 h = \int_0^{\delta_0} u\,dy. \tag{1.22}$$

Using Equation (1.22) in Equation (1.21) to eliminate h, one gets

$$D = \rho b \int_0^{\delta_0} u\,(U_0 - u)\,dy. \tag{1.23}$$

Note that the integrand on the right-hand side represents a measure of momentum deficit. If we represent the velocity distribution by a parabolic profile valid in the limit $0 \leq y \leq \delta$ (where δ is the boundary layer thickness, which is less than δ_0), then this is given by

$$\frac{u}{U_0} = \frac{2y}{\delta} - \frac{y^2}{\delta}. \tag{1.24}$$

The drag is obtained for this approximate velocity profile as

$$D = \frac{2}{15}\rho U_0^2 b\delta. \tag{1.25}$$

Note that for $\delta \leq y \leq \delta_0$: $u \equiv U_0$ and momentum deficit is zero. Despite such drastically simplified velocity profile being used, calculated drag for laminar flow value obtained incurs only 1% error.

1.4 Flow in Accelerating Control Volumes: Application of RTT

We are going to perform the analysis for the control volume which is fixed with the non-inertial frame (xyz axes system) in Figure 1.6, which translates at the rates given by the velocity and acceleration of its origin as $\frac{d\vec{R}}{dt}$ and $\frac{d^2\vec{R}}{dt^2}$, respectively. Additionally, the axes system performs

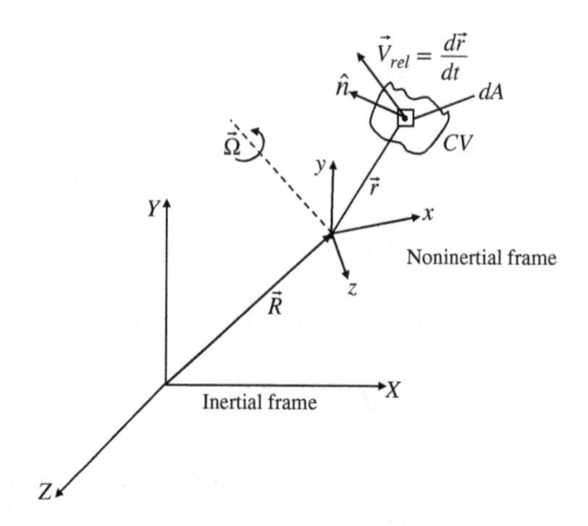

Figure 1.6 Control volume analysis of accelerating CV cases

a rotational motion given by $\vec{\Omega}$, with respect to the inertial frame of reference (XYZ axes system). Any arbitrary point inside the control volume is given by $\vec{r}$ in xyz-system.

Let the fluid velocity inside the control volume be given by $\vec{V}$ in the noninertial frame for an element of mass dm. The acceleration measured in the inertial ($\vec{a}_i$) and noninertial frames ($\frac{d\vec{V}}{dt}$) for this element mass are related by

$$\vec{a}_i = \frac{d\vec{V}}{dt} + \vec{a}_{rel} \tag{1.26}$$

From Newton's second law, sum of all the applied forces are given by $\sum \vec{F} = \int \vec{a}_i \, dm$ and this could be rearranged as

$$\sum \vec{F} - \int dm\, \vec{a}_{rel} = \int dm \frac{d\vec{V}}{dt}. \tag{1.27}$$

The second term on the left hand side incorporates all noninertial effects. The velocity in inertial frame ($\vec{V}_i$) is related to the velocity in the noninertial frame ($\vec{V}$) by

$$\vec{V}_i = \vec{V} + \frac{d\vec{R}}{dt} + \vec{\Omega} \times \vec{r}. \tag{1.28}$$

Similarly, the acceleration in these two frames of references are related by

$$\vec{a}_i = \frac{d\vec{V}}{dt} + \frac{d^2\vec{R}}{dt^2} + \frac{d\vec{\Omega}}{dt} \times \vec{r} + 2\vec{\Omega} \times \vec{V} + \vec{\Omega} \times (\vec{\Omega} \times \vec{r}). \tag{1.29}$$

Note that $\frac{d^2\vec{R}}{dt^2}$ is the acceleration of the origin of the noninertial frame; $\frac{d\vec{\Omega}}{dt} \times \vec{r}$ represents the angular acceleration effect or Euler acceleration term; $2\vec{\Omega} \times \vec{V}$ represents the Coriolis acceleration term and $\vec{\Omega} \times (\vec{\Omega} \times \vec{r})$ represents the centripetal acceleration. Using the Reynolds transport theorem on the right hand side of Equation (1.27) one gets

$$\sum \vec{F} - \int dm\, \vec{a}_{rel} = \frac{d}{dt} \int_{CV} \rho \vec{V} d\mathcal{V} + \int_{CS} \rho \vec{V} (\vec{V}_r \cdot \hat{n})\, dA, \tag{1.30}$$

where $\vec{a}_{rel} = \frac{d^2\vec{R}}{dt^2} + \frac{d\vec{\Omega}}{dt} \times \vec{r} + 2\vec{\Omega} \times \vec{V} + \vec{\Omega} \times (\vec{\Omega} \times \vec{r})$.

1.5 Atmosphere and Its Role in Aerodynamics

For aircraft aerodynamics, we consider its flight in atmosphere. Also the atmosphere may be regarded as an expanse of fluid substantially at rest. We can use hydrostatics and tools of reversible thermodynamics in relating pressure, temperature and other thermodynamic properties with altitude along with the assumption that the atmosphere is homogeneous.

Before we start discussing atmosphere, it is necessary to note the distinction between the height up to which the atmosphere exists to create aerodynamic forces and beyond which it does not exist. This is given by the Kármán line.

1.5.1 Von Kármán Line

The Kármán line is a fictitious dividing line used to demarcate the earths atmosphere and outer space into range of applicability of aeronautics and astronautics. This line is roughly about 100 kilometres above sea-level and named after Von Kármán who coined its first usage. The definition is used by Fédération Aéronautique Internationale (FAI), which maintains international standards and records for aeronautics and astronautics. As such there is no fixed height where Earth's atmosphere ends. On the contrary, the atmosphere gets progressively thinner with height and Von Kármán calculated that at approximately 100 kilometres from sea-level, an aircraft could maintain lift only if it flies at orbital velocity, at which a spacecraft can orbit around the Earth without losing altitude. Thus below 100 kilometres, lift can be maintained at suborbital velocities and this altitude constitute the realm of aeronautics and any other aerial activities above this altitude pertains to astronautics.

1.5.2 Structure of Atmosphere

Atmosphere consists of the following strata:

(a) The lowest region in the atmosphere is called the troposphere and extends over a range of 6 to 20 km depending upon the latitude of the place under consideration. However for aeronautical design and analysis purposes, an international standard atmosphere (ISA) is constructed in which troposphere is considered to extend up to a height of 11 km. ISA corresponds to temperate climate. Statistically speaking, the temperature is considered to drop linearly with height. This is the region where most of the aircrafts are noted to fly.

Table 1.1 Composition of atmosphere at sea level

Molecules	% by volume	Molecular weight	% by weight
N_2	78.088	28.016	75.525
O_2	20.950	32.00	23.143
Argon (A)	0.932	39.944	1.286
CO_2	0.030	44.011	0.046
+ small quantities of Neon, Helium, Krypton, Hydrogen, Xenon, Ozone and Radon.			

(b) The next region is the stratosphere where temperature remains statistically constant and is found up to a height of 50 km. The boundary between troposphere and stratosphere is called the tropopause. Stratosphere is studied with the help of the weather balloon.
(c) Above the stratosphere is the mesosphere which extends up to almost 85 km. Meteors are noted to burn in this region upon entering the atmosphere.
(d) The thermosphere extends beyond the mesosphere up to an altitude of 600 km. This is the region where the Kármán line is and the space shuttle orbit was in thermosphere. Auroras are noted in the lower reaches of thermosphere.
(e) Beyond the thermosphere, one notes the exosphere and is considered to extend up to 10 000 km.

Thus aerodynamics is related mostly to flights within the troposphere and stratosphere. In terms of composition, one notes the following variation with altitude:

1. 50% of total weight is accounted for by the first 18 000 ft.
2. 75% of total weight is accounted for by the first 36 000 ft.
3. Up to a height of 50 miles the composition of the air is more or less constant, except for variation of water vapour content.

Table 1.1 contains the composition of air at sea level, by volume and weight.

The following facts are noted furthermore for the consideration of operating aircrafts in the lower reaches of the atmosphere.

1. Up to 5–6 miles, the water vapour content is a function of ambient temperature. The higher the temperature, the more vapour can be held in a given volume.
2. At very high altitudes, the heavier gases fail to rise until around 50 miles, H_2 and He are predominant above this height.
3. Beyond 18 000 ft the human physiological limit is reached and extra O_2 must be supplied.
4. Beyond 100 000 ft sufficient O_2 is not available to support combustion of turbojets.

Among the above points, we note the role of human physiological limits are the limiting factors. The following line or limit is noted, which severely restricts operation of aircraft from physiological considerations.

1.5.3 Armstrong Line or Limit

This is the limiting altitude (also called the Armstrong line) where the ambient pressure (0.0618 times the sea level atmospheric pressure or 6.3 kPa) is so low that water boils at normal temperature of human body, i.e. at 37 degree Celsius or 98.6 degrees Fahrenheit (named after Harry G. Armstrong, founder of USAF's department of space medicine in 1947). This physiological limit implies a fictitious altitude which would require wearing a pressurizing suit. This is variously reported to be between 18.9 and 19.35 km (or roughly about 12 miles). At this height, bodily fluids in the exposed part will spontaneously boil due to body temperature. However, blood will not boil spontaneously!

One must remember that there would be other adverse physiological effects at significantly lower altitude which affects normal human functioning. Also, the variation of atmospheric composition also does not allow functioning of various components of aerospace vehicle. For example, at heights beyond 100 000 ft, sufficient oxygen is not available to support combustion of turbojets. This limit alerts us about the human physiological limits on design concept of aircrafts. It is just not simple optimization of engineering cost function(s)!

The human blood in a living body will not boil at Armstrong limit due to blood being always under pressure and the boiling point would be significantly higher than the normal body temperature. Blood pressure is a gauge pressure. For example, a person with a low diastolic blood pressure of 60 mm of Hg, will have absolute blood pressure which will be in addition to the ambient pressure. At the altitude of 19 km, if the ambient pressure is 47 mm of Hg, then the absolute pressure is more than double the ambient pressure and blood will not boil, while saliva, tears and liquids wetting the alveoli inside the lung will boil away.

Another major physiological issue of human sustenance and survival is related to availability of oxygen, lack of which causes what is known as hypoxia (confusion and loss of consciousness). Even a professionally well-trained pilot cannot operate on an unpressurized aircraft cabin at altitudes beyond 15 km and requires an oxygen mask. At 15 km, breathing pure oxygen through a mask has the same partial pressure with regular air at 4.7 km altitude. Commercial jetliners keep cabin pressure to an equivalent altitude of 8000 ft (2.438 km). For sustained operation for more than half an hour at altitudes above 12 500 ft or 3.81 km, a pilot must have access to supplemental oxygen supply. Passengers must also be provided with supplemental oxygen above 15 000 ft. Above this altitude, lungs expand like a balloon and tear off (condition known as pulmonary barotraumas), and this also requires pressurized suits. It is for this reason that the cabin is pressurized in commercial passenger planes.

1.5.4 International Standard Atmosphere (ISA) and Other Atmospheric Details

In Figure 1.7, the pressure, density and temperature profiles are plotted as function of altitude for ISA. Also shown are the cloud details which are essential for safe operation of aircraft. According to a general characteristic of atmosphere as released by ICAO, Nacreous and Cirrus clouds remain above the tropopause and are of less concern. However, the cumulonimbus clouds are extremely hazardous for aircraft operation. Noctilucent clouds occur near the mesopause and are caused by volcanic meteoric dust and ice crystals, the latter forming by reaction between atmospheric oxygen with outer space hydrogen.

The tropopause is not clear cut, varying from about 30 000 ft at the poles to about 54 000 ft at the equator. The high temperature noted in thermosphere is due to ionization of the atmospheric gases by cosmic radiation. Also, note the ozonosphere in Figure 1.7, which protects the atmosphere at lower heights. Maximum radiation occurs at 75 000 ft. Ozone is a toxic product of ionization and since the ozone cloud is found around 60 000 ft, the ambient air cannot be

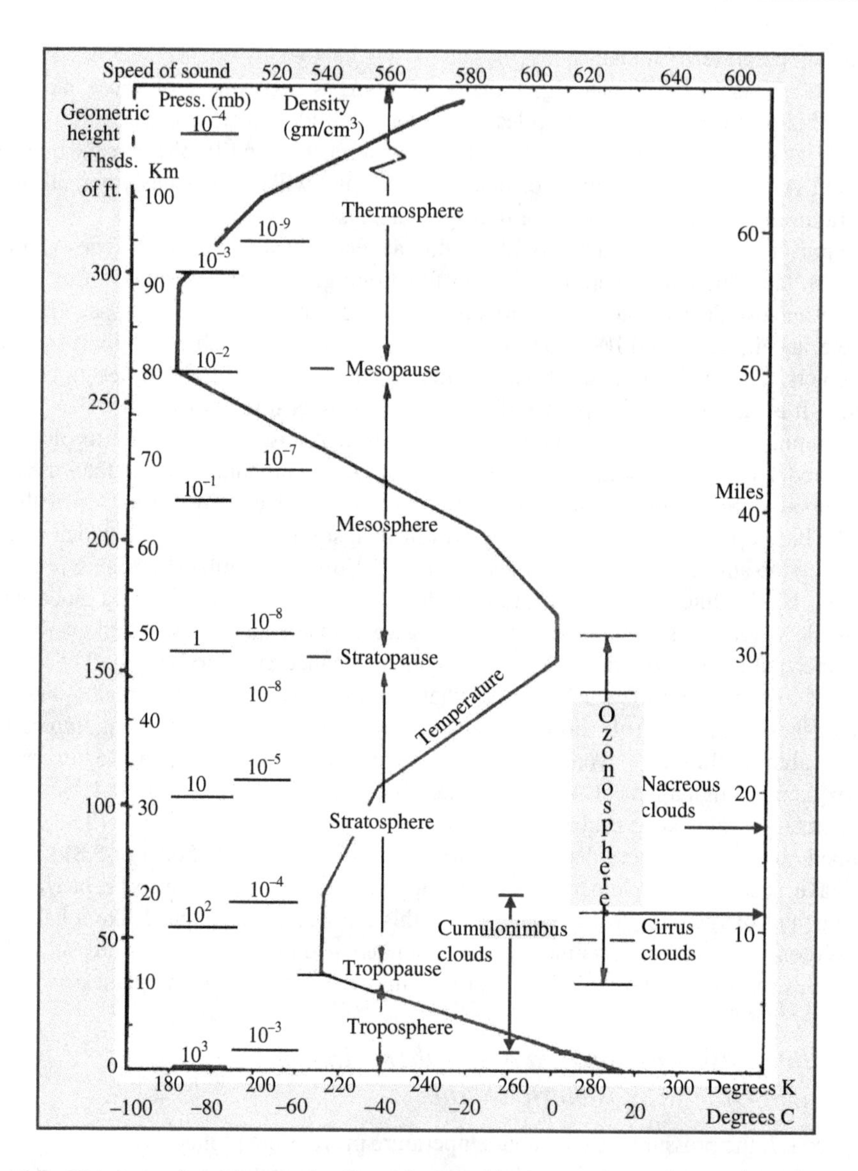

Figure 1.7 ISA atmosphere characterization and other altitude specific details useful for aeronautical operations

used for cabin pressurization. As the intensity of radiation is function of geometric height and latitude, this consideration is important for supersonic transport aircraft, which are perforce designed to fly at higher altitudes due to problems created by sonic boom.

Even when we assume the atmosphere to be static, the temperature variations within the atmosphere cause air mass to convect, resulting in wind, downdraft, wind shear and turbulence as a consequence of instabilities. Turbulence is the major hazard affecting the safe operation of aircraft. Thus, in this chapter we familiarize readers with static stability of atmosphere as

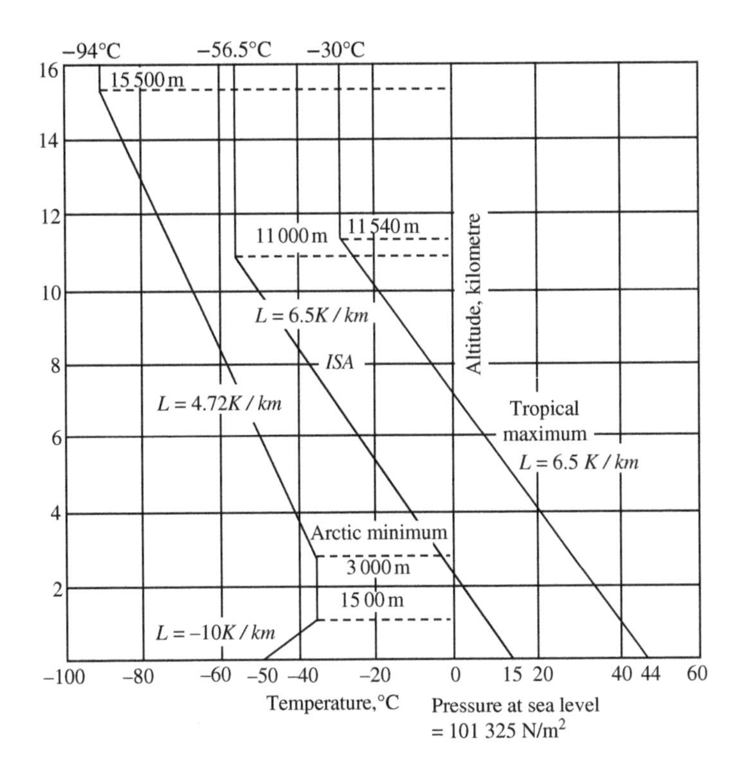

Figure 1.8 ISA atmosphere compared with that seen at arctic minimum and tropical maximum latitudes

a motivation to learn more about dynamic instability of atmosphere, which can be found in Sengupta (2012).

In Figure 1.8, the atmosphere as depicted by ISA condition is compared with those at the tropics and at the arctic.

In designing a new aircraft, it is necessary to note the extremes over which the vehicle is exposed in its lifetime. For example, operating in tropical latitudes the maximum sea level temperature endured is 44°C and in contrast, for minimum arctic climate the other extreme sea level temperature is about −50°C. For ISA, the sea level temperature is 15°C and the temperature falls at the rate of 6.5 K per km. The same lapse rate is noted for the tropical maximum temperature distribution. As the dynamic range of temperature is more at the tropics (from 44°C at sea level to −30°C at tropopause), than at temperate latitudes (from 15°C at sea level to −56.5°C at tropopause) – all low temperature flight trials are carried out in the tropics.

1.5.5 Property Variations in Troposphere and Stratosphere

In the following, the property variation for the ISA is obtained in troposphere and stratosphere by considering static atmosphere, so that conditions of hydrostatics apply. Consider a small cylindrical control volume of height dh and area of cross section A in the still atmosphere.

The temperature variation of the ambient air also causes the static pressure applied on A on two ends of a cylindrical control volume to differ by dp, so that from hydrostatic equilibrium one gets $dp = -\rho g\, dh$. If we treat the surrounding air as perfect gas, then $\frac{p}{\rho} = RT$ where R is the ideal gas constant equal to 287.26 J/(kg K). As noted in Figures 1.7 and 1.8, for the stratosphere the temperature remain constant as $T = T_s$. Using hydrostatic equilibrium and equation of state in the stratosphere

$$\frac{dp}{dh} = -\rho g = -\frac{gp}{RT} = -\frac{pg}{RT_s}.$$

Therefore by separation of variables one gets

$$\frac{dp}{p} = -\frac{g}{RT_s} dh.$$

Which upon integration between two heights h_1 and h_2 yields the relation between pressure at these two heights as

$$\mathrm{Ln}\frac{p_2}{p_1} = -\frac{g}{RT_s}(h_2 - h_1).$$

Thus, the pressure and density ratios are obtained as

$$\frac{p_2}{p_1} = Exp\left[\frac{g}{RT_s}(h_1 - h_2)\right] = \frac{\rho_2}{\rho_1} \tag{1.31}$$

In the troposphere, the temperature variation with altitude is given as

$$T = T_0 - Lh,$$

where L is the Lapse rate in $K/metre$. As before,

$$\frac{dp}{p} = -\frac{g}{R}\frac{dh}{T} = -\frac{g}{R}\frac{dh}{(T_0 - Lh)},$$

which allows us to integrate between two states indicated by subscripts '1' and '2' in the following

$$\mathrm{Ln}\frac{p_2}{p_1} = \frac{g}{LR}\Big[\mathrm{Ln}(T_0 - Lh)\Big]_{h_1}^{h_2}.$$

This can be alternately stated also as

$$\frac{p_2}{p_1} = \left(\frac{T_2}{T_1}\right)^{\frac{g}{LR}}, \tag{1.32}$$

where $T_i = T_0 - Lh_i$

Since $p/(\rho T)$ is constant, therefore using Equation (1.32) one gets

$$\frac{\rho_2}{\rho_1} = \frac{p_2}{p_1}\frac{T_1}{T_2} = \left(\frac{T_2}{T_1}\right)^{\frac{g}{LR}-1}.$$

Also, one can relate pressure and density as

$$\frac{p_2}{p_1} = \left(\frac{\rho_2}{\rho_1}\right)^{\frac{g}{g-LR}}.$$

So in general within troposphere, $p = \text{const}\ \rho^{\frac{g}{(g-LR)}}$.
In SI unit: L = 0.0065 K/m and R = 287.26 J/(kg K) and one has the following relations

$$\frac{p_2}{p_1} = \left(\frac{T_2}{T_1}\right)^{5.256}; \quad \frac{\rho_2}{\rho_1} = \left(\frac{T_2}{T_1}\right)^{4.256} \quad \text{and} \quad p = k\rho^{1.235}. \tag{1.33}$$

where k is constant.

Thus, movement in static air from one height to another is a polytropic process with polytropic constant $n = 1.235$.

A quantity often useful in aerodynamics is the relative density and is defined as

$$\sigma = \frac{(\rho)_{\text{alt}}}{(\rho)_{\text{SL and ISA}}},$$

where

$$(\rho)_{\text{SL and ISA}} = \frac{p_0}{RT_0} = \frac{101.325}{287.26 \times 288} \simeq 1.2256\ kg/m^3.$$

Thermodynamic properties as given in Equations (1.31)–(1.33) define the atmosphere, which is considered to be static. It is important to study the stability of this atmosphere for aerodynamic applications. For example, if the atmosphere is unstable, then there would be thermals which can be used for gliding, as the soaring birds do. If static instability is present in the local atmosphere, then that will ensure a necessary condition for updraft. It does not ensure the sufficient condition for instability, yet it provides indication for a need to study dynamic instability. Also, instability can cause adverse effects, as in the creation of turbulence and downdraft. In performing the study of static stability, we consider the atmosphere as an expanse of fluid that is at rest.

1.6 Static Stability of Atmosphere

In studying static stability of the atmosphere in the troposphere, we note that the static temperature decreases with altitude and is given by

$$T = T_0 - Lz$$

where T_0 is the sea-level temperature in K and L is the lapse rate in $K/metre$. To study static stability, consider a small element of atmosphere to be suddenly displaced a short distance upward, as shown in Figure 1.9. One is then interested in finding the tendency of the displaced element. Three possibilities exist:

(i) If the displaced mass takes up the new position, then the atmosphere is neutrally stable.
(ii) If the displaced parcel tends to return to its original undisturbed position, then the atmosphere is considered to be statically stable.

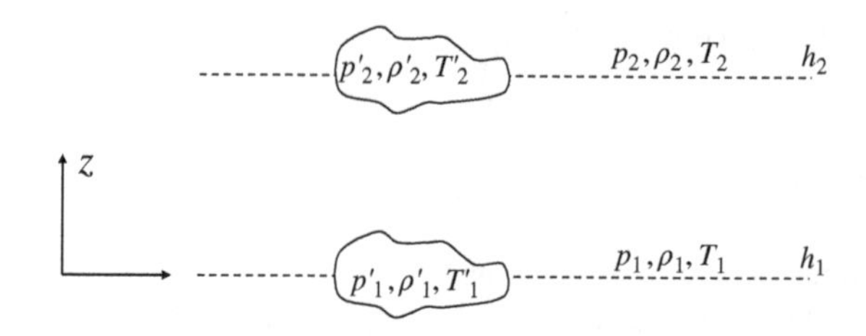

Figure 1.9 An element displaced from h_1 and h_2 with its conditions indicated by primed quantities and ambient condition indicated by unprimed quantities

(iii) If the displaced element tends to continue upward further, then the original equilibrium atmosphere is considered statically unstable.

Let a parcel of air be disturbed from the altitude h_1 to h_2, where the prevailing conditions at that altitude are denoted by unprimed quantities, while the primed quantities refer to the condition of the air-parcel, as shown in Figure 1.9.

For upward migration in the ambient air, we have shown that the pressure and density are related by

$$\left.\begin{array}{r} p = k\rho^{1.235} \text{ in troposphere of ISA} \\ p = k\rho \text{ for stratosphere in ISA} \end{array}\right\}.$$

Or in general

$$p = k\rho^n. \tag{1.34}$$

Hence for the ambient air

$$\frac{T_2}{T_1} = \frac{p_2}{p_1}\frac{\rho_1}{\rho_2} = \left(\frac{p_2}{p_1}\right)\left(\frac{p_1}{p_2}\right)^{1/n},$$

which can be simplified as

$$\frac{T_2}{T_1} = \left(\frac{p_2}{p_1}\right)^{\frac{n-1}{n}}. \tag{1.35}$$

If the upward migration of the air packet is very quick, then the process of moving the parcel can be considered adiabatic to be given by

$$p' = c\,\rho'^{\gamma}.$$

The temperature ratio of the air-parcel at h_1 and h_2 is related to the static pressure at these two heights by

$$\frac{T'_2}{T'_1} = \left(\frac{p'_2}{p'_1}\right)^{\frac{\gamma-1}{\gamma}} \tag{1.36}$$

Note that initially before the air-parcel is moved upward, it is in thermodynamic equilibrium with its surrounding. For mechanical, thermal and chemical equilibrium, one must have, $p_1' = p_1$, $T_1' = T_1$ and $\rho_1' = \rho_1$. Since we are studying the statics of the problem, then even after moving the parcel, it must be in mechanical equilibrium with its surrounding, i.e. the force balance demands $p_2' = p_2$, otherwise there will be expansion work done, along with acceleration of the parcel.

Thus from Equations (1.35) and (1.36) one gets

$$\frac{T_2'}{T_2} = \frac{T_2'}{T_1'}\left(\frac{T_1'}{T_1}\right)\frac{T_1}{T_2}.$$

As $T_1' = T_1$, this simplifies to

$$\frac{T_2'}{T_2} = \left(\frac{p_2}{p_1}\right)^{\frac{\gamma-1}{\gamma}}\left(\frac{p_1}{p_2}\right)^{\frac{n-1}{n}} = \left(\frac{p_2}{p_1}\right)^{\frac{\gamma-n}{\gamma n}}. \tag{1.37}$$

Also from the equation of state and dynamic equilibrium at the displaced condition

$$\frac{\rho_2'}{\rho_2} = \frac{p_2'}{T_2'}\frac{T_2}{p_2} = \frac{T_2}{T_2'} = \left(\frac{p_1}{p_2}\right)^{\frac{\gamma-n}{\gamma n}}. \tag{1.38}$$

Consider the following cases:

(1) For $n < \gamma$, one has $\frac{\gamma-n}{\gamma n} > 0$ and which leads to the following observation:

$$\frac{\rho_2'}{\rho_2} = \left(\frac{p_1}{p_2}\right)^{\alpha^2} > 1 \tag{1.39}$$

where α^2 is a positive quantity.

Thus, the displaced parcel is heavier than the ambient fluid which has been displaced and the parcel will sink back, indicating static stability.

(2) For $n = \gamma$, one obtains $\frac{\rho_2'}{\rho_2} = 1$ and this indicates neutral stability.

(3) For $n > \gamma$, then one obtains $\frac{\rho_2'}{\rho_2} < 1$ and as a consequence the parcel will continue in its motion upward, indicating static instability.

As $n = \frac{g}{g-LR}$, the condition for neutral stability $n = \gamma$ implies a lapse rate given by

$$(L)_{ns} = \left(-\frac{g}{\gamma} + g\right)\frac{1}{R} = \frac{g}{R}\left(\frac{\gamma-1}{\gamma}\right) \simeq \frac{9.807 \times 0.4}{1.4 \times 287.26} = 9.75K/km.$$

$(L)_{ns}$ is called the adiabatic lapse rate.

It is possible to study the dynamic stability of an air-parcel displaced from its equilibrium position. However, that is outside the scope of the present book. Interested readers can consult Sengupta (2012).

Bibliography

Anderson, John. D. Jr, (1997) *A History of Aerodynamics and Its Impact on Flying Machines.* Cambridge Univ. Press, New York, USA.

Kármán, T. von. (1921) Über laminare und turbulente reibung. *Z. Angew. Math. Mech.,* **1**, 233–252.

Sengupta, T.K. (2012) *Instabilities of Flows and Transition to Turbulence.* CRC Press, USA.

White, F.M. (2008) *Fluid Mechanics.* Sixth Edn., McGraw-Hill, New York, USA.

2

Basic Equations of Motion

2.1 Introduction

The fundamental physical laws are used to solve for the fluid motion in a general problem and these are given by the conservation principles of mass, momentum and energy. While these lead to the Navier–Stokes equation and the energy equations, early developments in aerodynamics are related to low-speed flights. In arriving at the basic equations of motion, one can either define the motion of each and every molecule or one can predict average behaviour of molecules in a control volume. The size of the elemental volume is important, as the number of molecules must be numerous in this volume to allow a statistical description of the flow. For low-speed flights, our concern in describing fluid motion is in large spaces containing very large number of particles, so that one can assume the fluid to be continuum. For example, air at normal temperature and pressure at sea level contains 2.5×10^{25} molecules in a volume of $1m^3$, while the mean free path is at $6.6 \times 10^{-8}m$. At an altitude of 130 km, $1m^3$ of air contains 1.6×10^{17} molecules, while the mean free increases to $10.2m$. As we have noted in the previous chapter that most of the conventional aircrafts (or even the high altitude reconnaissance planes) fly either in the troposphere or stratosphere, the continuum assumption can be used to define macroscopic properties of fluid flow meaningfully by statistical averages. Although one can use the concepts of kinetic theory of gases to define the thermodynamic properties of temperature (T) and pressure (p) by considering the motion to arise due to random molecular motion and collisions among molecules, we will adopt the classical viewpoint of these properties using continuum mechanics. We will also adopt the equation of state as the additional constitutive relation between these properties. For a thermally perfect gas, the equation of state is

$$p = \rho RT, \tag{2.1}$$

where ρ is the density and R is the gas constant. Despite this equation of state relating thermodynamic properties, for vehicles flying at speed less than approximately $100m/sec$, it is often assumed that the density of air flowing past the vehicle is constant. This seems more counter-intuitive that gas flow will be considered incompressible, as the modulus of

Theoretical and Computational Aerodynamics, First Edition. Tapan K. Sengupta.

Companion Website: www.wiley.com/go/sengupta

elasticity (E) of air is about 20 000 times smaller as compared to that of water. This is explained in the following subsection.

2.1.1 *Compressibility of Fluid Flow*

This property measures the ability of fluid to change its volume ($\mathcal{V}$) under the action of external force and a quantitative measure is given by the modulus of elasticity (E) given by the action of an applied pressure (Δp) and the relative change of volume by the following equation:

$$\Delta p = -E\frac{\Delta \mathcal{V}}{\mathcal{V}_0}, \tag{2.2}$$

where the relative change in volume ($\frac{\Delta \mathcal{V}}{\mathcal{V}_0}$) is brought about by Δp. The compressibility is negligible for liquid, e.g. an increase in pressure of 1 atmosphere causes relative change of volume of only about 0.005%. Thus, it is quite natural to consider liquid to be regarded as incompressible.

For flow of gases, it is possible to obtain modulus of elasticity directly from the process followed by the gas. For isothermal process gas follows Boyle's law,

$$(p_0 + \Delta p)(\mathcal{V}_0 + \Delta \mathcal{V}) = p_0 \mathcal{V}_0.$$

This shows that for isothermal flow of gases the modulus of elasticity is given by, $E = p_0$. This shows that at sea level natural temperature and pressure (14.7 psi) condition, isothermal flow of air is about 20 000 times more compressible than water (for which $E = 280\,000\, psi$). One can easily show that for adiabatic flow of gases the modulus of elasticity increases to $E = \gamma p_0$ and which can be alternately written as $E = a^2 \rho$, with a as the speed of sound. One can also use the continuity equation to relate

$$\frac{\Delta \mathcal{V}}{\mathcal{V}_0} = -\frac{\Delta \rho}{\rho_0}.$$

Thus, Equation (2.2) can also be written as

$$\Delta p = E\frac{\Delta \rho}{\rho_0}. \tag{2.3}$$

Consequently the flow of gases would be considered incompressible if $\frac{\Delta \rho}{\rho_0} \ll 1$. One can make a quick estimate when this is going to happen. For a gas flowing with speed, V and dynamic pressure, $q = \frac{1}{2}\rho V^2$, the maximum change in pressure that can be brought about is equal to the dynamic head and Equation (2.3) can be written as

$$\frac{\Delta \rho}{\rho} \equiv \frac{q}{E} = \frac{1}{2}\left(\frac{V}{a}\right)^2. \tag{2.4}$$

As one defines the Mach number by, $M = \frac{V}{a}$, the condition of incompressibility is equivalent to

$$\frac{1}{2}M^2 \ll 1.$$

For $M \leq 0.3$, it is noted that $\frac{\Delta\rho}{\rho}$ is less than equal to 0.05 or 5%. This is the rule of thumb often used in aerodynamics to treat the flow as incompressible if the Mach number of oncoming flow is less than equal to 0.3.

2.2 Conservation Principles

For low-speed flights, the energy equation may not be necessary to solve unless some explicit heat transfer problems are addressed. Additionally, conservation of linear momentum equations can be simplified by successively making inviscid and irrotational flow assumptions. These lead to significant ease of analysis and have led to development of simple, yet very elegant aerodynamics theories. One such theory is based on Laplace's equation for velocity potential and are solved computationally for incompressible irrotational flows, by the so-called panel methods. This along with boundary layer theory developed by Prandtl have made the preliminary design of low-speed aircraft very amenable. It is equally useful to remind the readers that for commercial flights, aircrafts may cruise at transonic speed, yet their landing, take-off and early climb segment can be analyzed using the same panel method supplemented by viscous corrections by boundary layer theory.

2.2.1 Flow Description Method: Eulerian and Lagrangian Approaches

There are two principal ways to describe fluid motion as governed by the conservation principles. These are enunciated for describing the kinematics and kinetics of the flows as

Lagrangian description:
Here the particles are followed in their motion around a body. It is as if the observer is in a ground-fixed or inertial reference frame.

Eulerian description:
Here the observer is with the body in its motion, i.e. the observer is in a vehicle-fixed coordinate moving with the vehicle.

Although Lagrangian descriptions appear to be simple, the Eulerian description assists in explaining various physical mechanisms better. Thus, we prefer Eulerian description of fluid flow in this book.

At points far from the body the fluid particles move towards the vehicle with the velocity U_∞, which in reality is the speed of the vehicle. If the vehicle is not accelerating, then it is immaterial whether the body is stationary and fluid is moving or fluid is stationary and the body is moving. This is the principle of relative motion among simply translating frames, on which aerodynamic bodies are experimentally tested in wind and water tunnels. The subscript ∞ is used to denote the undisturbed uniform flow approaching the body. Introduction of a body in the flow perturbs the uniform oncoming flow and the subject materials of this book aim to track the disturbances following the conservation equations.

2.2.2 The Continuity Equation: Mass Conservation

Conservation of mass is easily obtained by noting that the net inflow of mass is equal to the rate at which mass is created within the control volume. These two expressions must be identical and if an analysis is performed by considering three-dimensional flows in a control volume taken as a rectangular parallelepiped, then the result is

$$\frac{\partial \rho}{\partial t} + \frac{\partial}{\partial x}(\rho u) + \frac{\partial}{\partial y}(\rho v) + \frac{\partial}{\partial z}(\rho w) = 0. \tag{2.5}$$

Equation (2.5) is obtained by considering a Cartesian frame of reference and the same can be expressed in a vectorial coordinate system free form given by

$$\frac{\partial \rho}{\partial t} + \nabla \cdot (\rho \vec{V}) = 0. \tag{2.6}$$

We have already shown by using Reynolds transport theorem that for incompressible flows ρ can be treated as constant and the above simplifies to

$$\nabla \cdot \vec{V} = 0. \tag{2.7}$$

To describe the kinematics and kinetics of fluid flows, one defines the following.

Streamline: is defined as a line in the flow, whose tangent is everywhere parallel to the velocity field $\vec{V}$ instantaneously, i.e. $\vec{V} \times \vec{ds} = o$. If the line element $\vec{ds}$ is represented in the Cartesian frame by, $\vec{ds} = \hat{i}\, dx + \hat{j}\, dy + \hat{k}\, dz$, then this definition helps fixing the family of streamlines at any instant t, as the solutions of

$$\frac{dx}{u} = \frac{dy}{v} = \frac{dz}{w}, \tag{2.8}$$

when the flow is steady, the streamlines have the same form at all times. Also from the very definition, streamlines cannot intersect each other, otherwise at the point of intersection, a fluid particle will have more than one velocity vector and is simply not possible. However, streamlines will intersect at singular points, as at the stagnation point or at singular points. An example of the latter will be the point of separation or flow attachment.

Streamtube: is the surface formed instantaneously by all streamlines which pass through a given closed curve in the fluid.

Pathline: of a material element of fluid is the trajectory of the element. Thus path lines do not, in general, coincide with streamlines. It does so, only when the motion is steady.

Streakline: is a line on which all those fluid elements that at some earlier instant passed through a certain point of space; as noted in case of flow visualization by dye or marker.

A special aspect of fluid motion is noted that even in a steady flow, a material element experiences acceleration as it moves to a position where $\vec{V}$ has a different value. This is due to the fact that $\vec{V}$, apart from being an explicit function of time, is also an implicit function of time as (x, y, z) is a function of time. A fluid element experiences acceleration that can be obtained from

$$\frac{d\vec{V}}{dt} = \frac{\partial \vec{V}}{\partial t} + \frac{\partial \vec{V}}{\partial x}\frac{dx}{dt} + \frac{\partial \vec{V}}{\partial y}\frac{dy}{dt} + \frac{\partial \vec{V}}{\partial z}\frac{dz}{dt}.$$

$$\text{As,} \quad \frac{dx}{dt} = u, \; \frac{dy}{dt} = v, \; \frac{dz}{dt} = w, \; \text{therefore.}$$

$$\frac{d\vec{V}}{dt} = \frac{\partial \vec{V}}{\partial t} + u\frac{\partial \vec{V}}{\partial x} + v\frac{\partial \vec{V}}{\partial y} + w\frac{\partial \vec{V}}{\partial z}.$$

The derivative on the left-hand side is called the total or substantive derivative. The time derivative on the right-hand side is the local acceleration component, while the rest of the terms on the right-hand side constitute the convective acceleration.

2.3 Conservation of Linear Momentum: Integral Form

First, we derive it in integral form due to its simplicity and its appearance which allows relating to physical causes giving rise to different terms. For the derivation of integral form of momentum equation, one starts from Newton's second law of motion, i.e.

$$\begin{aligned}&\sum \text{Body forces} + \sum \text{Surface forces}\\ &= \frac{\partial}{\partial t}\iiint_{\mathcal{V}} \rho\vec{V}\, d\mathcal{V} + \text{net efflux of momentum through the control surfaces.}\end{aligned}$$

If one represents the forces for unit volume

$$\vec{f}_b = \sum \text{Body forces}$$

and

$$\vec{f}_s = \sum \text{Surface forces}$$

then

$$\vec{f}_b + \vec{f}_s = \frac{\partial}{\partial t}\iiint_{\mathcal{V}} \rho\vec{V} d\mathcal{V} + \oiint_{\mathcal{A}} \vec{V}(\rho\vec{V}\cdot\hat{n}\, d\mathcal{A}),$$

where the net efflux of momentum through the control surfaces is given by the last term on the right-hand side.

2.4 Conservation of Linear Momentum: Differential Form

This is obtained by applying Newton's second law on an infinitesimal control volume representing a fluid particle. The net force acting on the fluid particle is equal to the time rate of change of linear momentum of the fluid particle. As the fluid element moves in space, its shape and volume may change, but its mass is conserved. In an inertial coordinate system

$$\vec{F} = m\frac{d\vec{V}}{dt}. \tag{2.9}$$

Writing Equation (2.9) for a fluid element of unit volume

$$\rho\frac{d\vec{V}}{dt} = \vec{f}_b + \vec{f}_s, \tag{2.10}$$

where $\vec{f}_b$ is the mass dependent body force. Typical examples encountered in fluid dynamics are Coriolis force, gravitational force and other electrodynamic/magnetic forces appear due to electromagnetic effects. $\vec{f}_s$ represents all form of surface forces, which act on the surface of an immersed body due to various forms of stress distribution. Typical examples are the pressure and shear forces due to normal and shear stresses appearing as point forces. The surface forces depend on the rate at which the fluid is strained by the present velocity field in the flow. The system of applied forces defines a state of stress and one of the tasks is to relate stress with strain rate. We specifically note that this relationship is an empirical one. In this context, we consider fluids which are isotropic and which display a linear relationship between stress and strain, the so-called Newtonian fluids.

2.4.1 *General Stress System in a Deformable Body*

Consider an infinitesimal rectangular parallelepiped of volume $dx\, dy\, dz$. The coordinate of the lower left-hand corner O is (x, y, z). Consider two faces with area $(dy\, dz)$, which are perpendicular to x-axis, and over which two resultant stresses act, shown in Figure 2.1 as

$$p_x \quad \text{and} \quad p_x + \frac{\partial p_x}{\partial x}\, dx,$$

acting in a direction along which we define a unit vector, $\hat{i}_x$.

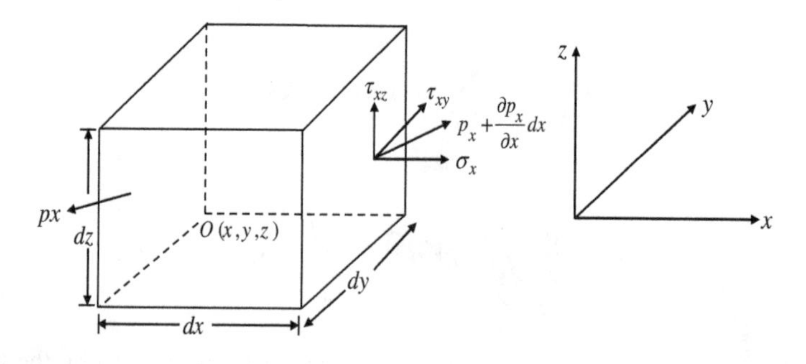

Figure 2.1 Control volume analysis for linear momentum of three-dimensional viscous flows

Similar terms can be obtained for the faces $(dx\,dz)$ and $(dx\,dy)$ which are perpendicular to y and z axes, respectively, in Figure 2.1 acting along the directions given by the unit vectors, $\hat{j}_y$ and $\hat{k}_z$, respectively. Net components of the surface force are obtained as

$$\text{On plane } in \text{ direction } \hat{i}_x : \frac{\partial p_x}{\partial x}\, dx\, dy\, dz.$$

$$\text{On plane } in \text{ direction } \hat{j}_y : \frac{\partial p_y}{\partial y}\, dx\, dy\, dz.$$

$$\text{On plane } in \text{ direction } \hat{k}_z : \frac{\partial p_z}{\partial z}\, dx\, dy\, dz.$$

The resultant surface force per unit volume is

$$\vec{f}_s = \frac{\partial p_x}{\partial x}\hat{i}_x + \frac{\partial p_y}{\partial y}\hat{j}_y + \frac{\partial p_z}{\partial z}\,\hat{k}_z. \tag{2.11}$$

Here, p_x, p_y and p_z are vectors which can be resolved into components along the co-ordinate directions acting on each face, i.e. into normal stress denoted by σ with subscript indicating the direction and into components parallel to each face, i.e. into shearing stresses denoted by τ with two subscripts. In the shear stress, the first subscript indicates the axis which is perpendicular to the face and the second subscript indicates the direction along which the shearing stress is parallel. With this notation we have

$$\begin{aligned} p_x &= \hat{i}\,\sigma_x + \hat{j}\,\tau_{xy} + \hat{k}\,\tau_{xz} \\ p_y &= \hat{i}\,\tau_{yx} + \hat{j}\,\sigma_y + \hat{k}\,\tau_{yz} \\ p_z &= \hat{i}\,\tau_{zx} + \hat{j}\,\tau_{zy} + \hat{k}\,\sigma_z. \end{aligned} \tag{2.12}$$

These nine quantities in Equation (2.12) constitute the stress tensor. One can show that the stress tensor is symmetric by noting angular motions about the coordinate directions. For the angular motion about y-axis with a rate $\dot{\omega}$, this is given by

$$\dot{\omega}_y\, dI_y = (\tau_{xz}\, dy\, dz)dx - (\tau_{zx}\, dx\, dy)\, dz,$$

with dI_y indicating the moment of inertia of the fluid element about the y-axis. This can be rewritten as

$$\tau_{xz} - \tau_{zx} = \dot{\omega}_y \frac{dI_y}{dx\, dy\, dz}.$$

Now the moment of inertia about the y-axis (dI_y) is proportional to $(dl)^5$, where dl is the typical length scale defining the elementary volume of the order of dx, dy and dz.

Thus,

$$\tau_{xz} - \tau_{zx} \simeq \dot{\omega}_y \, (dl)^2.$$

Thus, when the elementary volume shrinks to define a point property, the right-hand side would tend to zero. Hence the proof: $\tau_{xz} = \tau_{zx}$, provided $\dot{\omega}_y$ is not very large. The latter can arise due to distributed couple in the flow due to applied moment(s). Other two similar relations follow considering the angular motion about the other two axes.

Note that the stresses cease to be symmetric if the fluid develops a local moment which is proportional to the elemental volume – as is the case of a flow affected by imposed electromagnetic and/ or electrostatic field.

Finally, the surface force per unit volume is

$$\vec{f}_s = \hat{i}\left(\frac{\partial \sigma_x}{\partial x} + \frac{\partial \tau_{xy}}{\partial y} + \frac{\partial \tau_{xz}}{\partial z}\right) + \hat{j}\left(\frac{\partial \tau_{xy}}{\partial x} + \frac{\partial \sigma_y}{\partial y} + \frac{\partial \tau_{yz}}{\partial z}\right) + \hat{k}\left(\frac{\partial \tau_{xz}}{\partial x} + \frac{\partial \tau_{yz}}{\partial y} + \frac{\partial \sigma_y}{\partial z}\right). \tag{2.13}$$

If the components of the body force $\vec{f}_b$ are X, Y and Z, then the equations of motion are

$$\begin{aligned}
\rho \frac{Du}{Dt} &= X + \frac{\partial \sigma_x}{\partial x} + \frac{\partial \tau_{xy}}{\partial y} + \frac{\partial \tau_{xz}}{\partial z} \\
\rho \frac{Dv}{Dt} &= Y + \frac{\partial \tau_{xy}}{\partial y} + \frac{\partial \sigma_y}{\partial y} + \frac{\partial \tau_{yz}}{\partial z} \\
\rho \frac{Dw}{Dt} &= Z + \frac{\partial \tau_{zx}}{\partial x} + \frac{\partial \tau_{yz}}{\partial y} + \frac{\partial \sigma_z}{\partial z}.
\end{aligned} \tag{2.14}$$

These are also called the Cauchy equation. For a frictionless fluid, $\tau_{ij} = 0$ for $i \neq j$ and $\tau_{ii} = -p$ where p is the arithmetic mean of normal stresses.

$$\text{i.e.} \quad p = \frac{1}{3}(\sigma_x + \sigma_y + \sigma_z).$$

In general, Cauchy equations cannot be solved unless one has closure relations for the stress tensor. Next we relate the independent six stress components with the associated rate of strain field, as one of the ways this is achieved for a special class of fluids.

2.5 Strain Rate of Fluid Element in Flows

Using only the first term of a Taylor series expansion, the velocity at B in Figure 2.2, which is in relative motion with respect to the point A, is described by the following matrix of nine partial derivatives of the local velocity field

$$\begin{bmatrix} u_x & u_y & u_z \\ v_x & v_y & v_z \\ w_x & w_y & w_z \end{bmatrix}. \tag{2.15}$$

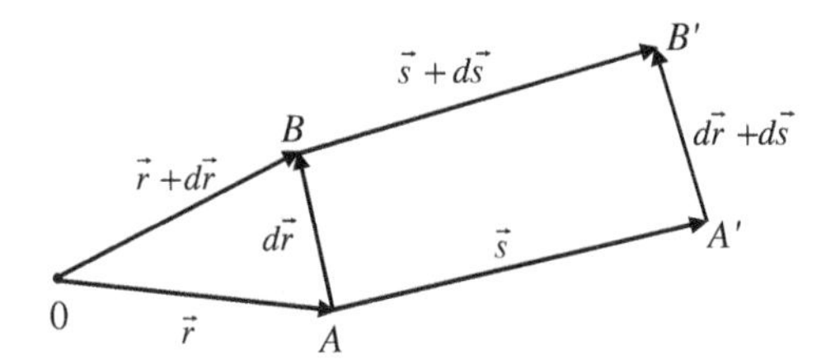

Figure 2.2 Relative motion of particles at A and B due to applied stresses

The relative velocity components du, dv and dw between A and B can also be rearranged as

$$\begin{aligned}du &= (\dot{\epsilon}_x\, dx + \dot{\epsilon}_{xy}\, dy + \dot{\epsilon}_{xz}\, dz) + (\eta\, d_z - \zeta\, dy)/2\\ dv &= (\dot{\epsilon}_{xy}\, dx + \dot{\epsilon}_y\, dy + \dot{\epsilon}_{yz}\, dz) + (\zeta\, d_x - \xi\, dy)/2\\ dw &= (\dot{\epsilon}_{zx}\, dx + \dot{\epsilon}_{zy}\, dy + \dot{\epsilon}_z\, dz) + (\xi\, d_y - \eta\, dx)/2.\end{aligned} \tag{2.15a}$$

where $\dot{\epsilon}_{ij} = \frac{1}{2}\left(\frac{\partial u_i}{\partial x_j} + \frac{\partial u_j}{\partial x_i}\right)$ for $i, j = 1$ to 3 represent the symmetric part of the strain tensor constituted from the partial derivatives of velocity components. Also, $\dot{\epsilon}_{ij}$'s are symmetric with respect to the subscripts.

One constructs the antisymmetric part of the strain tensor in terms of the vorticity vector defined by

$$\vec{\omega} = \nabla \times \vec{V} = \xi\, \hat{i} + \eta\, \hat{j} + \zeta\, \hat{k}.$$

2.5.1 *Kinematic Interpretation of Strain Tensor*

It is easy to see that $\dot{\epsilon}_{ii}$ represents the rate of elongation or contraction in the i-direction suffered by an element due to the applied strain. This will preserve the shape, while contracting or dilating the fluid element. Additional effects of applied strains could lead to alteration of the shape of the fluid element.

The change in volume imparted to a fluid element by simultaneous action of all three diagonal elements of the strain tensor results in the element expanding (considering all partial derivatives of velocity components with corresponding coordinates to be positive, without loss of generality) in all the three direction, and the change in the length of its three sides produces a change in volume at a relative rate

$$\dot{e} = \frac{\{dx + \frac{\partial u}{\partial x} dx\, dt\}\, \{dy + \frac{\partial v}{\partial y} dy\, dt\}\, \{dz + \frac{\partial w}{\partial z} dz\, dt\} - dx\, dy\, dz}{dx\, dy\, dz\, dt}$$

$$= \frac{\partial u}{\partial x} + \frac{\partial v}{\partial y} + \frac{\partial w}{\partial z} = \text{div}\, \vec{V}, \tag{2.16}$$

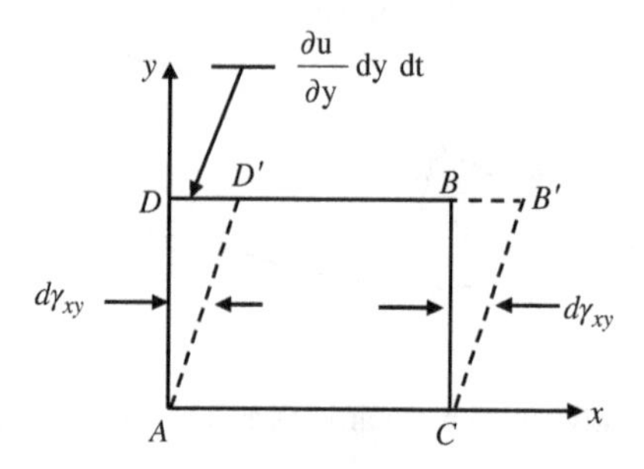

Figure 2.3 A case of pure strain applied to the fluid

retaining only the lowest order terms. During this change, the shape of the element described by the angles at its vertices remains unchanged. Thus, $\dot{e}$ describes the local instantaneous volumetric dilatation of a fluid element.

The relative velocity field is different when one of the off-diagonal term in the matrix of Equation (2.15) is nonzero. Consider, for example a case for which $u_y > 0$. The corresponding field is one of pure shear strain due to which a rectangular element of fluid distorts into a parallelogram, as shown in Figure 2.3. The original right angle at A in this figure changes at a rate measured by the angle

$$d\gamma_{xy} = \frac{\frac{\partial u}{\partial y}\, dy\, dt}{dy}$$

Thus, the included angle at A changes at a rate of $\frac{\partial u}{\partial y}$, i.e. the angular deformation rate due to pure strain is given by, $\dot{\gamma}_{xy} = \frac{\partial u}{\partial y}$

Consider next a case, when both $\frac{\partial u}{\partial y}$ and $\frac{\partial v}{\partial x} > 0$. The right angle at A will then distort owing to the superposition of these two motions, as shown in Figure 2.4. Figure 2.4 is drawn such that

$$\frac{\partial u}{\partial y} = \frac{\partial v}{\partial x}.$$

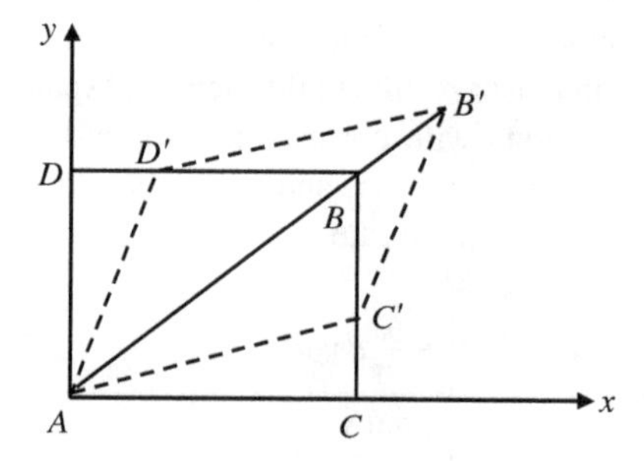

Figure 2.4 A case of irrotational flow. Note that the diagonal of the element does not rotate due to applied strain

This is a case when

$$\dot{\epsilon}_{xy} = \dot{\epsilon}_{yx} = \frac{1}{2}\left\{\left(\frac{\partial u}{\partial y}\right) + \left(\frac{\partial v}{\partial x}\right)\right\} > 0,$$

with all other terms being zero and this gives rise to distortion in shape.

For the imposed strain field, one clearly notices that the z component of the vorticity field is zero, i.e. $\zeta = \frac{1}{2}(-u_y + v_x) = 0$.

For the case of Figure 2.4, the right angle at A now distorts at the rate

$$\dot{\epsilon}_{xy} = \dot{\epsilon}_{yx} = \frac{1}{2}\,(u_y + v_x),$$

such that this distortion is volume preserving and affects only the shape. With B' now on the extension of AB, i.e. the diagonal do not rotate. Such is the case for irrotational flows for which vorticity is identically zero.

Finally, consider the case when $u_y = -v_x$, such that $\dot{\epsilon}_{xy} = \dot{\epsilon}_{yx} = 0$, i.e. the right angles at A remains a right angle. This is shown in Figure 2.5, where the fluid element rotates with respect to A. Instantaneously, this occurs without distortion and can be described as rigid body rotation, indicated by the rotation rate of the diagonal AB and is v_x or $-u_y$. This rigid body rotation can be generally represented by the vorticity ($\frac{1}{2}\nabla \times \vec{V}$) of the velocity field.

When all the elements of the matrix given by Equation (2.15) are nonzero, the overall relative motion is a combination of all the cases discussed above, whose mathematical representation is given in Equation (2.15a) by the relative motion of two fluid particles.

Both the stress tensor and the part of rate of strain tensor given by $\dot{\epsilon}_{ij}$ are symmetric. For such matrices, it is possible to find a transformation such that these are rendered diagonal, i.e. we move to the principal axes along which only normal stresses and rates of strain exist. We denote the values of these strictly diagonal components by a bar, i.e.

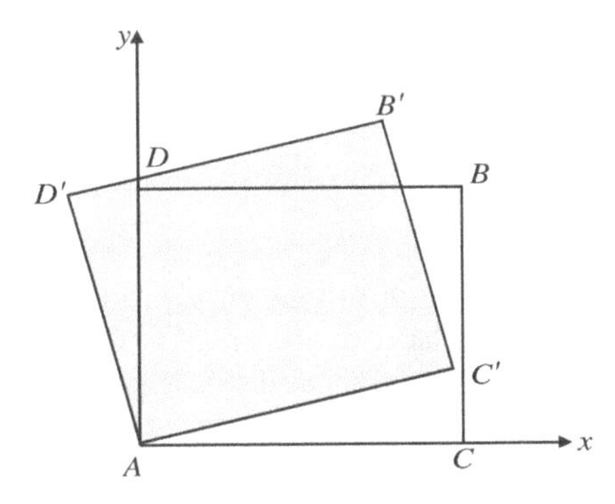

Figure 2.5 A case of rigid body rotation for which $u_y = v_x$

$(\bar{\sigma}_x, \bar{\sigma}_y$ and $\bar{\sigma}_z)$ and $(\bar{\mathring{\epsilon}}_x, \bar{\mathring{\epsilon}}_y, \bar{\mathring{\epsilon}}_z)$. From the invariant property of the trace of tensors one notes that

$$\sigma_x + \sigma_y + \sigma_z = \bar{\sigma}_x + \bar{\sigma}_y + \bar{\sigma}_z$$

and

$$\mathring{\epsilon}_x + \mathring{\epsilon}_y + \mathring{\epsilon}_z = \bar{\mathring{\epsilon}}_x + \bar{\mathring{\epsilon}}_y + \bar{\mathring{\epsilon}}_z.$$

2.6 Relation between Stress and Rate of Strain Tensors in Fluid Flow

Having discussed the strain field associated with the velocity of fluid flow, the task now remains to relate the stress with the rate of strain tensors, so that one can simplify the Cauchy equation. This was achieved by Navier and Stokes independently, in the first half of the nineteenth century.

When the fluid is at rest, it develops a uniform field of hydrostatic stress $(-p)$ which is identical to the thermodynamic pressure. When the fluid is in motion, the equation of state still determines a pressure at every point and one defines conveniently the deviatoric normal stresses by

$$\sigma_x' = \sigma_x + p, \quad \sigma_y' = \sigma_y + p \text{ and } \sigma_z' = \sigma_z + p, \tag{2.17}$$

along with the unchanged shear stresses. The six quantities thus obtained are now due to fluid motion. We realize that instantaneous translation and rigid body rotation of an element of fluid produces no surface forces on it and thus, the deviatoric stress components contain only velocity gradients. The relations between the two are postulated as linear and remain unchanged by rotation of the coordinate system to ensure isotropy. Isotropy also requires that the principle axis of strain and stress are identical, otherwise a preferred direction is introduced. This is the definition of Newtonian fluid for which stresses and rates of strains are related by linear relation.

Isotropy can be secured only if each one of the three normal stresses $\bar{\sigma}_x', \bar{\sigma}_y'$ and $\bar{\sigma}_z'$ are made to depend on the same component of rate of strain and on the trace of the strain tensor, each with a different factor of proportionality. Thus, we write one of the principal stresses as

$$\bar{\sigma}_x' = \lambda\left(\frac{\partial \bar{u}}{\partial \bar{x}} + \frac{\partial \bar{v}}{\partial \bar{y}} + \frac{\partial \bar{w}}{\partial \bar{z}}\right) + 2\mu \frac{\partial \bar{u}}{\partial \bar{x}} \tag{2.17a}$$

Here $\bar{x}, \bar{y}$ and $\bar{z}$ are the principal axes directions and $\bar{u}, \bar{v}$ and $\bar{w}$ are the velocity components along these directions. This relation has two properties of fluids, λ and μ, which should be treated as the elements of a fourth order tensor and the right hand side is essentially a contraction operation. Interested readers are referred to the book by Aris (1962) for application of tensors in fluid mechanics.

Also, relations like that is given in Equation (2.17a) can be written for any arbitrary coordinate systems obtained by a general rotation with appropriate linear transformation.

Such a derivation would show the stress system given in this arbitrary frame as

$$\sigma'_x = \lambda \operatorname{div} \vec{V} + 2\mu \frac{\partial u}{\partial x}$$

$$\sigma'_y = \lambda \operatorname{div} \vec{V} + 2\mu \frac{\partial v}{\partial y}$$

$$\sigma'_z = \lambda \operatorname{div} \vec{V} + 2\mu \frac{\partial w}{\partial z}$$

$$\tau_{xy} = \tau_{yx} = \mu \left(\frac{\partial v}{\partial x} + \frac{\partial u}{\partial y} \right) \tag{2.18}$$

$$\tau_{yz} = \tau_{zy} = \mu \left(\frac{\partial w}{\partial y} + \frac{\partial v}{\partial z} \right)$$

$$\tau_{zx} = \tau_{xz} = \mu \left(\frac{\partial u}{\partial z} + \frac{\partial w}{\partial x} \right).$$

The above relations can be written in general by $T_{ij} = -p\, \delta_{ij} + \lambda\, \partial_k v_k\, \delta_{ij} + 2\mu\, \partial_i\, (v_j)$, where $-p$ is the thermodynamic pressure and δ_{ij} is the Kronecker delta which takes a value equal to one, when $i = j$ and zero otherwise. If p_m is the mechanical pressure, then it should be given by

$$p_m = \frac{T_{ii}}{3}.$$

Therefore the difference between mechanical and thermodynamic pressure is given by

$$(-p) - p_m = -\left(\lambda + \frac{2}{3}\mu \right) \partial_k v_k = \left(\lambda + \frac{2}{3}\mu \right) \frac{1}{\rho} \frac{D\rho}{Dt}.$$

Stokes' hypothesis:

Stokes, from reversible work consideration of a system containing fluids, postulated thermodynamic and mechanical pressure to be identical and according to Stokes' hypothesis,

$$3\lambda + 2\mu = 0. \tag{2.19}$$

Now using Equation (2.17), we obtain the constitutive relation for an isotropic, Newtonian fluid as

$$\begin{aligned}
\sigma_x &= -p - \frac{2}{3}\mu \operatorname{div} \vec{V} + 2\mu\frac{\partial u}{\partial x} \\
\sigma_y &= -p - \frac{2}{3}\mu \operatorname{div} \vec{V} + 2\mu\frac{\partial v}{\partial y} \\
\sigma_z &= -p - \frac{2}{3}\mu \operatorname{div} \vec{V} + 2\mu\frac{\partial w}{\partial z} \\
\tau_{xy} &= \tau_{yx} = \mu\left(\frac{\partial v}{\partial x} + \frac{\partial u}{\partial y}\right) \\
\tau_{yz} &= \tau_{zy} = \mu\left(\frac{\partial w}{\partial y} + \frac{\partial v}{\partial z}\right) \\
\tau_{xz} &= \tau_{zx} = \mu\left(\frac{\partial u}{\partial z} + \frac{\partial w}{\partial x}\right).
\end{aligned} \tag{2.20}$$

Using Equation (2.20) in Equation (2.14), one gets

$$\begin{aligned}
\rho\frac{Du}{Dt} &= X - \frac{\partial p}{\partial x} + \frac{\partial}{\partial x}\left[\mu\left(2\frac{\partial u}{\partial x} - \frac{2}{3}\operatorname{div}\vec{V}\right)\right] + \frac{\partial}{\partial y}\left[\mu\left(\frac{\partial u}{\partial y} + \frac{\partial v}{\partial x}\right)\right] \\
&\quad + \frac{\partial}{\partial z}\left[\mu\left(\frac{\partial w}{\partial x} + \frac{\partial u}{\partial z}\right)\right] \\
\rho\frac{Dv}{Dt} &= Y - \frac{\partial p}{\partial y} + \frac{\partial}{\partial y}\left[\mu\left(2\frac{\partial v}{\partial y} - \frac{2}{3}\operatorname{div}\vec{V}\right)\right] + \frac{\partial}{\partial z}\left[\mu\left(\frac{\partial v}{\partial z} + \frac{\partial w}{\partial y}\right)\right] \\
&\quad + \frac{\partial}{\partial x}\left[\mu\left(\frac{\partial u}{\partial y} + \frac{\partial v}{\partial x}\right)\right] \\
\rho\frac{Dw}{Dt} &= Z - \frac{\partial p}{\partial z} + \frac{\partial}{\partial z}\left[\mu\left(2\frac{\partial w}{\partial z} - \frac{2}{3}\operatorname{div}\vec{V}\right)\right] + \frac{\partial}{\partial x}\left[\mu\left(\frac{\partial w}{\partial x} + \frac{\partial u}{\partial z}\right)\right] \\
&\quad + \frac{\partial}{\partial y}\left[\mu\left(\frac{\partial v}{\partial z} + \frac{\partial w}{\partial y}\right)\right].
\end{aligned} \tag{2.21}$$

These are the celebrated Navier–Stokes equation for compressible flows. The governing energy equation for compressible flows is given in Sections 2.17 to 2.21.

One can very easily simplify Equation (2.21) for incompressible flows in the following vector form

$$\rho \frac{D\vec{V}}{Dt} = \vec{f}_b - \nabla p + \mu \nabla^2 \vec{V}. \tag{2.22}$$

One can further reduce Navier–Stokes equation for inviscid flows by omitting the viscous stress terms on the right-hand side of the above to obtain the Euler's equation for incompressible flows

$$\rho \frac{D\vec{V}}{Dt} = \vec{f}_b - \nabla p.$$

The Euler's equation for compressible flow is directly obtained by omitting the viscous stress terms on the right-hand side of Equation (2.21).

While the Euler's equation governs inviscid fluid motion, it can be further simplified by considering irrotational flows. To distinguish between rotational and irrotational flows, we restrict our attention to incompressible flows and introduce important concepts associated with rotational flows. In general, any vector can be split into an irrotational and rotational part via what is known as Clebsch transform in Lamb (1945). We would shortly see that the velocity in general fluid flow can also be similarly split into rotational and irrotational parts. The rotational field is already introduced in the previous section through the vorticity vector, in defining relative motion between two points in a flow. Here, we introduce an associated concept of rotationality via circulation.

2.7 Circulation and Rotationality in Flows

This is defined as the closed contour integral of the velocity field given by, $\Gamma = \oint \vec{V} \cdot d\vec{r}$, which is considered as positive, when the contour is traced in a counter-clockwise direction. As the contour can be arbitrary, we consider the circulation around a small rectangular element in the (x, y)-plane, as shown in Figure 2.6. Integrating the velocity component along each of

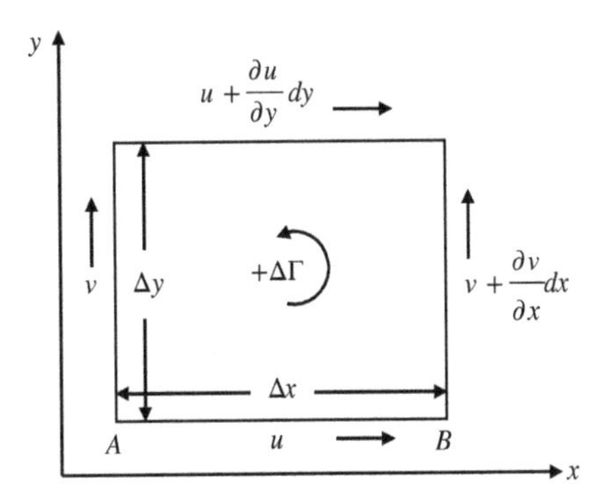

Figure 2.6 Defining circulation $(\Delta\Gamma)$ around an infinitesimal fluid element

the sides and proceeding counter-clockwise, we obtain the circulation for this infinitesimal element as

$$\Delta\Gamma = u\,\Delta x + \left(v + \frac{\partial v}{\partial x}\Delta x\right)\Delta y$$

$$-\left(u + \frac{\partial u}{\partial y}\Delta y\right)\Delta x - v\,\Delta y$$

$$= \left(\frac{\partial v}{\partial x} - \frac{\partial u}{\partial y}\right)\Delta x\,\Delta y.$$

Note:

(i) The variation of u along AB with x can also be considered, but the contributions mutually cancel out, leaving the above.
(ii) The right-hand side quantity of the above inside the parenthesis is nothing but the z-component of the vorticity defined already in Equation (2.15a).

This procedure can be extended to calculate the circulation around a general curve C in the (x, y)-plane. This result for the general curve in the (x, y)-plane is given by

$$\Gamma = \oint_C (u\,dx + v\,dy) = \iint_{\mathcal{A}} \left(\frac{\partial v}{\partial x} - \frac{\partial u}{\partial y}\right) dx\,dy. \tag{2.23}$$

Equation (2.23) is Green's theorem, relating a line integral to the corresponding area integral. The transformation from a line integral to a surface integral in three-dimensional space is governed by Stokes' theorem

$$\oint_C \vec{V}.\,d\vec{r} = \iint_{\mathcal{A}} (\nabla \times \vec{V}) \cdot \hat{n}\,d\mathcal{A}, \tag{2.24}$$

where $\hat{n}\,d\mathcal{A}$ is the area vector normal to the surface and is positive when it is pointing outward from the enclosed volume and equal in magnitude to the surface area.

Stokes theorem is valid when $\mathcal{A}$ represents a simply connected region in which $\vec{V}$ is continuously differentiable and Equation (2.24) is not valid if $\mathcal{A}$ contains regions where $\vec{V}$ or its derivatives are infinite.

2.8 Irrotational Flows and Velocity Potential

By Stokes' theorem, it is apparent that if the curl of $\vec{V}$, i.e. $(\nabla \times \vec{V})$ is zero at all points in the region bounded by C, then the circulation is zero.

If $(\nabla \times \vec{V}) = 0$, then

$$\Gamma = \oint \vec{V} \cdot d\vec{r} = 0 \tag{2.25}$$

and the flow contains no singularities, then the flow is said to be irrotational. Thus for irrotational flows, the line integral $\oint \vec{V} . d\vec{r}$ is independent of path. The line integral depends only on its end point. However, a line integral can be independent of the path of integration only if the integrand is an exact differential, i.e.

$$\vec{V} = \nabla\phi.$$

Extension of this to three-dimensional flows is trivial and is left as an exercise to the reader.

2.9 Stream Function and Vector Potential

We have already defined a streamline as a line drawn in the fluid such that the velocity vector is tangent to this line everywhere. Thus, there cannot be any flow across the streamline anywhere. Since this is identical to the condition at a solid boundary, it follows that any streamline can be replaced by a solid boundary – and this is true only for inviscid flow. In the same way as there is no outflow across a streamline, it is possible to prescribe a stream function ψ to each streamline and the mass flowing through any pair of streamlines is equal to the difference of stream function values of these streamlines. Strictly speaking, stream functions are only defined (mathematically) for two-dimensional flows and for three-dimensional flows this concept can be generalized and the resulting quantity is referred to as the vector potential.

First, we define the stream functions for two-dimensional flows. Examining the continuity equation for an incompressible, two-dimensional flow in the Cartesian frame

$$\nabla . \vec{V} = \frac{\partial u}{\partial x} + \frac{\partial v}{\partial y} = 0. \tag{2.26}$$

It is obvious that the equation is satisfied identically by a stream function ψ, such that

$$u = \frac{\partial \psi}{\partial y}$$

and

$$v = -\frac{\partial \psi}{\partial x}. \tag{2.27}$$

For polar coordinate system, we can obtain the radial and azimuthal components of velocity in terms of the stream function as

$$v_r = \frac{1}{r}\frac{\partial \psi}{\partial \theta}$$

and

$$v_\theta = -\frac{\partial \psi}{\partial r}. \tag{2.28}$$

Thus, the equation of continuity for an incompressible two-dimensional flow is the necessary and sufficient condition for the existence of a stream function. Since ψ is a point function, using Equation (2.27) we obtain

$$d\psi = \frac{\partial \psi}{\partial x}dx + \frac{\partial \psi}{\partial y}dy$$
$$= u\,dy - v\,dx. \tag{2.29}$$

We have already defined a streamline in two-dimensional flow by

$$\frac{dx}{u} = \frac{dy}{v} \quad \text{or} \quad u\,dy - v\,dx = 0.$$

Thus, along a streamline from Equation (2.29) we must have, $d\psi = 0$.

An extension of the concept of streamline from two-dimensional to three-dimensional flows can be easily obtained. We note that the conservation of mass for incompressible flows, $\nabla \cdot \vec{V} = 0$, is automatically satisfied if we postulate the existence of a vector potential, $\vec{\Psi}$ with components in a Cartesian frame given by, ψ_x, ψ_y and ψ_z, in the corresponding coordinate directions, such that

$$\vec{V} = \nabla \times \vec{\Psi}.$$

In Cartesian frame the velocity components are given by

$$u = \frac{\partial \psi_z}{\partial y} - \frac{\partial \psi_y}{\partial z}$$

$$v = -\frac{\partial \psi_z}{\partial x} + \frac{\partial \psi_x}{\partial z}$$

$$w = \frac{\partial \psi_y}{\partial x} - \frac{\partial \psi_x}{\partial y}.$$

2.10 Governing Equation for Irrotational Flows

Note that the two-dimensional flow case can be easily retrieved by noting that for such flows, $\psi_x = \psi_y = 0$ and there are no variations with respect to the third dimension, z.

As the circulation is defined as the line integral of the tangential velocity component around any closed curve, hence referring to Figure 2.7, the circulation is given by

$$\Gamma = \oint_C \vec{V} \cdot d\vec{r}.$$

Following the discussion given after Equation (2.25), one must have

$$\vec{V}.\,d\vec{r} = d\phi$$

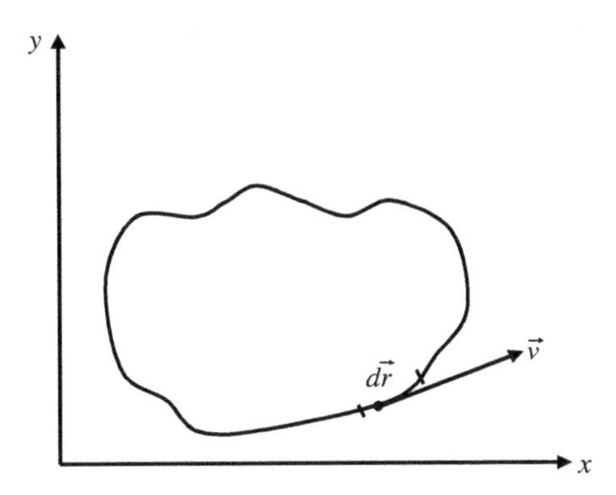

Figure 2.7 Physical implication of circulation as the line integral of tangential component of velocity vector

or

$$u\,dx + v\,dy = \frac{\partial \phi}{\partial x}dx + \frac{\partial \phi}{\partial y}dy$$

$$\vec{V} = \nabla \phi. \tag{2.30}$$

And the condition of irrotationality implies zero vorticity, i.e. $\nabla \times \vec{V} = 0$, which translates into $\nabla \times \nabla \phi = 0$ - an identity, if ϕ is a scalar potential and this is designated as the velocity potential. This is called the irrotational or the potential flow.

When $\vec{V}$ is given by Equation (2.30), then for incompressible flow the equation of continuity gives

$$\nabla . \vec{V} = 0.$$

That is,

$$\nabla^2 \phi = 0. \tag{2.31}$$

This is the governing equation for irrotational flow and one can easily show that satisfaction of this also satisfies the governing momentum conservation equation given by the Navier–Stokes equation. This equation is known as the Laplace's equation, named after the French mathematician who enunciated this for the first time.

A direct consequence of the condition of irrotationality is given by the important theorem due to Kelvin, used for analysis of inviscid flow experiencing conservative body force and the flow is such that the pressure can be expressed uniquely in terms of its density – a condition known as the barotropic flow condition. Incompressible flow is a special case and satisfies barotropic condition.

2.11 Kelvin's Theorem and Irrotationality

The theorem states that 'in an inviscid, homogeneous flow with conservative body forces, the circulation around a closed fluid line remains constant with respect to time.'

A homogeneous (or barotropic) fluid is one in which the density depends only on the pressure – as it is for isothermal flow described above.

From the definition of Γ, we have

$$\frac{d\Gamma}{dt} = \frac{d}{dt}\left(\oint_c \vec{V}\cdot d\vec{r}\right) = \oint_c \frac{d\vec{V}}{dt}\cdot d\vec{r} + \oint_c \vec{V}\cdot\frac{d}{dt}(d\vec{r}\,). \tag{2.32}$$

From Euler's equation

$$\frac{d\vec{V}}{dt} = \vec{f}_b - \frac{1}{\rho}\nabla p.$$

As the body force is considered here as conservative, so $\vec{f}_b = -\nabla F$ and Euler's equation is given by

$$\frac{d\vec{V}}{dt} = -\nabla F - \frac{1}{\rho}\nabla p.$$

Also note that

$$\frac{d}{dt}(d\vec{r}\,) = d\left(\frac{d\vec{r}}{dt}\right) = d\vec{V}.$$

Therefore

$$\frac{d}{dt}\oint_C \vec{V}\cdot d\vec{r} = -\oint_C dF - \oint \frac{dp}{\rho} + \oint_C V\cdot d\vec{V}. \tag{2.33}$$

Note, that for Cartesian coordinate,

$$\nabla F\cdot d\vec{r} = \left(\hat{i}\frac{\partial F}{\partial x} + \hat{j}\frac{\partial F}{\partial y} + \hat{k}\frac{\partial F}{\partial z}\right)\cdot(\hat{i}\,dx + \hat{j}\,dy + \hat{k}\,dz)$$

$$= \frac{\partial F}{\partial x}dx + \frac{\partial F}{\partial y}dy + \frac{\partial F}{\partial z}dz = dF.$$

This can also be generalized in coordinate system free context, by using vector calculus.

Since the density is a function of the pressure only, all the terms on the right-hand side of Equation (2.33) involve exact differentials. Also, the integral of an exact differential around a closed curve is zero and thus,

$$\frac{d}{dt}\left(\oint \vec{V}\cdot d\vec{r}\right) = 0 \tag{2.34}$$

That is, the circulation remains constant along the closed fluid line for the inviscid flow with conservative body force. Kelvin's theorem leads to the important conclusion that the entire flow remains irrotational in the absence of viscous stresses and of discontinuities, provided that the fluid is homogeneous and the body force is conservative and can be described by a potential function.

Another important consequence of irrotationality consequence is the integrability of the momentum equation and leads to the Bernoulli's equation which was derived by Euler and this is shown next.

2.12 Bernoulli's Equation: Relation of Pressure and Velocity

Bernoulli's equation relates pressure with velocity for inviscid, incompressible flow that is also steady and irrotational.

For inviscid steady flow, the governing Euler's equation simplifies to

$$(\vec{V}.\nabla)\vec{V} = \vec{f}_b - \frac{1}{\rho}\nabla p. \tag{2.35}$$

Let us consider body forces which are conservative (e.g., the gravity), then

$$\vec{f}_b = -\nabla F.$$

Also, we have the vector identity

$$(\vec{V}.\nabla)\vec{V} = \nabla\left(\frac{1}{2}|\vec{V}|^2\right) - \vec{V} \times (\nabla \times \vec{V}).$$

Then Equation (2.35) becomes

$$\nabla\left(\frac{1}{2}|\vec{V}|^2\right) + \nabla F + \frac{1}{\rho}\nabla p - \vec{V} \times (\nabla \times \vec{V}) = 0. \tag{2.36}$$

Let us calculate the change in magnitude of each of these terms along an arbitrary path $d\vec{r}$ for a barotropic fluid, i.e. we take a dot product of Equation (2.36) with $d\vec{r}$. The result is

$$d\left(\frac{1}{2}|\vec{V}|^2\right) + dF + \frac{dp}{\rho} - \vec{V} \times (\nabla \times \vec{V}).d\vec{r} = 0.$$

Also, the last term is zero for any arbitrary $d\vec{r}$ for irrotational flows ($\nabla \times \vec{V} = 0$). Note that $\vec{V} \times (\nabla \times \vec{V})$ is perpendicular to $\vec{V}$ and hence for rotational flows if $d\vec{r}$ is along a streamline, then also the last term is zero.

Therefore the first integral of Euler's equation for barotropic fluid is

$$\frac{|\vec{V}|^2}{2} + F + \frac{p}{\rho} = \text{constant}. \tag{2.37}$$

For an unsteady potential flow, one needs to retain the local acceleration term $(\frac{\partial \vec{V}}{\partial t}) = \nabla(\frac{\partial \phi}{\partial t})$, which leads to the general Bernoulli's equation for unsteady irrotational flow as

$$\frac{\partial \phi}{\partial t} + \frac{|\nabla \phi|^2}{2} + F + \frac{p}{\rho} = \text{constant}, \tag{2.38}$$

if the body force potential is due to gravity, then $|F| = gz$ and it is in z-direction. Therefore Equation (2.38) becomes

$$\frac{\partial \phi}{\partial t} + \frac{|\nabla \phi|^2}{2} + gz + \frac{p}{\rho} = \text{constant}. \tag{2.39}$$

Equation (2.37) or (2.39) is well known as the Bernoulli's equation and widely used in solving many fluid-flow problems. The working principle of many instruments in aircraft is governed by this equation. Thus, it is relevant to inquire whether such an equation, arising after making so many limiting assumptions, is still relevant for practical engineering devices. Of specific interest is the question regarding the neglect of viscous effects in deriving this equation. Although this equation has been used in measurement of air speeds, its justification came later via the brilliant work done by Ludwig Prandtl in developing a simple and elegant theory of boundary layer. This is described in Chapter 7, but a general introduction is given here in Section 2.14 for the justification for the use of potential flow theory and Bernoulli's equation.

2.13 Applications of Bernoulli's Equation: Air Speed Indicator

The instrument is based on the operating principle of Pitot static probe. The probe consists of two concentric tubes (P and Q, shown in Figure 2.8); one of which (P) has an open mouth facing the stream of air aligned to it, while the other tube is sealed on to P on all sides, barring some regularly spaced holes at a streamwise location. These holes communicate the ambient pressure to the air inside Q, and the ends of P and Q are connected to the two limbs of a U-tube manometer filled with liquid. When the probe is placed in the airstream, the air entering P stagnates and left limb of the U-tube senses the stagnation pressure of the oncoming air. Similarly the right limb senses the static pressure of the oncoming stream. For the steady oncoming incompressible flow, one can apply Equation (2.39) between the ports P and Q, by noting that both these are at the same height and do not have any potential difference due to

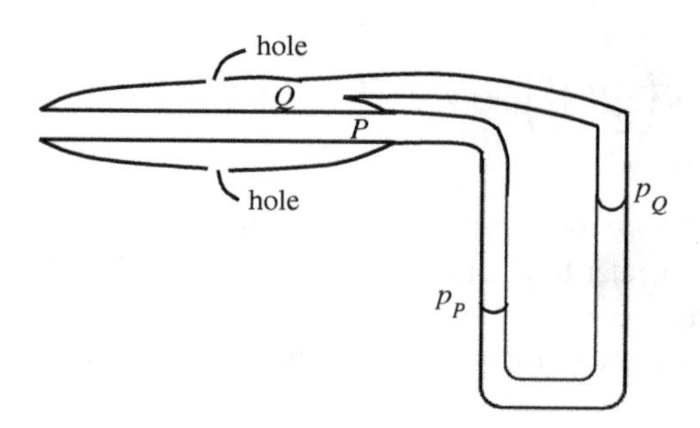

Figure 2.8 A Pitot pressure probe connected to a liquid manometer in a lab

gravity as the conservative body force (i.e. $F_P = F_Q$) and hence

$$p_P + \frac{1}{2}\rho v_P^2 = p_Q + \frac{1}{2}\rho v^2. \tag{2.40}$$

Noting that $\Delta p = p_P - p_Q$, is the pressure difference measured by the manometer and $v_P \equiv 0$, one obtains the measured flow speed from

$$v = \sqrt{2\Delta p/\rho}. \tag{2.41}$$

In a laboratory environment, it is easy to obtain the ambient constant density for the incompressible flow and use the above expression. Equation (2.40) states that the difference between the stagnation pressure and the static pressure as the dynamic pressure which is given by $\frac{1}{2}\rho v^2$ for incompressible flows. For compressible flows the difference between stagnation and static pressures is still given by the dynamic pressure, whose expression is the same obtained from that is given for incompressible flows.

2.13.1 Aircraft Speed Measurement

The air speed indicator works with identical principle for Pitot static probe, but one cannot use liquid manometer as noted above in Figure 2.8. Instead an aneroid barometric capsule or pressure transducer is used for the purpose of measuring air speed. The static pressure is experienced by the exterior of the capsule, while the stagnation pressure is transmitted to the interior. The expansion of the capsule is converted to the motion of an indicator on a dial, which requires a priori calibration with respect to standard condition. In aircraft applications, the standard condition refers to the sea level conditions in ISA. Thus, the reading on the indicator dial is referred to as indicated air speed (IAS). IAS is the speed as installed without correction for system error, but includes compressibility correction. Calibrated air speed (CAS) is the corrected IAS for instrument and position error. IAS without compressibility correction, referred with respect to ISA sea-level condition, is called the equivalent air speed (EAS) and indicated by v_E, i.e. the measured pressure differential is written as

$$\Delta p = \frac{1}{2}\rho_0 v_E^2, \tag{2.42}$$

where $\rho_0 = 1.226 kg\ m^{-3}$, is the sea-level air density for ISA. With respect to the actual flight speed or true air speed (TAS), v and ambient density, ρ, the same pressure differential would be written as

$$\Delta p = \frac{1}{2}\rho v^2. \tag{2.43}$$

Equating Equations (2.42) and (2.43), one obtains the true air speed as

$$v_E = v\sqrt{\sigma},$$

where the relative density is given by, $\sigma = \rho/\rho_0$.

2.13.2 The Pressure Coefficient

Application of Bernoulli's equation between any point with pressure p and velocity v with any point in the free stream condition is given by

$$p + \frac{1}{2}\rho v^2 = p_\infty + \frac{1}{2}\rho V_\infty^2, \tag{2.44}$$

where the condition at the free stream is indicated by the subscript ∞ in the above. Noting that the dynamic pressure of free stream is given by $\frac{1}{2}\rho V_\infty^2$, one can define the variation of local pressure with respect to the free stream pressure by the pressure coefficient by

$$C_p = \frac{(p - p_\infty)}{\frac{1}{2}\rho V_\infty^2}. \tag{2.45}$$

It is readily apparent that the pressure coefficient at the stagnation point is equal to unity. Also using Equation (2.44) in the above, one readily obtains

$$C_p = 1 - \left(\frac{v}{V_\infty}\right)^2. \tag{2.46}$$

This clearly indicates that a negative C_p implies local acceleration with respect to free stream speed and a positive value indicates deceleration with respect to free stream speed.

2.13.3 Compressibility Correction for Air Speed Indicator

First of all we note the property of compressible flows, when it is considered in its simplest one-dimensional unsteady flow condition, with the governing continuity and Euler equations given for a stream tube of area A by

$$\frac{\partial}{\partial x}(\rho u A) + \frac{\partial}{\partial t}(\rho A) = 0 \tag{2.47}$$

$$\frac{\partial u}{\partial t} + u\frac{\partial u}{\partial x} = -\frac{1}{\rho}\frac{\partial p}{\partial x}. \tag{2.48}$$

Multiplying Equation (2.47) by u and Equation (2.48) by ρA and adding, one gets

$$\frac{\partial}{\partial t}(\rho u A) + \frac{\partial}{\partial x}(\rho A u^2) = -\frac{\partial}{\partial x}(pA) + p\frac{\partial A}{\partial x}. \tag{2.49}$$

Which upon integration between stations '1' and '2' yields

$$\frac{\partial}{\partial t}\int_1^2 (\rho u A)\, dx + (\rho_2 u_2^2 A_2 - \rho_1 u_1^2 A_1) = (p_1 A_1 - p_2 A_2) + \int p\, dA, \tag{2.50}$$

where the subscripts indicate the values at respective stations. If we assume steady flow condition, then this can be further simplified by approximating the last integral on the right-hand side as $\frac{1}{2}(p_1 + p_2)(A_2 - A_1)$ to obtain

$$\rho_2 u_2^2 A_2 - \rho_1 u_1^2 A_1 - (p_1 A_1 - p_2 A_2) - \frac{1}{2}(p_1 + p_2)(A_2 - A_1) = 0. \tag{2.51}$$

This is supplemented with mass and energy conservation equations for steady flow as

$$\rho_2 u_2 A_2 = \rho_1 u_1 A_1 \tag{2.52}$$

$$c_p T_1 + \frac{u_1^2}{2} = c_p T_2 + \frac{u_2^2}{2}. \tag{2.53}$$

Noting that the speed of sound is given by $a = \sqrt{\gamma p/\rho} = \sqrt{\gamma RT}$ and $c_p = \frac{\gamma R}{\gamma - 1}$, the energy conservation equation, Equation (2.53) can be alternatively written as

$$\frac{u_1^2}{2} + \frac{\gamma}{\gamma - 1}\frac{p_1}{\rho_1} = \frac{u_2^2}{2} + \frac{\gamma}{\gamma - 1}\frac{p_2}{\rho_2}.$$

This can be further simplified to

$$\frac{u_2^2 - u_1^2}{2} = \frac{\gamma}{\gamma - 1}\left(\frac{p_1}{\rho_1} - \frac{p_2}{\rho_2}\right).$$

The states at stations '1' and '2' for the loss-less flow without heat transfer can be related by isentropic relation as $p_1/p_2 = (\rho_1/\rho_2)^\gamma$ so that the above equation can be written as

$$\frac{u_2^2 - u_1^2}{2} = \frac{\gamma}{\gamma - 1}\frac{p_2}{\rho_2}\left[\left(\frac{p_1}{p_2}\right)^{(\gamma-1)/\gamma} - 1\right]. \tag{2.54}$$

As $\gamma p_2/\rho_2 = a_2^2$, Equation (2.54) can be used to represent the pressure ratio as

$$\frac{p_1}{p_2} = \left[1 + \frac{\gamma - 1}{2}\frac{u_2^2 - u_1^2}{a_2^2}\right]^{\gamma/(\gamma-1)}. \tag{2.55}$$

If the station '1' is identified with the reservoir condition, i.e. $u_1 = 0$ and $p_1 = p_0$, then Equation (2.55) can be used to relate the stagnation condition with conditions at any station identified without any subscript by

$$\frac{p_0}{p} = \left[1 + \frac{\gamma - 1}{2}M^2\right]^{\gamma/(\gamma-1)}. \tag{2.56}$$

For air, one can consider it as primarily composed of diatomic gasses and use $\gamma = 7/5$ so that

$$\frac{p_0}{p} = \left[1 + \frac{1}{5}M^2\right]^{7/2}. \tag{2.57}$$

This relation is used to demonstrate compressibility correction for the air speed indicator by considering the pressure coefficient at the stagnation point, which by definition for incompressible flow is equal to one. Note that the free stream dynamic pressure is given by,

$$\frac{1}{2}\rho_\infty V_\infty^2 = \frac{1}{2}\left(\frac{\rho_\infty}{\gamma p_\infty}\right)\gamma p_\infty V_\infty^2.$$

As the speed of sound is defined as $a_\infty^2 = \frac{\gamma p_\infty}{\rho_\infty}$, the Mach number is $M_\infty = V_\infty / a_\infty$ and the dynamic pressure is written as $\frac{1}{2}\gamma p_\infty M_\infty^2$.

For flow of air with compressibility effects, the stagnation pressure coefficient is given by

$$C_{po} = \frac{1}{0.7M^2}\left[\frac{p_0}{p} - 1\right]. \tag{2.58}$$

Using the pressure ratio in Equation (2.57) in this equation, one gets

$$C_{po} = \frac{1}{0.7M^2}\left(\left[1 + \frac{1}{5}M^2\right]^{7/2} - 1\right). \tag{2.59}$$

2.14 Viscous Effects and Boundary Layers

Theoretical developments in aerodynamic studies took place in the late nineteenth and early twentieth century, utilizing irrotational flow models. The utility of such methods was supported strongly by the introduction of the concept of boundary layers by Ludwig Prandtl. This is enunciated based on phenomenology of the viscous action being confined very close to the boundary of thin aerodynamic shapes.

A qualitative view of the same can also be obtained directly from the Navier-Stokes equation. From Equation (2.22) one can look at an order of magnitude analysis for each of the terms represented in vectorial form as

$$\frac{\partial \vec{V}}{\partial t} + \vec{V} \cdot \nabla \vec{V} = -\frac{1}{\rho}\nabla p + \nu\, \nabla^2 \vec{V}. \tag{2.60}$$

In writing this equation we have omitted the body force without affecting the final outcome of the analysis. Consider an aircraft wing typically represented by an airfoil of chord c flying level and steady at a forward speed of U_∞. For such a cruise flight condition, the airfoil experiences a steady flow and the local acceleration term in Equation (2.60) is negligible. As stated above, the airfoil as the quintessential element displays two types of variation of flow variables. In the streamwise direction, the flow velocity (u) varies from zero value on the airfoil surface to U_∞ at a short wall-normal distance in a thin layer. Let us say that the dimension of this thin layer is indicated by δ, and then the order of magnitude of the wall-normal component of velocity (v) is obtained from the continuity equation as given by $U_\infty \delta / c$. This smaller order of magnitude of this component with respect to the streamwise component is the reason for the smaller thickness of the shear layer forming over the airfoil.

At any location over the airfoil, the convective acceleration term $\vec{V} \cdot \nabla \vec{V}$ of the streamwise momentum equation is seen to have the order of magnitude given by U_∞^2 / c, i.e.

$$\vec{V} \cdot \nabla \vec{V} \sim U_\infty^2 / c.$$

The pressure gradient term can at the most be of the same order of magnitude as the acceleration terms, when it is present in the streamwise momentum equation. One of the consequences of the thin boundary layer for the diffusion term $\nu\, \nabla^2 \vec{V}$ in the streamwise momentum equation is that the streamwise diffusion term is of the order of $\nu U_\infty / c^2$, which is significantly smaller than the wall-normal diffusion term given by $\nu U_\infty/\delta^2$. One of the consequence of this observation is that one can simplify the streamwise momentum equation, as will be shown in greater details in Chapter 7. The kinetics of the fluid flow demands that the convective acceleration terms on the left-hand side of Equation (2.60) must by balanced by viscous diffusion term on the right-hand side in terms of order of magnitude in this thin layer next to the airfoil, i.e. the viscous diffusion is of the same order of magnitude of the convective acceleration terms and cannot be neglected inside the boundary layer. Thus one must have,

$$U_\infty^2/c \sim \nu U_\infty/\delta^2.$$

This directly implies

$$\frac{\delta}{c} \sim \left(\frac{U_\infty c}{\nu}\right)^{-1/2}.$$

The quantity inside the bracket of the right-hand side is nothing but the Reynolds number defined in terms of the chord of the airfoil and free stream speed. Thus, the thickness of the shear layer nondimensionalzed by the chord is inversely proportional to the square root of the Reynolds number, implying the boundary layer thickness decreases with an increase in the Reynolds number.

One notes that in typical aeronautical applications, Reynolds number is of the order of 10^6 and then the boundary layer thickness is about one thousandth of the chord of the airfoil, i.e. there exists only a very thin shear layer next to the body where the flow cannot be treated as inviscid. Outside the boundary layer, the flow can be considered inviscid. As we will explain in detail in Chapter 7, the action of no-slip condition at the wall is to create vorticity, which convects and diffuses in general, following the Navier-Stokes equation. It is to be understood that the diffusion process transports the wall vorticity to the interior and it falls off to vanishingly small value at the edge of the boundary layer. The steady state of the flow implies a dynamic balance at all times and it is quite reasonable to assume the flow to be inviscid past the augmented body, which now represents the actual airfoil plus the boundary layer added to it. If such a body is started impulsively from rest, then following Kelvin's theorem, the flow can be treated as irrotational also, as the flow has zero circulation to begin with following the impulsive start. This is the basis of using inviscid, irrotational flow in the most part of the flow field, as we propose to do in the following few chapters up to Chapter 6. However, such an assumption will not allow obtaining any estimate of the viscous action, which can be obtained, by the simplification brought about by solving the boundary layer equation, as will be performed in Chapter 7.

2.15 Thermodynamics and Reynolds Transport Theorem

To analyze fluid flows or to study aerodynamics of high-speed flight heat transfer becomes important. For such flows we need to account for it by considering energy conservation. This is traditionally performed for a system with fixed mass (control mass system) with the help of first and second laws of thermodynamics. However for fluid flow, one cannot use a control mass system; instead we consider a control volume system for analysis, using Reynolds transport theorem, which is a general framework for all conservation principles. In control volume system analysis, we focus our attention on a fixed volume in space, distinguished by a control surface separating the system from the surrounding. Here, we mainly concentrate on conservation of energy.

2.16 Reynolds Transport Theorem

We can have control volume which is fixed, moving or deformable. However, we will only derive the theorem for a fixed control volume. Thereafter, we will state the same for moving or deformable control volume.

In Figure 2.9, we indicate the fixed control volume, CV by the solid thick-line, which also coincides with the control mass system at 't'. The control volume is distinguished from the surrounding by the control surface, CS. During the subsequent time interval, dt an amount of fluid enters CV, which is indicated by the area shaded by vertical lines. If we consider an elementary area for this purpose as dA_{in}, with outward normal drawn by $\hat{n}$, then the mass inflow through this element is given by

$$dm_{in} = -\vec{V} \cdot \hat{n}\, dA_{in} dt. \tag{2.61}$$

By convection, outflow is considered positive, while inflow is indicated by a negative sign. Actually, the sign will be determined by the angle θ between the velocity vector at the control surface, $\vec{V}$ and the outward drawn normal $\hat{n}$ to the area element. The control surface will be locally a streamline for $\theta = 90^o$. At the no-slip solid wall $\vec{V} = 0$ and such contributions are absent for inflow or outflow.

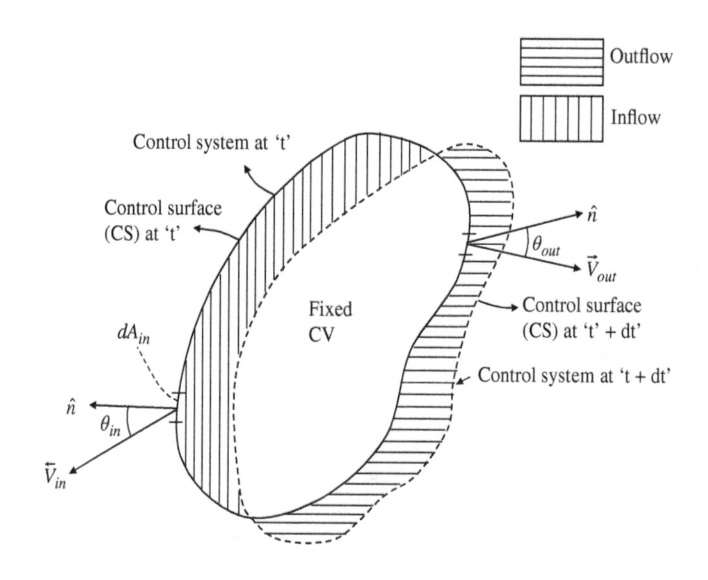

Figure 2.9 Arbitrary control volume within a flow domain

Let us consider Φ as same physical property and let ϕ_1 be the corresponding intensive or specifies property, i.e.

$$\phi_1 = \frac{d\Phi}{dm}$$

The control volume has the total property given by $\Phi_{CV} = \int_{CV} \rho\, \phi_1\, d\, \mathcal{V}$. The time rate of the property in the control volume arises from:

(i) Change in Φ_{CV} from $\frac{d}{dt}\left(\int_{CV} \phi_1\, \rho\, d\, \mathcal{V}\right)$. (2.62)

(ii) Inflow and outflow of the property from the control volume is given by

$$\int_{CS} \phi_1\, \rho\, \vec{V} \cdot d\vec{A}. \tag{2.63}$$

The Reynolds transport theorem equates two changes taken together as the net rate of change of property in the system to be given by

$$\frac{d}{dt}(\Phi_{sys}) = \frac{d}{dt}\left(\int_{CV} \phi_1\, \rho\, d\mathcal{V}\right) + \int_{CS} \phi_1\, \rho\, \vec{V} \cdot d\vec{A}. \tag{2.64}$$

For a fixed control volume, one can introduce the time derivative inside the integral sign, on the right-hand side of Equation (2.64).

If the control volume is moving (a) at a constant speed or (b) the control surfaces are deforming, then the Reynolds transport theorem is given by

$$\frac{d}{dt}(\Phi_{sys}) = \frac{d}{dt}\left(\int_{CV} \phi_1\, \rho\, \mathcal{V}\right) + \int_{CS} \phi_1\, \rho\, (\vec{V}_r \cdot d\vec{A}), \tag{2.65}$$

where $\vec{V}_r = \vec{V} - V_S$; V_S being the constant speed for case (a) and $\vec{V}$ is the fluid velocity relative to the reference coordinate system.

When the CS is deforming with time given by $\vec{V}_s(\vec{r}, t)$ and the fluid velocity at the control surface is given by $\vec{V}(\vec{r}, t)$, then in Equation (2.65)

$$\vec{V}_r = \vec{V}(\vec{r}, t) - \vec{V}_S(\vec{r}, t). \tag{2.66}$$

2.17 The Energy Equation

For fluid flow, we apply the Reynolds transport theorem as given by Equation (2.64) or (2.65) to first law of thermodynamics, with Φ_{sys} denoting energy E and the specific energy is give by $\phi_1 = e = \frac{dE}{dm}$. Writing the first law in rate form for the flow process as

$$\frac{dQ}{dt} - \frac{dW}{dt} = \frac{dE}{dt}, \tag{2.67}$$

with the right-hand side given by the Reynolds transport theorem by

$$\frac{dE}{dt} = \frac{d}{dt}\left(\int_{CV} e\,\rho\, d\,\mathcal{V}\right) + \int_{CS} e\,\rho\,(\vec{V}\cdot d\,\vec{A}), \tag{2.68}$$

we use the sign convection for Q as positive denoting heat added to the system and positive W indicates work done by the system.

The specific energy can be further subdivided as

$$e = e_{int} + e_{ke} + e_{pot}. \tag{2.69}$$

To the right-hand side of (2.69), one should add other components due to reactions (Chemical or nuclear) or body forces not accounted for in the potential energy component, $e_{pot} = gz$ where z is measured with respect to some datum. Thus,

$$e = e_{int} + \frac{1}{2}|\vec{V}|^2 + gz, \tag{2.70}$$

Similarity, the rate of work done term in Equation (2.67) can be subdivided by terms with dot indicating time derivative by

$$\mathring{W} = \mathring{W}_{sh} + \mathring{W}_p + \mathring{W}_V, \tag{2.71}$$

where $\mathring{W}_{sh}$ indicates shaft work done by active devices like fan, impeller, stream etc.; $\mathring{W}_p$ indicates work done by pressure via the movements of the control surface with internal pressure mutually cancelling effects (telescopically) and given by $\mathring{W}_p = \int_{CS} p(\vec{V}\cdot d\vec{A})$ finally, $\mathring{W}_V$ indicates the work due to viscous stresses acting on the control surfaces given in terms of the shear stress $\bar{\bar{\tau}}$, by the shear as

$$\mathring{W}_V = -\int_{CS} (\bar{\bar{\tau}}\cdot\vec{V})_{ss}\cdot d\vec{A}. \tag{2.72}$$

Note here that p and τ_{ss} are stress tensor components and the dot product in Equation (2.72) involving $\bar{\bar{\tau}}$ is a contraction of the tensor with a vector. On the solid surface, $\vec{V} \equiv 0$ and hence $\mathring{W}_V \equiv 0$ identically. If the flow at inflow and outflow is normal to the area element and only the normal stress terms contribution as $\mathring{W}_V = \int \tau_{nn} V_n |d\vec{A}|$. Such terms are usually negligible, except at the shock wave interior. If the control surface is part of a stream surface, then $\mathring{W}_V$ is nonzero depending upon the shear stress value. Thus, the rate of work term in Equation (2.71) consists of

$$\mathring{W} = \mathring{W}_{sh} + \int_{CS} p\vec{V}\cdot d\vec{A} - \int_{CS} (\bar{\bar{\tau}}\cdot\vec{V})_{ss}\cdot d\vec{A}, \tag{2.73}$$

where the subscript ss stands for stream surface. Including all these terms in Equation (2.68) one gets

$$\mathring{Q} - \mathring{W}_{sh} - \mathring{W}_V = \frac{\partial}{\partial t}\left(\int_{CV} e\,\rho\, d\,\mathcal{V}\right) + \int_{CS}\left(e + \frac{p}{\rho}\right)\rho\vec{V}\cdot d\vec{A}. \tag{2.74}$$

Noting that $(e_{int} + \frac{p}{\rho})$ is nothing but the specific enthalpy, h, the final form of first law far a flow process is given by

$$\mathring{Q} - \mathring{W}_{sh} - \mathring{W}_V = \frac{\partial}{\partial t}\left[\int_{CV}\left(e_{int} + \frac{|\vec{V}|^2}{2} + gz\right)\rho\, d\,\mathcal{V}\right] + \int_{CS}\left(h + \frac{|\vec{V}|^2}{2} + gz\right)\rho\vec{V}\cdot d\vec{A}. \tag{2.75}$$

In Chapter 11, while discussing compressible flows, we will discuss the one-dimensional version of Equation (2.75) with a series of one-dimensional inlets and outlets. The surface integral of Equation (2.75) then simplifies to

$$\int_{CS}\left(h + \frac{|\vec{V}|^2}{2} + gz\right)\rho\vec{V}\cdot d\vec{A} = \sum_{out}\left(h + \frac{|\vec{V}|^2}{2} + gz\right)_{out} \mathring{m}_{out} - \sum_{in}\left(h + \frac{|\vec{V}|^2}{2} + gz\right)_{in} \mathring{m}_{in}. \tag{2.76}$$

2.17.1 *The Steady Flow Energy Equation*

Consider the case of steady flow with one set of inlet and outlet treated as one-dimensional port; the energy equation takes the simplified form:

$$\mathring{Q} - \mathring{W}_{sh} - \mathring{W}_V = \mathring{m}_1(h_1 + \frac{1}{2}|\vec{V}_1|^2 + g\,z_1) - \mathring{m}_2(h_2 + \frac{1}{2}|\vec{V}_2|^2 + g\,z_2). \tag{2.77}$$

From mass conservation, $\mathring{m}_1 = \mathring{m}_2 = \mathring{m}$, and Equation (2.77) simplifies to

$$h_1 + \frac{1}{2}|\vec{V}_1|^2 + g\,z_1 = h_2 + \frac{1}{2}|\vec{V}_2|^2 + g\,z_2 - q + w_{sh} + w_V. \tag{2.78}$$

where $q = \frac{\mathring{Q}}{\mathring{m}} = \frac{dQ}{dm}$, the heat transferred per unit mass. Similarity

$$w_{sh} = \frac{dW_{sh}}{dm} \text{ and } w_V = \frac{dW_V}{dm}.$$

Noting that the stagnation enthalpy can be define as $H_i = h_i + \frac{1}{2}|\vec{V}_i|^2$ and the steady energy equation can be written as

$$H_1 + gz_1 = H_2 + gz_2 - q + w_{sh} + w_V. \tag{2.79}$$

where q is positive, if heat is added to CV and w_{sh} and w_V are positive, if work is done by fluid upon the surrounding.

2.18 Energy Conservation Equation

Here, the energy equation is derived as the first law of thermodynamics stated for a control volume system. In Figure 2.10, the *control volume* element has been shown with only the stress system acting in x-direction.

The first law of thermodynamics for a control system simply states: The rate of change of energy inside the *control volume* must be due to the heat interaction across the *control surface* plus the work done reversibly due to boundary displacement by body and surface forces. Each of the constituent terms are obtained next.

The work done term:

The rate of work done by body forces for the mass $\rho\ (dx\ dy\ dz)$ is

$$\rho\vec{F}\cdot\vec{V}(dx\ dy\ dz)$$

Contributions of surface forces acting on the element are due to normal and shear stresses. Only the contribution by x-component of forces are shown in Figure 2.10 and the rate of work done is the product of forces in the x-direction and the x-component of velocity (u). On face *abcd* of Figure 2.10, the work done term is, $\tau_{yx}\ dx\ dz\ u$. Similarly, the work done term due to shear stress acting on the face *efgh* is given by, $\left[u\tau_{yx} + \frac{\partial(u\tau_{yx})}{\partial y}dy\right]dx\ dz$.

Since the two shear forces are acting in the opposite direction, the net energy flux in the x-direction due to the shear stresses in the x-direction on faces *abcd* and *efgh* is $\frac{\partial(u\tau_{yx})}{\partial y}\ dx\ dy\ dz$.

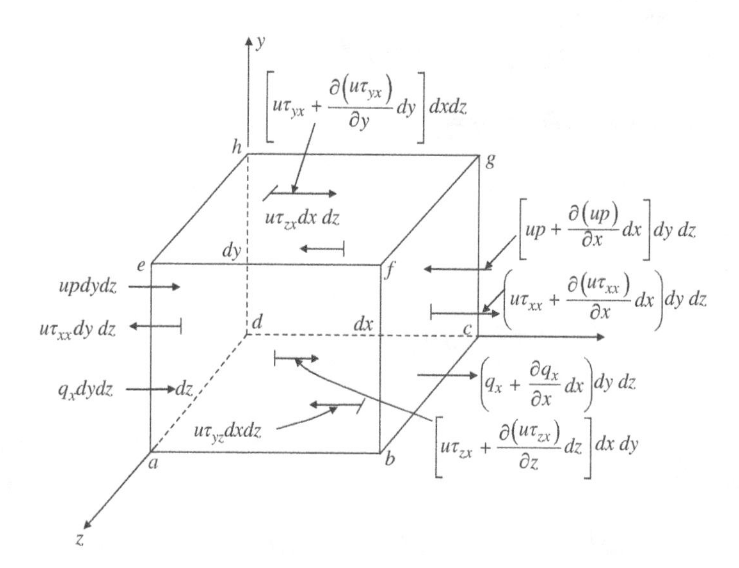

Figure 2.10 Energy fluxes associated to and from a control volume system. Only the fluxes in the x-direction are shown for clarity

Similarly, the pressure acting on faces *adhe* and *bcgf* give rise to the work done term

$$\left[up - (up + \frac{\partial(up)}{\partial x}dx)\right]dy\,dz = -\frac{\partial(up)}{\partial x}dx\,dy\,dz.$$

Now, considering all the forces shown in Figure 2.10, net energy flux for the *control volume* due to these forces is simply obtained as

$$\left[-\frac{\partial(up)}{\partial x} + \frac{\partial(u\tau_{xx})}{\partial x} + \frac{\partial(u\tau_{yx})}{\partial y} + \frac{\partial(u\tau_{zx})}{\partial z}\right]dx\,dy\,dz. \tag{2.80}$$

Expressions in Equation (2.80), are obtained by considering forces acting in x-direction only. When the surface forces in y- and z-directions are also included in deriving the work done term, we get the total contributions coming from body and surface forces as

$$\left[-\nabla \cdot (p\vec{V}) + \frac{\partial}{\partial x_j}(u_i\tau_{ji})\right]dx\,dy\,dz + \rho\vec{F}\cdot\vec{V}dx\,dy\,dz. \tag{2.81}$$

The heat transfer term:

The net heat flux is due to volumetric heating such as absorption/emission of radiation and heat transfer across the control surface due to thermal conduction. If $\dot{q}$ denotes rate of volumetric heat addition per unit mass, then the volumetric heating of the element is

$$= \rho\dot{q}\,dx\,dy\,dz. \tag{2.82}$$

In Figure 2.10, heat transferred due to conduction across the face *adhe* is: $\dot{q}_x\,dy\,dz$, where $\dot{q}_x$ is the heat flux in x-direction, per unit time, per unit area, by thermal conduction. Similarly, heat flux out of element across face *bcfg* is: $[\dot{q}_x + \frac{\partial\dot{q}_x}{\partial x}dx]\,dy\,dz$.

Net heat flux in x-direction, into the fluid element by thermal conduction is

$$= -\frac{\partial\dot{q}_x}{\partial x}\,dx\,dy\,dz. \tag{2.83}$$

Therefore, heating of the fluid element by thermal conduction is given by, considering all directional conduction contributions as

$$= -\frac{\partial(\dot{q}_j)}{\partial x_j}\,dx\,dy\,dz. \tag{2.84}$$

Once again, Einstein's summation convention is used in Equation (2.84). Using Newton's law, one can relate directional conductive heat transfer with temperature gradient, using the Fourier law by, $\dot{q}_j = -k\frac{\partial T}{\partial x_j}$, where k is the thermal conductivity. Hence, the total heat interaction term is obtained by using the Newton's law in Equation (2.84) and adding Equation (2.82) as

$$\left[\rho\dot{q} + \frac{\partial}{\partial x_j}\left(k\frac{\partial T}{\partial x_j}\right)\right]dx\,dy\,dz. \tag{2.85}$$

The rate of change of energy term:
The total energy of a moving fluid, per unit mass, is the sum of its internal energy per unit mass (e), and its kinetic energy per unit mass, $\frac{V^2}{2}$. Since we are following a moving fluid element, time rate of change of energy of unit mass is given by the substantive derivative as

$$\rho \frac{D}{Dt}\left(e + \frac{V^2}{2}\right) dx\, dy\, dz. \tag{2.86}$$

The final form:
The final form of energy equation is obtained by collating terms in Equations (2.80), (2.85) and (2.86) to obtain the *nonconservation* or *convective* form of the energy equation as

$$\rho \frac{D}{Dt}\left(e + \frac{V^2}{2}\right) = \left[\rho \dot{q} + \frac{\partial}{\partial x_j}\left(k \frac{\partial T}{\partial x_j}\right)\right] - \nabla \cdot (p\vec{V}) + \frac{\partial}{\partial x_j}(u_i \tau_{ji}) + \rho \vec{F} \cdot \vec{V}. \tag{2.87}$$

2.19 Alternate Forms of Energy Equation

Occasionally, energy equation is written in terms of the internal energy only. This is achieved by rewriting the momentum equation, Equation (2.14) after multiplying by u and multiplying y- and z-momentum equations by v and w, respectively. The component equations are obtained as

$$\rho \frac{D}{Dt}\left(\frac{u^2}{2}\right) = -u \frac{\partial p}{\partial x} + u\left(\frac{\partial \tau_{xx}}{\partial x} + \frac{\partial \tau_{yx}}{\partial y} + \frac{\partial \tau_{zx}}{\partial z}\right) + \rho u f_x \tag{2.88a}$$

$$\rho \frac{D}{Dt}\left(\frac{v^2}{2}\right) = -v \frac{\partial p}{\partial y} + v\left(\frac{\partial \tau_{xy}}{\partial x} + \frac{\partial \tau_{yy}}{\partial y} + \frac{\partial \tau_{zy}}{\partial z}\right) + \rho v f_y \tag{2.88b}$$

$$\rho \frac{D}{Dt}\left(\frac{w^2}{2}\right) = -w \frac{\partial p}{\partial z} + w\left(\frac{\partial \tau_{xz}}{\partial x} + \frac{\partial \tau_{yz}}{\partial y} + \frac{\partial \tau_{zz}}{\partial z}\right) + \rho w f_z. \tag{2.88c}$$

We have expressed the components of body force ($\vec{F}$) as f_x, f_y and f_z.

Adding these three equations one gets

$$\rho \frac{D}{Dt}\left(\frac{V^2}{2}\right) = -\vec{V} \cdot \nabla p + \rho \vec{F} \cdot \vec{V} + v_j \frac{\partial \tau_{ij}}{\partial x_i}. \tag{2.89}$$

Subtracting Equation (2.89) from Equation (2.87), one gets

$$\rho \frac{De}{Dt} = \rho \dot{q} + \frac{\partial}{\partial x_j}\left(k \frac{\partial T}{\partial x_j}\right) - p\nabla \cdot \vec{V} + \tau_{xx}\frac{\partial u}{\partial x} + \tau_{yx}\frac{\partial u}{\partial y} + \tau_{zx}\frac{\partial u}{\partial z} + \tau_{xy}\frac{\partial v}{\partial y}$$
$$+ \tau_{yy}\frac{\partial v}{\partial y} + \tau_{zy}\frac{\partial v}{\partial z} + \tau_{xz}\frac{\partial w}{\partial x} + \tau_{yz}\frac{\partial w}{\partial y} + \tau_{zz}\frac{\partial w}{\partial z}. \tag{2.90}$$

Equations (2.87) and (2.90) can be written in terms of flow variables by using the stress–strain relation, Equation (2.18) and also noting that the stress tensor is symmetric. Symmetry of stress tensor arises due to absence of distributed couples or moments in the flow field. Thus, Equation (2.90) simplifies to

$$\rho\frac{De}{Dt} = \rho\dot{q} + \frac{\partial}{\partial x_j}\left(k\frac{\partial T}{\partial x_j}\right) - p\nabla\cdot\vec{V} + \lambda(\nabla\cdot\vec{V})^2 + 2\mu\left[\left(\frac{\partial u}{\partial x}\right)^2 + \left(\frac{\partial v}{\partial y}\right)^2 + \left(\frac{\partial w}{\partial z}\right)^2\right]$$

$$+\mu\left[\left(\frac{\partial u}{\partial y} + \frac{\partial v}{\partial x}\right)^2 + \left(\frac{\partial u}{\partial z} + \frac{\partial w}{\partial x}\right)^2 + \left(\frac{\partial v}{\partial z} + \frac{\partial w}{\partial y}\right)^2\right]. \tag{2.91}$$

2.20 The Energy Equation in Conservation Form

This is readily obtained from Equation (2.91) by noting that

$$\rho\frac{De}{Dt} = \rho\frac{\partial e}{\partial t} + \rho(\vec{V}\cdot\nabla)e$$

and $\quad\nabla\cdot(\rho e\vec{V}) = e\nabla\cdot(\rho\vec{V}) + \rho(\vec{V}\cdot\nabla)e.$

Therefore $\quad\rho\frac{De}{Dt} = \frac{\partial(\rho e)}{\partial t} - e\left[\frac{\partial\rho}{\partial t} + \nabla\cdot(\rho\vec{V})\right] + \nabla\cdot(\rho e\vec{V}).$

The quantity in the square bracket is zero by Equation (2.6), and thus the left-hand side of Equation (2.91) is now expressed in the *conservation form*. The right-hand side of Equation (2.91) is already in the conservation form. This is also known as the divergence from of energy equation.

2.21 Strong Conservation and Weak Conservation Forms

The form of equations derived so far involves divergence of various flux terms and are said to be in weak conservation form. In a physical sense, flux terms are important contributors to the governing equations and hence the *divergence form* is often preferred for application in computing. All governing equations in fluid mechanics can also be written in the generic form

$$\frac{\partial U}{\partial t} + \frac{\partial F}{\partial x} + \frac{\partial G}{\partial y} + \frac{\partial H}{\partial z} = S. \tag{2.92}$$

This can be used to represent all the equations together (in CFD these are referred together as the Navier–Stokes equation) in the conservation form, where U, F, G and H are column vectors given by

$$U = \left[\rho \quad \rho u \quad \rho v \quad \rho w \quad \rho\left(e + \frac{V^2}{2}\right)\right]^T$$

$$F = \left[\rho u \quad (\rho u^2 + p - \tau_{xx}) \quad (\rho uv - \tau_{xy}) \quad (\rho uw - \tau_{xz}) \quad \rho u(e + \frac{V^2}{2})\right.$$

$$\left. + pu - k\frac{\partial T}{\partial x} - u\tau_{xx} - v\tau_{xy} - w\tau_{xz}\right]^T$$

$$G = \left[\rho v \quad (\rho uv - \tau_{yx}) \quad (\rho v^2 + p - \tau_{yy}) \quad (\rho vw - \tau_{yz}) \quad \rho v(e + \frac{V^2}{2}) \right.$$

$$\left. + pv - k\frac{\partial T}{\partial y} - u\tau_{yx} - v\tau_{yy} - w\tau_{yz}\right]^T$$

$$H = \left[\rho w \quad (\rho uw - \tau_{zx}) \quad (\rho vw - \tau_{zy}) \quad (\rho w^2 + p - \tau_{zz}) \quad \rho w(e + \frac{V^2}{2}) \right.$$

$$\left. + pw - k\frac{\partial T}{\partial z} - u\tau_{zx} - v\tau_{zy} - w\tau_{zz}\right]^T$$

and $$S = \left[0 \quad \rho f_x \quad \rho f_y \quad \rho f_z \quad \rho\vec{F}\cdot\vec{V} + \rho\dot{q}\right]^T.$$

In Equation (2.92), column vectors F, G and H are the flux terms and S is the source term. For an unsteady problem, U is called the state vector, as this is obtained numerically as a function of time.

When the conservation laws are written in the form given by Equation (2.92), we regard the equation to be in the strong conservation form, as opposed to the weak conservation form in Equation (2.91) for the energy equation. The reason for this nomenclature is simple, because in the strong conservation form, any symmetric discretization method does not affect the solution in the interior as the neighbouring cell terms 'telescope' into each other, leaving only the boundary terms, for a true boundary value problem. This is not the case with weak conservation form, where the numerical solution depends on the process of discretization. It is also possible to show that for compressible flows, in the presence of discontinuities for the physical variables, fluxes remain continuous, while the physical variables can be discontinuous. Continuous functions are easier to compute accurately rather than the discontinuous functions; the latter has larger bandwidth in wavenumber space. This is the second reason for using governing equations in strong conservation form. The third reason is that this form remains invariant even when the equations of motions are transformed in arbitrary curvilinear coordinate system.

2.22 Second Law of Thermodynamics and Entropy

Considering the extensive internal energy, E, one can define a thermodynamic pressure for a reversible change of state of an adiabatic process from

$$p = -\frac{\partial E}{\partial \mathcal{V}}. \tag{2.93}$$

This makes conceptual sense, as the change of energy for an elemental volume the adiabatic process is equal to work done by the pressure and is given by, $pd\mathcal{V}$. The negative sign indicating a stable system for which compressive force leads to reduction of volume. For one reversible mode of work, the system can be characterized by any two independent thermodynamic variables.

In conformity with the relation in Equation (2.93) one can define the temperature in terms of a new variable, which we will call the entropy. If this extensive property is denoted by S, then one can express the internal energy as

$$E = E(S, \mathcal{V}) \tag{2.94}$$

such that

$$p = -\left(\frac{\partial E}{\partial \mathcal{V}}\right)_S \tag{2.95a}$$

and

$$T = \left(\frac{\partial E}{\partial S}\right)_{\mathcal{V}}. \tag{2.95b}$$

Hence the incremental change in E for a reversible process can be expressed using Equation (2.94) as

$$dE = \left(\frac{\partial E}{\partial \mathcal{V}}\right)_S d\mathcal{V} + \left(\frac{\partial E}{\partial S}\right)_{\mathcal{V}} dS. \tag{2.96}$$

Using the relations Equations (2.95a) and (2.95b) in Equation (2.96), one obtains

$$dE = -pd\,\mathcal{V} + TdS. \tag{2.97}$$

Comparing this with the first law given in Equation (2.67) for a control mass by system

$$dE = -pd\,\mathcal{V} + dQ. \tag{2.98}$$

Thus, from Equations (2.97) and (2.98), we have the identity for a reversible process,

$$TdS = dQ. \tag{2.99}$$

It can be shown that entropy is a state property and its change from state '1' to '2' can be written as

$$S_2 - S_1 = \int_1^2 \frac{dQ}{T}, \tag{2.100}$$

using Equation (2.98)

$$S_2 - S_1 = \int_1^2 \frac{dE + pd\mathcal{V}}{T}. \tag{2.101}$$

From first law for a control mass system, $dQ = dE$, with $\mathcal{V}$ held constant. From the definition of specific heat obtained for a constant volume process

$$C_v = \left(\frac{dQ}{dt}\right)_{\mathcal{V}} = \left(\frac{\partial E}{\partial T}\right)_{\mathcal{V}}.$$

For a perfect gas, $E = E(T)$ and consequently

$$dE = C_v \, dT.$$

Also enthalpy can be written as

$$h = e(T) + p\,v, \tag{2.102}$$

where v is the specific volume $(1/\rho)$. As for a perfect gas, $p\,v = R\,T$, one notes that the enthalpy is also a function of T only. Such a flow is called thermally perfect. From Equation (2.102)

$$dh = de + pdv + vdp.$$

Using the first law, $dq = de + pdv$, one gets $dh = dq + vdp$.

Thus, for a constant pressure process, $dq \equiv dh$ and one can define the specific heat at constant pressure as

$$c_p = \left(\frac{dq}{dT}\right)_p$$

$$= \left(\frac{\partial h}{\partial T}\right)_p.$$

If the specific heat is also constant, we call it calorically perfect gas. Using Equation (2.102), this can be simplified as

$$c_p = \frac{de}{dT} + p\left(\frac{dv}{dT}\right)_p.$$

Thus, for a perfect gas

$$c_p - c_v = R \tag{2.103a}$$

$$e(T) = \int C_v dT + \text{Constant} \tag{2.103b}$$

$$h(T) = \int C_p dT + \text{Constant}. \tag{2.103c}$$

From Equation (2.101), we can calculate the change in entropy as

$$S_2 - S_1 = \int \frac{dE}{T} + \int \frac{pd\mathcal{V}}{T}.$$

Noting that $\mathcal{V} = Mv$ and $p/T = \rho R$, the above is simplified as

$$S_2 - S_1 = Mc_v \int \frac{dT}{T} + \int \frac{R}{v} \cdot M dv$$

$$= Mc_v \operatorname{Ln} \frac{T_2}{T_1} + MR \operatorname{Ln} \frac{v_2}{v_1}. \tag{2.104}$$

The same can be evaluated from

$$S_2 - S_1 = \int \frac{M dh - \mathcal{V}\, dp}{T}$$

as

$$\frac{S_2 - S_1}{M} = c_p \operatorname{Ln} \frac{T_2}{T_1} - R \operatorname{Ln} \frac{p_2}{p_1}. \tag{2.105}$$

Equations (2.104) and (2.105) can be further simplified using specific properties. For Equation (2.105)

$$s_2 - s_1 = c_p \operatorname{Ln} \frac{T_2}{T_1} - R \operatorname{Ln} \frac{p_2}{p_1}.$$

Also using $c_p/c_v = \gamma$, one notices from Equation (2.103a)

$$c_p = \frac{\gamma R}{\gamma - 1} \quad \text{and} \quad c_v = \frac{R}{\gamma - 1}. \tag{2.106}$$

Using this in the above

$$\frac{s_2 - s_1}{R} = \frac{\gamma}{\gamma - 1} \operatorname{Ln} \frac{T_2}{T_1} - \operatorname{Ln} \frac{p_2}{p_1}$$

$$= \operatorname{Ln} \frac{p_1}{p_2} \cdot \left(\frac{T_2}{T_1} \right)^{\frac{\gamma}{\gamma - 1}}.$$

Thus,

$$\frac{p}{T^{\frac{\gamma}{\gamma - 1}}} = \text{const.}\ e^{\frac{s_2 - s_1}{R}}. \tag{2.107}$$

Similarity from Equation (2.104)

$$\frac{s_2 - s_1}{R} = \frac{1}{\gamma - 1} \operatorname{Ln} \frac{T_2}{T_1} + \operatorname{Ln} \frac{v_2}{v_1} = \operatorname{Ln}\left(T_2^{\frac{1}{\gamma - 1}} v_2 \right) \Big/ \left(T_1^{\frac{1}{\gamma - 1}} v_1 \right).$$

As $v = 1/\rho$, one can also write the above

$$\frac{T^{\frac{1}{\gamma-1}}}{\rho} = \text{const } e^{\frac{s_2-s_1}{R}}. \tag{2.108}$$

If the process is reversible adiabatic, then

$$dq = ds = 0$$

and then for such process

$$\frac{p}{T^{\gamma/(\gamma-1)}} = \text{Constant} \tag{2.109a}$$

$$\frac{T^{1/(\gamma-1)}}{\rho} = \text{Constant}. \tag{2.109b}$$

Also

$$\frac{p}{\rho^{\gamma}} = \text{Constant}. \tag{2.109c}$$

For adiabatic one-dimensional flow of a perfect gas, if one neglects shaft work or work done due to viscous stresses, then from Equation (2.79) neglecting potential energy

$$h + \frac{1}{2}|\vec{V}|^2 = \text{Constant}. \tag{2.110}$$

For a calorically perfect gas flow process considered as adiabatic, the above equation can be simplified as

$$c_pT + \frac{1}{2}|\vec{V}|^2 = \text{Constant}. \tag{2.111}$$

2.23 Propagation of Sound and Mach Number

Sound propagates as a weak irrotational signal. Even if the ambient medium is quiescent, sound wave propagates due to compressibility of the fluid and its speed of propagation (a) is given by Liepmann and Roshko (1957)

$$a^2 = \left(\frac{\partial p}{\partial \rho}\right)_s$$

and the associated disturbance field propagates by an entropic process. Using Equation (2.109c), one obtain for a perfect gas

$$a^2 = \frac{\gamma p}{\rho} = \gamma RT, \tag{2.112}$$

when the ambient fluid flows with a speed u, one introduces a dimensionless parameter, the Mach number as

$$M = \frac{u}{a}. \tag{2.113}$$

For $M < 1$ we have subsonic condition, whereas for $M > 1$ one has locally a supersonic flow. By definition, transonic flow is one where one notices mixed condition in flow domain $M \gtrless 1$.

2.24 One-Dimensional Steady Flow

From Equation (2.47) for steady flow condition, $\rho Au =$ Constant, which can also be alternately stated as

$$\frac{d\rho}{\rho} + \frac{du}{u} + \frac{dA}{A} = 0. \tag{2.114}$$

Similarly, from momentum conservation equation in Equation (2.48) for steady flow condition one obtains

$$udu = -\frac{dp}{\rho} = -\frac{dp}{d\rho} \cdot \frac{d\rho}{\rho},$$

which can be further simplified, by noting $\frac{dp}{d\rho} = a^2$, as

$$udu = -a^2 \frac{d\rho}{\rho}.$$

Using Equation (2.113) for the Mach number provides

$$\frac{d\rho}{\rho} = -M^2 \frac{du}{u}. \tag{2.115}$$

Substituting this in Equation (2.114) and simplifying one gets the area-velocity relation for one-dimensional flow as

$$\frac{du}{u} = \frac{-1}{(1 - M^2)} \frac{dA}{A}. \tag{2.116}$$

As is evident from this relation that for subsonic condition ($M < 1$), increase in area is related to flow deceleration. However for supersonic flows, flow accelerates with increase in area normal to the flow direction. This counter-intuitive behaviour of flow at supersonic speed occurs due to the fact that density decreases at a larger rate in Equation (2.115), as compared to velocity change, as the proportionality factor is M^2, which is larger than one. To maintain mass conservation, one also has to satisfy Equation (2.116), which thus explains this phenomenon. The curious case of $M = 1$ shows from Equation (2.116) that

$$\frac{dA}{A} \equiv 0.$$

Such a location is called the throat in the flow (as in a wind tunnel) where flow will smoothly change from subsonic to supersonic and vice versa, as shown in Figure 2.11. Having a physical throat in the flow does not ensure that a flow will accelerate from subsonic condition to supersonic condition. Also this observation is based upon the fact that the flow is inviscid (we have used Euler equation) and isentropic. If the downstream portion of the throat where

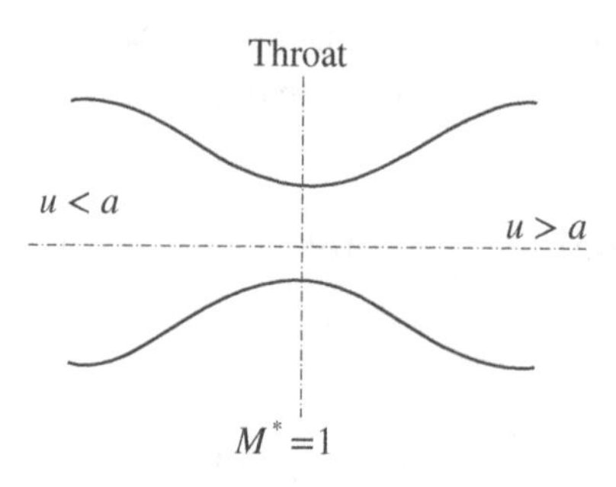

Figure 2.11 Flow in a convergent–divergent duct

the 1D flow terminates into ambient condition which does not match the local flow condition, then an abrupt change is in the form of a shock, which essentially is a nonisentropic process. Across shock, the total enthalpy remains the same, as a consequence of energy conservation principle stated in Equation (2.110). For a calorically perfect gas further simplifies to the condition that the total temperature remains the same across such discontinuities.

In Equation (2.105), if the states 1 and 2 are related to two reservoirs connected through a shock wave, the entropy must increase across the shock wave and is given by

$$s_2 - s_1 = R\,\mathrm{Ln}\frac{p_{01}}{p_{02}}. \tag{2.117}$$

As entropy must increase across a shock, it is readily apparent that $p_{02} < p_{01}$, i.e. the total pressure decrease across the shock. Thus, the total pressure must be viewed as the ability of the flow to be able to be converted to mechanical work. This ability reduces due to presence of shock.

2.25 Normal Shock Relation for Steady Flow

The shock (whose physical dimension is of the order of few mean free path) is a region of nonequilibrium. Even when the shock forms, one can consider the flow to be steady. We consider two adjacent streamlines and flow inside these streamlines can be approximated as one-dimensional constant-area flow. If we consider two stations (1 and 2) away from the nonequilibrium region, the conservation equations can be written as

$$\mathring{m} = \rho_1\, u_1 = \rho_2\, u_2 \tag{2.118a}$$

$$p_1 + \rho_1\, u_1^2 = p_2 + \rho_2\, u_2^2 \tag{2.118b}$$

$$h_1 + \frac{1}{2}\, u_1^2 = h_2 + \frac{1}{2}\, u_2^2. \tag{2.118c}$$

In these conservation equations for mass, momentum and energy, respectively, the subscripts indicate the location of stations. Dividing left and right-hand sides of Equation (2.118b) by the left and right-hand sides of Equation (2.118a) and rearranging, one obtains

$$u_1 - u_2 = \frac{p_2}{\rho_2\, u_2} - \frac{p_1}{\rho_1\, u_1}. \tag{2.119}$$

Considering the working fluid to be perfect gas, then

$$\frac{p_2}{\rho_2} = RT_2 = \frac{\gamma RT_2}{\gamma} = \frac{a_2^2}{\gamma} \text{ and } \frac{p_1}{\rho_1} = \frac{a_1^2}{\gamma}.$$

From Equation (2.119) one gets

$$u_1 - u_2 = \frac{a_2^2}{\gamma\, u_2} - \frac{a_1^2}{\gamma\, u_1}. \tag{2.120}$$

Also $h_1 = C_p\, T_1 = \frac{\gamma RT_1}{\gamma - 1}\, T_1 = \frac{a_1^2}{\gamma - 1}$

Hence from the energy equation, Equation (2.118c), one gets

$$\frac{u_1^2}{2} + \frac{a_1^2}{\gamma - 1} = \frac{u_2^2}{2} + \frac{a_2^2}{\gamma - 1}. \tag{2.121}$$

Denoting the condition at the throat by a superscript *, then $u^* = a^*$ (where $M^* = 1$).

Hence from Equation (2.121), one can relate the physical variables at stations 1 and 2, with respect to the throat condition as

$$\frac{u_1^2}{2} + \frac{a_1^2}{\gamma - 1} = \frac{u_2^2}{2} + \frac{a_2^2}{\gamma - 1} = \frac{a^{*2}}{2} + \frac{a^{*2}}{\gamma - 1}. \tag{2.122}$$

If the reservoir conditions be indicated by a subscript 0, then from Equation (2.118c)

$$\frac{u^2}{2} + \frac{a^2}{\gamma - 1} = \frac{a_0^2}{\gamma - 1}. \tag{2.123}$$

From Equations (2.122) and (2.123), one obtains

$$\frac{a^{*2}}{a_0^2} = \frac{T^*}{T_0} = \frac{2}{\gamma + 1} = 0.833 \quad \text{(for air)}. \tag{2.124}$$

Also in Equation (2.55), if we identify station 1 with reservoir and station 2 with the throat, then

$$\frac{p^*}{p_0} = \left(\frac{2}{\gamma + 1}\right)^{\frac{\gamma}{\gamma - 1}} = 0.528 \text{ (for air)}. \tag{2.125}$$

Similarly, one obtains the density ratio as

$$\frac{\rho^*}{\rho_0} = \left(\frac{2}{\gamma + 1}\right)^{\frac{1}{\gamma - 1}} = 0.634 \text{ (for air)}. \tag{2.126}$$

Note that there is no shock discontinuity between the reservoir and throat and hence isentropic relations hold (by neglecting viscous terms); secondly, the throat is used here as a reference condition, even though the throat condition is not achieved in a flow. Thus from Equation (2.122) one gets

$$1+\left(\frac{2}{\gamma-1}\right)\frac{1}{M^2}=\frac{\gamma+1}{\gamma-1}M^{*2}$$

or

$$M^{*2}=\frac{\gamma-1}{\gamma+1}\left[1+\frac{2}{\gamma-1}\frac{1}{M^2}\right]=\frac{\gamma-1}{\gamma+1}.$$

Thus,

$$\left[\frac{(\gamma-1)M^2+2}{(\gamma-1)M}\right]M^{*2}=\frac{1}{(\gamma+1)M^2}\,[2+(\gamma-1)M^2].$$

In transonic flows, one often refers the speed ratio u/a^* by M^*. From Equation (2.122) one has

$$\frac{u^2}{2}+\frac{a^2}{\gamma-1}=\frac{1}{2}\frac{\gamma+1}{\gamma-1}\,a^{*2}.$$

Dividing both side by u^2 and introducing the definitions of M and M^*, one gets

$$\frac{1}{2}+\frac{1}{(\gamma-1)M^2}=\frac{1}{2}\frac{\gamma+1}{\gamma-1}\frac{1}{M^{*2}} \tag{2.127}$$

2.26 Rankine–Hugoniot Relation

From energy conservation relation [Equation (2.121)]

$$\frac{u_1^2}{2}+\frac{\gamma p_1}{(\gamma-1)\rho_1}=C_pT_o=\frac{u_2^2}{2}+\frac{\gamma p_2}{(\gamma-1)\,\rho_2}$$

Or

$$\frac{\gamma}{\gamma-1}\left[\frac{p_1}{\rho_1}-\frac{p_2}{\rho_2}\right]=\frac{1}{2}\,(u_2-u_1)\,(u_2+u_1). \tag{2.128}$$

From mass conservation equation, Equation (2.118a), one can obtain

$$u_2+u_1=\mathring{m}\left(\frac{1}{\rho_2}+\frac{1}{\rho_1}\right). \tag{2.129}$$

And from momentum conservation relation, Equation (2.118b)

$$u_2-u_1=\frac{1}{\mathring{m}}\,(p_1-p_2). \tag{2.130}$$

Substitution of Equations (2.129) and (2.130) in Equation (2.128) yields

$$\frac{\gamma}{\gamma-1}\left(\frac{p_1}{\rho_1}-\frac{p_2}{\rho_2}\right)=\frac{1}{2}\left(p_1-p_2\right)\left(\frac{1}{\rho_1}+\frac{1}{\rho_2}\right).$$

One can simplify this by relating p_2/p_1 with ρ_2/ρ_1 as

$$\frac{p_2}{p_1}=\frac{\frac{\gamma+1}{\gamma-1}\frac{\rho_2}{\rho_1}-1}{\frac{\gamma+1}{\gamma-1}-\frac{\rho_2}{\rho_1}} \tag{2.131a}$$

or alternately,

$$\frac{\rho_2}{\rho_1}=\frac{\frac{\gamma+1}{\gamma-1}\frac{p_2}{p_1}+1}{\frac{\gamma+1}{\gamma-1}+\frac{p_2}{p_1}}. \tag{2.131b}$$

These pressure-density relations across shock are noted as Rankine–Hugoniot relations.

2.27 Prandtl or Meyer Relation

From energy conservation relation [Equation 2.121)], one can relate the conditions upstream and downstream of the shock by

$$\frac{p_1}{\rho_1}=\frac{\gamma-1}{\gamma}\left[C_p\,T_o-\frac{u_1^2}{2}\right] \tag{2.132a}$$

and

$$\frac{p_2}{\rho_2}=\frac{\gamma-1}{\gamma}\left[C_p\,T_o-\frac{u_2^2}{2}\right]. \tag{2.132b}$$

Dividing left- and right-hand sides of Equation (2.118b) by the corresponding terms of Equation (2.118a), one obtains

$$u_1-u_2=\frac{p_2}{\rho_2 u_2}-\frac{p_1}{\rho_1 u_1}.$$

Using Equations (2.132a) and (2.132b) in the right-hand side of above to get,

$$(u_1-u_2)=\frac{\gamma-1}{\gamma}(u_1-u_2)\left[\frac{1}{2}+\frac{C_pT_o}{u_1u_2}\right] \tag{2.133}$$

Disregarding the shock free solution for 1D inviscid flow condition, i.e. $u_1-u_2=0$, one obtains from Equation (2.133)

$$\frac{1}{2}+\frac{C_p\,T_o}{u_1\,u_2}=\frac{\gamma}{\gamma-1}$$

$$\text{or} \quad \frac{C_p\,T_o}{u_1\,u_2} = \frac{\gamma+1}{2(\gamma-1)}$$

$$\text{or} \quad u_1\,u_2 = \frac{2(\gamma-1)}{\gamma+1}C_p\,T_o = \frac{2(\gamma-1)}{\gamma+1}\frac{\gamma R}{(\gamma-1)}T_o = \frac{2}{\gamma+1}\,a_0^2.$$

From Equation (2.124), $\frac{a^{*2}}{a_0^2} = \frac{2}{\gamma+1}$ and hence

$$u_1\,u_2 = a_*^{*2}. \tag{2.134}$$

This is the Prandtl or Meyer relation and expresses the fact that the throat condition is the geometric mean between upstream and downstream condition of the shock. Relationship between other physical properties of the flow, before and after the shock can be obtained and are expressed below in terms of Mach number ahead of the shock, M_1.

(a) Static pressure ratio:

From momentum conservation relation (Equation (2.118b)),

$$\frac{p_2-p_1}{p_1} = \frac{\rho_1 u_1^2 - \rho_2 u_2^2}{p_1} = \frac{\rho_1 u_1^2}{p_1}\left[1-\frac{u_2}{u_1}\right].$$

As

$$\frac{\rho_1\,u_1^2}{p_1} = \frac{\rho_1\,u_1^2}{\rho_1 R\,T_1} = \frac{\gamma u_1^2}{a_1^2} = \gamma M_1^2.$$

Hence

$$\frac{p_2}{p_1} - 1 = \gamma M_1^2\left[1-\frac{\rho_1}{\rho_2}\right].$$

Using the Rankine–Hugoniot relation [Equation (2.131b)] in the above, one gets

$$\frac{p_2}{p_1} - 1 = \gamma M_1^2\left\{1-\left[\frac{\gamma+1}{\gamma-1}\frac{p_1}{p_2}+1\right]\Big/\left[\frac{\gamma+1}{\gamma-1}+\frac{p_1}{p_2}\right]\right\}.$$

It is left as an exercise for the readers to simplify the above to yield

$$\frac{p_2}{p_1} = \frac{2\gamma M_1^2}{\gamma+1} - \frac{\gamma-1}{\gamma+1}. \tag{2.135}$$

Using the Prandtl or Meyer relation, one can obtain the static pressure ratio in terms of M_2 (exit Mach number) as

$$\frac{p_1}{p_2} = \frac{2\gamma M_2^2 - (\gamma-1)}{\gamma+1}. \tag{2.136}$$

One can relate M_1 and M_2 from Equations (2.135) and (2.136) from

$$\frac{2\gamma M_1^2}{\gamma+1}-\frac{\gamma-1}{\gamma+1}=\frac{\gamma+1}{2\gamma M_2^2-(\gamma-1)}.$$

Simplification of this yields

$$M_2^2=\frac{(\gamma-1)M_1^2+2}{2\gamma M_1^2-(\gamma-1)}. \tag{2.136a}$$

(b) Density jump across a normal shock:

Using Rankine–Hugoniot relation, Equation (2.131b) in Equation (2.135), one obtains the density ratio

$$\frac{\rho_2}{\rho_1}=\left[\frac{\gamma+1}{\gamma-1}\left(\frac{2\gamma M_1^2-(\gamma-1)}{(\gamma+1)}\right)+1\right]\Bigg/\left[\frac{\gamma+1}{\gamma-1}+\frac{2\gamma M_1^2-(\gamma-1)}{\gamma+1}\right],$$

which can be simplified to obtain

$$\frac{\rho_2}{\rho_1}=\frac{(\gamma+1)\,M_1^2}{2+(\gamma-1)\,M_1^2}. \tag{2.137}$$

(c) Temperature ratio across shock:

This is readily obtained from the equation of state as

$$\frac{T_2}{T_1}=\frac{p_2}{p_1}\Big/\frac{\rho_2}{\rho_1}$$

and using Equations (2.135) and (2.137) one obtains

$$\frac{T_2}{T_1}=\left[\frac{2\gamma M_1^2-(\gamma-1)}{\gamma+1}\right]\left[\frac{2+(\gamma-1)M_1^2}{(\gamma+1)M_1^2}\right]. \tag{2.138}$$

(d) Entropy change across a normal shock:

From Equation (2.105),

$$s_2-s_1=c_p\,\mathrm{Ln}\,\frac{T_2}{T_1}-R\,\mathrm{Ln}\,\frac{p_2}{p_1}.$$

As $c_p=\frac{\gamma R}{\gamma-1}$ and $c_v=\frac{R}{\gamma-1}$, therefore.

$$\frac{s_2-s_1}{R}=\frac{\gamma}{\gamma-1}\,\mathrm{Ln}\,\frac{p_2\,\rho_1}{p_1\rho_2}-\mathrm{Ln}\,\frac{p_2}{p_1}=\frac{\gamma-\gamma+1}{\gamma-1}\mathrm{Ln}\,\frac{p_2}{p_1}-\frac{\gamma}{\gamma-1}\,\mathrm{Ln}\,\frac{\rho_2}{\rho_1}.$$

This is further simplified to

$$\frac{s_2 - s_1}{R/(\gamma - 1)} = \text{Ln}\,\frac{p_2}{p_1} - \gamma\,\text{Ln}\,\frac{\rho_2}{\rho_1}.$$

Using Equations (2.135) and (2.137) in the above and noting $c_v = R/(\gamma - 1)$ one gets

$$\frac{s_2 - s_1}{c_v} = \text{Ln}\left[\frac{2\gamma M_1^2 - (\gamma - 1)}{\gamma + 1}\right] - \gamma\,\text{Ln}\left[\frac{(\gamma + 1)M_1^2}{2 + (\gamma + 1)M_1^2}\right]$$

$$= \text{Ln}\left[1 + \frac{2\gamma}{\gamma + 1}(M_1^2 - 1)\right] - \gamma\,\text{Ln}\,(M_1^2) + \gamma\,\text{Ln}\left[1 + \frac{\gamma - 1}{\gamma + 1}(M_1^2 - 1)\right].$$

Expressing $M_1^2 - 1 = \epsilon$, above can be rewritten as

$$\frac{s_2 - s_1}{c_v} = \text{Ln}\left[1 + \frac{2\gamma\epsilon}{\gamma + 1}\right] - \gamma\,\text{Ln}\,(1 + \epsilon) + \gamma\,\text{Ln}\left[1 + \frac{\gamma - 1}{\gamma + 1}\epsilon\right].$$

Noting that $\text{Ln}\,(1 + x) = x - \frac{x^2}{2} + \frac{x^3}{3} \cdots$ for $|x| \ll 1$, the above can be simplified to

$$\frac{\Delta s}{c_v} = \frac{2\gamma}{\gamma + 1}\,\epsilon - \frac{1}{2}\left(\frac{2\gamma}{\gamma + 1}\,\epsilon\right)^2 + \frac{1}{3}\left(\frac{2\gamma\epsilon}{\gamma + 1}\right)^3 + \gamma\left\{\frac{\gamma - 1}{\gamma + 1}\epsilon - \frac{1}{2}\right.$$

$$\left.\left(\frac{\gamma - 1}{\gamma + 1}\,\epsilon\right)^2 + \frac{1}{3}\left(\frac{\gamma - 1}{\gamma + 1}\,\epsilon\right)^3\right\}$$

$$-\gamma\left\{\epsilon - \frac{\epsilon^2}{2} + \frac{\epsilon^3}{3}\right\} + \text{Higher Order Terms}$$

$$= \gamma\,\epsilon\left\{\frac{2}{\gamma + 1} + \frac{\gamma - 1}{\gamma + 1} - 1\right\} + \frac{\epsilon^2}{2}\left\{\gamma - \gamma\left(\frac{\gamma - 1}{\gamma + 1}\right)^2 - \frac{4\gamma^2}{(\gamma + 1)^2}\right\} +$$

$$\frac{\epsilon^3}{3}\left\{\frac{8\gamma^3}{(\gamma + 1)^3} + \gamma\left(\frac{\gamma - 1}{\gamma + 1}\right)^3 - \gamma\right\}.$$

The coefficients of ϵ and ϵ^2 are identically zero and one obtains by substituting $\epsilon = M_1^2 - 1$

$$\frac{\Delta s}{c_v} = \frac{(M_1^2 - 1)^3}{3(\gamma + 1)^3}\left[8\gamma^3 + \gamma(\gamma - 1)^3 - \gamma(\gamma + 1)^3\right] = \frac{(M_1^2 - 1)^3}{3(\gamma + 1)^3}\,2\gamma\,(\gamma^2 - 1).$$

Thus

$$\frac{\Delta s}{c_v} = \frac{2}{3}\frac{\gamma(\gamma^2 - 1)}{(\gamma + 1)^3}\,(M_1^2 - 1)^3. \tag{2.139}$$

If M_1 does not exceed one by a large value, then the entropy change is negligibly small, as compared to variation of the other physical variables like p, ρ and T.

2.28 Oblique Shock Waves

Here we consider the formation of oblique stationary shock wave in a two-dimensional flow. The shock is inclined at an angle β with respect to oncoming flow ($\vec{V}_1$), while the velocity after the shock ($\vec{V}_2$) is deflected by an angle θ, with respect to $\vec{V}_1$. The property of the oblique shock is analyzed with the help of the following physical observation, via the decomposition of $\vec{V}_1$ and $\vec{V}_2$ into tangential and normal to shock components. For an inviscid analysis, it is natural to assume that the tangential component remains the same, before and after the shock. However, the normal component before the shock (u_1) is reduced to (u_2), decided upon the normal shock relation for the normal component of the oncoming flow Mach number ($u_1/a_1 = M_1 \sin\beta$).

The Mach number after the shock (M_2) has the normal component given by $u_2/a_2 = M_2 \sin(\beta - \theta)$. Substitution of these in Equation (2.136a) yields

$$M_2^2 \sin^2(\beta - \theta) = \frac{(\gamma - 1)M_1^2 \sin^2\beta + 2}{2\gamma M_1^2 \sin^2\beta - (\gamma - 1)}. \tag{2.140}$$

Density ratio across an oblique shock is obtained from Equation (2.137) as

$$\frac{\rho_2}{\rho_1} = \frac{(\gamma + 1)M_1^2 \sin^2\beta}{2 + (\gamma - 1)M_1^2 \sin^2\beta}. \tag{2.141}$$

Pressure ratio across the oblique shock is obtained from Equation (2.135) as

$$\frac{p_2}{p_1} = \frac{2\gamma(M_1^2 \sin^2\beta) - (\gamma - 1)}{(\gamma + 1)}. \tag{2.142}$$

Temperature ratio across the oblique shock is obtained from Equation (2.138) as

$$\frac{T_2}{T_1} = \left[\frac{2\gamma M_1^2 \sin^2\beta - (\gamma - 1)}{\gamma + 1}\right]\left[\frac{2 + (\gamma - 1)M_1^2 \sin^2\beta}{(\gamma + 1)M_1^2 \sin^2\beta}\right]. \tag{2.143}$$

The entropy jump is similarly obtained from

$$\frac{\Delta s}{c_v} = \frac{2}{3}\frac{\gamma(\gamma^2 - 1)}{(\gamma + 1)^3}\left[M_1^2 \sin^2\beta - 1\right]^3. \tag{2.144}$$

Relationship between oblique shock angle (β) and deflection angle:
From Figure 2.12, one notes the following relations

$$\tan\beta = \frac{u_1}{v} \quad \text{and} \quad \tan(\beta - \theta) = \frac{u_2}{v}. \tag{2.145}$$

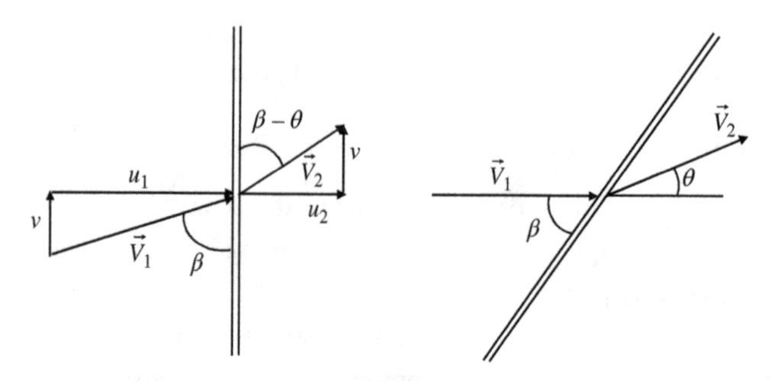

Figure 2.12 Plain oblique shock

Using Equation (2.141) in

$$\frac{\rho_2}{\rho_1} = \frac{u_1}{u_2} = \frac{\tan\ \beta}{\tan(\beta - \theta)}$$

one gets

$$\frac{(\gamma + 1)\ M_1^2 \sin^2 \beta}{2 + (\gamma - 1) M_1^2 \sin^2 \beta} = \frac{\tan\ \beta}{\tan\ (\beta - \theta)}.$$

We rewrite this as

$$\frac{\tan\ (\beta - \theta)}{\tan\ \beta} = \frac{(\gamma + 1)\ M_1^2 \sin^2 \beta}{2 + (\gamma - 1) M_1^2 \sin^2 \beta} = \alpha \text{ (say)}.$$

Then

$$\frac{\tan \beta - \tan \theta}{(1 + \tan \beta \tan \theta) \tan \beta} = \alpha$$

or

$$\alpha\ (1 + \tan \beta \tan \theta) = 1 - \tan\ \theta\ \cot \beta$$

$$\tan \theta\ (\alpha \tan \beta + \cot\ \beta) = 1 - \alpha.$$

Hence

$$\tan \theta = \frac{(1 - \alpha)\ \cot\ \beta}{(\alpha + \cot^2 \beta)}. \tag{2.146}$$

Now,

$$1 - \alpha = \frac{(\gamma + 1) M_1^2 \sin^2 \beta - 2 - (\gamma - 1) M_1^2 \sin^2 \beta}{(\gamma + 1) M_1^2 \sin^2 \beta} = \frac{2(M_1^2 \sin^2 \beta - 1)}{(\gamma + 1) M_1^2 \sin^2 \beta}. \tag{2.147a}$$

Also,

$$\alpha + \cot^2 \beta = \frac{2 + (\gamma - 1)M_1^2 \sin^2 \beta}{(\gamma + 1)M_1^2 \sin^2 \beta} + \frac{\cos^2 \beta}{\sin^2 \beta} = \frac{2 + (\gamma - 1)M_1^2 \sin^2 \beta + (\gamma + 1)M_1^2 cos^2 \beta}{(\gamma + 1)M_1^2 \sin^2 \beta}$$

$$= \frac{2 + (\gamma - 1)M_1^2 \sin^2 \beta + (\gamma + 1)M_1^2(1 - \sin^2 \beta)}{(\gamma + 1)M_1^2 \sin^2 \beta}$$

$$= \frac{2 + (\gamma + 1)M_1^2 - 2M_1^2 \sin^2 \beta}{(\gamma + 1)M_1^2 \sin^2 \beta} = \frac{2 + \gamma M_1^2 + M_1^2(1 - 2\sin^2 \beta)}{(\gamma + 1)M_1^2 \sin^2 \beta}.$$

Thus

$$\alpha + \cot^2 \beta = \frac{2 + M_1^2(\gamma + \cos^2 \beta)}{(\gamma + 1)M_1^2 \sin^2 \beta}. \tag{2.147b}$$

Substituting Equation (2.147a) and (2.147b) in Equation (2.146), one gets

$$\tan \theta = 2 \cot \beta \frac{M_1^2 \sin^2 \beta - 1}{M_1^2(\gamma + \cos^2 \beta) + 2}. \tag{2.148}$$

In the relation of β with θ, we note the variations possible for the former to be dictated upon the fact that the normal component of Mach number before the shock must be supersonic, i.e. $M_1 \sin \beta > 1$ or $\beta > sin^{-1}(\frac{1}{M_1})$. Also the maximum value of shock angle corresponds to the normal shock for which $\beta = \pi/2$. Variations of θ with β is shown plotted in Figure 2.13, for which one notes the following.

(i) In the range $\sin^{-1}(\frac{1}{M_1}) \leq \beta \leq \pi/2$, the deflection angle θ is positive and attains a maximum value (θ_{max}) for different M_1. The global maximum is 46° for the case $M_1 \to \infty$.
(ii) For every $\theta < \theta_{max}$, there are two possible solutions:
   (a) the weak oblique shock solution corresponding to $M_2 > 1$ and
   (b) the strong oblique shock solution for which $M_2 < 1$.

2.29 Weak Oblique Shock

For the weak solution ($M_2 > 1$), when θ decreases to zero, the limiting value of β is indicated by a value μ, which can be obtained from (2.148) as

$$M^2 \sin^2 \mu - 1 = 0 \quad \text{or} \quad \mu = \sin^{-1} \frac{1}{M}. \tag{2.149}$$

This is called the Mach angle and is a property determined by the local Mach number. It is readily apparent that the corresponding process can be idealized as isentropic (from Equation (2.144)).

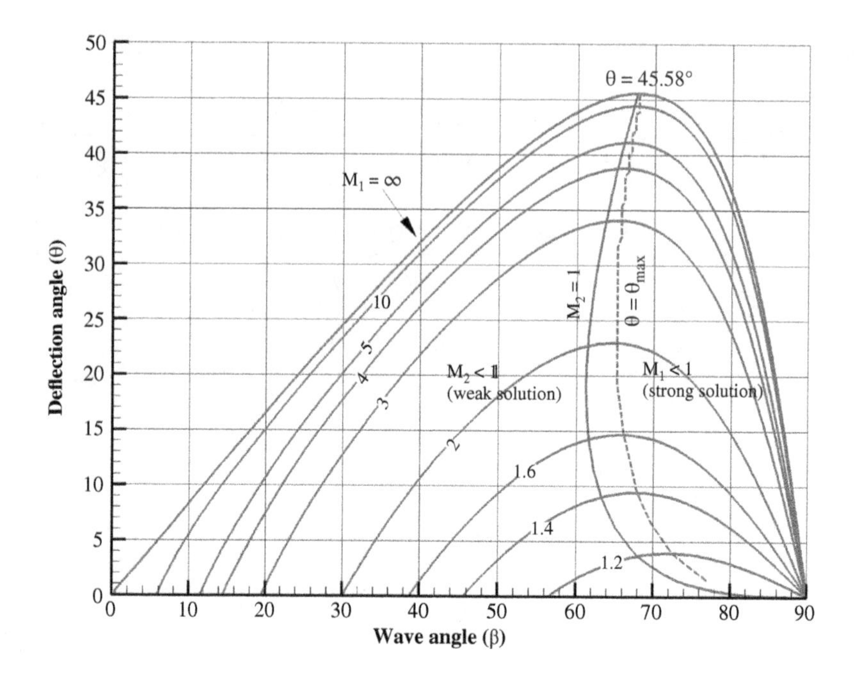

Figure 2.13 Plain oblique shock

For nonuniform flow, μ varies with local Mach number and the corresponding Mach lines (two values given by Equation (2.149) as $\pm\mu$, with respect to the streamline passing through the point. For a 2D flow, one therefore talks about right and left running Mach waves through any point.

For nonzero, but small deflection angle, $\tan\theta \simeq \theta$ and $\tan\beta \simeq \tan\mu = \frac{1}{\sqrt{M_1^2-1}}$. From Equation (2.148) one can write

$$\theta \simeq 2\cot\beta \frac{M_1^2\sin^2\beta - 1}{M_1^2(\gamma + \cos^2\beta) + 2}. \tag{2.150}$$

Also, for this small deflection, $\sin\beta \simeq \frac{1}{M_1}$

Therefore

$$\cos 2\beta = 1 - 2\sin^2\beta = \frac{M_1^2 - 2}{M_1^2}.$$

Thus form Equation (2.150)

$$M_1^2\sin^2\beta - 1 = \frac{M_1^2(\gamma + \frac{M_1^2-2}{M_1^2}) + 2}{2\sqrt{M_1^2 - 1}}\theta = \frac{M_1^2}{\sqrt{M_1^2 - 1}}\frac{\gamma + 1}{2}\theta \tag{2.151}$$

From Equation (2.142) the pressure jump across weak shock can be obtained as

$$\frac{\Delta p}{p_1} = \frac{p_2 - p_1}{p_1} = \frac{2\gamma M_1^2 \sin^2 \beta_1 - \gamma + 1}{\gamma + 1} - 1$$

$$= \frac{2\gamma (M_1^2 \sin^2 \beta_1 - 1)}{\gamma + 1}.$$

Using Equation (2.101) in this expression one can quantify the strength of the wave as

$$\frac{\Delta p}{p_1} = \frac{2\gamma}{\gamma + 1} \frac{\gamma + 1}{2} \frac{M_1^2}{\sqrt{M_1^2 - 1}} \theta = \frac{\gamma M_1^2}{\sqrt{M_1^2 - 1}} \theta. \tag{2.152}$$

The shock strength can be expressed in terms of other physical variables, ρ and T. We have already noted that entropy change is proportional to that given by Equation (2.144) as

$$\frac{\Delta s}{c_v} = \frac{2}{3} \frac{\gamma(\gamma^2 - 1)}{(\gamma + 1)^3} \left[\frac{\gamma + 1}{2} \frac{M_1^2}{\sqrt{M_1^2 - 1}} \theta \right]^3 = \frac{\gamma(\gamma^2 - 1)}{12} \frac{M_1^6}{(M_1^2 - 1)^{3/2}} \theta^3. \tag{2.153}$$

Results in (2.152) and (2.153) indicate an interesting possibility of turning a flow smoothly, rather than abruptly, as in a compression corner.

This is shown in Figure 2.14, where the same flow turning is achieved by two strategies. In Figure 2.14(a), the turning is achieved by a single compression ramp of angle θ. While in Figure 2.14(b), same turning angle is achieved by continuous turning. For the purpose of illustration, let us say that this is idealized by n such turning, with $\theta = n\Delta\theta$. In both the cases, the shock strength as represented by $\frac{\Delta p}{p_1}$ is the same. However for case (b), entropy change is proportional to

$$\Delta s \sim n(\Delta\theta)^3 = n\Delta\theta(\Delta\theta)^2 = \theta(\Delta\theta)^2 = \frac{\theta^3}{n^2}.$$

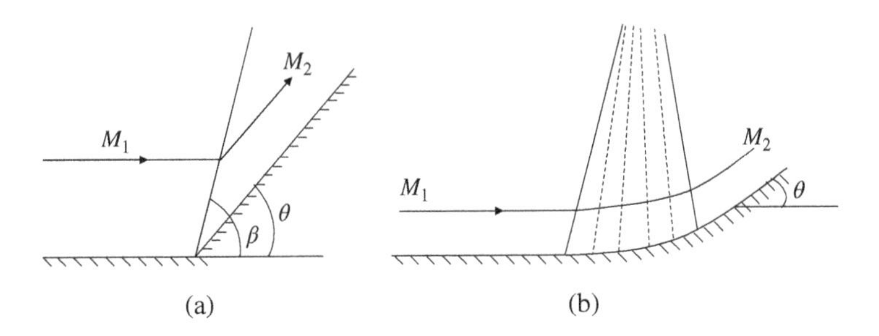

Figure 2.14 Compression of a supersonic flow ($M_1 > 1$) by (a) single ramp and (b) by continuous turning (only a flow turning is shown for the purpose of illustration)

Thus the entropy increase in case (b) is reduced by a factor of $\frac{1}{n^2}$. In the limit of n taking a very large value, the turning may be considered almost isentropic and individual shock can be viewed as Mach waves. Far out in the flow, these Mach lines intersect to form a single shock.

2.30 Expansion of Supersonic Flows

When a supersonic flow moves around a convex corner by a single turn by a single wave $(-\theta)$, then there arises a conceptual problem. For example, Equation (2.153) indicates that the entropy will decrease with flow turning by a single wave, even though the flow will obey the conservation laws. This problem can be circumvented, if the flow turns via a large number Mach waves, bracketed between the Mach wave corresponding to the upstream flow at M_1 and the Mach wave corresponding to the downstream flow Mach number M_2. Each of the unit process is isentropic and the overall ensemble of these unit processes do not lead to entropy increase.

The schematic of the expansion process, also known as turning via expansion fan, is shown in Figure 2.15. This fan is continuous, with infinite number of Mach waves whose extent is given by the angles

$$\mu_1 \leq \mu \leq \mu_2, \text{ where } \mu_1 = \sin^{-1}\left(\frac{1}{M_1}\right) \text{ and } \mu_2 = \sin^{-1}\left(\frac{1}{M_2}\right).$$

This is also called a centred fan around the turning point, O.

Let us consider a single Mach wave across which the velocity increase from V to $V + dV$, as indicated in Figure 2.16, via a turn of angle $d\theta$. The velocity diagram is shown in this figure, with the horizontal segment PR representing the velocity vector, V, while PQ is inclined by $d\theta$ with respect to this, indicating an incremental velocity vector $V + dV$. As the velocity does not change in the tangential direction, the change in the velocity (dV) along QR occurs normal to the Mach wave. From the property of triangle applied on PQR yields

$$\frac{V + dV}{V} = \frac{\sin(\pi/2 + \mu)}{\sin(\pi/2 - \mu - d\theta)}. \tag{2.154}$$

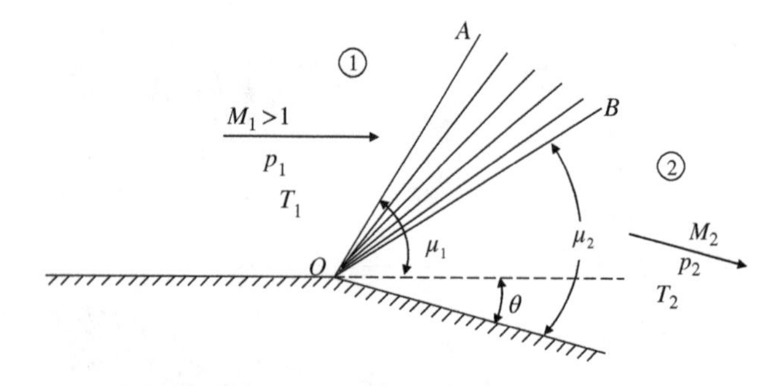

Figure 2.15 Prandtl-Meyer expansion

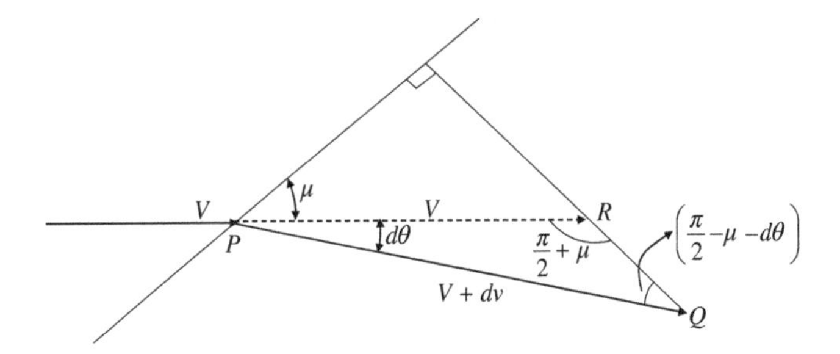

Figure 2.16 Geometrical construction for the infinitesimal changes across an infinitesimally weak wave (in the limit of a Mach wave)

For small deflection, $d\theta$, the denominator on the right-hand side simplifies to

$$\sin\left[\frac{\pi}{2}-(\mu+d\theta)\right]=\cos(\mu+d\theta)=\cos\mu-d\theta\sin\mu.$$

Thus

$$1+\frac{dV}{V}=\frac{\cos\mu}{\cos\mu-d\theta\sin\mu}=\frac{1}{1-d\theta\tan\mu}.$$

The right-hand side can be expanded in a power series and linearized to obtain

$$\frac{dV}{V}=d\theta\tan\mu. \tag{2.155}$$

As we have already established $\tan\mu=\frac{1}{\sqrt{M^2-1}}$, hence the turning angle can be expressed as

$$d\theta=\sqrt{M^2-1}\,\frac{dV}{V}. \tag{2.156}$$

This is the differential turning describing the flow expansion inside the fan. The right side of Equation (2.156) can be expressed as a function of M only, by noting $V=Ma$, and therefore

$$\frac{dV}{V}=\frac{dM}{M}+\frac{da}{a}. \tag{2.157}$$

Since from Equation (2.123)

$$a=a_o\left(1+\frac{\gamma-1}{2}M^2\right)^{-1/2}.$$

Therefore

$$\frac{da}{a}=-\frac{1}{2}\left(1+\frac{\gamma-1}{2}M^2\right)^{-3/2+1/2}\frac{\gamma-1}{2}(2M)\,dM=-\frac{\gamma-1}{2}\left(1+\frac{\gamma-1}{2}M^2\right)^{-1}M\,dM.$$

Substituting this in Equation (2.157) one obtains

$$\frac{dV}{V} = \frac{1}{1 + \frac{\gamma-1}{2}M^2} \frac{dM}{M}. \tag{2.158}$$

Thus, from Equation (2.156) one obtains

$$\theta_{12} = \int_{M_1}^{M_2} \frac{\sqrt{M^2 - 1}}{1 + \frac{\gamma-1}{2}M^2} \frac{dM}{M}. \tag{2.159}$$

The right-hand side integral is usually denoted as $\nu(M)$.
That is

$$\nu(M) = \int \frac{\sqrt{M^2 - 1}}{1 + \frac{\gamma-1}{2}M^2} \frac{dM}{M}, \tag{2.160}$$

which is also called the Prandtl-Meyer function and can be simplified as:

$$\nu(M) = \sqrt{\frac{\gamma+1}{\gamma-1}} \tan^{-1} \sqrt{\frac{\gamma-1}{\gamma+1}(M^2 - 1)} - \tan^{-1} \sqrt{M^2 - 1}, \tag{2.161}$$

such that $\theta_{12} = \nu(M_2) - \nu(M_1)$

The Prandtl-Meyer function can be found tabulated in Liepmann and Roshko (1957), Shapiro (1953) and other books on elementary gas dynamics. This is always associated with supersonic flow, such that $\nu(M = 1) = 0$ and

$$\nu(M \to \infty) = \nu_{max}, \text{ where } \nu_{max} = \frac{\pi}{2}\left[\sqrt{\frac{\gamma+1}{\gamma-1}} - 1\right]. \tag{2.162}$$

The above procedure of associating a Mach number with $\nu(M)$ can be used for both compression and expansion by considering the absolute value of turning. For compression $\nu(M)$ reduces after turning, while it increases during an expansion.

Bibliography

Aris, R. (1962) *Vectors, Tensors, and the Basic Equation of Fluid Mechanics*. Prentice-Hall Inc., USA.

Lamb, H. (1945) *Hydrodynamics*. 375 Dover Publications, USA.

Liepmann, H.W. and Roshko, A. (1957) *Elements of Gas Dynamics*. John Wiley & Sons, Inc., USA.

Shapiro, A.H. (1953) *The Dynamics and Thermodynamics of Compressible Fluid Flow*. **1**, John Wiley & Sons, Inc., USA.

3

Theoretical Aerodynamics of Potential Flows

3.1 Introduction

In the last chapter, we have noted that for incompressible flows the governing equations are obtained from the continuity and momentum conservation equations. It was furthermore noted that for an aircraft in cruise condition the wing section is set at a low angle of attack, for which the flow remains attached and the viscous effects are confined inside the boundary layer. For typical operational Reynolds numbers in the range of millions, the boundary layer thickness is a small fraction of the chord of the wing. Due to this observation of thinness of the boundary layers, it is noted that the rest of the flow can be viewed as inviscid. Furthermore, if the wing is started impulsively then using the Kelvin's theorem in Section 2.11, one can reason that the resulting flow is irrotational and the governing equation is obtained from continuity equation as given in Equation (2.31). Also, from the kinematic definition of vorticity

$$\vec{\omega} = \nabla \times \vec{V}. \tag{3.1}$$

The continuity equation for incompressible flow is given by, $\nabla \cdot \vec{V} = 0$ and it is readily satisfied if one postulates the existence of vector function, $\vec{\Psi}$, such that $\vec{V} = \nabla \times \vec{\Psi}$ automatically satisfies mass conservation. For a general three-dimensional flow field, this function $\vec{\Psi}$ is known as the vector potential, as already defined in Section 2.9. Thus, from Equation (3.1), one gets

$$\vec{\omega} = \nabla \times \nabla \times \vec{\Psi}. \tag{3.2}$$

Since we are dealing with physical quantities, such variables must be unique and this requires the quantities to be divergence-free or *solenoidal*. Hence, using the *solenoidality condition* for the vector potential, Equation (3.2) can be simplified to

$$\vec{\omega} = -\nabla^2\vec{\Psi}. \tag{3.3}$$

Theoretical and Computational Aerodynamics, First Edition. Tapan K. Sengupta.

Companion Website: www.wiley.com/go/sengupta

These are three equations for the three components of vorticity vector, in terms of the three components of vector potential for 3D flows. For 2D flows, considerable saving is derived, as the vorticity has single component (in the out-of-flow plane, i.e. if the flow is in (x, y)-plane, then the nonzero component of vorticity is in z-direction) and so will also be the vector potential. The vector potential for 2D flow is the stream function (ψ) and has also the only nonzero component in the z-direction. Thus, for 2D irrotational flow (with zero vorticity), Equation (3.3) reduces to the Laplace's equation given by

$$\nabla^2 \psi = 0. \tag{3.4}$$

Thus for 2D irrotational flows, the governing equation is either Equation (2.31) or (3.4). This similar appearance of the governing equations indicate the existence of some generality between the two solutions obtained in term of ϕ or ψ. This is obtained in the general framework of potential theory derived using independent variables expressed as complex numbers. This is discussed next. This approach and various other theoretical tools discussed in this and subsequent chapters are also described quite effectively in Currie (2010), Kundu and Cohen (2008), Houghton and Carpenter (1993) and Milne-Thomson (1958).

3.2 Preliminaries of Complex Analysis for 2D Irrotational Flows: Cauchy–Riemann Relations

Any complex variable can be represented as $z = x + iy$ or $z = re^{i\theta}$, with $i^2 = -1$. The complex conjugate of the same variable can be written as $\bar{z} = x - iy$ or $re^{-i\theta}$, so that $r = (z\bar{z})^{1/2}$. The angle, θ is the argument, and one takes the principal value of it to lie within $\pm\pi$.

Quite often in analysis, one comes across analytic functions (also called holomorphic function) of complex variable. Let $f(z)$ be one such function defined inside a closed contour C, then it must have the following properties:

(i) $f(z)$ is finite and unique within C and

(ii) for any z within C, it has a single valued finite derivative defined by

$$f'(z) = \underset{z' \to z}{Lim} \frac{f(z') - f(z)}{z' - z}.$$

Example of analytic functions are $\sin z$, $\cos z$, e^z and z^n, with n positive. The function $(z - z_0)^{-n}$, with n positive is not analytic at $z = z_0$. The real and imaginary parts of an analytic function are called *conjugate functions*. With the help of velocity potential and stream function, we can construct an analytic function and call it the complex potential as

$$W = \phi + i\psi.$$

As ϕ and ψ are continuous functions of their argument, so is W of its argument z. Thus, one can obtain the following partial derivatives of the complex potential

$$\frac{\partial \phi}{\partial x} + i\frac{\partial \psi}{\partial x} = \frac{\partial W}{\partial x} = \frac{dW}{dz}\frac{\partial z}{\partial x} = W'(z) \tag{3.5}$$

and

$$\frac{\partial \phi}{\partial y} + i\frac{\partial \psi}{\partial y} = \frac{\partial W}{\partial y} = \frac{dW}{dz}\frac{\partial z}{\partial y} = iW'(z). \tag{3.6}$$

From Equations (3.5) and (3.6), one obtains the following relations

$$\frac{\partial \phi}{\partial x} = \frac{\partial \psi}{\partial y} \quad and \quad \frac{\partial \phi}{\partial y} = -\frac{\partial \psi}{\partial x}. \tag{3.7}$$

These are the well known Cauchy–Riemann relations. Elimination of ϕ and ψ from the Cauchy–Riemann relations yields once again the governing equation, as the Laplace's equations for ϕ and ψ given in Equations (2.31) and (3.4).

A corollary of Cauchy–Riemann relations is that the $\phi = constant$ and $\psi = constant$ lines meet at right angles to each other. A proof of this follows from the observation that along $\phi = constant$ and $\psi = constant$ lines one can write

$$d\phi = 0 = \frac{\partial \phi}{\partial x} + \frac{\partial \phi}{\partial y}\left(\frac{dy}{dx}\right)_1 \tag{3.8}$$

and

$$d\psi = 0 = \frac{\partial \psi}{\partial x} + \frac{\partial \psi}{\partial y}\left(\frac{dy}{dx}\right)_2. \tag{3.9}$$

From Equations (3.8) and (3.9), one obtains the product of the slopes $\left(\frac{dy}{dx}\right)_1$ and $\left(\frac{dy}{dx}\right)_2$ of constant velocity potential and stream function lines as

$$\left(\frac{dy}{dx}\right)_1\left(\frac{dy}{dx}\right)_2 = \frac{\phi_x\psi_x}{\phi_y\psi_y},$$

where the subscripts in the numerator and denominator of the right-hand side above indicate partial derivatives with respect to that variable. By the Cauchy–Riemann relation, the right-hand side is equal to -1, implying the proposition that at any point in the flow, the $\phi = constant$ and $\psi = constant$ lines meet at right angles to each other.

Cauchy's Integral Theorem: states that for an analytic function defined inside a closed curve C, $\oint f(z)dz = 0$.

In all contour integrals considered, the counter-clockwise direction will be considered as positive. The proof of this theorem also follows from Cauchy–Riemann relations for, if $f(z) = \phi + i\psi$, then one notes that

$$\oint f(z)dz = \oint (\phi + i\psi)(dx + idy) = \oint (\phi\, dx - \psi\, dy) + i\oint (\psi\, dx + \phi\, dy).$$

Using Stokes' theorem: $\oint (P\, dy - Q\, dx) = \iint \left(\frac{\partial P}{\partial x} + \frac{\partial Q}{\partial y}\right) dx\, dy$, one obtains

$$\oint (\phi\, dx - \psi\, dy) = -\iint \left(\frac{\partial \psi}{\partial x} + \frac{\partial \phi}{\partial y}\right) dx\, dy = 0, \text{ by Cauchy–Riemann relation}$$

and

$$\oint (\psi \, dx + \phi \, dy) = \iint \left(\frac{\partial \phi}{\partial x} - \frac{\partial \psi}{\partial y} \right) dx \, dy = 0 \text{ also by Cauchy–Riemann relation.}$$

Singularities:
Are the points in the Argand diagram, where the complex function, $f(z)$, ceases to be analytic. In the neighbourhood of such a point, indicated here by z^*, the function can be expressed in terms of positive and negative powers of $(z - z^*)$ by Laurent series expansion

$$f(z) = ... + a_3(z - z^*)^3 + a_2(z - z^*)^2 + a_1(z - z^*) + a_0 \\ + b_1(z - z^*)^{-1} + b_2(z - z^*)^{-2} + ... \tag{3.10}$$

If the number of terms with negative exponent in the above Laurent series expansion is finite in number, then $z = z^*$ is called the pole. Furthermore, if all b_j's are zero except b_1, then the point $z = z^*$ is called the simple pole. This coefficient b_1 is called the residue of the function at z^*. The residue is very important, as it can only produce nonzero contribution for a contour integral performed for the function given in Equation (3.10) over the closed curve that includes the pole. To show this, consider the contour integral of the powers of $(z - z^*)$ on a circle of radius R centred at z^* by

$$I = \oint (z - z^*)^n dz = \int_0^{2\pi} R^{n+1} e^{i(n+1)\theta} \, i \, d\theta$$

as one can express $(z - z^*) = Re^{i\theta}$ along the contour of the circle. The above integral can be further simplified to

$$I = \frac{R^{n+1}}{(n+1)} \left[e^{(n+1) \, i\theta} \right]_0^{2\pi} = 0 \text{ if n} \neq -1.$$

However, for $n = -1$

$$I = \oint \frac{dz}{(z - z^*)} = \int_0^{2\pi} i \, d\theta = 2\pi i.$$

Thus the contour integral of the function given in Equation (3.10) can be simply written as

$$\oint f(z) dz = 2\pi i b_1. \tag{3.11}$$

This result is further generalized in the following theorem.

3.2.1 Cauchy's Residue Theorem

Let C be a closed contour inside and upon which $f(z)$ is analytic, except at finite number of singular points within C, for which if the residues are given as $r_1, r_2, r_3, \dots r_n$, then

$$\oint f(z)dz = 2\pi i(r_1 + r_2 + r_3 + \dots + r_n). \tag{3.12}$$

3.2.2 Complex Potential and Complex Velocity

We have defined the complex potential as one for which the real part is the velocity potential and stream function is the imaginary part. As this is an analytic function everywhere except at the singularities, ϕ and ψ satisfy the Cauchy–Riemann relation. Conversely, any analytical function can be viewed to represent a fluid flow, whose real part is a valid velocity potential and imaginary part is a potential stream function. Now from Equation (3.5) and the Cauchy–Riemann relation given in Equation (3.7), we get

$$\frac{dW}{dz} = \frac{\partial\phi}{\partial x} + i\frac{\partial\psi}{\partial x} = u - iv. \tag{3.13}$$

Thus, the derivative $\frac{dW}{dz}$ is called the complex velocity. If we represent $\bar{W}$ as the complex conjugate of W, then it is easy to see that

$$\frac{dW}{dz}\frac{d\bar{W}}{dz} = (u - iv)(u + iv) = u^2 + v^2 \tag{3.14}$$

with the right-hand side representing the speed squared.

For uniform flow with velocity U_∞ in the x-direction, the complex potential can be obtained as

$$W_\infty = U_\infty Z.$$

3.3 Elementary Singularities in Fluid Flows

We have already noted that any complex function ceases to be analytic at singular points. As the derivative does not remain finite at the singular points for the complex potential, then from Equation (3.13) one notices that velocity components are infinite there. We have noted that the contribution of such singularities can be obtained using Cauchy's residue theorem. In fluid mechanics, such primary singularities are the source, sink and vortices. In the context of two-dimensional flows, one considers only line vortex. There are two ways of analyzing such problems: either by posing these in the physical plane and obtain the solutions in a heuristic manner or assuming some analytic complex potential function and obtain the streamlines to infer possible flow fields. Both these approaches are used here to have better appreciation of the roles of these elementary singularities.

We have already obtained the complex potential due to uniform flow, which implies a source and sink pair positioned at infinite distances from the object under investigation. Next, we obtain ϕ and ψ for elementary singularities like source/sink and line vortex from the first principle.

Flow field due to source/sink:

Consider a source placed at the origin of a Cartesian system in an infinite expanse, which will emit fluid in all directions without bias. This will therefore create fluid flow that will be purely radial and symmetric. The strength of the source is given by the volume flow rate of the source and let the source under consideration have the strength as m. Using polar coordinate system, the radial velocity component can be written in terms of ϕ as

$$v_r = \frac{\partial \phi}{\partial r}.$$

Constructing a control volume as a circle at a distance r from the source, as shown in Figure 3.1(a), we note that the velocity at all points of the control surface is given by the same radial velocity component, v_r. Hence the mass flow through the control surface is given by, $2\pi r v_r$ and by definition, this is the strength of the source, i.e.

$$\frac{\partial \phi}{\partial r} = \frac{m}{2\pi r}. \tag{3.15}$$

This equation can be readily integrated to obtain

$$\phi = \frac{m}{2\pi} \text{Ln} \frac{r}{r_0}, \tag{3.16}$$

where $r = r_0$ is the radius of the equi-potential line, $\phi = 0$. Since the radial velocity can also be written down in terms of the stream function by

$$v_r = \frac{1}{r}\frac{\partial \psi}{\partial \theta}.$$

This expression along with Equations (3.15) and (3.16) provides the stream function as

$$\psi = \frac{m}{2\pi}(\theta - \theta_0), \tag{3.17}$$

where θ_0 indicates the orientation of the streamline for which $\psi = 0$. The $\psi = constant$ lines are drawn as solid lines in Figure 3.1(a) and the equi-potential lines are circular and shown by

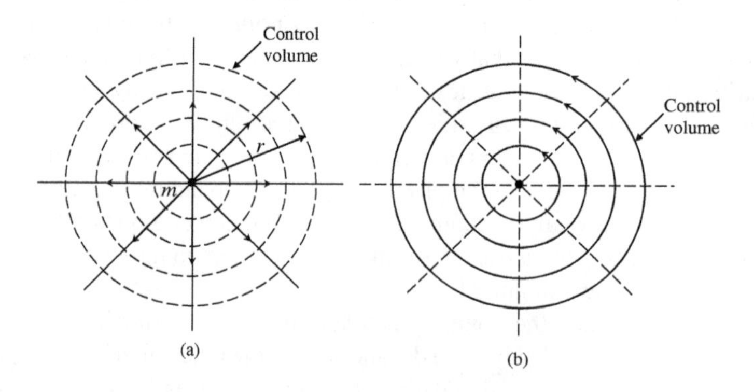

Figure 3.1 Streamlines and equipotential lines created by (a) a source and (b) a line vortex

dashed lines. With the help of ϕ and ψ in Equations (3.15) and (3.16), one can construct the complex potential as

$$W = \frac{m}{2\pi}\left(\text{Ln}\frac{r}{r_0} + i(\theta - \theta_0)\right) = \frac{m}{2\pi}\text{Ln}\frac{z}{z_0}, \tag{3.18}$$

where $z_0 = r_0 e^{-i\theta_0}$ represents the datum for W. For a sink the complex potential will be of same magnitude, but of opposite sign. We also note from Equation (3.15) that $v_r \to \infty$ as $r \to 0$ – the location of the source/sink. In a mathematical sense, the source/sink can be viewed as a Dirac delta function, such that $\text{Lim}_{r\to 0}\, 2\pi r v_r = m$.

Flow field due to a line vortex:
The line vortex is a singularity, which has an associated constant circulation, as defined in Section 2.7. The circulation is considered positive, when the path is traversed in counter-clockwise direction and the line integral is positive. Although the streamlines describe closed circular paths, vorticity in the flow is zero everywhere, excluding the singular line vortex element. Thus, the induced velocity is strictly in azimuthal direction, with zero radial component. Let Γ be the circulation associated with the line vortex. Considering the control volume of radius r in Figure 3.1(b) shown by any circle along which one can perform a contour integral in using the definition of Γ in Equation (2.25) to get

$$v_\theta = \frac{1}{r}\frac{\partial \phi}{\partial \theta} = \frac{\Gamma}{2\pi r}, \tag{3.19}$$

which can be integrated to obtain

$$\phi = \frac{\Gamma}{2\pi}(\theta - \theta_0). \tag{3.20}$$

As $v_\theta = -\frac{\partial \psi}{\partial r}$, one can obtain using Equation (3.19) the stream function as

$$\psi = -\frac{\Gamma}{2\pi}\text{Ln}\frac{r}{r_0}. \tag{3.21}$$

Using Equations (3.20) and (3.21) one can obtain the complex potential as

$$W = \phi + i\psi = \frac{\Gamma}{2\pi}\left(\theta - \theta_0 - i\text{Ln}\frac{r}{r_0}\right) = -\frac{i\Gamma}{2\pi}\left[i(\theta - \theta_0) + \text{Ln}\frac{r}{r_0}\right] = -\frac{i\Gamma}{2\pi}\text{Ln}\frac{z}{z_0}. \tag{3.22}$$

Once again z_0 defines the datum for the complex potential.

We also note from Equation (3.19) that $v_\theta \to \infty$ as $r \to 0$ – the location of the vortex. In a mathematical sense, the vortex also is a Dirac delta function, such that $\text{Lim}_{r\to 0}\, 2\pi r v_\theta = \Gamma$.

3.3.1 Superposing Solutions of Irrotational Flows

We have noted that the governing equation for incompressible, irrotational flows are given by the Laplace's equation for ϕ and ψ, given in Equations (2.31) and (3.4). The flow field due to elementary singularities are valid solutions of Equations (2.31) and (3.4). Noticing that these governing equations are linear, it is natural to superpose these solutions of elementary singularities and construct more complex flow fields. We will consider some of these compound solutions.

(a) Flow field due to a pair of source and sink:
Here, a pair of source and sink of identical strength m is kept at a distance of $2s$ from each other, as shown in Figure 3.2. The sink is placed to the left of the source. The flow field is obtained at any arbitrary field point P whose coordinate is given by (x, y). The lines joining P with the source and sink subtend angles θ_1 and θ_2, respectively, with the x-axis. The included angle between these two lines is indicated by θ.

The stream function at the point P is obtained by superposing the solution due to source and sink as

$$\psi = \frac{m}{2\pi}(\theta_1 - \theta_2) = \frac{m\theta}{2\pi}.$$

This could also be written in terms of Cartesian coordinates as

$$\psi = \frac{m}{2\pi}\left(\tan^{-1}\frac{y}{x-s} - \tan^{-1}\frac{y}{x+s}\right),$$

which using the trigonometric identity

$$(\theta_1 - \theta_2) = \tan^{-1}\frac{\tan\theta_1 - \tan\theta_2}{1 + \tan\theta_1 \tan\theta_2}$$

yields the simplified expression for ψ as

$$\psi = \frac{m}{2\pi}\tan^{-1}\frac{2sy}{x^2 + y^2 - s^2}. \tag{3.23}$$

The streamlines are obtained as ψ = constant lines given as

$$\tan\frac{2\pi\psi}{m} = \frac{2sy}{x^2 + y^2 - s^2}.$$

This can be further simplified as

$$x^2 + \left(y - s\cot\frac{2\pi\psi}{m}\right)^2 = s^2 \text{cosec}^2\frac{2\pi\psi}{m}.$$

This represents circles for different values of ψ with centre on the y-axis at $s\cot\frac{2\pi\psi}{m}$ and radius as $s\,\text{cosec}\frac{2\pi\psi}{m}$.

Using the value of ϕ from Equation (3.16) for the source, one can obtain the velocity potential for the combination of source and sink of Figure 3.2 as

$$\phi = \frac{m}{2\pi}\text{Ln}\frac{r_1}{r_2} = \frac{m}{4\pi}\text{Ln}\frac{x^2 + y^2 + s^2 - 2xs}{x^2 + y^2 + s^2 + 2xs}. \tag{3.24}$$

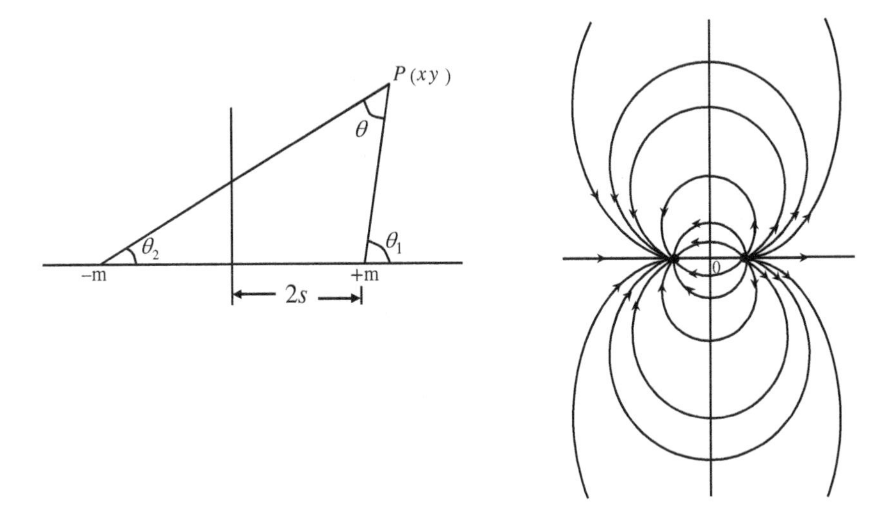

Figure 3.2 A source placed in the upstream of a sink, both having the same strength and separated by a distance $2s$

The equi-potential lines are obtained by considering ϕ = constant in

$$\frac{x^2 + y^2 + s^2 - 2xs}{x^2 + y^2 + s^2 + 2xs} = e^{4\pi\phi/m} = \lambda,$$

which represents a circle whose equation is given by

$$x^2 + y^2 + 2xs\left(\frac{\lambda + 1}{\lambda - 1}\right) + s^2 = 0.$$

The centres are located on the x-axis at $-s\left(\frac{\lambda+1}{\lambda-1}\right)$ and whose radius is given by

$$s\sqrt{\left(\frac{\lambda + 1}{\lambda - 1}\right)^2 - 1} = \frac{2s\sqrt{\lambda}}{\lambda - 1} = 2s \text{ cosech } \frac{2\pi\phi}{m}.$$

(b) Flow field due to a doublet:
This is a special case of the above, when the distance between the source and sink vanishes, i.e. $2s \to 0$. In such a case, from Figure 3.2 one can note that $\theta \to 0$, so that $\tan\theta \simeq \theta$. Thus, from Equation (3.23) one simplifies to obtain

$$\psi_{doublet} = \lim_{2s \to 0} \frac{m}{2\pi} \frac{2sy}{x^2 + y^2 - s^2}. \tag{3.25}$$

The strength of the doublet is defined as $\mu = \lim\limits_{2s\to 0} 2ms$ and in this limit the denominator of Equation (3.25) tends to r^2 and in the numerator one can substitute $y = r\sin\theta$ to obtain

$$\Psi_{doublet} = \frac{\mu y}{2\pi r^2} = \frac{\mu \sin\theta}{2\pi r}. \tag{3.26}$$

From the definition of the radial velocity

$$\frac{1}{r}\frac{\partial \psi}{\partial \theta} = \frac{\partial \phi}{\partial r}.$$

Thus, the velocity potential can be obtained by integration as

$$\phi_{doublet} = -\frac{\mu\cos\theta}{2\pi r}. \tag{3.27}$$

The complex potential is obtained from Equations (3.26) and (3.27) as

$$W = -\frac{\mu\cos\theta}{2\pi r} + \frac{i\mu}{2\pi r}\sin\theta = -\frac{\mu}{2\pi r}(\cos\theta - i\sin\theta) = -\frac{\mu}{2\pi r e^{i\theta}} = -\frac{\mu}{2\pi z}. \tag{3.28}$$

The streamlines created by the doublet is shown in Figure 3.3. The doublet has a specific direction and this is indicated to be from the source to sink, as indicated by the thick line with the arrowhead marking the direction. The doublet is equivalent to the derivative of a delta function, and is an elementary singularity, which will be used to describe flow field past complex body shapes.

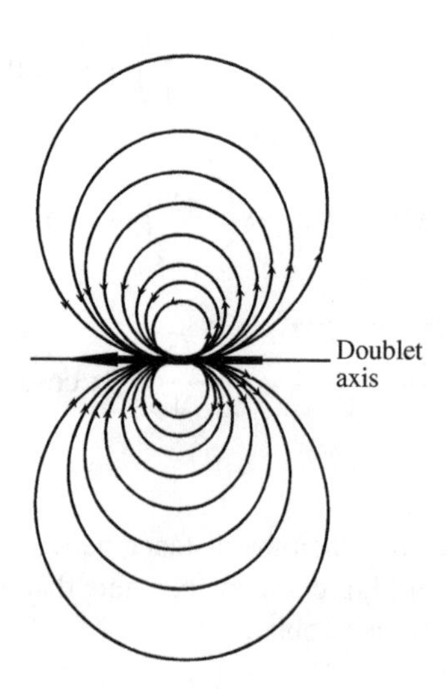

Figure 3.3 Streamlines created by the doublet. The axis is pointed from the source to the sink, as indicated by the thick line with the arrowhead

(c) Flow field due to a source-sink pair in uniform horizontal stream:
The stream function for a uniform horizontal stream (U) from right to left is given by $-Uy$ and due to the source-sink pair is given by Equation (3.23) and the combination of the two gives

$$\psi = \frac{m}{2\pi}\tan^{-1}\frac{2sy}{x^2+y^2-s^2} - Uy \tag{3.29}$$

and the velocity potential is similarly obtained as

$$\phi = \frac{m}{4\pi}\,\mathrm{Ln}\,\frac{x^2+y^2+s^2-2xs}{x^2+y^2+s^2+2xs} - Ux. \tag{3.30}$$

Note that the streamline $\psi = 0$ is obtained from Equation (3.29) as

$$\tan\left(\frac{2\pi U}{m}y\right) = \frac{2sy}{x^2+y^2-s^2},$$

which is symmetric about the x and y-axes and represents a closed oval-shaped geometry. Thus, the stream functions given by Equation (3.29) for which $\psi > 0$ represents flow at zero angle of attack past this oval. The x-axis (i.e. $y = 0$ line) outside the oval also corresponds to $\psi = 0$ and its intersections with the oval define the front and rear stagnation points, as indicated in Figure 3.4.

Also, $\psi = 0$ represents the closed double-symmetric oval with maximum thickness located at $x = 0$ given by $y = 2t_0$, as indicated in Figure 3.4. This maximum thickness is obtained from

$$\psi = 0 = \frac{m}{2\pi}\tan^{-1}\frac{2s\,t_0}{t_0^2-s^2} - Ut_0$$

or

$$\tan\frac{2\pi Ut_0}{m} = \frac{2t_0 s}{t_0^2-s^2}.$$

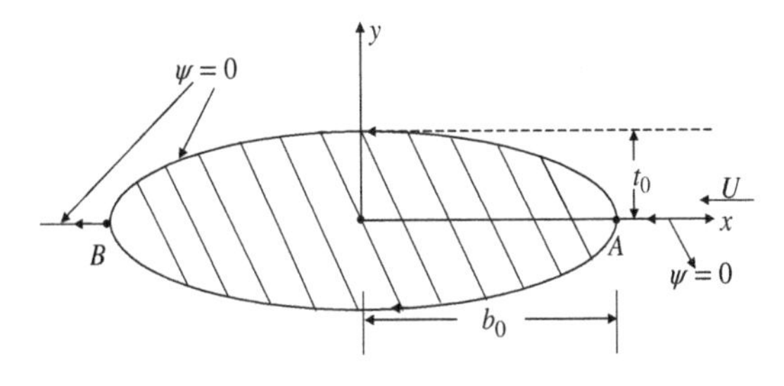

Figure 3.4 A pair of source and sink of strength (m) placed at a distance of $2s$ from each other, placed in a uniform horizontal stream from right to left. Note the front and rear stagnation points marked on the oval, which is the body streamline, $\psi = 0$

The stagnation points are at the tip of the oval, i.e. at $x = b_0$ and at $y = 0$, as marked in Figure 3.4. This can be obtained from the definition, i.e. from $u = 0$ (note that for potential flows, the wall-normal component of velocity is always zero at all points on the surface) as

$$\frac{\partial \psi}{\partial y} = 0.$$

That is

$$\frac{\partial \psi}{\partial y} = \frac{m}{2\pi} \frac{1}{1 + \left(\frac{2sy}{x^2+y^2-s^2}\right)^2} \frac{(x^2 + y^2 - s^2)2s - 4sy^2}{(x^2 + y^2 - s^2)^2} - U.$$

Thus, the coordinate of the stagnation point is ($x = b_0$ $y = 0$), and is obtained from

$$\frac{\partial \psi}{\partial y} = 0,$$

which gives

$$0 = \frac{m}{2\pi} \frac{(b_0^2 - s^2)2s}{(b_0^2 - s^2)^2} - U.$$

This expression relates the strength of the source-sink with the oncoming flow speed and the separation distance between the source and sink given by

$$m = \pi U \frac{b_0^2 - s^2}{s}.$$

(d) Flow field due to a doublet in uniform horizontal stream:
We analyze the flow field by looking at the combination of complex potentials. It is shown that the complex potential due to uniform horizontal velocity as $-Uz$, when the flow is from right to left. From Equation (3.28) the complex potential for the doublet is given by $-\mu/(2\pi z)$. Thus, the complex potential for the combination is given by

$$W = -Uz - \frac{\mu}{2\pi\, z} = -U\left(z + \frac{a^2}{z}\right), \tag{3.31}$$

where $a = \sqrt{\frac{\mu}{2\pi U}}$. From Equation (3.31) one can also write the stream function as

$$\psi = \frac{\mu}{2\pi r} \sin\theta - U r \sin\theta$$

or

$$\psi = \frac{\mu}{2\pi} \frac{y}{x^2 + y^2} - U y.$$

Thus for $\psi = 0$, one must have

$$y\left(\frac{\mu}{2\pi(x^2+y^2)} - U\right) = 0,$$

which implies that

$$\psi = 0 \text{ is for } y = 0 \text{ and } x^2 + y^2 = \frac{\mu}{2\pi U}.$$

The same condition of $\psi = 0$ in polar coordinate provides

$$\left(\frac{\mu}{2\pi r} - Ur\right)\sin\theta = 0.$$

Thus, $\psi = 0$ represents the following possibilities

$$\sin\theta = 0, \; i.e. \; \theta = 0 \text{ and } \pm\pi \text{ (the } x-\text{axis).}$$

The first factor defines a circle of radius a obtained from

$$\frac{\mu}{2\pi a} - Ua = 0 \Longrightarrow a = \sqrt{\frac{\mu}{2\pi U}}.$$

Therefore, the stream function $\psi = 0$ is given by the circle of the radius a and the x-axis. Thus, this represents a flow past a circular cylinder of radius a and where it intersects the x-axis, are the stagnation point. This stream function is given in polar representation as

$$\psi = U\sin\theta\left(-r + \frac{a^2}{r}\right). \tag{3.32}$$

The streamlines of the flow field is depicted in Figure 3.5, with the front and rear stagnation points marked as A and B. The readers are informed that such a flow field will not be seen experimentally due to massive separation that the flow will suffer beyond the shoulder of the cylinder in the leeward side due to adverse pressure gradient. In fact, this distinguishes flows past streamline shape with the bluff body flows. Flow past a circular cylinder is a canonical example of bluff body flow and the solution given by Equation (3.32) is quite useful for solving impulsively started viscous flow as initial condition for numerical solution of the Navier-Stokes equation.

Using Equation (3.32), velocity components are obtained at any point in the flow domain as

$$v_r = \frac{1}{r}\frac{\partial\psi}{\partial\theta} = U\left(-1 + \frac{a^2}{r^2}\right)\cos\theta \tag{3.33}$$

and

$$v_\theta = -\frac{\partial\psi}{\partial r} = U\left(1 + \frac{a^2}{r^2}\right)\sin\theta. \tag{3.34}$$

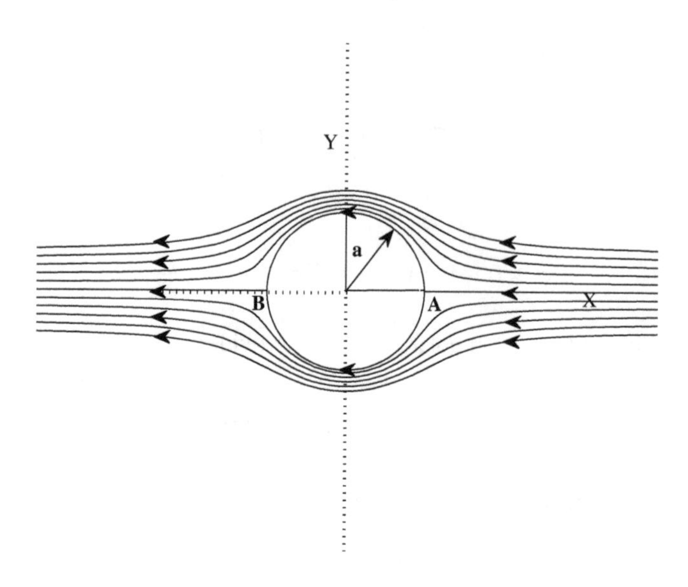

Figure 3.5 Flow past a stationary cylinder modelled as a doublet in uniform horizontal stream from right to left. Note the front and rear stagnation points marked on the circle as A and B

On the surface of the cylinder ($r = a$): $v_r = 0$ and $v_\theta = 2U \sin\theta$.

This enables us to obtain the pressure (p) distribution on the surface of the cylinder by using Bernoulli's equation. The speed of the inviscid flow (q) on the surface of the cylinder is given by v_θ at $r = a$. Far away from the cylinder, the effect of it will disappear and the free stream speed is therefore U. If the free stream pressure is represented as p_∞, then an application of Bernoulli's equation for steady irrotational flow provides

$$p_\infty + \frac{1}{2}\rho U^2 = p + \frac{1}{2}\rho q^2.$$

Thus, the coefficient of pressure is obtained as

$$C_p = \frac{p - p_\infty}{\frac{1}{2}\rho U^2} = 1 - 4\sin^2\theta. \tag{3.35}$$

The force acting on the cylinder by the irrotational flow can be obtained by integrating the pressure distribution given in Equation (3.35). However, there is an elegant theorem due to Blasius, which helps in obtaining the force components and pitching moment directly, once the complex potential of the flow past any arbitrary body is known.

3.4 Blasius' Theorem: Forces and Moment for Potential Flows

Let a fixed body be placed in an irrotational steady flow and let X, Y and M denote the Cartesian components of force and moment about the origin due to pressure distribution, then neglecting external forces one obtains

$$X - iY = \frac{i\rho}{2} \oint \left(\frac{dW}{dz}\right)^2 dz \tag{3.36}$$

$$M = \text{Real}\left[-\frac{\rho}{2} \oint z\left(\frac{dW}{dz}\right)^2 dz\right]. \tag{3.37}$$

Proof:
Consider an arbitrary body shown in Figure 3.6 (in this case an aerofoil is shown for reference), kept in an irrotational flow with a small element shown on which the differential force components are shown to act due to the pressure distribution.

The force components on the surface element ds (with projection as dx and dy in x- and y-directions, respectively) along the coordinate axes are given by

$$dX = -p\,dy \ \text{ and } \ dY = p\,dx.$$

Similarly the moment of these force components about the origin is given by

$$dM = p(x\,dx + y\,dy),$$

where the counter-clockwise direction is considered as positive. Using the complex conjugate of z as $\bar{z}$, one can write the force components together as

$$d(X - iY) = -ip\,d\bar{z}. \tag{3.38}$$

Similarly the pitching moment about the origin can also be written as

$$dM = \text{Real}\,[pz\,d\bar{z}]. \tag{3.39}$$

One can write the pressure using Bernoulli's equation as

$$p = C_1 - \frac{1}{2}\rho\,q^2,$$

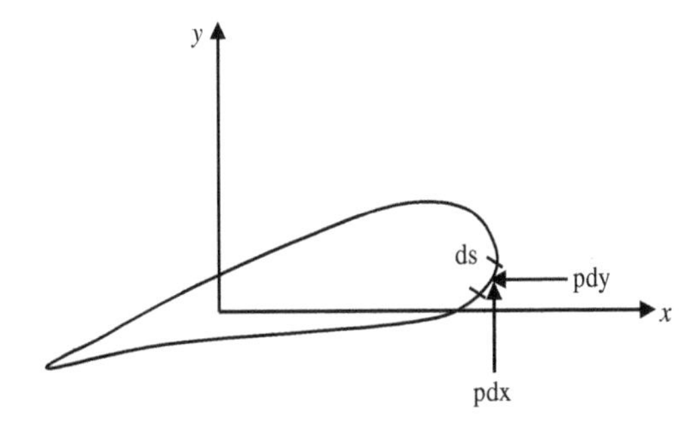

Figure 3.6 An arbitrary body shown with differential force acting on a small element for the application of Blasius' theorem

where C_1 is the Bernoulli head and the speed on any point on the surface of the body is obtained from

$$q^2 = (u^2 + v^2) = (u + iv)(u - iv) = \frac{dW}{dz}\frac{d\bar{W}}{d\bar{z}}.$$

Using these simplifications in terms of the complex potential and its derivative, one can further write in compact form the complex force and moment in Equations (3.38) and (3.39) as

$$(X - iY) = \frac{i\rho}{2}\oint \frac{dW}{dz}d\bar{W} + i\oint C_1 d\bar{z}.$$

The last contour integral on right-hand side of the above is identically zero and hence does not contribute to forces and moment integrals. Also

$$M = \text{Real}\left[\oint -\frac{\rho z}{2}\frac{dW}{dz}d\bar{W}\right].$$

As the surface of the body represents $\psi =$ constant on it, one notes $d\psi = 0$ and as a consequence $dW = d\bar{W}$ on the surface of the body in Figure 3.6. Thus

$$d\bar{W} = \frac{dW}{dz}dz$$

and one can further simplify Equations (3.38) and (3.39) as

$$(X - iY) = \frac{i\rho}{2}\oint \left(\frac{dW}{dz}\right)^2 dz$$

and

$$M = \text{Real}\left[-\oint \frac{\rho z}{2}\left(\frac{dW}{dz}\right)^2 dz\right].$$

The theorem follows upon performing the contour integral following the surface. Here, we also invoke the observation that this contour can be furthermore deformed or enlarged, until and unless we encounter any singularities. We also note that the flow is composed of singularities which are already in the interior of the body. For any general case, we would consider the contour to be obtained by expanding the body streamline without encountering any other singularities.

3.4.1 Force Acting on a Vortex in a Uniform Flow

Here, we use the Blasius' theorem to calculate the force acting on the vortex placed in a horizontal stream flowing from left to right. Consider the vortex to be of strength, Γ placed in a horizontal flow U with the circulation in the clockwise direction. The complex potential of this combination is given by

$$W = Uz + \frac{i\Gamma}{2\pi}\mathrm{Ln}z. \tag{3.40}$$

Thus, the derivative of this complex potential is given by

$$\frac{dW}{dz} = U + \frac{i\Gamma}{2\pi z}. \tag{3.41}$$

Hence

$$\left(\frac{dW}{dz}\right)^2 = U^2\left[1 + \frac{i\Gamma}{\pi U z} - \frac{\Gamma^2}{4\pi^2 U^2 z^2}\right]. \tag{3.42}$$

For the complex potential, the singularity lies at the origin of the Cartesian or polar coordinate system. Hence, for the application of Blasius' theorem we consider a closed contour given by a unit circle $z = e^{i\theta}$. Therefore from Equations (3.36) and (3.42), one can write

$$\begin{aligned}
X - iY &= \int_0^{2\pi} \frac{i\rho}{2}\left(\frac{dW}{dz}\right)^2 dz \\
&= \frac{i\rho U^2}{2}\int_0^{2\pi}\left[1 + \frac{i\Gamma}{\pi U}e^{-i\theta} - \frac{\Gamma^2}{4\pi^2 U^2}e^{-2i\theta}\right] ie^{i\theta}d\theta \\
&= -\frac{\rho U^2}{2}\left[-ie^{i\theta}\Big|_0^{2\pi} + \frac{i\Gamma}{\pi U}\theta\Big|_0^{2\pi} - \frac{\Gamma^2}{4\pi^2 U^2}ie^{-i\theta}\Big|_0^{2\pi}\right].
\end{aligned}$$

The first and third terms do not contribute to the integral, with the middle term contributing as

$$X - iY = -i\rho U\Gamma. \tag{3.43}$$

Thus, $X = 0$ and $Y = \rho U\Gamma$. The same results will be obtained if one uses contour integral and uses Cauchy's residue theorem using Equation (3.42) in Equation (3.36) for the pole at $z = 0$. In Equation (3.42), we note that the second term constitute a simple pole and the third term representing a pole of second order. Equation (3.43) is also known as the **Kutta-Jukowski theorem**, which states that the creation of a transverse force (in y-direction) in a uniform horizontal flow (in x-direction) is by keeping a vortex whose strength is Γ and is in the z-direction. Thus, this acting force ($\vec{F}$) could also be written down for a general flow ($\vec{V}$) acting on the vortex ($\vec{\Gamma} = \Gamma\hat{k}$) in vectorial notation by

$$\vec{F} = \rho\vec{V} \times \vec{\Gamma}. \tag{3.44}$$

Physically we cannot conceive of a force acting without a physical body and the above derivation may appear as abstract. This result could be attributed to a physical body like a cylinder and the vortex can be thought to be lumped at its centre, created by spinning the cylinder. This is established next.

3.4.2 *Flow Past a Translating and Rotating Cylinder: Lift Generation Mechanism*

Let us consider a cylinder of radius a translating with speed U and spinning about its axis, so that one can consider a lumped vortex of circulation Γ at its centre. The flow is from left to right and the circulation is in clockwise direction. Considering the flow to be Galilean invariant, one can construct the flow past translating cylinder from the complex potential by

$$W_1 = U\left(z + \frac{a^2}{z}\right). \tag{3.45}$$

For the spinning cylinder with radius a and with circulation Γ, the complex potential is given by

$$W_2 = \frac{i\Gamma}{2\pi}\mathrm{Ln}\frac{z}{a}. \tag{3.46}$$

Thus the cylinder ($z = ae^{i\theta}$) represents the streamline, $\psi = 0$ in this case and the singularities, doublet for the translating cylinder and the vortex for the spinning action, reside at the centre. Thus, one can apply Blasius' theorem by considering a contour along the surface of the cylinder. The complex potential for the translating and spinning cylinder is obtained by superposing the solution given by Equations (3.45) and (3.46) as

$$W = U\left(z + \frac{a^2}{z}\right) + \frac{i\Gamma}{2\pi}\mathrm{Ln}\frac{z}{a}. \tag{3.47}$$

To obtain general form of the streamlines, one can first obtain the stagnation points obtained from the complex velocity being equal to zero, i.e.

$$\frac{dW}{dz} = 0 = U\left(1 - \frac{a^2}{z^2}\right) + \frac{i\Gamma}{2\pi z} \tag{3.48}$$

or

$$\frac{z^2}{a^2} + \frac{z}{a}\frac{i\Gamma}{2\pi U a} - 1 = 0.$$

Whence

$$\frac{z_{1,2}}{a} = -\frac{i\Gamma}{4\pi U a} \pm \sqrt{1 - \frac{\Gamma^2}{16\pi^2 a^2 U^2}}. \tag{3.49}$$

Depending upon the sign of the quantity under the radical sign, few subcases are considered next.

(a) The circulation magnitude is such that, $\Gamma < 4\pi U a$: Here, one can introduce an auxiliary variable β, such that

$$\frac{\Gamma}{4\pi U a} = \sin\beta.$$

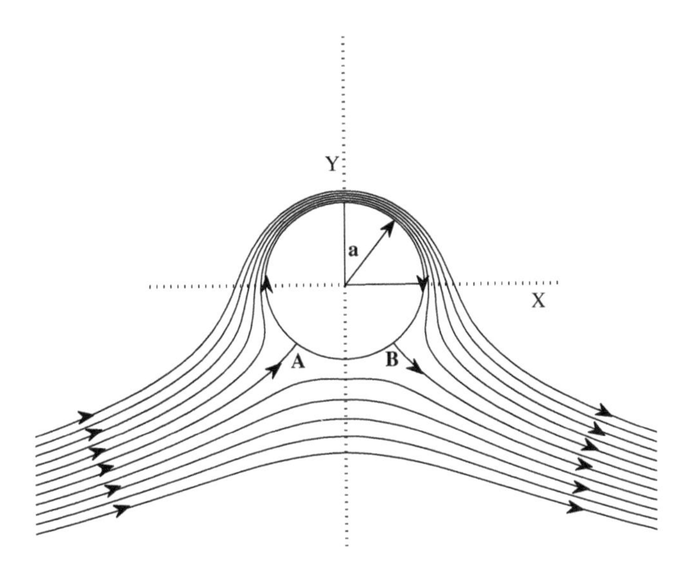

Figure 3.7 Streamlines past the rotating and translating cylinder, with the subcritical circulation

This is used to simplify Equation (3.49) one obtains

$$\frac{z_{1,2}}{a} = -i \sin \beta \pm \cos \beta = e^{-i\beta} \text{ and } e^{i\beta+\pi}. \tag{3.50}$$

The stagnation points and few representative streamlines are shown in Figure 3.7 for this subcase. Note that β represents an angle and the figure shows two stagnation points symmetrically located about the y-axis and depressed by the angle β with respect to the x-axis or flow direction. These two points are marked as A and B in the figure. One notes the existence of a limiting circulation for which $\beta = \pi/2$ and the two stagnation points coalesce. We will consider the value of circulation for which $\beta < \pi/2$ as belonging to subcritical cases. Criticality would correspond to when A and B will merge together. This is also related to the maximum lift which can be generated by this inviscid irrotational model of the flow.

It is easy to see that in the absence of rotation, $\beta = 0$ and the stagnation points would be at $(-a,\ 0)$ and $(a,\ 0)$, i.e. along the x-axis and the origin is at the centre of the circle. The existence of circulation forces these stagnation points to move downwards. Thus, the presence of circulation depresses the stagnation points, and considering steady flow, one notices that the fluid particles above A and B will traverse a much longer distances as compared to particles which remain below A and B. This implies immediately that the flow velocity will be higher for the streamlines staying above A and B, as compared to U. In the same way, the flow velocity will be lower for the streamlines lying below A and B. Now from Bernoulli's equation, this state of affair would indicate pressure to be lower above AB (suction) and higher below, with respect to free stream pressure, p_∞. This would create a lift force in the positive y-direction. However, the drawn streamlines also indicate a perfect fore-aft symmetry with respect to the y-axis, implying zero drag experienced. For irrotational flow this is a consequence of D'Alembert's paradox (see White 2008).

The lift experienced by the cylinder can be obtained from Kutta-Jukowski theorem as

$$L = \rho U\Gamma = 4\pi\rho U^2 a \sin\beta.$$

Choosing the diameter as the length scale, one can obtain the lift coefficient as

$$C_l = \frac{L}{\frac{1}{2}\rho U^2 2a} = 4\pi \sin\beta. \tag{3.51}$$

We will indicate the two-dimensional lift coefficient with the subscript l and the corresponding three-dimensional value will be denoted by C_L. One can try to relate the rotation rate (Ω^*) with circulation by calculating the latter from the basic definition as $\Gamma = 2\pi a^2\Omega^*$. If we call the corresponding surface speed as $U_s = a\Omega^*$, then $\Gamma = 2\pi a U_s$.

(b) The circulation magnitude is such that, $\Gamma = 4\pi U a$: This case represents a critical case for which $\beta = \pi/2$ and then the stagnation points A and B coincide. From Equation (3.51), one would expect the lift generated to be maximum, i.e. $C_{l,max} = 4\pi$. This was noted by Prandtl and is referred to as the Prandtl's limit. Going by the state of art in aeronautics, this appears to be a very generous restriction. However, the readers are warned about the above derivation, in which the circulation is referred to parametrically, apart from the qualitative statement that this is generated by spinning the cylinder. It is to be understood that the generated circulation is a consequence of application of conservation laws and is not an independent variable. One cannot increase circulation in an unbounded manner. Thus, it has to be sought from the solution of complete Navier-Stokes equation and the readers' attention is drawn to the exhaustive discussion about it in Section 8.4 of White (2008). The flow field is depicted in Figure 3.8 for a qualitative understanding. It is noted that the extent over which suction is now acting covers the full cylinder, with the stagnation points coincident. This will be referred to

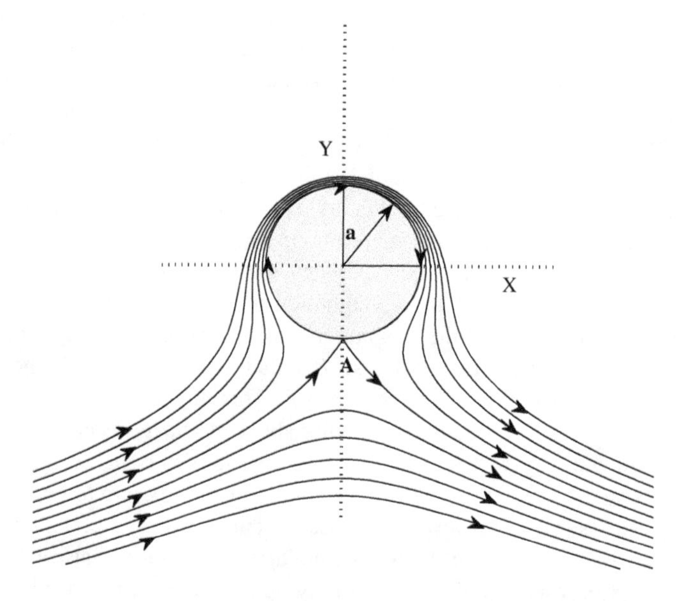

Figure 3.8 Streamlines past the rotating and translating cylinder, with the critical circulation ($\Omega = 2$)

as the critical case. If we call this as Γ_{cr} and corresponding oncoming flow is termed U_{cr}, then one can equate this circulation with $\Gamma = 2\pi a U_s$ to get $U_{cr} = 2U_s$ and the nondimensional rotational rate, $\Omega = \Omega^* a/U = 2$.

Before we proceed further, we note classification of singular points and their effects. From the definition of streamline, one observes that the lines must be nonintersecting and the slope variation must be smooth. Yet in the above two subcases we have noted that the body streamlines do not have continuous slopes/derivatives. Does this imply that the velocity at those points are not defined, as in the case of the stagnation points A and B? While it is understood that viscous actions can smooth out such discontinuities, still these will be noted for viscous flows at the stagnation, separation and reattachment points. When this points are on the surface of a physical body, these are referred to as half-saddle points. One of the redeeming features of these stagnation points are that at these points stream function are continuous and its derivatives are zero. Thus, these causes no ambiguity in prescribing the flow velocity at such points. For the subcase discussed above, these two half-saddle points coincide. However, the following subcase is discussed which corresponds to supercritical circulation.

(c) The circulation magnitude is given by, $\Gamma > 4\pi U a$: For this case, we now define,

$$\frac{\Gamma}{4\pi U a} = \cosh\beta.$$

Here, β does not represent any angle and is a simple parameter. From Equation (3.49) one gets

$$\frac{z_{1,2}}{a} = i(-\cosh\beta \pm \sinh\beta) = -ie^{\beta} \text{ and } -ie^{-\beta}. \tag{3.52}$$

This case also can be classified as $U > 2U_s$ or by the nondimensional rotation rate given by $\Omega > 2$. Now for this case, both the stagnation points are on the y-axis and moreover, $z_1 z_2 = a^2$, implying that these two are inverse points. If one of the points is outside the cylinder $|z| > a$, then the other point must necessarily be inside the cylinder, so that the product remains always equal to a^2. Thus, one of the stagnation points now descends along the y-axis and this is called the full saddle point. The streamlines are shown plotted in this case in Figure 3.9, for the supercritical circulation, with the saddle point (B) clearly visible.

3.4.3 *Prandtl's Limit on Maximum Circulation and its Violation*

For circulation exceeding critical case creates a closed recirculating streamline, as shown in Figure 3.9. Prandtl reasoned that as there is no flow across the streamline, so there cannot be any momentum transfer across this limiting streamline and this will limit the lift experienced by the cylinder when $U > 2U_s$. As mentioned before, this appears as an optimistic limit for aeronautical applications for generating lift by a rotating cylinder. This effect is now known as Magnus-Robins effect (see White (2008) and Sengupta and Talla (2004) for further details of the historical accounts). While it is possible to calculate the actual lift generated on a rotating cylinder by solving Navier-Stokes equation, and it has been noted that for subcritical case of rotation the actual lift generated falls short of this inviscid, irrotational flow model. However, research continues on the topic of super-critical rotation rate and it has been shown both experimentally by Tokumaru and Dimotakis (1993) and computationally by Sengupta *et al.* (2003a) results, which clearly demonstrate that for rotation rates significantly higher than the

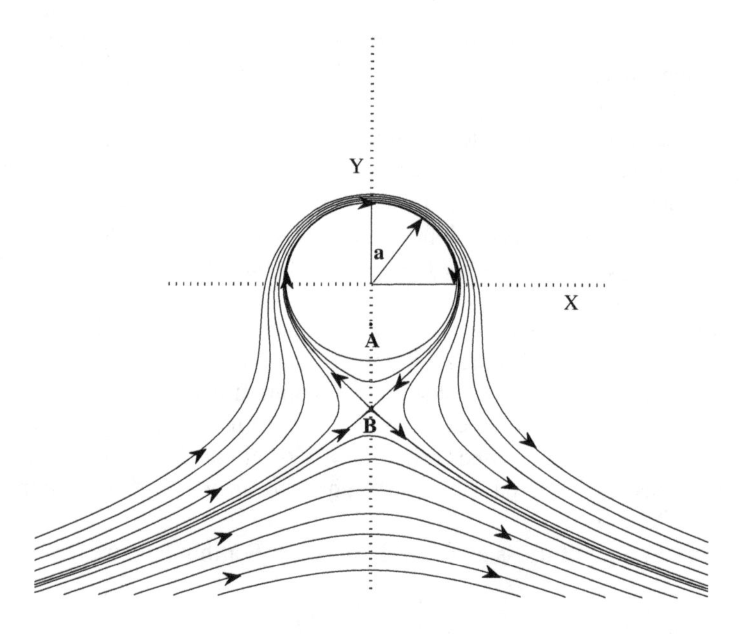

Figure 3.9 Streamlines past the rotating and translating cylinder, with the circulation super-critical ($\Omega > 2$)

critical rate, Prandtl's limit is violated! It has been explained in Sengupta and Talla (2004) and Sengupta *et al.* (2003a) that this is due to viscous diffusion across the limiting streamline and the process of continuous generation of vorticity at the wall, due to no slip condition.

3.4.4 Pressure Distribution on Spinning and Translating Cylinder

Given the complex potential

$$W = U\left(z + \frac{a^2}{z}\right) + \frac{i\Gamma}{2\pi}\text{Ln}\frac{z}{a}. \tag{3.53}$$

The stream function can be obtained from this as

$$\psi = Ur\left(\frac{a^2}{r^2} - 1\right)\sin\theta - \frac{\Gamma}{2\pi}\text{Ln}\frac{r}{a}. \tag{3.54}$$

From which the radial and azimuthal components of velocity are obtained as

$$u_\theta = -\frac{\partial\psi}{\partial r} = U\left(\frac{a^2}{r^2} + 1\right)\sin\theta - \frac{\Gamma}{2\pi r} \tag{3.55}$$

$$u_r = \frac{1}{r}\frac{\partial\psi}{\partial\theta} = U\left(\frac{a^2}{r^2} - 1\right)\cos\theta. \tag{3.56}$$

Thus, on the surface of the cylinder ($r = a$)

$$u_r = 0 \text{ and } u_\theta = 2U\sin\theta - \frac{\Gamma}{2\pi a}.$$

This can be used in the Bernoulli's equation to relate local pressure (p) with free stream pressure (p_∞) as

$$p_\infty + \frac{1}{2}\rho U^2 = p + \frac{\rho}{2}\left(2U\sin\theta - \frac{\Gamma}{2\pi a}\right)^2.$$

This is used to obtain the pressure coefficient as

$$C_p = \frac{p - p_\infty}{\frac{1}{2}\rho U^2} = 1 - \left(2\sin\theta - \frac{\Gamma}{2\pi aU}\right)^2. \quad (3.57)$$

(a) For the subcritical rotation case ($\Gamma < 4\pi aU$), as noted before

$$\frac{\Gamma}{2\pi aU} = 2\sin\beta$$

and the pressure coefficient is then given by

$$C_p = 1 - 4(\sin\theta - \sin\beta)^2.$$

(b) For critical rotation rate ($\Gamma = 4\pi aU$): The stagnation points move such that $\beta = \pi/2$ and then

$$C_p = 1 - 4(\sin\theta - 1)^2.$$

(c) When the rotation rate is super-critical, i.e. $\Gamma > 4\pi aU$, then we defined

$$\frac{\Gamma}{4\pi aU} = \cosh\beta$$

and then

$$C_p = 1 - 4(\sin\theta - \cosh\beta)^2.$$

These pressure distribution can be integrated from $\theta = 0$ to 2π to obtain the lift generated and establish that $L = \rho U\Gamma$ in all the cases.

3.5 Method of Images

So far we have superposed elementary flow solutions to obtain flows of some practical interest. This has been made possible due to the linear governing equation for the irrotational flows given by Laplace's equation. This process can be further aided by method of images, where the presence of multiple walls or bodies can be routinely implemented, again with the principle of superposition of solutions. The methodology originated in electromagnetic problems solved by Gauss, Kirchhoff, Kelvin and others. Its implementation in fluid mechanics is due to Kelvin, Helmholtz, Stokes among many others.

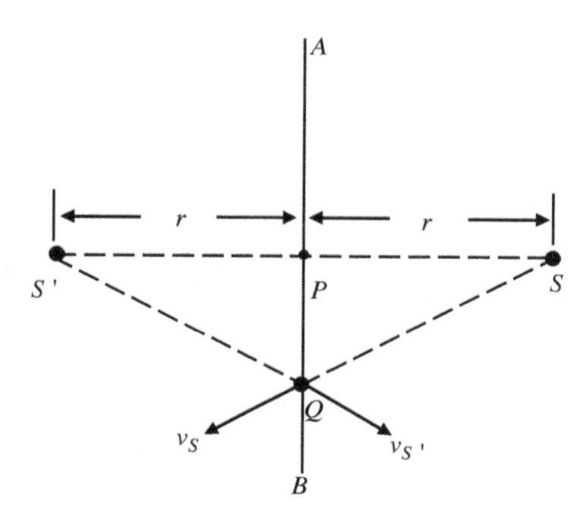

Figure 3.10 Method of images used for a source in front of a plane wall

The working principle of method of images is demonstrated with the help of elementary singularities placed in front of a solid wall. As there cannot be any flow across the wall, this wall must by itself be a streamline. This condition is used as a working constraint for the flow field, in which apart from the singularity or a combination of many of these placed in front of the wall must be modelled. Let us demonstrate it with the help of Figure 3.10, where we place a source of strength m in front of a plane wall AB. Now, the wall AB must by itself be a streamline. At the same time, the source placed at a normal distance r from the wall will induce a velocity at the foot of this normal at P.

Presence of the source will create a normal component of velocity at P, which will violate the requirement that AB be a streamline. To nullify this induced normal velocity component, we place another source of same strength (m), at the same distance from the wall (r) at S', on the other side of the wall. This is the image of the original source placed in front of the wall. Notice that this image source at S' will induce a wall-normal velocity component, which is same in magnitude to which was created by S at P, but of opposite sign. A simple superposition of these two velocities will ensure that there is no wall-normal component of velocity at P. One can notice any other points on the wall also, to see that this zero wall-normal velocity condition along AB is satisfied everywhere. Consider a general point Q, for which the velocity due to the source and its image are drawn as v_S and $v_{S'}$ in Figure 3.10. One can resolve these two velocities into two components: one along the wall and and another perpendicular to the wall. One notices that the wall-normal components of v_S and $v_{S'}$ are equal and opposite, while the component along the wall adds up. These later component does not violate the definition of the streamline.

Next, we consider a vortex in front of the plane wall. In this case, created streamlines by the vortex at the location of the plane wall can be nullified with another vortex (image vortex) which is located behind the plane wall at the same normal distance from it. However, the image vortex will now be of the opposite sign. The readers are cautioned to not treat this case of a single vortex in front of a wall as a trivial mathematical exercise. As the dynamics of a single isolated vortex in front of a finite plane can lead to a sequence of events, beginning with the formation of a boundary layer and its subsequent destabilization by the vortex over it, a topic which has been discussed in great detail in Sengupta *et al.* (2012). It is worthwhile to remember that placement of vortex is different from placement of source or sink. As the image system by itself would impart a motion to the actual vortex, even if there is no mean flow, the vortex and its image system imparts a motion to the fluid at the surface. In real fluid, such a slip velocity at the surface is suppressed via the creation of wall vorticity, whose diffusion and convection leads to formation of a transient boundary layer. Subsequently this boundary vorticity will interact with the free stream vortex and can lead to vortex-induced instability, a topic of fluid mechanics described in Sengupta *et al.* (2003) and Lim *et al.* (2004).

Also, readers' attention is drawn to various cases for method of images discussed in Robertson (1965) for plane and curved walls. In passing, we remark that the uniform flow itself may be viewed as putting a source and and sink at $\pm\infty$. Now placing a cylinder in uniform flow is equivalent to considering a flow where the images of the source and sink at $\pm\infty$ will have to be at the centre of the circle. This is precisely equivalent to placing the doublet at the centre of the circle. Also, in Robertson (1965), positioning of a doublet in front of a wall is described and the curious effect of shape distortion of the streamline representing the circular cylinder placed in front of the wall is discussed, when the distance between the two is decreased.

3.6 Conformal Mapping: Use of Cauchy–Riemann Relation

This is one of the common techniques, by which the flow past a complicated body is analyzed, by studying flow past a simpler body and linking these two flow fields via convenient transformation. For example, in the previous sections we have studied analytically in great detail, the flow past a circular cylinder. Now from an aerodynamic perspective it would be extremely convenient if we could link this flow with the flow around an aerofoil. However, we have to first establish that the governing equation in the physical z-plane retains the same form in the transformed ζ-plane so that the complex potential described in the physical plane can be gainfully employed in the transformed plane.

Let us define the plane defining the complex body as ζ, while the simpler body is represented in the z-plane. Thus in essence, we are looking for a complex function

$$\zeta = f(z), \tag{3.58}$$

such that the geometric shape in the z-plane whose coordinates are x and y is mapped to a different shape in the ζ-plane, whose coordinates are given by ξ and η, respectively. It is assumed that the transformation represented by Equation (3.58) is analytic. If it is so, then it is possible to establish a simple relationship between the physical and transformed plane flow properties.

For any point z_0 in z-plane, we call the point $\zeta_0 = f(z_0)$ as the image of z_0 with respect to the map, $f(z)$. In this way, one maps lines and planes via the same mapping function. In conformal mapping, one tries to relate special shapes in the z-plane which will transform to simpler shapes in the ζ-plane.

Given the transform in Equation (3.58), it is easy to see that

$$d\zeta = f'(z)\, dz.$$

If $f'(z) = re^{i\theta}$, then $|d\zeta| = r|dz|$, i.e. the transformed element equals the the original element rotated through θ and multiplied by r.

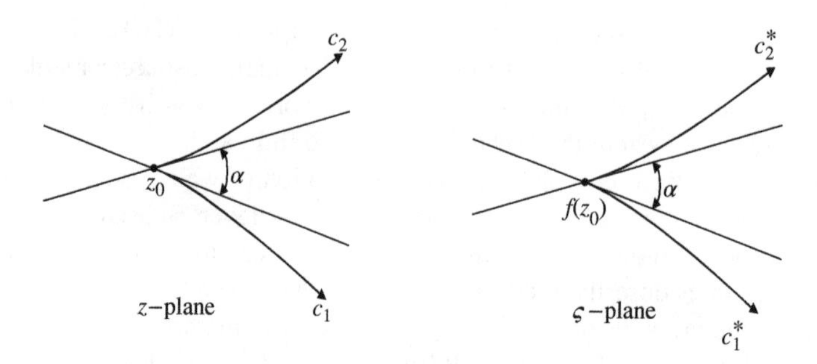

Figure 3.11 Conformal mapping between z- and ζ-plane explained through oriented curves

Consider the two planes shown in Figure 3.11, which shows that the lines C_1 and C_2 passing through z_0 in z-plane is mapped to C_1^* and C_2^*, respectively, in ζ-plane passing through ζ_0. Now, if the included angle between C_1 and C_2 is α at z_0 in the z-plane, then the included angle at ζ_0 between C_1^* and C_2^* is also α. It is to be noted that a conformal mapping is one which preserves angles between any curves both in magnitude and orientation.

3.6.1 Laplacian in the Transformed Plane

Using chain rule

$$\frac{\partial}{\partial x} = \frac{\partial}{\partial \xi}\xi_x + \frac{\partial}{\partial \eta}\eta_x.$$

Therefore

$$\frac{\partial^2}{\partial x^2} = \xi_{xx}\frac{\partial}{\partial \xi} + \xi_x\frac{\partial}{\partial x}\left(\frac{\partial}{\partial \xi}\right) + \eta_{xx}\frac{\partial}{\partial \eta} + \eta_x\frac{\partial}{\partial \eta}\left(\frac{\partial}{\partial x}\right),$$

where

$$\frac{\partial^2}{\partial \xi \partial x} = \frac{\partial}{\partial x}\left(\frac{\partial}{\partial \xi}\right) = \left(\xi_x\frac{\partial}{\partial \xi} + \eta_x\frac{\partial}{\partial \eta}\right)\frac{\partial}{\partial \xi} = \xi_x\frac{\partial^2}{\partial \xi^2} + \eta_x\frac{\partial^2}{\partial \xi \partial \eta}$$

and

$$\frac{\partial^2}{\partial \eta \partial x} = \xi_x\frac{\partial^2}{\partial \xi \partial \eta} + \eta_x\frac{\partial^2}{\partial \eta^2}.$$

Using these two relations, we finally obtain

$$\frac{\partial^2}{\partial x^2} = \xi_{xx}\frac{\partial}{\partial \xi} + \eta_{xx}\frac{\partial}{\partial \eta} + \xi_x^2\frac{\partial^2}{\partial \xi^2} + \eta_x^2\frac{\partial^2}{\partial \eta^2} + 2\xi_x\eta_x\frac{\partial^2}{\partial \xi \partial \eta}.$$

Similarly

$$\frac{\partial^2}{\partial y^2} = \xi_{yy}\frac{\partial}{\partial \xi} + \eta_{yy}\frac{\partial}{\partial \eta} + \xi_y^2\frac{\partial^2}{\partial \xi^2} + \eta_y^2\frac{\partial^2}{\partial \eta^2} + 2\xi_y\eta_y\frac{\partial^2}{\partial \xi \partial \eta}.$$

We can therefore write down the Laplacian operator as

$$\nabla^2 = [\xi_{xx} + \xi_{yy}]\,\frac{\partial}{\partial \xi} + [\eta_{xx} + \eta_{yy}]\,\frac{\partial}{\partial \eta} + [\xi_x^2 + \xi_y^2]\,\frac{\partial^2}{\partial \xi^2}$$
$$+[\eta_x^2 + \eta_y^2]\frac{\partial^2}{\partial \eta^2} + 2[\xi_x\eta_x + \xi_y\eta_y]\frac{\partial^2}{\partial \xi \partial \eta}. \tag{3.59}$$

As discussed in Section 3.2, if we consider $\zeta = \xi + i\eta$ as an analytic function of z, then ξ and η satisfy Cauchy–Riemann relations given by

$$\frac{\partial \xi}{\partial x} = \frac{\partial \eta}{\partial y} \quad \text{and} \quad \frac{\partial \xi}{\partial y} = -\frac{\partial \eta}{\partial x}. \tag{3.60}$$

Immediately one notices from Equation (3.60) that $\xi_x\eta_x + \xi_y\eta_y = 0$. Also from Equation (3.60), one can deduce from Cauchy–Riemann relations in a straightforward manner

$$\frac{\partial^2 \xi}{\partial x^2} + \frac{\partial^2 \xi}{\partial y^2} = 0 \quad \text{and} \quad \frac{\partial^2 \eta}{\partial x^2} + \frac{\partial^2 \eta}{\partial y^2} = 0.$$

Equation (3.59) therefore simplifies to

$$\nabla^2 = [\xi_x^2 + \xi_y^2]\,\frac{\partial^2}{\partial \xi^2} + [\eta_x^2 + \eta_y^2]\frac{\partial^2}{\partial \eta^2}. \tag{3.61}$$

Another consequence of Cauchy–Riemann relations is that

$$[\xi_x^2 + \xi_y^2] = [\eta_x^2 + \eta_y^2],$$

the so-called equality of scale factors, used in conformal grid mapping. Hence, the Laplacian operator in the transformed ζ-plane is simply given by

$$\nabla^2 = [\xi_x^2 + \xi_y^2]\left(\frac{\partial^2}{\partial \xi^2} + \frac{\partial^2}{\partial \eta^2}\right) \tag{3.62}$$

or

$$\nabla^2 = [\eta_x^2 + \eta_y^2]\left(\frac{\partial^2}{\partial \xi^2} + \frac{\partial^2}{\partial \eta^2}\right). \tag{3.63}$$

Thus the governing equation for incompressible, irrotational flow in terms of ϕ is given by

$$\nabla^2\phi = \left(\frac{\partial^2 \phi}{\partial \xi^2} + \frac{\partial^2 \phi}{\partial \eta^2}\right) = 0, \tag{3.64}$$

which is given in the same form, as it is in the z-plane. The same observation holds for the governing equation involving ψ.

Mapping defined by an analytic function $f(z)$ is conformal except at critical points, i.e. at points where $f'(z)$ is zero or infinity. Take for example the Jukowski transformation given by

$$\zeta = z + \frac{b^2}{z} \tag{3.65}$$

Note that for this transformation

$$\frac{d\zeta}{dz} = 1 - \frac{b^2}{z^2}.$$

Thus, the derivative is zero at $z = \pm b$. The concept of criticality will become apparent when one relates velocities in the two plane, as shown next.

3.6.2 *Relation between Complex Velocity in Two Planes*

The complex velocity in the ζ-plane is given by

$$(u - iv)_\zeta = \frac{dW}{d\zeta} = \frac{dW}{dz}\frac{dz}{d\zeta} = (u - iv)_z \frac{dz}{d\zeta}. \tag{3.66}$$

The complex velocities in the two planes are thus related via $\frac{dz}{d\zeta}$. The velocities in z-plane are magnified by $|\frac{dz}{d\zeta}|$ and rotated by $\arg(\frac{dz}{d\zeta})$. In general, $|\frac{dz}{d\zeta}|$ has finite value differing from zero, except at singular or critical points.

One of the fundamental ideas behind conformal mapping is that the two flow fields must be the same, far away from the respective bodies in the two planes. This is equivalent to the condition

$$\operatorname*{Lim}_{\zeta, z \to \infty} \frac{dz}{d\zeta} \to 1.$$

This is readily seen for the Jukowski transformation given in Equation (3.65) as

$$\operatorname*{Lim}_{\zeta, z \to \infty} \left|\frac{dz}{d\zeta}\right| = \operatorname*{Lim}_{z \to \infty} \frac{1}{1 - \frac{b^2}{z^2}} = 1.$$

3.6.3 *Application of Conformal Transformation*

As noted before, the main use of conformal transformation is to reduce complex flows to simpler ones, which can be analyzed easily. In particular, we study Jukowski transformation, which is capable of producing family of airfoils with associated flow patterns. For the Jukowski transformation let us use the polar representation in Equation (3.65) as

$$\zeta = \xi + i\eta = re^{i\theta} + \frac{b^2}{re^{i\theta}}. \tag{3.67}$$

Thus

$$\xi = \left(r + \frac{b^2}{r}\right)\cos\theta \quad \text{and} \quad \eta = \left(r - \frac{b^2}{r}\right)\sin\theta. \tag{3.68}$$

(a) Transforming a circular cylinder to a flat plate:

Before we begin derivations, we note that the point of singularity brings in a slope discontinuity, even when it is located on the surface of a body. On the contrary, presence of singularity inside a closed smooth curve transforms to another closed smooth curve. This also points to the fact that a body with sharp discontinuity can be transformed to a smooth geometry, if we perform the transform such that a point of singularity falls exactly on the smooth curve in the z-plane to obtain a transformed shape with sharp corner in the ζ-plane.

Thus in using the Jukowski transformation of the flat plate to a circular cylinder, we must use $b = a$ in Equation (3.65), where a is the radius of the circle in the z-plane with its centre at the origin. Thus both the singular points $z = \pm b$ falls on the surface of the cylinder. Now, from Equations (3.67) and (3.68), the circle $r = a$ transforms to

$$\xi = 2a\cos\theta \quad \text{and} \quad \eta = 0. \tag{3.69}$$

Thus the chord of the flat plate in ζ-plane is $4a$, as $-2a \leq \xi \leq +2a$. The singularities at $z = \pm a$ produce the sharp edges at $\xi = \pm 2a$, respectively. In Figure 3.12, we note the mapping between the circle and flat plate planes.

(b) Transforming a circular cylinder to an ellipse:

Noting from above, we want to transform a smooth curve to another smooth curve by the Jukowski transform and thus, we will place the singular points $z = \pm b$ inside the circular cylinder, i.e. $a \geq b$ in Equation (3.65). The transformation of the circle, $r = a$, produces

$$\xi = \left(a + \frac{b^2}{a}\right)\cos\theta \quad \text{and} \quad \eta = \left(a - \frac{b^2}{a}\right)\sin\theta. \tag{3.70}$$

The equation of the curve in ζ-plane is obtained by eliminating θ from Equation (3.70) as

$$\left(\frac{\xi}{a + \frac{b^2}{a}}\right)^2 + \left(\frac{\eta}{a - \frac{b^2}{a}}\right)^2 = 1. \tag{3.71}$$

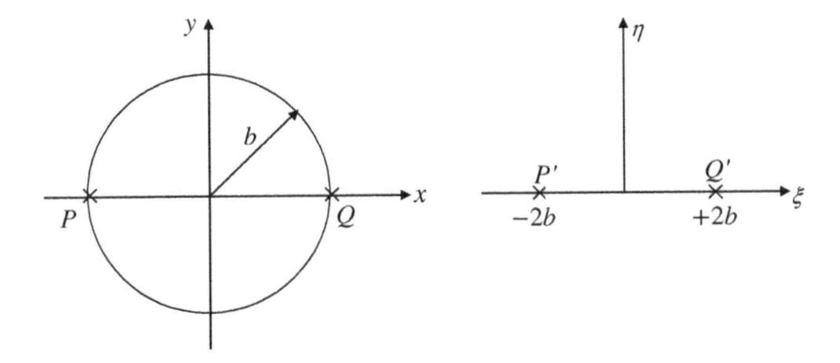

Figure 3.12 Mapping a circle to a flat plate by Jukowski transform

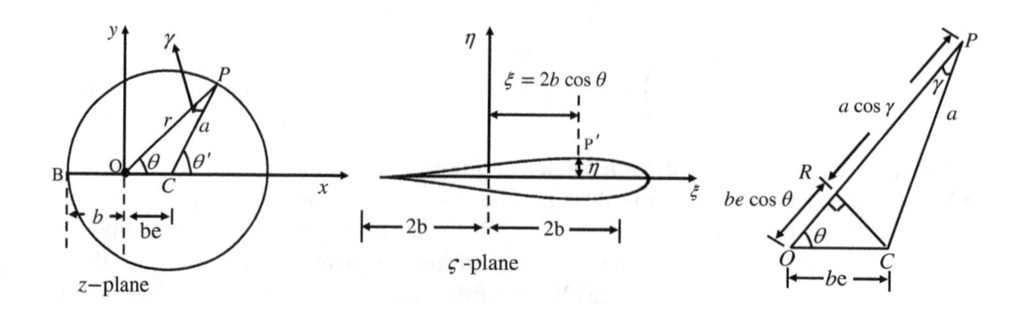

Figure 3.13 Mapping between circle and symmetric airfoil planes by Jukowski transform

The following geometric dimensions define the ellipse:

The chord of the ellipse: $2\left(a+\frac{b^2}{a}\right)$

The thickness of the ellipse: $2\left(a+\frac{b^2}{a}\right)$

And the fineness ratio: $(a^2+b^2)/(a^2-b^2)$.

(c) Transforming a circular cylinder to a symmetric airfoil:
Here, the circle and the airfoil planes are as shown in Figure 3.13, with one of the singular points of the Jukowski transformation at $z=-b$ lies on the circle of radius a. The centre of the circle at C is offset from the origin O, by an amount be from the origin in the z-plane. The offset is very small, i.e. $e \ll 1$. Consider a point P whose polar coordinate is given by (r, θ). The line CP subtends an angle θ' with respect to the x-axis. In Figure 3.13, we also show an enlarged view of the triangle OCP, where $\angle OPC$ is given by γ. A perpendicular line from C is dropped on the line OP which meets the line at R.

In Figure 3.13, we note the following:

$$OR = be\cos\theta \text{ and } PR = a\cos\gamma \simeq a.$$

Therefore, $OP = r = a + be\cos\theta$ and the radius of the circle can be approximated as $a = b(1+e)$. Thus, the radial distance can be written as

$$\frac{r}{b} = 1 + e + e\cos\theta, \tag{3.72}$$

which also could be alternately written

$$\frac{b}{r} = [1+e+e\cos\theta]^{-1} = 1 - e(1+\cos\theta) + \frac{e^2}{2}(1+\cos\theta)^2,$$

which could be linearized (for $e \ll 1$) to simplify further as

$$\frac{b}{r} = 1 - e - e\cos\theta. \tag{3.73}$$

Note that linearization is perfectly consistent, as our governing equation is linear. As

$$\xi = b\left(\frac{r}{b} + \frac{b}{r}\right)\cos\theta \ \text{ and } \ \eta = b\left(\frac{r}{b} - \frac{b}{r}\right)\sin\theta. \tag{3.74}$$

We can use Equations (3.72) and (3.73) in these equations to get

$$\xi = 2b\cos\theta \quad \text{and} \quad \eta = 2be(1+\cos\theta)\sin\theta. \tag{3.75}$$

It is clearly evident that this represent a symmetrical shape with chord $4b$ [as $-2b \le \xi \le +2b$ and $\eta(\theta) = -\eta(-\theta)$]. The maximum thickness can be obtained from

$$\frac{d\eta}{d\theta} = 0,$$

which simplifies as follows:

$$\cos^2\theta - \sin^2\theta + \cos\theta = 0.$$

This quadratic has the following solutions:

$$\cos\theta = \frac{1}{2} \ \text{ and } \ -1.$$

The value of $\cos\theta = \frac{1}{2}$ corresponds to $\theta = \frac{\pi}{3}$ for which $\xi = b$; i.e. the quarter chord point represents the maximum thickness location. Whereas $\cos\theta = -1$ occurs when $\theta = \pi$ and this is the trailing edge location, $\xi = -2b$. The circle and the symmetrical airfoil planes are shown in Figure 3.13. The maximum thickness is $2\eta_{max} = 4be\left(1+\cos\frac{\pi}{3}\right)\sin\frac{\pi}{3} = 3\sqrt{3}\,be$. The thickness to chord ratio of the symmetric airfoil is given by

$$\frac{t}{c} = \frac{3\sqrt{3}be}{4b} = 1.299e.$$

Thus the parameter e in the Jukowski transformation of Equations (3.71)–(3.75) defines the t/c ratio of the airfoil. While it is easy to show that the slope at the leading edge is a right angle $\left[\text{as} \frac{d\eta}{d\xi}|_{\theta=0} = \frac{\pi}{2}\right]$, the slope at the trailing has a very special feature.

The slope at the trailing edge is given by

$$\left.\frac{d\eta}{d\xi}\right|_{\xi=-2b} = -\left.\frac{d\eta/d\theta}{d\xi/d\theta}\right|_{\theta=\pi} = \left.\frac{2be[2\cos^2\theta + \cos\theta - 1]}{2b\sin\theta}\right|_{\theta=\pi}.$$

At first sight this appears in the indeterminate form, but it can be evaluated by L'Hospital's rule as zero. This implies that the airfoil has a cusp shape. This is a general feature of all Jukowski airfoils.

(d) Transforming a circular cylinder to a circular arc:

In performing this Jukowski transformation, we position the circle in such a way in the z-plane that the centre is shifted along the y-axis and the singular points A and B are on the circle, i.e.

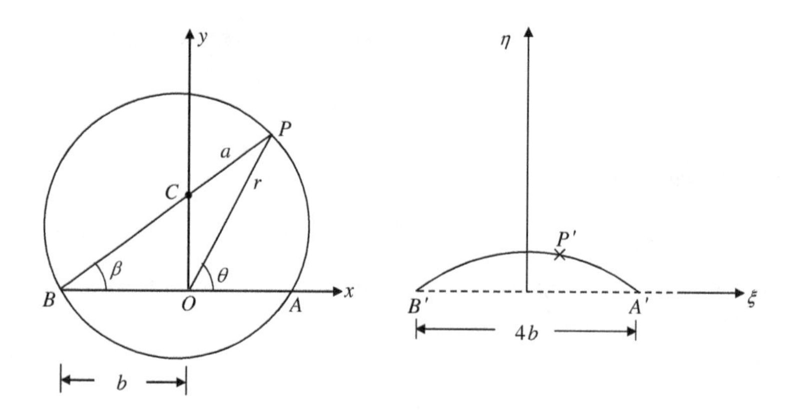

Figure 3.14 Transforming a circular cylinder to a circular arc airfoil of zero thickness

$OB = OA = b$ in Figure 3.14. The radius of the circle (CP) is given by a as before, so that $CB = a = b\ \sec\beta$ and $OC = b\tan\beta$.

In Figure 3.14 from the triangle OCP,

$$CP^2 = OC^2 + OP^2 - 2\ OC.\ OP\ \cos(\pi/2 - \theta)$$

or

$$b^2\sec^2\beta = b^2\tan^2\beta + r^2 - 2br\tan\beta\ \sin\theta,$$

which upon simplification provides

$$\frac{r^2 - b^2}{r} = 2b\tan\beta\ \sin\theta. \quad (3.76)$$

Now from Equation (3.68) we have

$$\xi = (r + b^2/r)\cos\theta \ \text{ and } \ \eta = (r - b^2/r)\sin\theta. \quad (3.77)$$

Using Equation (3.76) in these expressions, we get the simplified form of one of the coordinate as

$$\eta = 2b\tan\beta\ \sin^2\theta. \quad (3.78)$$

This immediately shows that two points symmetrically located about AB and indicated by $\pm\theta$, map to a single point in the ζ-plane. This establishes that the body in the ζ-plane has zero thickness. We also note that A and B maps to A' and B', respectively, and corresponds to $\theta = 0$ and π. Thus, $A'B' = 4b$ and the circle transforms to a cambered plate. Maximum value of η occurs for $\theta = \pi/2$ and is given by

$$\eta_{max} = 2b\tan\beta.$$

The camber ratio is defined as

$$\frac{\eta_{max}}{\text{chord}} = \frac{2b\tan\beta}{4b} = \frac{1}{2}\tan\beta. \tag{3.79}$$

From Equation (3.77) one gets

$$\frac{\xi^2}{\cos^2\theta} - \frac{\eta^2}{\sin^2\theta} = (r^2 + \frac{b^4}{r^2} + 2b^2) - (r^2 + \frac{b^4}{r^2} - 2b^2) = 4b^2. \tag{3.80}$$

From Equation (3.78), $\sin^2\theta = \eta/(2b\ \tan\beta)$ and using this in Equation (3.80), one gets

$$\xi^2 - \eta^2 \frac{\left(1 - \frac{\eta}{2b\tan\beta}\right)}{\frac{\eta}{2b\tan\beta}} = 4b^2\left(1 - \frac{\eta}{2b\tan\beta}\right), \tag{3.81}$$

which can be further simplified as

$$\xi^2 - 2b\eta\ \tan\beta\left(1 - \frac{\eta}{2b\tan\beta}\right) = 4b^2\left(1 - \frac{\eta}{2b\tan\beta}\right)$$

or

$$\xi^2 + \eta^2 - 2b\eta\ \tan\beta = 4b^2 - 2b\ \cot\beta$$

or

$$\xi^2 + \eta^2 - 4b\eta\ \cot 2\beta = 4b^2$$

or

$$\xi^2 + (\eta + 2b\ \cot 2\beta)^2 = 4b^2 + 4b^2\ \cot^2 2\beta.$$

Thus, the circle transforms to an arc in the ζ-plane whose governing equation is given by

$$\xi^2 + (\eta + 2b\ \cot 2\beta)^2 = 4b^2\ \text{cosec}^2 2\beta. \tag{3.82}$$

This is a circular arc with centre at $(0,\ -2b\ \cot 2\beta)$ and with radius $2b$ cosec 2β. The amount of the camber is decided by the offset of the centre of the cylinder from the origin.

(e) Transforming circular cylinder to a cambered airfoil:

Having already seen that if the singular point of the Jukowski transformation passes through the cylinder, it creates a transformed shape in ζ-plane with a sharp edge at the corresponding point. Also, the offset of the centre in the horizontal direction gives rise to thickness distribution and offset in the vertical direction gives rise to camber distribution; a combination of the two will give rise to a general cambered airfoil with finite thickness.

The z- and ζ-planes are shown in Figure 3.15, with the circle in the former plane mapped to a general airfoil in the latter. In this case, the centre is offset both horizontally, as well

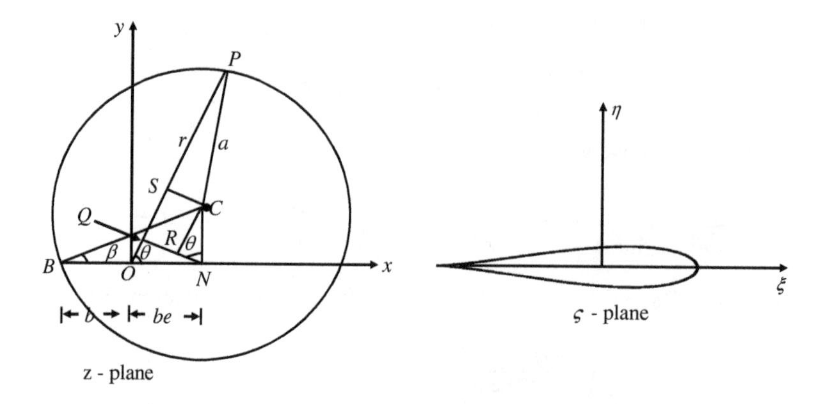

Figure 3.15 Transforming a circular cylinder to a general cambered airfoil of finite thickness

as vertically from the origin. In the circle plane, consider an arbitrary point P whose polar coordinate is given by (r, θ), with $CP = a$ and $OB = b$.

The horizontal shift (ON) is given by be, whereas the vertical shift of the centre (CN) is given by h. The corresponding angular offset, $\angle CBO = \beta$ is considered as small and $\angle CPO = \gamma$, so that $\cos\beta \simeq 1 = \cos\gamma$. From the figure, $CN = BC\ \sin\beta$, i.e. $h = a\ \sin\beta$. Also, $BN = BC\ \cos\beta$, which implies

$$b + be = a\cos\beta \simeq a. \tag{3.83}$$

Therefore

$$h = b(1+e)\beta. \tag{3.84}$$

From the part of the circle plane shown on the left of Figure 3.15

$$OP = OQ + CR + PS = ON\ \cos\theta + CN\ \sin\theta + CP\ \cos\gamma.$$

Thus

$$\begin{aligned} r &= be\ \cos\theta + h\ \sin\theta + a\ \cos\gamma \\ &= be\ \cos\theta + b\beta\ (1+e)\ \sin\theta + b(1+e). \end{aligned}$$

Since $b\beta e$ is a product of two small terms, it can be neglected in the linear analysis and one obtains

$$\frac{r}{b} = 1 + e + e\cos\theta + \beta\sin\theta \tag{3.85}$$

$$\frac{b}{r} = 1 - (e + e\cos\theta + \beta\sin\theta). \tag{3.86}$$

From Jukowski transform given in Equation (3.74), one gets

$$\xi = 2b\ \cos\theta \ \text{ and } \ \eta = 2b\ (e + e\cos\theta + \beta\sin\theta)\ \sin\theta. \tag{3.87}$$

The ordinate of the airfoil can also be written as

$$\eta = 2be\ (1 + \cos\theta)\ \sin\theta + 2b\beta\ \sin^2\theta. \tag{3.88}$$

One can readily see that the β-term gives rise to asymmetry, as this is always positive. Thus β represent camber, as for $\beta = 0$ one recovers the symmetric airfoil of Case (c). By differentiating Equation (3.88) with respect to θ, one can locate the maximum of η to be at the quarter chord, as shown below.

The thickness at any section is given by $t = \eta_1 - \eta_2$, where η_1 and η_2 are for the same ξ coordinate, i.e. $\theta_2 = -\theta_1$. Therefore the section thickness is given by

$$4be\ (1 + \cos\theta_1)\ \sin\theta_1.$$

As the chord of the airfoil is $c = 4b$, the thickness to chord distribution is given by

$$\frac{t}{c} = e\ (1 + \cos\theta)\ \sin\theta. \tag{3.89}$$

Once again, t/c is maximum at $\theta = \pi/3$. In general, camber of an airfoil is maximum deviation of the mean camberline from the chord, which is given by $\frac{1}{2}(\eta_1 + \eta_2)$. Thus the camber as defined before is given by

$$\text{Camber} = \frac{(\eta_1 + \eta_2)_{max}}{2\ \text{chord}} = \frac{(4b\beta\ \sin^2\theta)_{max}}{8b} = \frac{\beta}{2}, \tag{3.90}$$

which occurs at $\theta = \pi/2$. These general cambered sections with thickness, obtained by Jukowski transformation, are called the Jukowski airfoil. Note that these airfoils are obtained by the successive movement of the centre of the circle in horizontal and vertical directions, to create the thickness and camber, respectively. Thus in constructing the airfoil, the camber line is drawn as a function of x, to which the thickness is added as a sheared distribution. In an alternative practice, the thickness is added normally over the camber line. We recommend the former approach of constructing the airfoils.

3.7 Lift Created by Jukowski Airfoil

In Section 1.3, we have probed the origin of the aerodynamic forces and it was noted that an ideal aerodynamic device (in the absence of viscous diffusion) would be such that the action of the device would be to turn a large volume of air by the smallest possible angle, with respect to the oncoming flow. In the above, we have noted the means by which one can get what is the quintessential element of an aircraft wing, namely the airfoil obtained by Jukowski transformation. The flow past this airfoil can be analytically related to flow past a circular cylinder via the Jukowski transform. Here, we study the ability of such a section to create lift force.

Lift force is generated here by angle of attack and camber. The oncoming flow to this airfoil creates a steady circulation, whose origin is strictly viscous, yet it can be parametrized by phenomenological observation and heuristics. The application of transformation technique rests on the fact that far away from the bodies in z- and ζ-planes, the flow field remains identical, as shown in Section 3.6.2. This is with respect to the Jukowski transform given in Equation (3.65) applied to an airfoil discussed in the previous subsection. The other streamlines

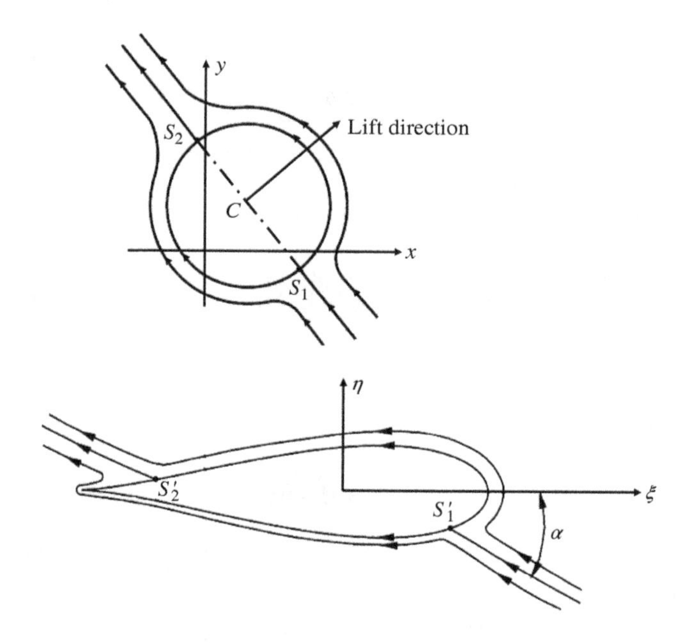

Figure 3.16 Flow past a Jukowski airfoil without circulation. Also shown is the corresponding circle plane flow streamlines

are also generated via the same transformation for the flow approaching the airfoil at an angle of attack, as shown in Figure 3.16.

The flow field shown in Figure 3.16 is an ideal flow field which does not create any circulation or lift on the airfoil. The flow field in both the planes can be developed and understood simultaneously. In the z-plane, we construct a circle whose centre (C) is offset from the origin of the complex plane in both the horizontal and the vertical directions. The angle of attack (α) of the flow in the airfoil plane is indicated in the figure by the front and rear stagnation points being located in the circle plane at S_1 and S_2, displaced by α with respect to the x-axis. Corresponding stagnation points on the airfoil are marked at S_1' and S_2', respectively. This would be the flow field if there is no circulation in the airfoil. However for any real flow, vorticity will be continually generated at the no-slip wall, which will be diffused and convected, as governed by the vorticity transport equation. A steady state will eventually be reached following an impulsive start and the steady state wall vorticity value accounts for the circulation in the airfoil plane, about which we are interested.

Corresponding circulation in the circle plane will be as given by the complex potential. It is easy to show that $\oint_{C_z} (u - iv)_z \, dz = \oint_{C_\zeta} (u - iv)_\zeta \, d\zeta$, which shows that the singularity strength in one plane remains the same, in the other plane. Flow past a circular cylinder, having a bound circulation, has been discussed in Section 3.4.2, which showed that the presence of circulation will cause the stagnation points for flow past a stationary cylinder (as shown in Figure 3.16 as S_1 and S_2), will be symmetrically depressed. For subcritical circulation, these two stagnation points will be distinct and depressed from the no-circulation case of Figure 3.16 to that shown in Figure 3.7. The pertinent question is: how much circulation is created after the steady state is attained, around the airfoil? This can be answered by invoking physical nature of the flow, as follows.

3.7.1 Kutta Condition and Circulation Generation

This is done by noting that the body streamline and the neighbouring ones are strained enormously for the flow field shown in Figure 3.16, as these are forced to turn abruptly around the trailing edge. The curvature of streamlines is indicative of strain rates suffered by the fluid flow. One observes that such highly turning flows are possible in the steady state, if the energy supplied by the flow is sufficiently large. The energy dissipation rate is the product of stress and rates of strain and rapidly turning streamline imply very large strain rate. Thus in case of finite energy fluid flows, such abrupt turning is not feasible. This is prevented, if the rear stagnation point S_2' is moved backwards to the trailing edge, so that the flow will leave the trailing edge smoothly without suffering large strain rate. This physical consideration, in effect, fixes the circulation created by the airfoil. We note that such a movement of rear stagnation point from S_2' to M would depress the rear stagnation point in the circle plane from S_2 to m, as indicated in the figure. We have already noted that the movement of rear stagnation to m in the circle plane will also symmetrically depress the forward stagnation point from S_1 to n. This will also cause the forward stagnation point to move from S_1' to N in the airfoil plane, as a consequence of the movement of the rear stagnation point from S_2' to M. The altered flow field via the movement of stagnation points will correspondingly fix the circulation created on the airfoil. This generated circulation (Γ) in a uniform fluid flow (U_∞) will create a lift force given by the Kutta-Jukowski theorem by $L = \rho U_\infty \Gamma$, as described in Section 3.4.1. The direction of the lift force is perpendicular to S_1CS_2 in the circle plane. In the airfoil plane also, lift is perpendicular to the free stream direction. Note that the free stream (measured by streamlines far away from the body) remains identically inclined as the complex velocities in the two planes are in the same direction, as $\frac{d\zeta}{dz} \rightarrow 1$, as $z, \zeta \rightarrow \infty$.

We also note that the Jukowski transform creates an airfoil with a cusped trailing edge, i.e. the slope of the airfoil viewed from the top and bottom surfaces is zero. Thus, fixing of the rear stagnation point at the trailing edge causes the flow to emerge smoothly, which will cause the pressure to be also identical, i.e. the load at the trailing edge is zero. These observations of flow emerging smoothly from the trailing edge and zero loading at the trailing edge was proposed by Kutta to calculate the generated lift on the Jukowski airfoil and is known as the **Kutta condition at the trailing edge**. This heuristic argument is quite powerful, yet the reader should keep in mind that this is essentially a model, whose validity has been noted for many airfoils for very small angles of attack. However in an actual flow, the flow past the airfoil need not be steady, specially when the flow is turbulent at high Reynolds numbers and such flows are inherently unsteady. As a consequence in actual experiments, one can expect to notice that the rear stagnation points would move about the trailing edge. Many of our accepted theories are based on steady-state aerodynamics and they are quite useful for high Reynolds number application for steady-state devices. For conventional aircrafts at high Reynolds number cruise flight condition such models yield quite realistic engineering results. For unsteady flows, or flows at reduced Reynolds number applications, as in micro air vehicle (MAV), one may be required to appeal for an answer from a more realistic solution of the Navier-Stokes equation.

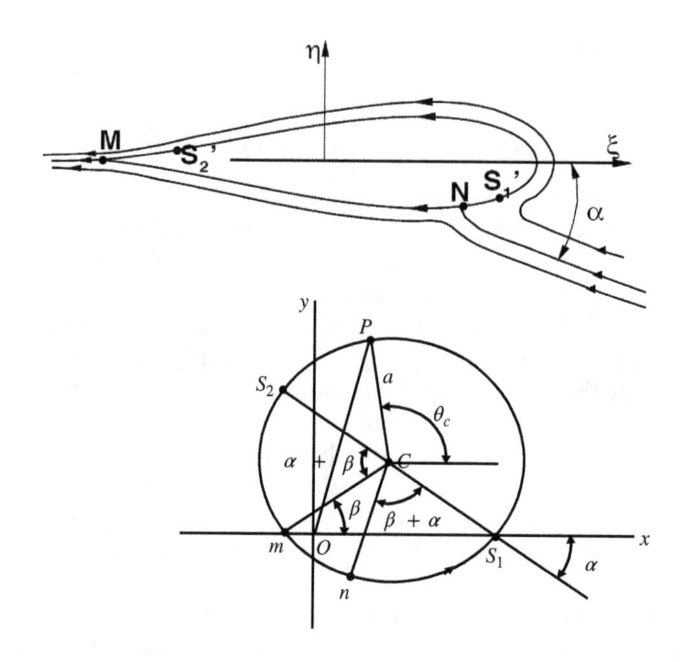

Figure 3.17 Flow past a Jukowski airfoil with circulation studied in the corresponding circle plane

3.7.2 Lift on Jukowski Airfoil

The analysis of the flow field and calculation of the lift acting on the airfoil is performed with the help of Figure 3.17, showing the various important points along the body streamline, for the case with circulation.

We define an auxiliary azimuthal variable (θ_c) in Figure 3.17, such that $\theta_c = \theta + \alpha$, to define an arbitrary point P with coordinate $(r,\ \theta)$ on the circle of radius a, i.e. $OP = r$, and $\angle POS_1$. Note that the line along the diameter S_1CS_2 is set at the angle of attack of the airfoil (α). The Jukowski transform which relates the airfoil to this circle plane has one of the singular points at m, so that the aerofoil camber is defined by the angle β which the line Cm subtends with the x-axis and this angle is considered to be very small. From the geometry shown in Figure 3.17, we note that $\angle S_2Cm = \angle S_1Cn = \alpha + \beta$. In the circle plane, the complex potential is given by

$$W = U_\infty\left(z + \frac{a^2}{z}\right) + \frac{i\Gamma}{2\pi}\,\mathrm{Ln}\frac{z}{a}. \tag{3.91}$$

If we now define the circle by $z = a\,e^{i\theta_c}$, then the complex velocity on the circle is given by

$$\frac{dW}{dz} = U_\infty\left(1 - e^{-2i\theta_c}\right) + \frac{i\Gamma}{2\pi a}\,e^{-i\theta_c} = e^{-i\theta_c}\left[2iU_\infty \sin\theta_c + \frac{i\Gamma}{2\pi a}\right]. \tag{3.92}$$

If the speed of the flow at P is q, then

$$q^2 = \left(2U_\infty \sin\theta_c + \frac{\Gamma}{2\pi a}\right)^2.$$

Note that as of now the circulation Γ is undetermined and which will be fixed by Kutta condition. We have noted from above discussion that applying Kutta condition is equivalent to forcing the stagnation points to be located at m and n, respectively. At the point n, $\theta_c = -(\alpha + \beta)$ and $q = 0$. Thus from this information

$$0 = -2U_\infty \sin(\alpha + \beta) + \frac{\Gamma}{2\pi a}$$

we fix the circulation as

$$\Gamma = 4\pi U_\infty a \sin(\alpha + \beta). \tag{3.93}$$

Using this one can also fix the surface speed distribution in the circle plane as

$$q = 2U_\infty[\sin\theta_c + \sin(\alpha + \beta)]. \tag{3.94}$$

Also having fixed the circulation, one can calculate the lift experienced by the Jukowski airfoil from Kutta-Jukowski theorem as

$$L = 4\pi\rho U_\infty^2 a \sin(\alpha + \beta). \tag{3.95}$$

As the chord of the airfoil is $c = 4b$, where $a = b(1 + e)$, one can obtain the lift coefficient from $L = \frac{1}{2}\rho U_\infty^2 c C_l$ as

$$C_l = 2\pi(1 + e) \sin(\alpha + \beta). \tag{3.96}$$

Neglecting e in comparison, the maximum lift coefficient is given by $C_{l,max} = 2\pi$, when $\alpha + \beta = \pi/2$. It is a very unrealistic value, as the developed theory is with linearized assumption for which β is small from the conditions set above and α should also be similarly restricted to negligibly small value to avoid viscous effects and flow separation to reduce the lift to smaller value. However, despite this the lift curve slope is obtained which is quite good and is a distinct contribution of this theory. This analytical expression for lift was obtained within a few decades for an aerofoil and is a singular achievement, after the foundation of the Royal Aeronautical Society with the initial goal of finding lift and drag acting on a flat plate.

The lift curve slope is given by

$$C_{l\alpha} = \frac{\partial C_l}{\partial \alpha} \simeq 2\pi. \tag{3.97}$$

This is often quoted also as equivalent to producing a lift coefficient of value of 0.1 per degree, approximately. Having obtained the total lift generated, one would also be interested to obtain the velocity and pressure distribution on the Jukowski aerofoil.

3.7.3 *Velocity and Pressure Distribution on Jukowski Airfoil*

Speed distribution in the circle plane is given by Equation (3.94). From the Jukowski transformation

$$\frac{d\zeta}{dz} = 1 - \frac{b^2}{z^2} = 1 - \frac{b^2}{r^2}(\cos 2\theta - i \sin 2\theta)$$

$$\left|\frac{d\zeta}{dz}\right| = \left[1 - \frac{2b^2}{r^2}\cos 2\theta + \frac{b^4}{r^4}\right]^{1/2}.$$

Therefore the speed distribution in the airfoil plane is given by

$$v_a = 2U_\infty \frac{[\sin\theta_c + \sin(\alpha+\beta)]}{\left[1 - \frac{2b^2}{r^2}\cos 2\theta + \frac{b^4}{r^4}\right]^{1/2}}. \tag{3.98}$$

For the point P in Figure 3.17,

$$y = r\ \sin\theta = a\ \sin(\theta_c - \alpha) + a\ \sin\beta.$$

Therefore

$$\sin(\theta_c - \alpha) = \frac{r}{a}\sin\theta - \beta$$

or

$$\theta_c = \alpha + \sin^{-1}\left[\frac{r}{a}\sin\theta - \beta\right]. \tag{3.99}$$

Also as

$$r = b + be + be\ \cos\theta + b\beta\ \sin\theta.$$

One can obtain r/b from this relation and use it in Equation (3.98) to obtain the speed distribution on the surface of the airfoil.

The pressure coefficient over the airfoil is thereafter obtained from

$$C_p = 1 - \left(\frac{v_a}{U_\infty}\right)^2.$$

3.8 Thin Airfoil Theory

In the previous sections we obtained the section characteristics of airfoils, by using the inviscid irrotational flow model governed by Laplace's equation for velocity potential. This was done by conformal mapping and is restricted to Jukowski airfoil. Recognizing that most aerofoils could be classified as thin, an aerofoil theory was propounded by Munk, which uses regular perturbation theory. In studying this part, the reader is not expected to have knowledge of perturbation theory, as the necessary elements are described here.

Taking account of the fact that the transformation $\zeta = z + \frac{a^2}{z}$ applied to a circle in a uniform stream gave a flat plate or cambered plate as a rudimentary aerofoil, the thin airfoil theory assumes that a general cambered airfoil with finite thickness can be replaced by its camber line, which can be further assumed to be a slight distortion of a flat plate. Consequently, the shape from which it was to be transformed would be a similar slight distortion from the original circle. The essential elements of the general theory is presented below, with the assumptions of size and shape retained, i.e. aerofoil is thin and can be represented by its camber line, which is slightly different from the chord line.

This theory consists in considering the camber line to have a string of line vortices of infinitesimal variable strengths, so that the total circulation about the chord is the sum of the vortex elements' strength.

Thus, if we measure the length along the camber line by s, then the circulation carried by the aerofoil is given by

$$\Gamma = \int \gamma \, ds. \tag{3.100}$$

Based on the assumption that the chord and camber line are the same, Equation (3.100) can also be rewritten as

$$\Gamma = \int_0^c \gamma \, dx \tag{3.101}$$

and using Kutta-Jukowski theorem, the lift experienced by the airfoil is given by

$$L = \rho \, U_\infty \int_0^c \gamma \, dx.$$

Similarly the pitching moment experienced by the airfoil section about its leading edge can be obtained from

$$M_{LE} = -\rho \, U_\infty \int_0^c \gamma \, x \, dx. \tag{3.102}$$

Once again, nose-up pitching moment is considered as positive. One can estimate by considering inviscid irrotational flow, the vortex element strength by applying Bernoulli's equation on an infinitesimal element of length dx shown in Figure 3.18. The camberline represents a line of discontinuity with pressure above and below indicated by p_1 and p_2 and local perturbation velocity indicated by u_1 and u_2, respectively.

The streamlines below and above the camberline originate from the same reservoir condition and hence one can apply Bernoulli's equation above and below the elementary vortex ($\gamma \, dx$) and obtain

$$p_2 - p_1 = \frac{1}{2} \rho \, U_\infty^2 \left[2\left(\frac{u_1}{U_\infty} - \frac{u_2}{U_\infty} \right) + \left(\frac{u_1}{U_\infty} \right)^2 - \left(\frac{u_2}{U_\infty} \right)^2 \right].$$

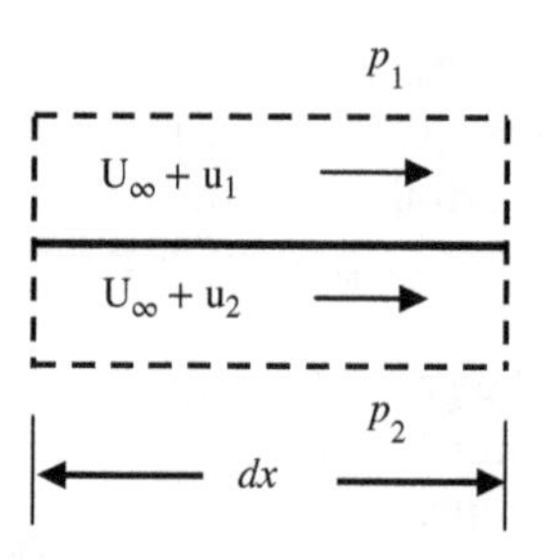

Figure 3.18 An element of vorticity distributed along the camber line and the flow variables surrounding the element

For the general thin airfoil, the perturbation velocities are small and one can perform linearization to obtain the loading as

$$\Delta p = p_2 - p_1 \simeq \rho\, U_\infty (u_1 - u_2). \tag{3.103}$$

Hence

$$C_p = \frac{2}{U_\infty}(u_1 - u_2).$$

One can also perform a contour integral over the dotted contour in Figure 3.18 to produce the circulation on the element ($\gamma\, dx$) as

$$\gamma\, dx = (U_\infty + u_1)dx - (U_\infty + u_2)dx.$$

Thus the loading distribution over the camber line can be obtained easily by that is given above. Next, the shape of the camberline is related to the loading distribution from the boundary condition, as applied for the airfoil placed at an angle of attack α, as shown in Figure 3.19.

With reference to the diagram in Figure 3.19, the camberline must itself be a streamline. Thus, the slope of the camberline must match the flow direction and this condition is given by

$$\left(U_\infty + u\right)\frac{dy}{dx} = v + U_\infty\, \alpha.$$

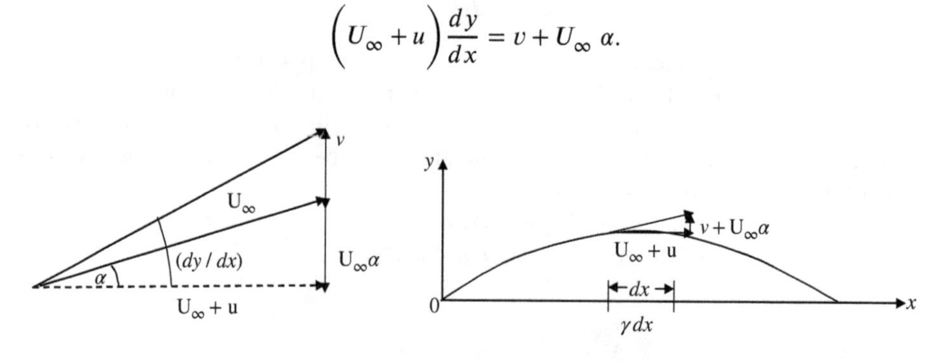

Figure 3.19 A thin airfoil shown represented by the camber line placed at an angle of attack, α and the corresponding velocity diagram for a point at a location, $x = x_1$

The term $u \frac{dy}{dx}$ is of lower order and can be neglected upon linearization to obtain

$$v \cong U_\infty \left(\frac{dy}{dx} - \alpha \right). \tag{3.104}$$

Now, the left-hand side of Equation (3.104) can be obtained as the induced velocity at the point x_1, created by a line vortex element of strength $\gamma\, dx$ by

$$v(x_1) = \frac{1}{2\pi} \int_0^c \frac{\gamma\, dx}{x - x_1}. \tag{3.105}$$

Equating Equations (3.104) and (3.105), on the camber line one gets

$$U_\infty \left(\frac{dy}{dx} - \alpha \right)_{x=x_1} = \frac{1}{2\pi} \int \frac{\gamma\, dx}{x - x_1}. \tag{3.106}$$

Equation (3.106) is the key element of thin airfoil theory, whereby the airfoil camber shape denoted by $\left(\frac{dy}{dx}\right)$ is related to the vorticity distribution. This is an integral equation which forms the basis for studying various types of airfoil, as described next.

3.8.1 *Thin Symmetric Flat Plate Airfoil*

In this case the camber line is straight along the x-axis, i.e. $\frac{dy}{dx} = 0$, and when this is used in Equation (3.106) one gets

$$U_\infty\, \alpha = \frac{1}{2\pi} \int_0^c \frac{\gamma\, dx}{x_1 - x}.$$

This equation can be simplified by applying a change of variable

$$x = \frac{c}{2}(1 - \cos\theta)$$

and therefore

$$dx = \frac{c}{2} \sin\theta\, d\theta.$$

Using these in the above equation one gets

$$U_\infty\, \alpha = \frac{1}{2\pi} \int_0^\pi \frac{\gamma \sin\theta\, d\theta}{(\cos\theta - \cos\theta_1)}. \tag{3.107}$$

This is called the Glauert's integral. We demonstrate the solution of the integral equation by using the results of the conformal mapping introduced here.

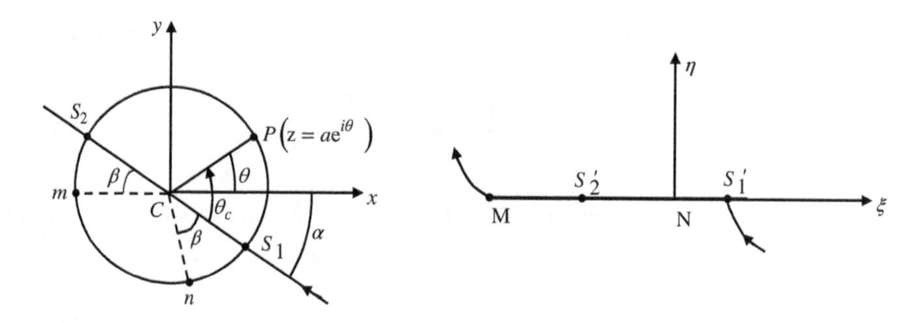

Figure 3.20 The circle plane and its map in the flat plate plane, indicating the front and rear stagnation points

Direct method of finding the loading distribution, γ:
With reference to Figure 3.20, one obtains the velocity in the circle plane first from Equation (3.92). Note that θ_c is measured with respect to the diameter S_1S_2 as, $\theta_c = (\theta + \alpha)$. Thus,

$$v_c = 2U_\infty \sin(\theta + \alpha) + \frac{\Gamma}{2\pi a},$$

where Γ is the total circulation in the airfoil plane considered lumped at the centre of the circle and which can be fixed by noting that the stagnation point is at m, i.e. at $\theta = \pi$.

Thus, $v_c = 0$ at m produces

$$\Gamma = 4\pi a\, U_\infty \sin \alpha.$$

Therefore the velocity in the circle plane is given by

$$v_c = 2\, U_\infty\, [\sin(\theta + \alpha) + \sin \alpha].$$

In the airfoil plane the velocity is given by

$$v_a = \frac{v_c}{\left|\frac{d\zeta}{dz}\right|},$$

where the denominator evaluated on the surface of the cylinder is given by

$$\left|\frac{d\zeta}{dz}\right| = \left|1 - \frac{a^2}{z^2}\right| = \left|1 - e^{-i\,2\theta}\right| = \left[(1 - \cos 2\theta)^2 + \sin^2 2\theta\right]^{1/2} = 2\sin\theta.$$

Thus, the velocity distribution in the airfoil plane simplifies to

$$v_a = U_\infty \frac{\sin(\theta + \alpha) + \sin \alpha}{\sin \theta}$$

or

$$\frac{v_a}{U_\infty} = \cos\alpha + \frac{\sin\alpha}{\sin\theta}(1 + \cos\theta). \tag{3.108}$$

For small angle of attack, $\cos\alpha \simeq 1$ and $\sin\alpha \simeq \alpha$, so that one obtains the nondimensional velocity in the airfoil plane as

$$\frac{v_a}{U_\infty} \simeq 1 + \alpha\frac{1 + \cos\theta}{\sin\theta}$$

or

$$\frac{v_a}{U_\infty} = 1 + \alpha\cot\theta/2. \tag{3.109}$$

Having obtained the velocity distribution, one can obtain the loading for the flow from $\gamma = v_{a_1} - v_{a_2}$, where the velocities are as shown in Figure 3.18, for which $\theta_1 = -\theta_2 = \theta$. Thus,

$$\gamma = 2\,U_\infty\,\alpha\,\frac{1 + \cos\theta}{\sin\theta}. \tag{3.110}$$

One can integrate the loading distribution to obtain the lift acting on the section as

$$\text{Lift} = \rho\,U_\infty \int_0^\pi 2\,U_\infty\alpha\,\frac{1 + \cos\theta}{\sin\theta}\,\frac{c}{2}\sin\theta\,d\theta.$$

Thus, the lift acting on the flat plate is given by

$$\text{Lift} = \rho\,U_\infty^2 c\,\alpha \int_0^\pi (1 + \cos\theta)\,d\theta = \pi\alpha\rho\,U_\infty^2 c. \tag{3.111}$$

And the lift coefficient is given by

$$C_l = 2\pi\alpha. \tag{3.112}$$

Using the loading distribution given by Equation (3.110) in Equation (3.102), one gets the pitching moment about the leading edge as

$$\begin{aligned} M_{LE} &= -\rho\,U_\infty \int 2\,U_\infty\,\alpha\,\frac{1 + \cos\theta}{\sin\theta}\,\frac{c}{2}(1 - \cos\theta)\,\frac{c}{2}\sin\theta\,d\theta \\ &= -\frac{1}{2}\rho\,U_\infty^2\,\alpha\,c^2 \int_0^\pi (1 - \cos^2\theta)\,d\theta = C_{M_{LE}}\,\frac{1}{2}\rho\,U_\infty^2 c^2. \end{aligned}$$

Thus the pitching moment coefficient about the leading edge can be written as

$$C_{M_{LE}} = -\frac{\pi}{2}\alpha = -\frac{C_l}{4}. \tag{3.113}$$

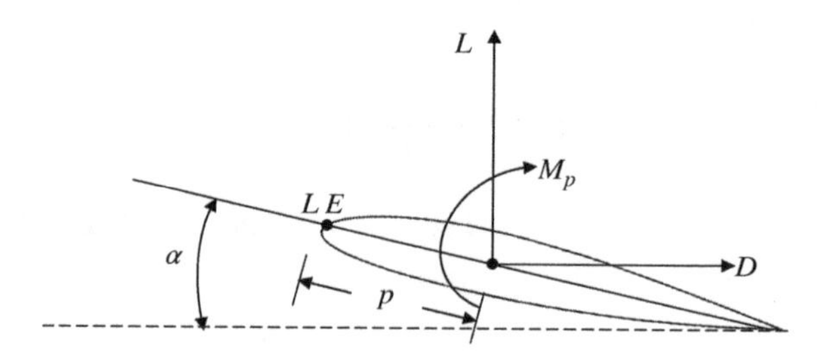

Figure 3.21 Representation of section properties of lift, drag and pitching moment referred to different points

Before proceeding further we will discuss two important aerodynamic parameters, which will help us to understand the section properties better and relate aerodynamic results with flight dynamics.

3.8.2 Aerodynamic Centre and Centre of Pressure

Consider an airfoil at an angle of attack α, as shown in Figure 3.21, with the sectional force components and pitching moment referred to a datum point which is at a distance p from the leading edge of the airfoil. Material in this subsection is not related to potential flow which has been described in the rest of this chapter and applies to any general case, for which the force and moment coefficients are considered to have been obtained by any means, theoretically, computationally or experimentally. The lift and drag force components are indicated as L and D, respectively. The pitching moment about this datum point is indicated as M_p, with clockwise moment (nose-up moment) treated as positive.

The pitching moment about the leading edge (M_{LE}) can be written down in terms of M_p and the pitching moment about any other arbitrary point located at a distance x from the leading edge by

$$M_{LE} = M_p - L\,p\,\cos\alpha - D\,p\,\sin\alpha = M_x - L\,x\,\cos\alpha - D\,x\,\sin\alpha. \tag{3.114}$$

Therefore

$$M_x = M_p - (L\cos\alpha + D\sin\alpha)(p - x).$$

Defining the pitching moment coefficient as before by

$$C_M = \frac{M}{\frac{1}{2}\rho U_\infty^2 c^2},$$

one can write

$$C_{M_x} = C_{M_p} - (C_l\cos\alpha + C_d\sin\alpha)\left(\frac{p}{c} - \frac{x}{c}\right).$$

When p is at the leading edge

$$C_{M_x} = C_{M_{LE}} + (C_l \cos\alpha + C_d \sin\alpha)\frac{x}{c}. \tag{3.115}$$

Aerodynamic centre:
Along the chord there is one point for which C_{M_x} is independent of C_l and this point is called the aerodynamic centre. Let x_{ac} be the position of the aerodynamic centre behind the leading edge, then

$$C_{M_p} = C_{M_{ac}} - (C_l \cos\alpha + C_d \sin\alpha)\left(\frac{x_{ac}}{c} - \frac{p}{c}\right).$$

For small angles of attack, $\cos\alpha \simeq 1$ and $\sin\alpha \simeq \alpha$ and for all aeronautical applications $C_l \gg C_d$. Therefore

$$C_{M_p} = C_{M_{ac}} - C_l\left(\frac{x_{ac}}{c} - \frac{p}{c}\right).$$

It is readily apparent that for $C_l = 0 : C_{M_p} = C_{M_{ac}}$ and for this reason, one often indicates $C_{M_{ac}} = C_{M_0}$. According to the definition of aerodynamic centre C_{M_p} is independent of C_l, i.e.

$$\frac{dC_{M_p}}{dC_l} = \frac{dC_{M_{ac}}}{dC_l} - \left(\frac{x_{ac}}{c} - \frac{p}{c}\right).$$

Therefore the aerodynamic centre can be located from the above equation as

$$\frac{x_{ac}}{c} = \frac{p}{c} - \frac{dC_{M_p}}{dC_l}. \tag{3.116}$$

Thus, the aerodynamic centre can be found by measuring lift and pitching moment coefficients about any point p and use Equation (3.116) to obtain the aerodynamics centre. For many high Reynolds number applications, if the angle of attack of an airfoil is below which the flow does not separate, then the aerodynamic centre is seen experimentally to be located in the range $0.23 \le \frac{x_{ac}}{c} \le 0.25$. Thus, it is customary for incompressible flow applications to state that the aerodynamic centre is at the quarter chord point. This is specifically true for flat or curved plate of negligible thickness. It is noted that thickness and viscous effects push the aerodynamic centre forward. It is to be noted that for supersonic flow past thin airfoils the aerodynamic centre is at the mid-chord.

For the case of the thin symmetric flat plate airfoil, from Equation (3.113) one notes that the aerodynamic centre can be obtained from Equation (3.116) by using $p = 0$ as

$$\frac{x_{ac}}{c} = -\frac{dC_{M_{LE}}}{dC_l} = \frac{1}{4}. \tag{3.117}$$

Thus the aerodynamic centre is located at the quarter chord point for this case.

Centre of pressure:
For every lift coefficient value, there is one point along the chord line about which the pitching moment is found to be zero. This point is referred to as the centre of pressure. It is important to note that it is not necessary for the centre of pressure to lie inside the airfoil.

From this definition of centre of pressure and from Equation (3.114), if we indicate the location of the centre of pressure to be given by x_{cp}, then

$$M_{LE} = M_{ac} - (L \cos\alpha + D \sin\alpha)\, x_{ac} = -(L \cos\alpha + D \sin\alpha)\, x_{cp}. \tag{3.118}$$

Therefore from this equation,

$$\frac{x_{cp}}{c} = \frac{x_{ac}}{c} - \frac{C_{M_{ac}}}{C_l \cos\alpha + C_d \sin\alpha}. \tag{3.119}$$

Once again, as the developed theory is for small perturbation, we will keep our attention focused on small angle of attack, for which $\cos\alpha = 1$; $\sin\alpha = \alpha$ and $C_d \sin\alpha$ is negligibly small, so that the denominator in Equation (3.119) can be approximated by C_l only. Based on these observations and Equation (3.118), one can also locate the centre of pressure from

$$\frac{x_{cp}}{c} = \frac{x_{ac}}{c} - \frac{C_{M_{ac}}}{C_l} = -\frac{C_{M_{LE}}}{C_l}. \tag{3.120}$$

For traditional airfoils, $C_{M_{ac}} < 0$ and so $x_{cp} > x_{ac}$, i.e. the centre of pressure is behind the aerodynamic centre. Specifically, one notes that for zero lift condition the centre of pressure goes to $-\infty$.

For the symmetric flat plate using the results of Equation (3.113) in Equation (3.120), it is easy to see that

$$\frac{x_{cp}}{c} = -\frac{C_{M_{LE}}}{C_l} = \frac{1}{4}. \tag{3.121}$$

Hence the aerodynamic centre and centre of pressure are at identical location of quarter chord, for the symmetric flat plate aerofoil.

3.8.3 The Circular Arc Airfoil

In Section 3.6.3, we have shown how a circular arc can be obtained by Jukowski transform from a circular cylinder and these two planes have been shown in Figure 3.14. The Jukowski transformation was applied to the circle, with its centre offset along the y-axis. Consider the transformation of a circle of radius a with centre at the point $y = h = a \sin\beta$ in the circle plane to the aerofoil plane via the transformation $\zeta = z + \frac{b^2}{z}$ where $b = a\cos\beta \simeq a$. The circle transforms to a circular arc of camber $\frac{1}{2}\tan\beta$.

The flow field is analyzed by looking at the circle plane shown in Figure 3.22. Velocity on the circle plane is given by

$$v_c = 2\, U_\infty\, [a \sin\theta_c + \sin(\alpha + \beta)], \tag{3.122}$$

where the circulation is obtained by applying the Kutta condition at m in Equation (3.92), i.e. at $\theta_c = \pi + (\alpha + \beta)$.

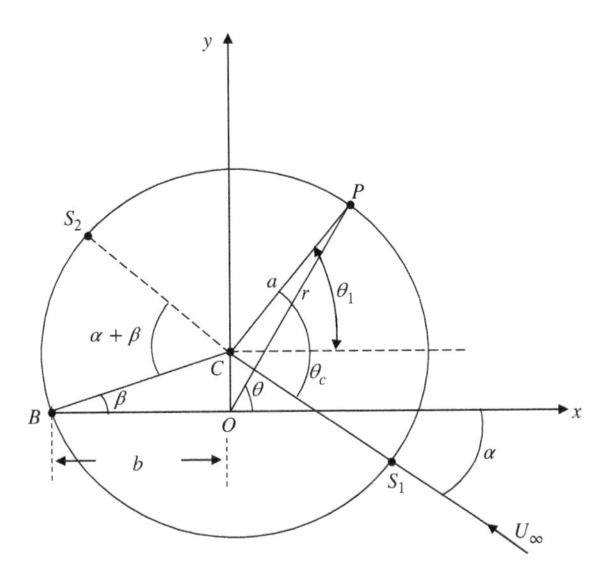

Figure 3.22 Flow in the circle plane for flow past a circular arc airfoil. Note the rear stagnation point is now located at B. Also, note that the front stagnation point is not as shown at S_1, which will also be correspondingly depressed

From Figure 3.22, $\theta_c = \theta_1 + \alpha$. Therefore

$$\sin\theta_c = \sin\theta_1 + \alpha\cos\theta_1.$$

Using this in Equation (3.122) one obtains

$$v_c = 2\,U_\infty(\sin\theta_1 + \alpha\cos\theta_1 + \alpha + \beta).$$

Also for the point P in Figure 3.22,

$$y = a\sin\theta_1 + b\beta = r\sin\theta.$$

Thus

$$\sin\theta_1 = \frac{r}{a}\sin\theta - \frac{b}{a}\,\beta \simeq \frac{r}{b}\sin\theta - \beta.$$

Similarly for the point P

$$x = a\cos\theta_1 = r\cos\theta.$$

Therefore

$$\cos\theta_1 = \frac{r}{a}\cos\theta = \frac{r}{b}\cos\theta$$

$$v_c = 2\,U_\infty \left(\overbrace{\left(\frac{r}{b}\sin\theta - \beta\right)}^{\sin\theta_1} + \alpha\, \overbrace{\frac{r}{b}\cos\theta}^{\cos\theta_1} + \alpha + \beta \right)$$

and

$$\frac{r}{b} \simeq 1 + \beta \sin\theta \text{ and } \alpha\beta = \beta^2 \simeq 0.$$

From Figure 3.22,

$$\begin{aligned} CP^2 &= OC^2 + OP^2 - 2OC.\,OP\cos\ \angle COP \\ a^2 &= a^2\beta^2 + r^2 - 2a\beta r\cos(\pi/2 - \theta) \\ a^2 &= a^2\beta^2 + r^2 - 2a\beta r\sin\theta. \end{aligned} \tag{3.123}$$

Also from Figure 3.22

$$CB^2 = CO^2 + OB^2.$$

Therefore

$$a^2 = a^2\beta^2 + b^2. \tag{3.124}$$

From Equations (3.123) and (3.124) one gets

$$b^2 = r^2 - 2ra\beta\sin\theta.$$

Solving this quadratic equation for r, one obtains

$$\begin{aligned} r &= \frac{2a\beta\sin\theta \pm \sqrt{4a^2\beta^2\sin^2\theta + 4b^2}}{2} \\ &= a\beta\sin\theta \pm \sqrt{b^2 + a^2\beta^2\sin^2\theta}. \end{aligned}$$

Since r is always positive, we will consider only the positive sign before the radical to get

$$r = a\beta\sin\theta + b\sqrt{1 + \left(\frac{a\beta}{b}\right)^2 \sin^2\theta}.$$

Since also $\beta \ll 1$, so, $a\beta \ll b$, and therefore

$$r \simeq a\beta\sin\theta + b. \tag{3.125}$$

Thus $\frac{r}{b} \simeq 1 + \frac{a}{b}\,\beta\,\sin\theta \simeq 1 + \beta\sin\theta.$

These simplifications result in the circle plane velocity as

$$v_c = 2\,U_\infty\,[\sin\theta\,(1+\beta\sin\theta) + \alpha(1+\cos\theta)]. \tag{3.126}$$

For this transformation to a circular arc airfoil, we have

$$\begin{aligned}
\left|\frac{d\zeta}{dz}\right| &= \left|1-\frac{b^2}{z^2}\right| = \left|1-\frac{b^2}{r^2}e^{-i2\theta}\right| \\
&= \left|1-\frac{\cos 2\theta - i\sin 2\theta}{(1+\beta\sin\theta)^2}\right| \\
&= \left[1-\frac{2\cos 2\theta}{(1+\beta\sin\theta)^2}+\frac{1}{(1+\beta\sin\theta)^4}\right]^{1/2}.
\end{aligned}$$

One can write the following based on the observation that β is small,

$$\left|\frac{d\zeta}{dz}\right| = \left[(1+\beta\sin\theta)^2 - 2\cos 2\theta + (1+\beta\sin\theta)^{-2}\right]^{1/2}(1+\beta\sin\theta)^{-1}.$$

Also using the following simplifications

$$(1+\beta\sin\theta)^2 \simeq 1+2\beta\sin\theta$$

and

$$(1+\beta\sin\theta)^{-2} \simeq 1-2\beta\sin\theta,$$

one obtains the following simplified result for the transformation

$$\left|\frac{d\zeta}{dz}\right| = [2-2\cos 2\theta]^{1/2}(1+\beta\sin\theta)^{-1} = \frac{2\sin\theta}{1+\beta\sin\theta}.$$

With help of the simplified expression for the transformation, one obtains the speed on the surface of the airfoil as

$$\begin{aligned}
v_a = \frac{v_c}{\left|\frac{d\zeta}{dz}\right|} &= \frac{U_\infty(1+\beta\sin\theta)}{\sin\theta}\left[\sin\theta(1+\beta\sin\theta)+\alpha(1+\cos\theta)\right] \\
&= U_\infty\left[(1+\beta\sin\theta)^2+\alpha\,\frac{1+\cos\theta}{\sin\theta}(1+\beta\sin\theta)\right] \\
&\simeq U_\infty\left[(1+2\beta\sin\theta)+\alpha\,\frac{1+\cos\theta}{\sin\theta}\right].
\end{aligned}$$

From the speed distribution on the surface of the airfoil, one can obtain the circulation distribution as before from $\gamma = v_{a_1} - v_{a_2}$, where for these points on top and bottom of the camberline one has $\theta_1 = -\theta_2$

$$\gamma = 2\,U_\infty\left[2\beta\sin\theta_1 + \alpha\,\frac{1+\cos\theta_1}{\sin\theta_1}\right]. \tag{3.127}$$

Comparing with the flat plate result, the camber increases γ by $4\,U_\infty\,\beta\sin\theta$.
For the circular arc airfoil, using Equation (3.127) one obtains the lift generated as

$$\begin{aligned}
\text{Lift} &= 2\rho\,U_\infty^2\int_0^\pi\left[2\beta\sin\theta + \alpha\,\frac{1+\cos\theta}{\sin\theta}\right]\frac{c}{2}\sin\theta\,d\theta\\
&= \rho\,U_\infty^2 c\int_0^\pi[\alpha(1+\cos\theta)+2\beta\sin^2\theta]d\theta\\
&= \pi\rho\,U_\infty^2 c\,(\alpha+\beta).
\end{aligned}$$

Therefore the lift coefficient is obtained as

$$C_l = 2\pi(\alpha+\beta) \tag{3.128}$$

and

$$\frac{dC_l}{d\alpha} = 2\pi.$$

As before, one calculates the pitching moment about the leading edge as given by

$$\begin{aligned}
M_{LE} &= -2\rho\,U_\infty^2\int_0^\pi\left[\alpha\frac{1+\cos\theta}{\sin\theta} + 2\beta\sin\theta\right]\frac{c}{2}(1-\cos\theta)\frac{c}{2}\sin\theta\,d\theta\\
&= -\frac{1}{2}\,\rho\,U_\infty^2c^2\int_0^\pi[\alpha(1-\cos^2\theta)+2\beta\sin^2\theta(1-\cos\theta)]d\theta\\
&= \frac{1}{2}\,\rho\,U_\infty^2c^2\,C_{M_{LE}}.
\end{aligned}$$

Therefore

$$C_{M_{LE}} = -\frac{\pi}{2}\,(\alpha+2\beta) \tag{3.129}$$

and one obtains the centre of pressure using

$$\frac{x_{cp}}{c} = -\frac{C_{M_{LE}}}{C_l} = \frac{1}{4}+\frac{\pi\beta}{C_l}. \tag{3.130}$$

Form the lift coefficient value in Equations (3.128) and (3.129), one notices that

$$C_{M_{LE}} = -\frac{\pi}{2}\,(\alpha+2\beta) = -\frac{C_l}{4}-\frac{\pi\beta}{2}.$$

Thus,

$$\frac{dC_{M_{LE}}}{dC_l} = 1/4,$$

which also shows that the aerodynamic centre is at the quarter chord. At zero lift, the centre of pressure is at an infinite distance behind the aerofoil, which means that there can always be a moment without any lift force.

The above analysis was based on conformal mapping for Jukowski airfoil which is considered to be very thin. In the limit of zero thickness, we have noted here that the load distribution for a circular arc airfoil is given by Equation (3.127). This expression shows that the load distribution is directly dependent on the camber and angle of attack. Also for any general airfoil with finite thickness distribution, one can superpose circulation distribution for the flat plate airfoil (indicated here by γ_a) and that due to the circular arc airfoil (indicated here by γ_b) to obtain

$$\gamma = \gamma_a + \gamma_b.$$

In the following, we try to develop a theory for any general airfoil which has similar features noted in Jukowski airfoils.

3.9 General Thin Airfoil Theory

Here we write the circulation at any point to consists of two parts

$$\gamma = \gamma_a + \gamma_b,$$

where γ_a represents the load due to angle of attack, in the absence of the camber. This is represented by

$$\gamma_a = 2\, U_\infty\, A_0 \left(\frac{1 + \cos\theta}{\sin\theta} \right),$$

with A_0 as a constant to be evaluated and which absorbs any differences between the 'equivalent' flat plate aerofoil and the actual chord line. The flat plate load distribution is given in Equation (3.110), which is created due to angle of attack.

The other component γ_b, is due to camber effect, as noted in Equation (3.127) and can be conveniently represented by a Fourier series

$$\gamma_b = 2\, U_\infty \sum_1^\infty A_n \sin n\theta.$$

Therefore the general load distribution is given by

$$\gamma = 2\, U_\infty \left\{ A_0 \frac{1 + \cos\theta}{\sin\theta} + \sum_1^\infty A_n \sin n\theta \right\}. \tag{3.131}$$

The coefficients can be found by substituting Equation (3.131) in Equation (3.106)

$$U_\infty\left[\frac{dy}{dx}-\alpha\right]=\frac{1}{2\pi}\int_0^c\frac{\gamma\, dx}{x-x_1}.$$

This equation can be simplified by applying the change of variable, $x=\frac{c}{2}(1-\cos\theta)$ and $dx=\frac{c}{2}\sin\theta\, d\theta$ in the above to get the right-hand side as

$$=-\frac{1}{2\pi}\int_0^\pi\frac{\gamma\sin\theta\, d\theta}{\cos\theta-\cos\theta_1}.$$

Using Equation (3.131) in this expression yields

$$=-\frac{U_\infty}{\pi}\int_0^\pi\left\{\frac{A_0(1+\cos\theta)}{\sin\theta}+\sum_1^\infty A_n\sin n\theta\right\}\frac{\sin\theta\, d\theta}{\cos\theta-\cos\theta_1}.$$

Therefore

$$\frac{dy}{dx}-\alpha=-\frac{A_0}{\pi}\int_0^\pi\frac{1+\cos\theta}{\cos\theta-\cos\theta_1}\,d\theta-\frac{1}{\pi}\int_0^\pi\sum\frac{A_n\sin n\theta\sin\theta}{\cos\theta-\cos\theta_1}\,d\theta. \tag{3.132}$$

Using the identity

$$A_n\sin n\theta\sin\theta=\frac{A_n}{2}\,[\cos(n-1)\theta-\cos(n+1)\theta].$$

Equation (3.132) can be represented symbolically as

$$\frac{dy}{dx}-\alpha=-\left\{\frac{A_0}{\pi}\,G_0+\frac{A_0}{\pi}\,G_1+\sum\frac{A_n}{2\pi}\,G_{n-1}-\sum\frac{A_n}{2\pi}\,G_{n+1}\right\},$$

where G_n is the Glauert integral given by

$$G_n=\int_0^\pi\frac{\cos n\theta}{\cos\theta-\cos\theta_1}\,d\theta=\frac{\pi\sin n\theta_1}{\sin\theta_1}.$$

Proof of the above result follows from the complex integral $\oint f(z)dz$ round the circle of unit radius centred at the origin ($z=e^{i\theta}$), for the function

$$f(z)=\frac{z^n}{z^2-2z\cos\theta_1+1}.$$

The Glauert's integral is thus rewritten as

$$G_n=\int_0^\pi\frac{ie^{in\theta}\,d\theta}{e^{i\theta}-2\cos\theta_1+e^{-i\theta}}=\int_0^\pi\frac{ie^{in\theta}\,d\theta}{2(\cos\theta-\cos\theta_1)}.$$

The denominator of the above integrand has zeros given by

$$
\begin{aligned}
z_{1,2} &= \frac{2\cos\theta_1 \pm \sqrt{4\cos^2\theta_1 - 4}}{2} \\
&= \cos\theta_1 \pm i\sin\theta_1 = e^{\pm i\theta_1}.
\end{aligned}
$$

The denominator can be written as

$$(e^{i\theta} - e^{i\theta_1})(e^{i\theta} - e^{-i\theta_1})$$

and the poles are at $e^{\pm i\theta_1}$. Residue for the pole at $z = e^{i\theta_1}$ is

$$\frac{ie^{in\theta_1}}{e^{i\theta_1} - e^{-i\theta_1}} = \frac{ie^{in\theta_1}}{2i\sin\theta_1} = \frac{e^{in\theta_1}}{2\sin\theta_1}.$$

Thus, the residue at these poles are

$$\frac{\sin n\theta_1 \pm \; i\cos n\theta_1}{2\sin\theta_1},$$

the sum of these residues is

$$\frac{\sin n\theta_1}{\sin\theta_1}.$$

Therefore

$$\frac{i}{2}\int_{-\pi}^{\pi} \frac{\cos n\theta + i\sin n\theta}{\cos\theta - \cos\theta_1}\, d\theta = \pi i\,\frac{\sin n\theta_1}{\sin\theta_1}.$$

Equating the imaginary part of this equation one gets

$$\frac{1}{2}\int_{-\pi}^{\pi} \frac{\cos n\theta}{\cos\theta - \cos\theta_1}\, d\theta = \pi\frac{\sin n\theta_1}{\sin\theta_1}.$$

By the symmetry of the integrand, one reduces the above to the following with changed interval

$$\int_{0}^{\pi} \frac{\cos n\theta}{\cos\theta - \cos\theta_1}\, d\theta = \frac{1}{2}\int_{-\pi}^{\pi} \frac{\cos n\theta}{\cos\theta - \cos\theta_1}\, d\theta.$$

Hence follows the proof

$$\int_{0}^{\pi} \frac{\cos n\theta}{\cos\theta - \cos\theta_1}\, d\theta = \pi\frac{\sin n\theta_1}{\sin\theta_1}.$$

Using this result in Equation (3.132), one gets

$$\frac{dy}{dx} - \alpha = -\frac{A_0}{\pi}\pi - \sum A_n \frac{\sin(n-1)\theta_1 - \sin(n+1)\theta_1}{2\sin\theta_1}$$
$$= -A_0 + \sum A_n \cos n\theta.$$

Thus

$$\frac{dy}{dx} = -A_0 + \alpha + \sum A_n \cos n\theta. \tag{3.133}$$

This equation can be integrated over the full range of θ to get

$$A_0 = \alpha - \frac{1}{\pi}\int_0^{\pi} \frac{dy}{dx}\, d\theta. \tag{3.134}$$

Multiplying Equation (3.133) by $\cos m\theta$, where m is an integer, and integrating with respect to θ, one gets

$$\int_0^{\pi} \frac{dy}{dx} \cos m\theta\, d\theta = \int_0^{\pi} (\alpha - A_0)\cos m\theta\, d\theta + \int_0^{\pi} \sum_n A_n \cos n\theta \cos m\theta\, d\theta$$
$$= 0 + \int_0^{\pi} A_m \cos^2 m\theta\, d\theta = \frac{\pi}{2} A_m.$$

Thus,

$$A_m = \frac{2}{\pi}\int_0^{\pi} \frac{dy}{dx} \cos m\theta\, d\theta. \tag{3.135}$$

For symmetric airfoil the camber is zero, i.e. $\frac{dy}{dx} = 0$. Thus, all A_m's are zero except A_0, which gives one back the results for the flat plate case, for which $A_0 = \alpha$, from Equations (3.134) and (3.135).

Also from the γ-distribution given in Equation (3.131) one can obtain the lift in terms of the coefficients as

$$\text{Lift} = \int_0^{c} \rho\, U_\infty\, \gamma\, dx = 2\rho\, U_\infty^2\, \frac{c}{2}\int_0^{\pi}\left[A_0(1+\cos\theta) + \sum_1^{\infty} A_n \sin n\theta \sin\theta\right] d\theta$$
$$= 2\,\rho\, U_\infty^2\, \frac{c}{2}\left[\pi\, A_0 + \frac{\pi}{2}\, A_1\right].$$

The lift coefficient is given by

$$C_l = 2\pi\left[A_0 + \frac{A_1}{2}\right]. \tag{3.136}$$

Thus the lift coefficient is solely determined by A_0 and A_1.

Zero lift angle:
From Equations (3.134) and (3.135) one obtains

$$A_0 = \alpha - \frac{1}{\pi}\int \frac{dy}{dx}\, d\theta$$

and

$$A_1 = \frac{2}{\pi}\int_0^{\pi} \frac{dy}{dx}\cos\theta\, d\theta.$$

Substituting these in Equation (3.136) one gets

$$C_l = \frac{dC_l}{d\alpha}\left[\alpha - \frac{1}{\pi}\int \frac{dy}{dx}\, d\theta + \frac{1}{\pi}\int \frac{dy}{dx}\cos\theta\, d\theta\right],$$

where we have used $\frac{dC_l}{d\alpha} = 2\pi$ and therefore

$$C_l = \frac{dC_l}{d\alpha}\,(\alpha - \alpha_0)$$

with

$$\alpha_0 = \frac{1}{\pi}\int \frac{dy}{dx}(1 - \cos\theta)\, d\theta \tag{3.137}$$

as the zero lift angle.

Similarly from the load distribution given in Equation (3.131) one can obtain the pitching moment about the leading edge as

$$-M_{LE} = -C_{M_{LE}}\,\frac{1}{2}\rho U_\infty^2\, c^2 = \rho\, U_\infty \int_0^c \gamma\, x\, dx.$$

Thus

$$C_{M_{LE}} = -\int_0^{\pi}\left[A_0\frac{1+\cos\theta}{\sin\theta} + \sum_1^{\infty} A_n \sin n\theta\right]\sin\theta(1-\cos\theta)d\theta.$$

Using various trigonometric identities, this can be simplified to

$$C_{M_{LE}} = -\frac{\pi}{2}\,A_0 - \frac{\pi}{2}\,A_1 + \frac{\pi}{4}\,A_2. \tag{3.138}$$

Using Equation (3.136) in Equation (3.138) one can write

$$C_{M_{LE}} = -\frac{C_l}{4}\left[1 + \frac{A_1 - A_2}{C_l}\pi\right]. \tag{3.139}$$

Thus the centre of pressure is obtained from

$$\frac{x_{cp}}{c} = -\frac{C_{M_{LE}}}{C_l} = \frac{1}{4} + \frac{\pi}{4C_l}(A_1 - A_2). \tag{3.140}$$

From Equation (3.115)

$$C_{M_x} = C_{M_{LE}} + \frac{x}{c}(C_l \cos\alpha + C_d \sin\alpha).$$

For small angles of attack, $\cos\alpha \simeq 1$ and $\sin\alpha \simeq \alpha$ and also $C_d \ll C_l$. Using these information and substituting $x = c/4$ one gets

$$\begin{aligned} C_{M_{c/4}} &= C_{M_{LE}} + \frac{C_l}{4} \\ &= -\frac{\pi}{2}A_0 - \frac{\pi}{2}A_1 + \frac{\pi}{4}A_2 + \frac{\pi}{2}(A_0 + A_{1/2}) \\ &= \frac{\pi}{4}(A_2 - A_1). \end{aligned}$$

For symmetrical airfoil, it is easy to establish that all A_m's are zero. Hence, $C_{M_{c/4}} = 0$. One can also establish this same result by using direct method.

One notes the location of the aerodynamic centre from Equation (3.139) as

$$\frac{x_{ac}}{c} = -\frac{dC_{M_{LE}}}{dC_l} = \frac{1}{4}. \tag{3.141}$$

3.10 Theodorsen Condition for General Thin Airfoil Theory

For thin airfoil theory, the circulation distribution is given by Equation (3.131) as

$$\frac{\gamma}{2\,U_\infty} = A_0\frac{1+\cos\theta}{\sin\theta} + \sum_1^\infty A_n \sin n\theta,$$

where the first part on the right-hand side is due to angle of attack and the second part is due to camber. The first part also show a curious feature that there is an infinite suction at the leading edge ($\theta = 0$) due to the behaviour of $\frac{1+\cos\theta}{\sin\theta}$. This can only be alleviated if $A_0 = 0$, for which this singularity disappears. Physically also, presence of an infinite suction at leading edge will cause an adverse pressure gradient following this, which will cause twin problems of separation and premature transition. These can be avoided by choosing $A_0 = 0$. As

$$A_0 = \alpha - \frac{1}{\pi}\int_0^\pi \frac{dy}{dx}d\theta.$$

This amounts to choosing an α for which the intensity of circulation at the leading edge is zero and the stream flows smoothly on to the camber line, with leading edge also as a stagnation point. This is the so-called Theodorsen condition and the corresponding C_l is the

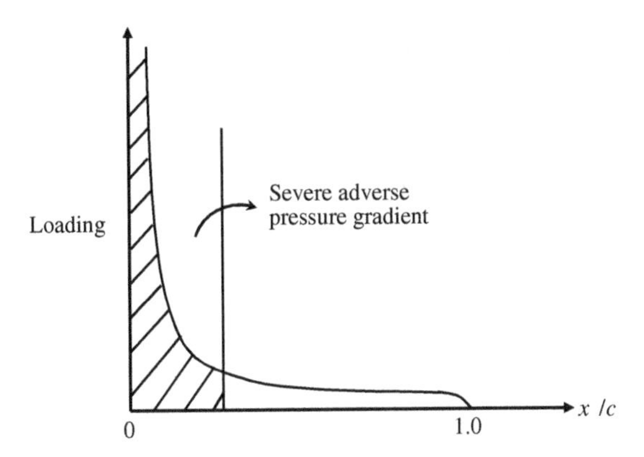

Figure 3.23 Effect of leading edge singularity at the leading edge of thin airfoil. Note the severe adverse pressure gradient immediately following the leading edge suction singularity

ideal optimum or design lift coefficient $C_{L\text{opt}}$. Thus, the optimum angle of attack and optimum or design lift coefficient is obtained as

$$\alpha_{\text{opt}} = \frac{1}{\pi} \int_0^{\pi} \frac{dy}{dx} d\theta \tag{3.142}$$

$$\text{and } C_{L\text{opt}} = \pi A_1 = \pi \times \frac{2}{\pi} \int_0^{\pi} \frac{dy}{dx} \cos\theta \, d\theta.$$

Thus the optimum lift coefficient, for which there is no leading edge singularity, is given by

$$C_{L\text{opt}} = 2 \int \frac{dy}{dx} \cos\theta \, d\theta. \tag{3.143}$$

The main reason for avoiding the leading edge singularity is to avoid the associated massive adverse pressure gradient over the top surface of the airfoil. This is depicted in Figure 3.23, which shows the resultant adverse pressure gradient near the leading edge of the airfoil, which can trigger flow separation and hence cause severe drag increase, even for very small angles of attack.

Bibliography

Currie, I.G. (2010) *Fundamental Mechanics of Fluids*. CRC Press, India.

Houghton, E.L. and Carpenter, P.W. (1993) *Aerodynamics for Engineering Students*. 4th Edn., Edward Arnold (Publishers) Ltd., UK.

Kundu, P.K. and Cohen, I.M. (2008) *Fluid Mechanics*. Academic Press, New York, USA.

Lim, T.T., Sengupta, T.K. and Chattopadhyay, M. (2004) A visual study of vortex-induced subcritical instability on a flat plate laminar boundary layer. *Expts. Fluids*, **37**, 47–55.

Milne-Thomson, L.M. (1958) *Theoretical Aerodynamics*. Dover Publication Inc., New York, USA.

Robertson, J.M. (1965) *Hydrodynamics in Theory and Application*. Prentice Hall International, New Jersey, USA.

Sengupta, T.K., Bhumkar, Y.G. and Sengupta, S. (2012) Dynamics and instability of a shielded vortex in close proximity of a wall. *Computers and Fluids*, **70**, 166–175.

Sengupta, T.K., De, S. and Sarkar, S. (2003) Vortex-induced instability of an incompressible wall-bounded shear layer. *J. Fluid Mech.,* **493**, 277–286.

Sengupta, T.K., Kasliwal, A., De, S. and Nair, M. (2003a) Temporal flow instability for Magnus-Robins effect at high rotation rates. *J. Fluids and Struct.* **17**, 941–953.

Sengupta, T.K. and Talla, S.B. (2004) Robins-Magnus effect: A continuing saga. *Current Science*, **86**(7), 1033–1036.

Tokumaru, P.T. and Dimotakis, P.E. (1993) The lift of a cylinder rotary motion in a uniform flow. *J. Fluid Mech.,* **255**, 1–10.

White, F.M. (2008) *Fluid Mechanics.* Sixth Edn., McGraw-Hill, New York, USA.

4

Finite Wing Theory

4.1 Introduction

This theory was developed by Lanchester (1908) and Prandtl (1921) in the late nineteenth and early twentieth century. The essential contribution of this theory is to replace the wing by a system of vortices of particular geometries which create a flow field similar to the actual flow in terms of the lift sustained by the wing. The vortex system representing the wing can be divided into three main parts: (i) a bound vortex system, which essentially mimics the sectional lift property; (ii) a starting vortex system necessary to satisfy a system of theorems for vortical field and (iii) the trailing vortex system to complement the other two elements into a single coherent description of the load distribution on a finite wing. Thus to understand the role of each element, we need to develop the basics of elementary vortex motion as described next.

4.2 Fundamental Laws of Vortex Motion

Vortex in a flow system has been as defined in the previous chapter, which represents a singularity in ideal irrotational fluid flow. In real flow this occupies a finite area in a section placed normally and can be viewed as a distribution of vorticity. The reader should realize that this is an idealization, as there are flows which have distributed vorticity, yet do not have concentrated vortices, as in a boundary layer, while wakes and free shear layer have regions of large concentrated vorticity distribution, which can be viewed distinctly as vortices. The unity of both of this type of description can be brought about by the single descriptor, as the circulation. In general, the vortex is a curve in space and has a finite area normal to its axis given by S. It may be convenient to consider the finite area vortex to be made up of several infinitesimal elements, in the same framework that a singularity can be described by a distribution. Alternatively, the vortex consists of a bundle of elemental vortex lines or filaments. Such a bundle is often called a vortex tube, being a tube consisting of many vortex filaments. For inviscid flows, the properties of vortex in motion is governed by the following theorems due to Helmholtz and Kelvin.

Theoretical and Computational Aerodynamics, First Edition. Tapan K. Sengupta.

Companion Website: www.wiley.com/go/sengupta

4.3 Helmholtz's Theorems of Vortex Motion

These are essentially four fundamental laws which help in building up the properties of vortices in describing various fluid flow phenomena.

Theorem 1:
This refers to a fluid particle in general motion possessing some of the following kinematic properties: linear velocity, vorticity and distortion.

Theorem 2:
This demonstrates constancy of strength of vortex along its length, which at times is referred to as the equation of vortex continuity. It is shown that the strength of a vortex cannot grow or diminish along its axis or length.

We have already defined circulation as the contour integral of the velocity vector, defined as $\Gamma = \oint \vec{V} \cdot d\vec{r}$. As the velocity vector can be expressed in terms of velocity potential as, $\vec{V} = \nabla\phi$, one can represent the circulation as

$$\Gamma = \oint \nabla\phi \cdot d\vec{r} = \oint d\phi.$$

Thus for a flow with circulation, ϕ is not single-valued and is not defined.

Proof of Theorem 2:
Since $\Gamma = \oint \vec{V} \cdot d\vec{r}$, its substantial derivative is given by

$$\frac{d\Gamma}{dt} = \frac{d}{dt}\oint \vec{V} \cdot d\vec{r} \tag{4.1}$$

$$= \oint \frac{d\vec{V}}{dt} \cdot d\vec{r} + \oint \vec{V} \cdot d\left(\frac{d\vec{r}}{dt}\right).$$

For inviscid flow, the governing Euler equation is

$$\frac{d\vec{V}}{dt} = -\frac{\nabla p}{\rho} - \nabla f, \tag{4.2}$$

where f is a conservative body force. Therefore

$$\frac{d\vec{V}}{dt} \cdot d\vec{r} = -\frac{\nabla p}{\rho} \cdot d\vec{r} - \nabla f \cdot d\vec{r}.$$

As

$$\nabla p \cdot d\vec{r} = dp \;\; and \;\; \nabla f \cdot d\vec{r} = df.$$

Therefore

$$\frac{d\vec{V}}{dt} \cdot d\vec{r} = -\frac{dp}{\rho} - df.$$

Hence follows the proof

$$\frac{d\Gamma}{dt} = \oint -\frac{dp}{\rho} - \oint df + \oint d\,(|\vec{V}|^2) \equiv 0. \tag{4.3}$$

Corollary of Theorem 2:
The strength of a vortex is the magnitude of the circulation around it and this is also equal to

$$\Gamma = \iint_S \omega \cdot dS,$$

with ω defined as the vorticity, as before.

Since infinite vorticity is unacceptable, so $S \to 0$ is not a possibility. Thus a vortex cannot abruptly end in the fluid. In practice any vortex must form a closed loop or originate (or terminate) in a discontinuity in the fluid such as a solid body or a surface of separation.

Corollary of constancy of circulation:
Since Γ remains constant in an inviscid flow, then upon an impulsive start, the circulation generated by a lifting body must be equally balanced by a starting vortex leaving the body, so that the net circulation is zero. This is shown in Figure 4.1. As at $t = 0^-$, there was no motion and hence there was no circulation, it must remain so thereafter, according to theorem 2 above. This is true for two-dimensional, as well as, three-dimensional bodies. This is also the foundation of finite wing theory.

Theorems 3 and 4:
These two theorems demonstrate, respectively, that a vortex tube consists of the same particles of fluid, i.e. that there is no fluid interchange between the vortex tube and surrounding fluid and that the strength of a vortex remains the same as the vortex moves through fluid confines.

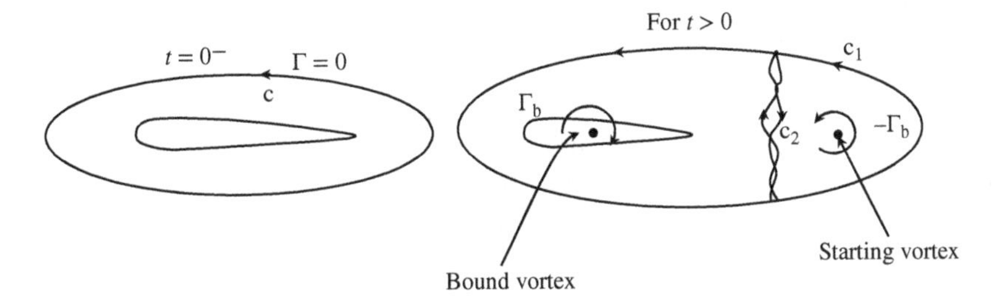

Figure 4.1 The necessary vortex system for an infinite aspect ratio wing, which requires only the bound and the starting vortex. Both the vortices are of same strength and opposite in sign and extend to infinity in the spanwise direction

Having defined the fundamental vortex laws, we can describe various elements of the vortex system forming on a wing of finite span in steady flight condition.

4.4 The Bound Vortex Element

This is associated with the circulation developing over a finite wing. The finite aspect ratio of the wing necessitates that the loading must decay to zero at the wing tip, where there is no physical surface to sustain any loading. Thus, one expects to see grading of the spanwise load distribution and the bound vortex system is so designed to help construct this spanwise load distribution. This is a hypothetical arrangement of vortices, which replaces the wing to support the lift distribution in every details, except the thickness distribution. It is concerned with developing the equivalent bound vortex element which explains the sectional loading along the span, to reproduce the lift experienced by the wing realistically.

Consider an aircraft wing in steady level flight, which requires the spanwise load distribution to be symmetric. Such symmetric load distribution necessarily must have the maximum load experienced at the wing root, which is at the mid-span. It is also necessary to represent this load distribution in a systematic manner which explains the load variation with basic wing planform variations, namely the span, planform and aerodynamic or geometric twist etc. As the load in such steady flight motion remains invariant with time, the bound vortex system must reflect the same feature and this model must qualitatively explain the variation of load distribution with wing planform variations, while the sectional property can be obtained from the methodologies of Chapter 3.

The grading of lift in spanwise direction can also be interpreted as a consequence of the wing to equalize pressures on the top and bottom surfaces of an airfoil as one approaches the wing tips. Thus to represent this spanwise distribution, the bound vortex system should have spanwise vortex elements whose cumulative strength will be represented by a requisite number of such elements. This portrait is not complete yet.

4.5 Starting Vortex Element

In the context of obtaining the sectional properties of an airfoil in Chapter 3, we have also used the Kelvin's second theorem in showing constancy of circulation of a barotropic inviscid flow (see Equation (4.3)). There we stated that the created and sustained circulation around a section following an impulsive start must also be associated with a starting vortex, so that the sum total of bound and starting vortices is zero. This is an absolute necessity, as before the impulsive start the quiet ambiance did not have any circulation. To maintain the zero circulation, the starting vortex counterbalances the bound vortex in strength, but opposite in sign. This is pictorially shown in Figure 4.1. From a two-dimensional flow point of view, the lift system consist of a spanwise line vortex extending from $-\infty$ to $+\infty$ positioned over the infinite aspect ratio wing. To counterbalance this circulation, the starting vortex is considered to be present at $x \to +\infty$, again infinitely long in the spanwise direction with opposite sign circulation. Thus, these two vortex system meet with each other at $\pm\infty$ and forming a loop, without violating Kelvin's second theorem.

The situation changes qualitatively for a finite wing and one requires a model circulation system, which is more subtle and elaborate in construction for two reasons. First, the loading is graded to fall off to zero value at the wing tip. The falling off of load towards tips can be replicated by taking vortex elements of differing spanwise lengths. However, we cannot have vortices of finite length and ending abruptly, according to second Helmholtz's theorem. This necessitates a third element of the vortex system, as defined next.

4.6 Trailing Vortex Element

Consider an aircraft of finite aspect ratio wing flying level and steady, with symmetric loading. Such a wing will produce a graded lift as discussed above, in terms of the bound vortex system. We reasoned that this must consist of bundle of line vortices of varying length, so that any closed contour at any spanwise station will account for the sectional load as the sum total strength of the enclosed line vortices. We have also reasoned that such line vortices cannot end abruptly. A plausible model was proposed by Lanchester and Prandtl, as depicted in Figure 4.2.

We noted above that the spanwise vortices have finite length in spanwise direction to account for the falling lift, as one moves from wing root to tip section. At the same time to satisfy Kelvin's vortex theorems – mainly the second theorem – these finite spanwise line vortices must all be turned backwards at each end symmetrically to form pair of the trailing vortex elements. This is one plausible solution, as the second Helmholtz's theorem states that vortices cannot end abruptly. This explains the formation of the trailing vortex system as shown in Figure 4.2. This trailing vortex system simulates the wake seen behind the wing and thus, two physical attributes of a real wing are represented in terms of lift variation in spanwise direction and the wake behind the wing. This turning of line vortices from spanwise to streamwise direction is the key feature of the finite wing theory. While such turning does not violate conservation of translational momentum, what happens to the conservation of angular momentum? The reader's attention is drawn to this fact. Even if the strength of the spanwise and streamwise vorticity may remain same, i.e. $|\omega_z| \equiv |\omega_x|$ during the turning, moment of inertia about these two axes are dissimilar! This is a point to ponder.

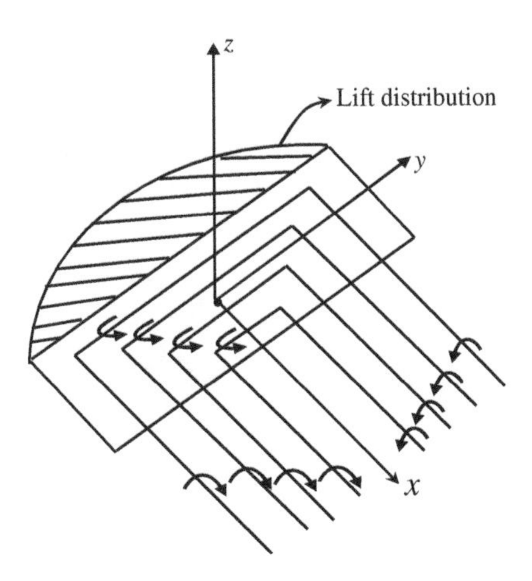

Figure 4.2 Vortex system over a finite aspect ratio, as put forward by Lanchester and Prandtl. This vortex system has three elements as shown in the figure and discussed in the text

Also, we noted in Chapter 2, that the nonlinear action of vortex stretching term in Navier-Stokes equation is needed to transform one component of vorticity to contribute to two orthogonal directions to satisfy the conservation of angular momentum for nondissipative flows. This has been used to explain the energy cascade from large to small scales in actual turbulent flows. This is beyond the scope of the present book.

We must note, however, that this finite wing theory is merely a model and further refinement of model is a distinct possibility, specially in the context of computational aerodynamics now feasible with powerful computers and high accuracy methods, as advocated here.

4.7 Horse Shoe Vortex

The final vortex system consisting of three elements replacing the finite wing by bound, trailing and starting vortices again forms multiple loops, consistent with Helmholtz's vortex theorems. We specially note from third and fourth theorems of Helmholtz, associated vortex tubes also bends likewise from bound spanwise vortex to streamwise trailing vortex and again the spanwise starting vortex. The starting vortex is left behind and the trailing pair stretches effectively to infinity as steady flight proceeds. This three-sided vortex has been called the horse shoes vortex.

Having described qualitatively the vortex system in terms of the three elements, it is necessary to quantify the effects of such loading distribution. This is in essence the major contribution of finite wing theory. One specifically notes that the finite aspect ratio of the wing causes the bound vortex to be of finite spanwise length and the trailing vortices on either side of the wing is of semi-infinite length. Such semi-infinite vortices will affect the flow field over the wing, which was completely absent for a wing of infinite aspect ratio. The quantitative measure of such interaction between bound and trailing vortices is found by the following law, which originated in electromagnetism.

4.8 The Biot-Savart Law

Consider a vortex tube of strength Γ, which can be viewed as a collection of infinite number of line vortices and as it cannot end abruptly, we allow this stem to terminate on a spherical surface centred around A, of radius R in Figure 4.3. As a consequence of vortex continuity, the total strength of the vortex filaments is allowed to spread uniformly over the spherical surface, to maintain symmetry and isotropy in all directions. The vorticity is then distributed over the spherical surface, whose sum must add up to Γ, according to this model.

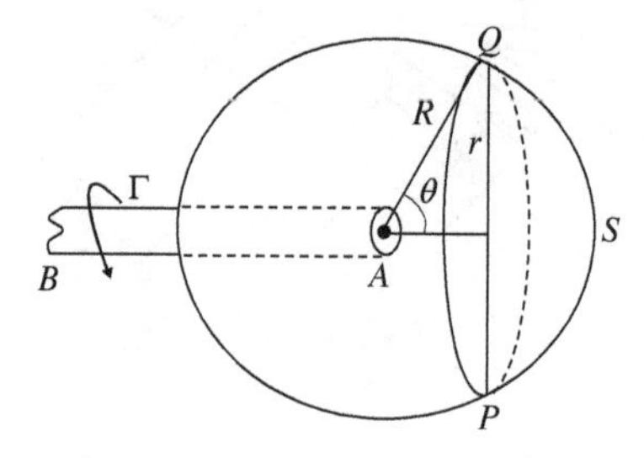

Figure 4.3 A sketch showing a line vortex terminating onto a sphere and its effect to calculate velocity distribution by Biot-Savart Law

The symmetry of the vorticity distribution over the spherical surface causes the velocity field also to be in azimuthal direction, if we draw a section perpendicular to the axis of the vortex tube. Consider the section PQ shown in Figure 4.3. Induced velocity along the periphery of PQ will be in tangential direction in conformity with the directionality of the circulation of the vortex tube. The figure shows the circle of radius r, which forms a conical angle 2θ at A.

If the velocity at the point Q defined by (R, θ) is v, then the circulation round the circuit PQ is Γ' and is obtained by the contour integral, by noting that the velocity along the circle is v and is perpendicular to the radius r $(= R\sin\theta)$. Thus,

$$\Gamma' = 2\pi r v = 2\pi \; R\sin\theta \; v.$$

The alternate way of defining this circulation is integrating the vorticity over the spherical cap PQS as an area integral, where the strength per unit area is given by $\frac{\Gamma}{4\pi R^2}$. Thus

$$\Gamma' = \frac{\text{surface area of cap PQS}}{\text{surface area of the sphere}}\Gamma = \frac{2\pi \; R^2(1-\cos\theta)}{4\pi R^2}\Gamma.$$

Hence

$$\Gamma' = \frac{\Gamma}{2}(1-\cos\theta).$$

Thus the induced velocity on the circular circuit is given by

$$v = \frac{\Gamma'}{2\pi r} = \frac{\Gamma}{4\pi r}(1-\cos\theta). \tag{4.4}$$

Now suppose the length of the vortex AB is decreased, until it is a very short element, AA_1 as shown in Figure 4.4. The circle PQ is now influenced by the complete element and the induced velocity dv can be very easily obtained by differentiation of Equation (4.4) as

$$dv = \frac{\Gamma}{4\pi r}\sin\theta \; d\theta. \tag{4.5}$$

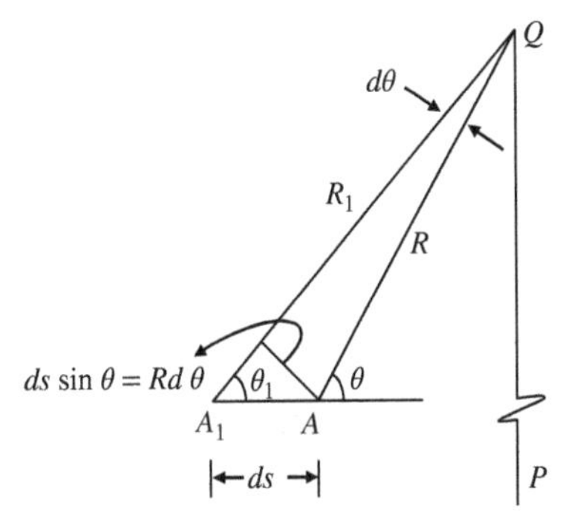

Figure 4.4 Calculating velocity distribution at the field point Q due to an elementary length of vortex, ds. With reference to the element, the field point is at $(R, \; \theta)$

This is the induced velocity at the point Q in the field, due to an elementary vortex of length ds and strength per unit length Γ, which subtends an angle $d\theta$ at Q located by the coordinates (R, θ) from the element. The above incremental induced velocity is written in terms of the angle as the independent variable. This can be restated in terms of the length as the independent variable (since $r = R\sin\theta$ and $R\, d\theta = ds \sin\theta$) by

$$dv = \frac{\Gamma}{4\pi R^2} \sin\theta\, ds. \tag{4.6}$$

We will use this representation of Biot-Savart law for further developments in finite wing theory. First, we will consider a few spacial cases, keeping in view of the geometries of the bound and trailing vortex elements. Note that the starting vortex being at infinite distance away from the lifting wing, will have zero induced velocity, as one can readily see from Equation (4.6) with $R \to \infty$ in the denominator.

4.8.1 Biot-Savart Law for Simplified Cases

In finite wing theory of unswept planform, the adopted model of vortex system consists of elements which are linear. While the bound vortex filaments are of finite length, the trailing vortex consists of semi-infinite elements originating over the planform and extends to infinite distance downstream. In further simplification, we can replace this vortex system consisting of countably infinite number of filaments to a single bound vortex turned backwards to two trailing vortices extending to infinity.

We show a schematic diagram in Figure 4.5, to calculate induced velocity at a field point P by a line vortex of finite length and strength per unit length is Γ. The velocity at P induced by the elemental length $d\bar{s}$ on AB. The field point P is located with respect to this infinitesimal element $d\bar{s}$ given by the polar coordinate of (R, θ). Also the field point subtends angles θ_L and θ_R with the finite length vortex AB. From Equation (4.6) one can calculate the induced velocity as

$$dv = \frac{\Gamma}{4\pi R^2} \sin\theta\, d\bar{s}. \tag{4.7}$$

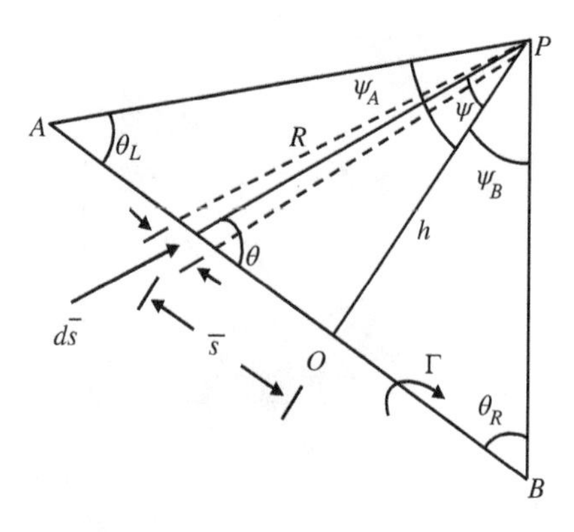

Figure 4.5 Schematic to calculate induced flow field due to a straight line vortex of constant strength

This induced velocity is in the direction of the normal to the plane APB.

The total induced velocity at P due to the complete line element AB, is the summation of effects of such successive elements obtained by integrating Equation (4.7) for the geometry shown in Figure 4.5. The normal PO drawn upon AB from P measures a distance of h. Defining an auxiliary angle ψ in the figure, we note that this varies from ψ_A and ψ_B. The angle ψ is given by $\psi = \frac{\pi}{2} - \theta$. Also,

$$\psi_A = -\left(\frac{\pi}{2} - \theta_L\right) \text{ and } \psi_B = +\left(\frac{\pi}{2} - \theta_R\right).$$

Note that ψ_A is oriented in clockwise direction with respect to PO and hence is negative and ψ_B is in anti-clockwise direction with respect to PO and hence is positive.

Since

$$\sin\theta = \cos\psi, \quad R = h\sec\psi \text{ and } \bar{s} = h\tan\psi. \tag{4.8}$$

Therefore

$$d\bar{s} = d\,(h\tan\psi) = h\sec^2\psi\, d\psi.$$

Using these in Equation (4.7), we get

$$v = \int_{\psi_A}^{\psi_B} \frac{\Gamma}{4\pi h}\cos\psi\, d\psi = \frac{\Gamma}{4\pi h}\left[\sin(\pi/2 - \theta_R) + \sin(\pi/2 - \theta_L)\right].$$

Thus, the induced velocity at P is obtained in terms θ_L and θ_R as

$$= \frac{\Gamma}{4\pi h}(\cos\theta_L + \cos\theta_R). \tag{4.9}$$

Influence of a semi-infinite vortex:

Consider that in Figure 4.5 one end of the vortex stretches to $+\infty$ (say the end B), then $\theta_R = 0$ and $\cos\theta_R = 1$ and Equation (4.9) gives

$$v = \frac{\Gamma}{4\pi h}(\cos\theta_L + 1). \tag{4.10}$$

Furthermore if $\theta_L = \pi/2$, i.e. for semi-infinite vortex, then the induced velocity of this special subcase is given by

$$v = \frac{\Gamma}{4\pi h}. \tag{4.11}$$

Influence of an infinite vortex:

This is the case of an infinite line vortex which we considered in the previous chapter. Thus, in this case A and B tends to infinite distance in the opposite directions, so that $\theta_L = \theta_R = 0$

and from Equation (4.9) or Equation (4.10) one gets

$$v = \frac{\Gamma}{2\pi h}. \tag{4.12}$$

4.9 Theory for a Finite Wing

The wing is now replaced by a bundle of bound vortex filaments each of different length in the spanwise direction and turned back into two trailing vortex filaments.

4.9.1 Relation between Spanwise Loading and Trailing Vortices

From Helmholtz's second law, circulation round any section of the bundle of vortices is the sum of the strengths of the vortex filaments cut by the sectional plane.

In this process it is easy to see that the spanwise loading drops off towards the wing tip, if different elements have different spanwise length and turned backwards. Now as the sectional plane is moved outwards from the midspan, the bound vortex filaments and corresponding sum of the strength of the bundle decreases.

In a wing of infinite aspect ratio, all the sections must have the same vortex strength. Finite aspect ratio of a wing causes the lift to fall off and this is possible that the lost circulation between two successive spanwise stations, should be shed in the wake of the wing in the form of trailing vortices. Thus the strength of the vortex filaments shed in the wake must increase, as the rate of fall in circulation strength falls rapidly in the vicinity of the wing tip. Thus, the general rule of change in circulation from section to section is equal to the strength of the vortices shed between the sections, for example, for the spanwise load distribution shown in Figure 4.6

At section y: The chordwise circulation around the section is Γ.

At an adjoining section at $y + dy$: The circulation has fallen to $\Gamma - d\Gamma$.

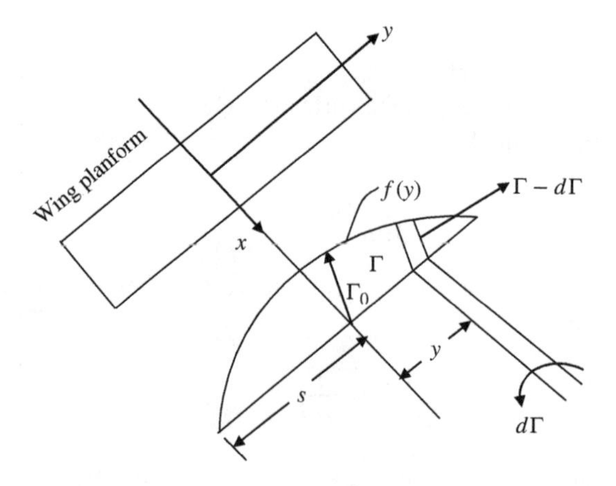

Figure 4.6 Spanwise load distribution for a straight wing and shed vortices

Therefore between the two stations at y and $y + dy$, trailing vortices of strength $d\Gamma$ have to be shed according to this steady state model. If the circulation variation is as shown in the drawn curve in Figure 4.6 and which can be described as some function of y, say $f(y)$, then the strength of circulation shed between the above-mentioned two sections dy apart can be written as

$$d\Gamma = -f'(y)\, dy,$$

where the prime indicates derivative with respect to the argument.

Now at any station the lift per unit span is given by the Kutta-Jukowski theorem as

$$L = \rho\, U_\infty\, \Gamma.$$

As noted in Figures 4.2 and 4.6, the induced velocity by the trailing vortices causes a downward component of velocity at any other station, which is inboard of the trailing vortices. As the strength of trailing vortices increases towards the wing-tip, the net contribution of such induced effects would be to introduce a downward component of velocity over the complete planform. When this component is calculated, this is thus justifiably called the **downwash**. Consider now the influence of the trailing vortex filaments of strength $d\Gamma$ shed from the section at y. At some other point y_1 along the span an induced velocity is created equal to

$$dw(y_1) = -\frac{f'(y)dy}{4\pi(y - y_1)}. \tag{4.13}$$

In Equation (4.13) use is made of Equation (4.11). Also the direction of this induced velocity is downwards, if y_1 is inboard of y and is given by

$$w(y_1) = -\frac{1}{4\pi}\int_{-s}^{s} \frac{f'(y)dy}{(y - y_1)}. \tag{4.14}$$

In computing this integral, we have considered the spanwise load distribution over the complete wing, indicated by the range of integral.

4.10 Consequence of Downwash: Induced Drag

The downwash for any spanwise station y_1 is felt to a lesser extent ahead of the airfoil than on the wing and downstream of it. The role of this downwash is very important as this can add to the component of a drag, which is inviscid in origin and is entirely due to the finite aspect ratio of the wing creating the above-mentioned trailing vortices.

At any spanwise station, the consequence is depicted in Figure 4.7, showing the airfoil section at a geometric angle of attack of α, which is the angle subtended by the chord line with the oncoming flow velocity (U_∞) direction. The downwash component w is vectorially added to U_∞ to provide the resultant velocity, U_R. Now notice that this resultant velocity vector makes a reduced angle of attack of α_∞ with chord line. This is the equivalent angle of attack made by an infinite aspect ratio wing section. The difference between these two angles

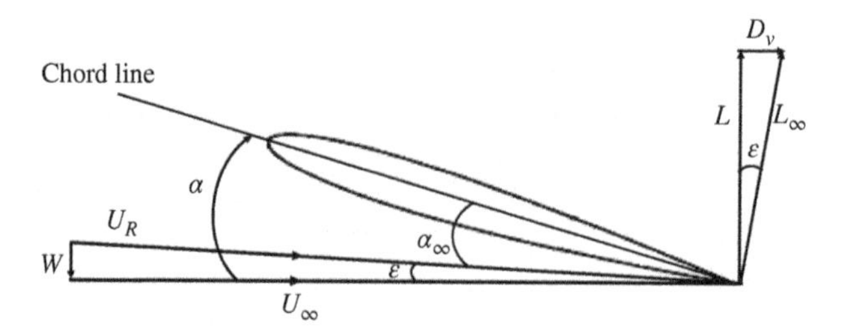

Figure 4.7 The effect of induced downwash by trailing vortices and creation of induced drag

of attack is indicated by ϵ, which is given for small downwash angles by

$$\epsilon = \tan^{-1} \frac{w}{U_\infty} \simeq \frac{w}{U_\infty}. \tag{4.15}$$

There is another interesting consequence of this downwash angle for this inviscid flow. The lift acting on the wing section (L_∞) will be perpendicular to U_R and there is no drag for this potential flow. However, going by our sign convention, the actual lift should be referred to the actual free stream direction, as indicated by the component L of L_∞. The other component of L_∞ is indicated by D_v and is in the direction of drag and is called the induced or trailing vortex drag.

There is an alternative viewpoint about the lift experienced by the finite aspect ratio wing. The induced velocity, w in effect reduces the effective incidence of the wing section, so that to generate the same lift as created by an infinite aspect ratio wing with the same airfoil section at incidence α_∞, the corresponding angle of incidence for the finite aspect ratio wing at the same section must be set at an incidence of $\alpha = \alpha_\infty + \epsilon$. Needless to say that for small downwash, $L = L_\infty \cos\epsilon \simeq L_\infty$, i.e. the lift remains virtually the same for the finite aspect ratio wing section.

However, the other component, $D_v = L_\infty \sin\epsilon \simeq L_\infty \epsilon$ is normal to lift direction and against the direction of the forward velocity of the airfoil, and justifiably called the induced drag.

One can quantify the lift and induced drag of a finite aspect ratio wing as following. If the wing experiences variable circulation distribution along the span, as shown in Figure 4.6, then the total lift experienced by the wing of semispan s is

$$L_\infty = L = \int_{-s}^{s} \rho\, U_\infty\, \Gamma\, dy \tag{4.16}$$

and the induced drag is

$$D_v = \int_{-s}^{s} \rho\, U_\infty \left(\frac{w}{U_\infty}\right) \Gamma\, dy,$$

which can be conveniently written in terms of the downwash velocity and circulation distribution only as

$$D_v = \int_{-s}^{s} \rho\, w\, \Gamma\, dy. \tag{4.17}$$

In Chapter 10, we will provide consolidated data which show that for a long haul transport aircraft flying in turbulent flow condition, the aircraft experiences induced drag which is more than 30% of the total drag. Thus, induced drag is not really insignificant and should be understood in proper perspective. It has been estimated that a commercial aircraft company charges in million dollars for a device called the winglet to reduce induced drag. What is interesting to note however, that such increased price is recouped by commercial airliners in couple of years' saving in reduced fuel consumption via reduced induced drag.

4.11 Simple Symmetric Loading: Elliptic Distribution

For a given load distribution described in Figure 4.6, one can obtain the lift and induced drag from Equations (4.16) and (4.17). Let us consider the load distribution given for a aircraft flying level and steady, for which the loading must by symmetric. For such a loading the maximum circulation would be experienced at the root chord (let us say given by, Γ_0) and which progressively falls off to zero at the wing tips at $y = \pm s$. The simplest possible variation can be a quadratic variation given as

$$\left(\frac{\Gamma}{\Gamma_0}\right)^2 + \left(\frac{y}{s}\right)^2 = 1. \tag{4.18}$$

Note that this represents an equation for an ellipse and such loading is therefore called the elliptic loading. Out of two possible signs, we only consider the possible positive sign for the circulation, Γ.

For such a loading distribution, the lift experienced by the wing is given by

$$L = \int_{-s}^{s} \rho\, U_\infty\, \Gamma\, dy = \rho\, U_\infty\, \Gamma_0 \int_{-s}^{s} \sqrt{1 - \left(\frac{y}{s}\right)^2}\, dy.$$

Using $y = -s\cos\phi$, from Equation (4.18) one gets $\Gamma = \Gamma_0 \sin\phi$ with $0 \leq \phi \leq \pi$. Plugging these expressions in Equation (4.16) one gets

$$L = \rho\, U_\infty \Gamma_0 \pi\, s/2. \tag{4.19}$$

Using the expression of lift coefficient for the finite wing (C_L) with planform area S as

$$L = C_L \frac{1}{2} \rho U_\infty^2 S$$

gives the circulation at the midspan as

$$\Gamma_0 = \frac{C_L\, U_\infty S}{\pi s}. \tag{4.20}$$

Note that the lift coefficient for the finite wing has been represented by C_L, instead of C_l used for 2D wing. For the circulation distribution given by Equation (4.18), one can calculate the rate of spanwise variation of loading as

$$\frac{d\Gamma}{dy} = -\frac{\Gamma_0}{s}\frac{y}{\sqrt{s^2-y^2}}.$$

Using this expression in Equation (4.14), one gets the downwash as

$$w(y_1) = \frac{\Gamma_0}{4\pi s}\int_{-s}^{s}\frac{y\,dy}{\sqrt{s^2-y^2}(y-y_1)}.$$

This can be simplified further to

$$= \frac{\Gamma_0}{4\pi s}\left[\int_{-s}^{s}\frac{dy}{\sqrt{s^2-y^2}} + y_1\int_{-s}^{s}\frac{dy}{(s^2-y^2)^{1/2}(y-y_1)}\right].$$

The first integral on the right-hand side can be easily evaluated and representing the second integral on the right-hand side by I, we can write the downwash as

$$w(y_1) = \frac{\Gamma_0}{4\pi s}\left[\pi + y_1 I\right].$$

For any symmetric flight, the loading will be symmetric on the wing and hence the induced downwash on two stations located symmetrically on either side of the midspan must be the same, i.e. $w(y_1) = w(-y_1)$.

This implies then

$$\frac{\Gamma_0}{4\pi s}\left[\pi + y_1\, I\right] = \frac{\Gamma_0}{4\pi s}\left[\pi - y_1\, I'\right].$$

We note that the two integrals are not the same and therefore the above identity of symmetric downwash distribution on either side of the wing is satisfied only if $I \equiv 0 = I'$.

And thus

$$w(y_1) = \frac{\Gamma_0}{4s}. \tag{4.21}$$

One of the key features of the elliptic load distribution in Equation (4.18) is that the induced downwash is constant along the span. We will shortly establish that such loading distribution is optimum in terms of induced drag, which is calculated next.

4.11.1 Induced Drag for Elliptic Loading

The induced drag for the elliptic loading given by Equation (4.18) is obtained from Equation (4.17) as

$$D_v = \int_{-s}^{s} \rho\, w\, \Gamma\, dy = \rho \frac{\Gamma_0}{4s}\, \Gamma_0 \int_{-s}^{s} \sqrt{1 - \left(\frac{y}{s}\right)^2}\, dy$$

$$= \frac{\pi}{8} \rho\, \Gamma_0^2 = C_{D_v} \frac{1}{2}\, \rho\, U_\infty^2\, S.$$

Thus, the induced drag coefficient, C_{D_v} can be obtained in terms of the wing lift coefficient, by substituting the midspan circulation expression of $\Gamma_0 = \frac{C_L\, U_\infty S}{\pi s}$ in the above to get

$$C_{D_v} = \frac{C_L^2}{\pi AR}, \tag{4.22}$$

where

$$\frac{4s^2}{S} = \frac{(\text{wing span})^2}{\text{wing planform area}} = AR \text{ (aspect ratio)}.$$

The elliptic loading represented by Equation (4.18) is also given by

$$\Gamma = \Gamma_0 \sin \phi. \tag{4.23}$$

Here, $0 \leq \phi \leq \pi$ and the argument is not allowed outside this limit. This is so, as the angle is allowed corresponding to $-s$ to $+s$. However, the circulation is represented by a periodic function described over half the wavelength. This lacks mathematical rigour, as the expression for the circulation implies corresponding periodic extension of the same. But, we must remember that this is a model for the load distribution and we do not expect mathematical rigour from such a model. Interested readers are encouraged to work out the rigorous derivation of the same by multiplying the load distribution in Equation (4.23) by appropriate unit step functions and this is left as an exercise.

It is pointed out that the elliptic loading distribution provides the optimum with respect to least induced drag and hence is of great interest. Such loading can be obtained in an actual wing by various means. For example, if one considers a plane wing without any wing twist distribution and having identical airfoil section along the full span, such distribution is simply obtained by spanwise chord length distribution given by, $c(y) = c_0 \sin \phi$, where c_0 is the wing root chord. This type of distribution can be obtained by a wing with planform shape representing a half-ellipse, represented by a planform shown in the loading distribution in Figure 4.6. However, with different wing sections and geometric twist, a whole range of planform will provide elliptic loading. An example of this is the famous spitfire aircraft wing shape, which used the same wing section and the chord length distribution was elliptic.

4.11.2 *Modified Elliptic Load Distribution*

We note from Equation (4.23) that the loading is symmetric about the midplane. Such symmetric loading, with the constraints of zero loading at the wing tip can be obtained by all odd superharmonics of the fundamental given in Equation (4.23). While we will describe most general load distribution in a later section, here we consider a modified load distribution given by

$$\Gamma = \Gamma_0 \sqrt{1-\left(\frac{y}{s}\right)^2}\left[1+a\left(\frac{y}{s}\right)^2\right] \tag{4.24}$$

$$= \Gamma_0 \sqrt{1-\left(\frac{y}{s}\right)^2} + a\Gamma_0 \sqrt{1-\left(\frac{y}{s}\right)^2}\left(\frac{y}{s}\right)^2,$$

where the first factor represents the elliptic variation and the second factor represents modification to it. These two components are shown in Figure 4.8. The first component has a maximum value at the mid-span, while the correction term is zero at the wing-root and wing tips. While the maximum correction is noted at half way from the wing root to wing tip. Substituting this expression of Equation (4.24) directly in to Equation (4.16), one can obtain the lift on the wing as

$$L = \int_{-s}^{s} \rho\, U_\infty\, \Gamma_0 \sqrt{1-\left(\frac{y}{s}\right)^2}\left[1+a\left(\frac{y}{s}\right)^2\right] dy$$

$$= 2\,\rho\, U_\infty\, \Gamma_0 \int_0^s \sqrt{1-\left(\frac{y}{s}\right)^2}\left[1+a\left(\frac{y}{s}\right)^2\right] dy.$$

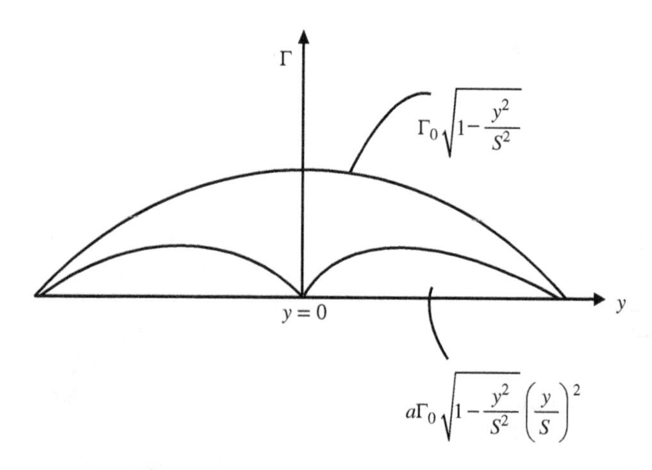

Figure 4.8 The individual harmonics of modified elliptic load distribution

As before, if we use the transformation $y = s\sin\phi$, with $-\pi/2 \le \phi \le \pi/2$ and $dy = s\cos\phi\, d\phi$, then from the above we get

$$L = 2\rho\, U_\infty\, \Gamma_0\, s \int_0^{\pi/2} \cos^2\phi(1 + a\sin^2\phi)\, d\phi.$$

By writing $a = 4\lambda$, one can simplify the above lift as

$$L = 2\rho\, U_\infty\, \Gamma_0\, s \int_0^{\pi/2} \left[\cos^2\phi + 4\lambda\cos^2\phi\sin^2\phi\right] d\phi$$

$$= 2\rho\, U_\infty\, \Gamma_0\, s \left[\frac{\pi}{4} + 4\lambda\,\frac{\pi}{16}\right] = \rho\, U_\infty\, \Gamma_0\, s\,\frac{\pi}{2}(1 + \lambda).$$

Thus, the lift coefficient for the modified loading is obtained as

$$C_L = \frac{\pi\, \Gamma_0\, s}{U_\infty S}(1 + \lambda). \tag{4.25}$$

4.11.3 *The Downwash for Modified Elliptic Loading*

Using Equation (4.24) in the expression for the downwash in Equation (4.14) and upon simplification it is easy to show that

$$w = \frac{\Gamma_0}{4s}\left\{1 - 2\lambda + 12\lambda\left(\frac{y}{s}\right)^2\right\}. \tag{4.26}$$

From this expression of downwash, one can obtain the induced drag from Equation (4.17) as

$$D_v = \int_{-s}^{s} \rho\, w\, \Gamma\, dy$$

$$= \frac{\pi\rho\,\Gamma_0^2}{8}\left[1 + 2\lambda + 4\lambda^2\right].$$

Hence it is easy to show that

$$C_{D_v} = \frac{C_L^2}{\pi AR}[1 + \delta] \tag{4.27}$$

where we have used the formula $\int \sin^4\phi\cos^2\phi\, d\phi = \pi/32$

There is an alternate viewpoint of identifying the modified elliptic loading by noting that $y = -s\cos\phi$ leads to representing Equation (4.24) by

$$\Gamma = \Gamma_0 \sin\phi\left[1 + a(1 - \sin^2\phi)\right].$$

That is the nondimensional circulation is written as

$$\frac{\Gamma}{\Gamma_0} = (1 + a) \sin \phi - a \sin^3 \phi.$$

As

$$\sin^3 \phi = \frac{1}{4}\left[3 \sin \phi - \sin 3\phi\right]$$

$$\frac{\Gamma}{\Gamma_0} = \left(1 + \frac{a}{4}\right) \sin \phi + \frac{a}{4} \sin 3\phi.$$

Thus one can write the compact notation as

$$A_1 = (1 + \lambda)\Gamma_0 \text{ and } A_3 = \lambda\Gamma_0. \tag{4.28}$$

And then the modified elliptic loading is given by

$$\Gamma = A_1 \sin \phi + A_3 \sin 3\phi. \tag{4.29}$$

These odd harmonics of sine function provide symmetric loading.

4.12 General Loading on a Wing

We have noted above that the loading represented by Equation (4.29) provides symmetric loading. In contrast, even harmonics of sine function provide anti-symmetric loading. Thus, the two terms in Equation (4.29) are not the same contributions shown in Figure 4.8. However, this provides us the way to represent any arbitrary loading over the wing by

$$\Gamma = 4\, s\, U_\infty \sum_{n=01}^{N} A_n \sin n\theta, \tag{4.30}$$

where the spanwise coordinate is defined as $y = -s \cos \theta$, with $0 \leq \theta \leq \pi$. The starboard wing-tip corresponds to $\theta = \pi$ and the port wing-tip is given by $\theta = 0$. The circulation scale has been taken as $4\, s\, U_\infty$. Depending upon the prescribed loading pattern, one can appropriately choose the number of terms (N) in the general loading function. Thus, the lift on a section of width $dy = s \sin \theta d\theta$ at $y = -s \cos \theta$ is obtained by application of Kutta-Jukowski theorem as

$$dL = 4\rho\, s^2\, U_\infty^2 \sum_{n=01}^{N} A_n \sin n\theta\, \sin \theta\, d\theta. \tag{4.31}$$

Thus, the lift acting on the wing is given by

$$L = 4\rho\, s^2\, U_\infty^2 \int_0^\pi \sum_{n=01}^{N} A_n \sin n\theta\, \sin \theta\, d\theta \tag{4.32}$$

or

$$L = 2\rho\, s^2\, U_\infty^2 \int_0^\pi \sum A_n[\cos(n-1)\theta - \cos(n+1)\theta]d\theta$$

$$= 2\rho\, s^2\, U_\infty^2 \sum \left[A_n\left(\frac{\sin(n-1)\theta}{n-1} - \frac{\sin(n+1)\theta}{n+1}\right)\right]_0^\pi.$$

It is noted that both the terms on the right-hand side do not contribute for all n, except for $n = 1$, for which one gets

$$L = 2\rho\, s^2\, U_\infty^2 \pi A_1. \tag{4.33}$$

Thus it is important to realize that for a general loading distribution, the lift is decided by the first odd harmonic of the loading distribution. The lift coefficient is obtained as

$$C_L = \pi A_1\, AR. \tag{4.34}$$

4.12.1 Downwash for General Loading

The general loading distribution given by Equation (4.30) can be used in Equation (4.14) to obtain the corresponding downwash. By transforming to θ as the independent variable using $y = -s\cos\theta$, the downwash is given by

$$w(\theta_1) = \frac{1}{4\pi s}\int_0^\pi \frac{\frac{d\Gamma}{d\theta}d\theta}{\cos\theta - \cos\theta_1}.$$

From Equation (4.30)

$$\frac{d\Gamma}{d\theta} = 4s\, U_\infty \sum n\, A_n \cos n\theta.$$

Thus,

$$w(\theta_1) = \frac{4s\, U_\infty}{4\pi s}\int_0^\pi \frac{\sum n\, A_n \cos n\theta}{\cos\theta - \cos\theta_1}d\theta = \frac{U_\infty}{\pi}\sum n\, A_n\, G_n, \tag{4.35}$$

with Glauert's integral already defined in Chapter 3 as

$$G_n = \frac{\pi \sin n\theta_1}{\sin\theta_1}.$$

Example: Obtain the downwash for the modified elliptic loading as given in Equation (4.26).

Answer: Here, the general loading distribution reduces by using Equation (4.28)

$$A_1 = (1+\lambda)\Gamma_0 \;\text{ and }\; A_3 = \lambda\Gamma_0,$$

with $\Gamma_0 = 4sU_\infty$ and all other terms as zero. Furthermore using Glauert's integral in Equation (4.35), one obtains the downwash as

$$w(\theta_1) = \frac{\Gamma_0 \pi}{4\pi s}\left[A_1 + 3A_3 \frac{\sin 3\theta_1}{\sin \theta_1}\right] = \frac{\Gamma_0}{4s}\left[(1+\lambda) + 3\lambda\left\{3 - 4\sin^2\theta\right\}\right]$$

$$= \frac{\Gamma_0}{4s}\left[1 + \lambda + 9\lambda - 12\lambda\left(1 - \frac{y^2}{s^2}\right)\right] = \frac{\Gamma_0}{4s}\left[1 - 2\lambda + 12\lambda\left(\frac{y}{s}\right)^2\right].$$

4.12.2 Induced Drag on a Finite Wing for General Loading

For the general loading of the wing given by Γ and the downwash $w(\theta)$ acting on the finite wing, the induced drag is obtained from

$$D_v = \int_{-s}^{s} \rho\, w\, \Gamma\, dy.$$

Using the expression of Γ from Equation (4.30) and $w(\theta)$ from Equation (4.35) in the above, one gets the induced drag from

$$D_v = \int_0^\pi \rho \frac{U_\infty \sum n\, A_n \sin n\theta}{\sin\theta} 4s\, U_\infty \sum A_m \sin m\theta\; s\; \sin\theta\, d\theta.$$

Or

$$D_v = 4\rho\, U_\infty^2\, s^2 \int_0^\pi \sum_n A_n \sin n\theta \sum_m A_m \sin m\theta\, d\theta$$

$$= 2\rho\, U_\infty^2\, s^2 \sum n\, A_n^2\, \pi.$$

Writing the induced drag in terms of a coefficient as

$$D_v = C_{D_v} \frac{1}{2} \rho\, U_\infty^2\, S,$$

one gets

$$C_{D_v} = \pi AR \sum n\, A_n^2. \tag{4.36}$$

From Equation (4.34), $A_1^2 = \frac{C_L^2}{\pi^2 AR^2}$, and hence the induced drag coefficient can be expressed as

$$C_{D_v} = \frac{C_L^2}{\pi AR} \sum n \left(\frac{A_n}{A_1}\right)^2.$$

Thus, the induced drag for general loading distribution can be obtained for symmetric loading from

$$C_{D_v} = \frac{C_L^2}{\pi AR}\left[1 + \left(\frac{3A_3^2}{A_1^2} + \frac{5A_5^2}{A_1^2} + \frac{7A_7^2}{A_1^2} + \cdots\right)\right] = \frac{C_L^2}{\pi AR}\left[1 + \delta\right]. \quad (4.37)$$

As an example, consider the modified elliptic loading for which one gets the following simplified induced drag coefficient expression by retaining A_1 and A_3 terms only as

$$C_{D_v} = \frac{C_L^2}{\pi AR}\left[1 + \frac{3A_3^2}{A_1^2}\right] = \frac{C_L^2}{\pi AR}\left[1 + \frac{3\lambda^2}{(1+\lambda)^2}\right].$$

4.12.3 Load Distribution for Minimum Drag

For the general loading distribution, we have obtained the induced drag coefficient as

$$C_{D_v} = \pi AR\left[A_1^2 + 2A_2^2 + 3A_3^2 + 4A_4^2 + \cdots\right]. \quad (4.38)$$

This drag is a minimum, when $A_2 = A_3 = \cdots = A_n = \cdots = 0$ and $A_1 \neq 0$.
For such a distribution,

$$\Gamma = 4s\, U_\infty\, A_1 \sin\theta \quad \text{with} \quad y = -s\cos\theta.$$

Thus

$$\Gamma = 4s\, U_\infty\, A_1 \sqrt{1 - \left(\frac{y}{s}\right)^2}.$$

which is nothing but the elliptic distribution.

4.13 Asymmetric Loading: Rolling and Yawing Moment

Consider the asymmetric loading indicated in Figure 4.9, for the irrotational potential flow model. From previous sections, we noted that one gets an associated induced drag, which also takes an asymmetric form, whose distribution is also sketched in the figure. Due to the asymmetric loading, there will be a rolling moment indicated by L_R in the figure. At the same time, due to asymmetric drag grading, there will be a yawing moment, N, as also indicated in the figure.

4.13.1 Rolling Moment (L_R)

Due to the asymmetric loading, for any strip of width dy, located at a distance y, the differential lift acting is given by

$$dL = \rho\, U_\infty\, \Gamma\, dy,$$

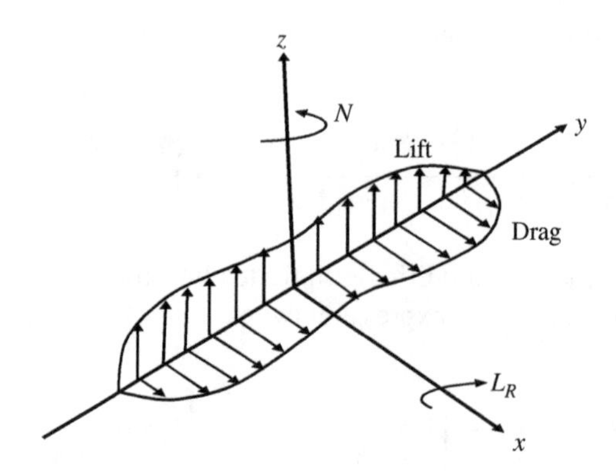

Figure 4.9 Asymmetric spanwise loading giving rise to rolling and yawing moments

which will contribute to the differential rolling moment. With the sign convention that starboard wing going down as positive, this rolling moment is given by

$$dL_R = -dL\ y.$$

The total rolling moment acting on the wing is obtained as

$$L_R = -\int_{-s}^{s} \rho\, U_\infty\, \Gamma\, y\, dy. \tag{4.39}$$

Using the loading distribution given in Equation (4.30), this simplifies to

$$L_R = 4\rho s\, U_\infty^2 \int_0^\pi \sum s^2\, A_n \sin n\theta \cos\theta\ \sin\theta\ d\theta.$$

Upon simplification, one gets

$$\begin{aligned} L_R &= 2\rho s^3\, U_\infty^2 \int_0^\pi \sum A_n \sin n\theta\ \sin 2\theta\ d\theta \\ &= \rho\, U_\infty^2\, s^3 \left[\sum A_n \left(\frac{\sin(n-2)\theta}{n-2} - \frac{\sin(n+2)\theta}{n+2}\right)\right]_0^\pi. \end{aligned}$$

Writing the rolling moment in terms of the rolling moment coefficient C_{lR} as

$$\rho s^3\, U_\infty^2\, A_2 \pi = C_{lR} \frac{1}{2}\, \rho\, U_\infty^2\, S\, s.$$

Thus

$$C_{lR} = \frac{\pi}{2} AR\, A_2. \tag{4.40}$$

Thus, the rolling moment is given by the first even harmonic of the general loading distribution.

4.13.2 Yawing Moment (N)

As noted in Figure 4.9, the asymmetric loading is associated with asymmetric induced drag. This asymmetric drag grading will create a net yawing moment (N) which can be calculated from D_v as

$$dN = dD_v\, y,$$

where $dD_v = \rho\, w\, \Gamma\, dy$. This can be integrated over the whole wing span to get

$$N = \int_{-s}^{s} \rho\, w\, \Gamma\, y\, dy,$$

which can be simplified as

$$N = -4s^3 \rho\, U_\infty^2 \int_0^\pi \frac{\sum n\, A_n \sin n\theta}{\sin\theta} \sum A_m \sin m\theta\, \cos\theta \sin\theta\, d\theta.$$

Equating this with the yawing moment expressed in terms of moment coefficient one gets

$$N = C_N \frac{1}{2} \rho\, U_\infty^2\, Ss.$$

Thus

$$C_N = 2AR \int_0^\pi \sum n\, A_n \sin n\theta \sum A_m \sin m\theta \cos\theta\, d\theta.$$

The integral on the right hand side has nonzero value only for $m = n + 1$, thus

$$C_N = \frac{\pi}{2}\, AR \left[3A_1\, A_2 + 5A_2\, A_3 + 7A_3\, A_4 + \cdots + (2n+1)\, A_n\, A_{n+1} \right]. \tag{4.41}$$

4.13.3 Effect of Aspect Ratio on Lift Curve Slope

Consider the (C_L, α) plot shown in Figure 4.10, where the lift curve slope of the finite wing is defined by $a = \frac{dC_L}{d\alpha}$. Additional subscript ∞ refers to conditions for an infinite aspect ratio wing. It is understood from the previous discussion that the infinite aspect ratio wing will be more efficient aerodynamically in generating lift. Hence, the lift curve slope of a 3D wing will also be lower as compared to a 2D wing.

Thus from Figure 4.10, to generate the same lift coefficient, C_L, we must have

$$C_L = a_\infty[\alpha_\infty - \alpha_0] = a[\alpha - \alpha_0],$$

where α_0 is the zero lift angle. We have already noted that the finite aspect ratio wing induces a downwash characterized by the tilting of the lift backwards by an angle ϵ so that $\alpha_\infty = \alpha - \epsilon$

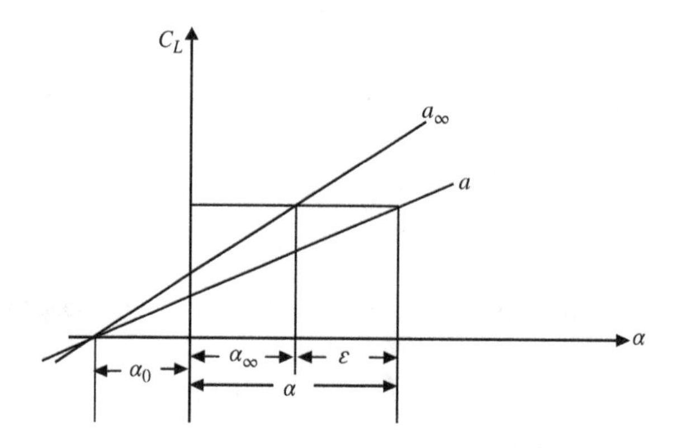

Figure 4.10 Lift curves for the finite AR wing and the equivalent infinite aspect ratio wing

and

$$C_L = a_\infty\,[(\alpha - \alpha_0) - \epsilon] = a(\alpha - \alpha_0).$$

Thus,

$$a = a_\infty\left[1 - \frac{\epsilon}{(\alpha - \alpha_0)}\right]. \tag{4.42}$$

Now for elliptic load distribution $\epsilon = \frac{w}{U_\infty} = \frac{\Gamma_0}{4s\,U_\infty}$ with $\Gamma_0 = \frac{C_L\,U_\infty\,S}{\pi s}$.
Therefore

$$\epsilon = \frac{C_L S}{\pi 4 s^2} = \frac{C_L}{\pi AR}.$$

Using this in Equation (4.42), we get

$$a = a_\infty\left[1 - \frac{C_L}{\pi AR(\alpha - \alpha_0)}\right] = a_\infty\left[1 - \frac{a}{\pi AR}\right]. \tag{4.43}$$

Or

$$a = \frac{a_\infty}{1 + a_\infty/\pi AR}. \tag{4.44}$$

For a finite wing of general planform, the above equation is modified as

$$a = \frac{a_\infty}{1 + \frac{a_\infty}{\pi AR}(1 + \tau)}. \tag{4.44a}$$

The value of τ as a function of A_n in the loading distribution has been calculated by Glauert (1926) and usually this value lies between 0.05 and 0.25.

4.14 Simplified Horse Shoe Vortex

The chordwise loading distribution can be obtained by thin airfoil theory developed in Chapter 3. This sectional information can be used for the finite wing theory considered so far by bound, trailing and starting vortices, out of which the first two elements are important to calculate the lift of the finite wing, downwash and induced drag. Further significant simplification to the model can be brought about if one can replace the bundle of bound vortex by a single bound spanwise vortex of constant strength which is turned through 90° at each end to form the semi-infinite trailing vortices behind the wing. We note that this simplification is only proposed and possible for symmetric loading.

Such a replacement must satisfy that the simplified bound vortex to account for the same total lift and the same must be true of the equivalent trailing vortex. To satisfy the additional condition that both the systems have the same circulation at mid span, the span of the simplified horse shoe vortex must have a different span. The schematics of the actual and equivalent vortex system are shown in Figure 4.11.

Note that both the systems have the same mid-span circulation Γ_0. The span of the simplified horse shoe system has a span of $2s'$, which can be obtained from

$$\Gamma_0\, 2s' = \frac{\text{Total lift}}{\rho\, U_\infty}.$$

Thus, the ratio of the actual semi-span to the semi-span of the equivalent system is given by

$$\frac{s'}{s} = \frac{\text{Total lift}}{2\rho s\, U_\infty\, \Gamma_0}. \tag{4.45}$$

We have noted earlier for general loading distribution from Equation (4.33) that the total lift is given by

$$L = 2\pi\rho\, U_\infty^2\, s^2\, A_1. \tag{4.33}$$

The mid-span circulation is obtained from Equation (4.30) by substituting $\theta = \pi/2$ to obtain

$$\Gamma_0 = 4s\, U_\infty \sum A_n \sin n\, \pi/2. \tag{4.46}$$

Substituting Equations (4.33) and (4.46) in Equation (4.45) one gets

$$\frac{s'}{s} = \frac{2\pi\rho\ U_\infty^2\ s^2\ A_1}{2\rho\, U_\infty^2\, 4s^2 \sum A_n \sin n\, \pi/2} = \frac{\pi}{4}\,\frac{A_1}{A_1 - A_3 + A_5 - A_7 \cdots}. \tag{4.47}$$

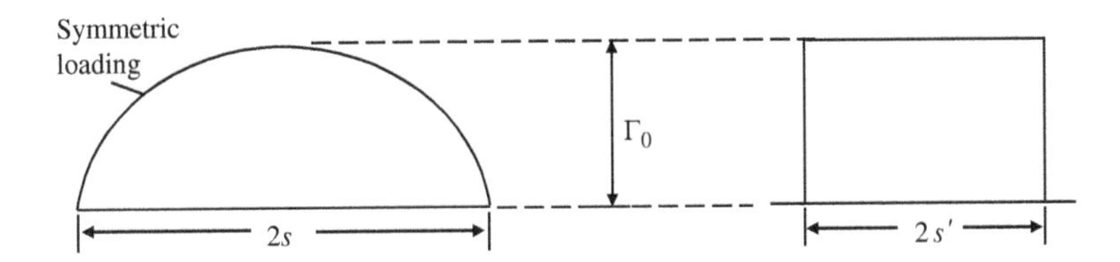

Figure 4.11 Actual spanwise loading of circulation (left) and its equivalent horse shoe vortex (right)

For elliptic loading distribution

$$A_3 = A_5 = A_7 = \cdots = 0.$$

Therefore the equivalent description of elliptic loading by a constant bound vortex is of span given by the ratio

$$\frac{s'}{s} = \frac{\pi}{4}.$$

4.15 Applications of Simplified Horse Shoe Vortex System

In the following subsections, we discuss the utility of the simplified horse shoe vortex system, which assists us in quickly estimating the consequences of finite wing theory. We note, however, that the simplified horse shoe vortex system has exaggerated trailing vortex system, which is constructed based on vortex continuity and is not on the same logical ground the general distribution is developed by Prandtl-Lanchester theory. Nobody would notice such a strong trailing vortex as used in this simplified model. While the model of Prandtl-Lanchester depicts a varying load distribution with the more intense shedding near the wing-tip, as can be seen behind the wake of an wing of an aircraft taking off on a winter morning. A vortex sheet is a discontinuity and such discontinuity is more prominent on a wing near the tip, than near the root.

4.15.1 Influence of Downwash on Tailplane

This is often needed for calculating the effects of downwash on the horizontal stabilizer. Created downwash from the main wing affects the performance of the horizontal tail plane making its control effectiveness less desirable. To calculate such effects we replace the lifting wing by the equivalent simplified horse shoe vortex of span $2s'$ and let us say that the tailplane is located at a distance l, aft of the wing, which is represented by the bound vortex. The effect on the tailplane is calculated on the midplane of the aircraft at the quarter chord of the horizontal tailplane, located at P in Figure 4.12. In this figure, the lifting wing is replaced by its equivalent horse shoe vortex system. Let the point P makes an angle θ_L with the end of the trailing vortex meeting the bound vortex.

The induced downwash at P is due to two contributions. The first contribution is from the bound vortex and the second contribution comes from the two trailing vortex elements, with both of the trailing vortex element producing identical effects and the net downwash is given by

$$w_p = \frac{\Gamma_0}{4\pi l} 2 \sin\theta_L + \frac{2\Gamma_0}{4\pi s'}(1 + \cos\theta_L)$$

$$= \frac{\Gamma_0}{2\pi}\left[\frac{\sin\theta_L}{l} + \frac{1 + \cos\theta_L}{s'}\right].$$

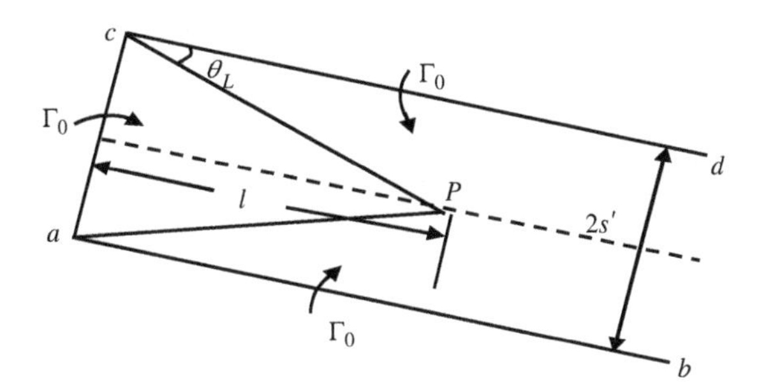

Figure 4.12 Effects of wing downwash on tailplane located at a distance l behind the bound vortex of the wing

From the figure $l = s' \cot\theta_L$ and $s' = s\pi/4$.
Thus the induced downwash at P is given by

$$w_p = \frac{2\Gamma_0}{\pi^2 s}[1 + \sec\theta_L].$$

But as

$$\Gamma_0 = \frac{C_L U_\infty S}{\pi s},$$

the downwash tilts the lift by the angle obtained for this simplified model as

$$\epsilon = \frac{w_p}{U_\infty} = \frac{8C_L}{\pi^3 AR}[1 + \sec\theta_L].$$

In studying stability and control of an aircraft, the quantity that appears in estimating the effectiveness of the horizontal tail plane is given by the following derivative

$$\frac{\partial\epsilon}{\partial\alpha} = \frac{\partial\epsilon}{\partial C_L}\frac{dC_L}{d\alpha}.$$

Here, the lift curve slope of the main wing is needed in the second factor, which we identify as a. Thus, the tailplane effectiveness factor is given by

$$\frac{\partial\epsilon}{\partial\alpha} = a\frac{\partial\epsilon}{\partial C_L} = \frac{8a}{\pi^3 AR}(1 + \sec\theta_L). \quad (4.48)$$

4.15.2 Formation-flight of Birds

This is an effect which shows the benefit of vortex system, often used in the natural world by birds, while flying in formation during long flights. In fact, this is a case when the birds use the upwash effect of the vortex system of birds flying in unison, in specific formation.

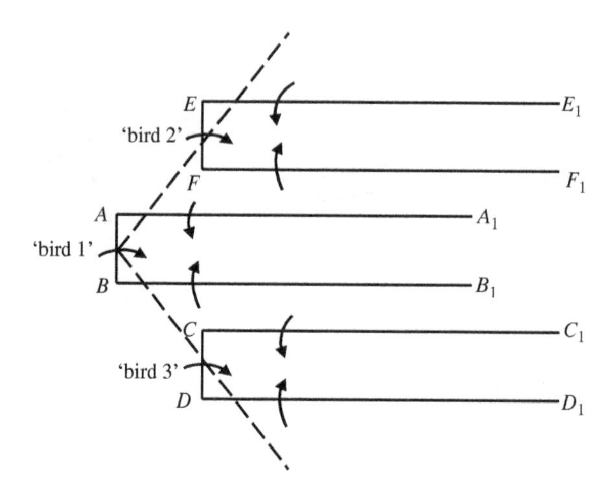

Figure 4.13 Three birds flying in V-formation, represented by simplified horse shoe vortex model

Earlier we have noted that downwash causes the lift vector to tilt backward with respect to the oncoming flow direction, thereby reducing the lift marginally. But more importantly, the backward lift also has the component against the wind direction, causing the induced drag. Now if a lifting surface works in the presence of upwash, then the lift instead of tilting backward, will be made to tilt forward. Such tilt in the opposite direction will not only decrease drag, it can even create a forward thrust.

When birds are flying in a group, in a situation where the neighbouring birds' equivalent vortex systems (considered as a finite wing model) create a net upwash, this upwash reduces induced drag of the individual birds in the group. This is achieved by flying in a 'V'-formation. An interesting study of the phenomenon is reported in Lissaman and Shollenberger (1970), which provides details of the formation's spacing and resultant drag reduction. It is noted that more benefit accrues with the number of birds increasing in the formation. Heuristically, more benefit also accrues with the spacing reduced.

This is explained qualitatively below by just showing the flight of three birds in formation, in Figure 4.13. Replace each bird by its equivalent horse shoe vortex system. as shown for the three birds in 'V'-formation, in this figure. The arrows on the bound and trailing vortices indicate the circulation direction.

Thus, it is very apparent that an individual vortex system creates a downwash in a region enclosed by the bound and trailing vortex system. For example, looking at the bound vortex of any particular bird, one will notice the streamlines and the induced velocity by the trailing vortex elements, as shown in Figure 4.14, if it was flying solo. The effective regions of upwash and downwash are marked in this figure, in the front and plan views of any one bird flying solo.

Now concentrating on 'bird 1' in the formation, represented by AB in Figure 4.13, the effect of the bound vortices CD and EF is to produce an upward velocity or upwash. Again at AB, the effect of trailing vortices CC_1 and FF_1 is to create an upwash, while DD_1 and EE_1 create downwash. However, induced downwash is lower compared to upwash for AB because of distances involved in Biot-Savart law. As a result, the 'bird 1' experiences a net upward velocity, the result of which will reduce the rearward tilt of the lift (as experienced by

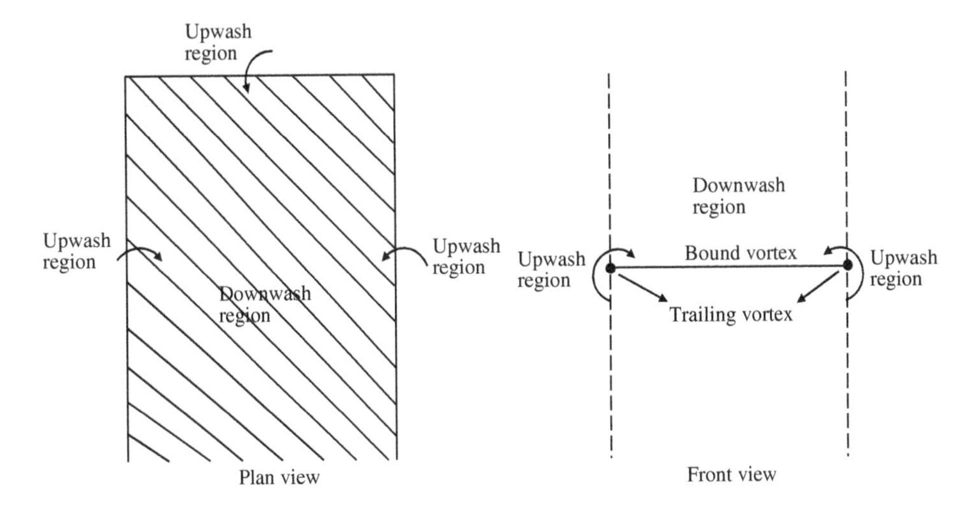

Figure 4.14 Zone of upwash and downwash due to an individual horse shoe vortex system

the bird flying solo), due to the presence of 'bird 2' and 'bird 3'. Thus, the lead bird will be affected constructively by the presence of all the other birds following it.

The scenario is similar for other individual birds in the 'V'-formation. In Figure 4.13, both the 'bird 2' and 'bird 3' will experience identical beneficial effects. We can reason in a similar manner to account for this with the only difference being that the bound vortices of 'bird 2' and 'bird 3'will have no effects on each other, as they are collinear. However, the bound vortex AB of 'bird 1' will cause downwash on 'bird 2' and 'bird 3'. However, we note that the trailing vortices play more vital role in creating upwash on each member of the flock, rather than the bound vortices. Due to closer proximity of the trailing vortex elements to any bound vortices, the Biot-Savart law dictates the intensity of such interactions.

4.15.3 Wing-in-Ground Effect

This is another example of beneficial effects of the vortex system of a finite aspect ratio wing, flying in close proximity of the ground. If we represent the flying wing by an equivalent vortex system, then the aircraft flying close to the ground (at a height of about two to three wing span distance) will experience a constructive interference of image horse shoe vortex system of the actual wing horse shoe vortex system. This is discussed here with reference to Figure 4.15, where the aircraft is at a height H from the ground. The front view shows the bound vortex AB representing the aircraft wing, whose image system is indicated by $A'B'$, also at the distance H. Strength of the original and the image vortex system is given by Γ_0 and width of the bound vortex is given by $2s'$, which is lower compared to the wing span, $2s$.

To account for the mitigating effect of the image system, consider a general point P on the bound vortex, which is at a distance y along the span of the bound vortex of the wing. The line joining P with A' and B' subtend angles, θ_L and θ_R with the y-axis. If PA' and PB' are $r_L = \sqrt{4H^2 + (y + s')^2}$ and $r_R = \sqrt{4H^2 + (s' - y)^2}$, respectively, then the induced azimuthal velocities at P by the image of the trailing vortices are obtained from Equation (4.11)

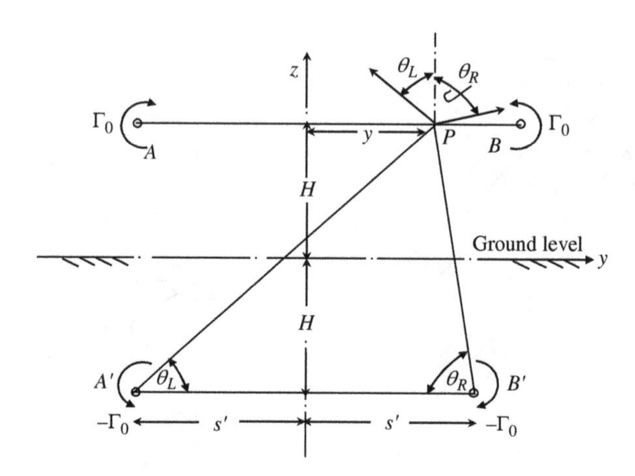

Figure 4.15 Sketch showing the ground effect for an aircraft flying close to ground

as $\frac{\Gamma_0}{4\pi r_L}$ and $\frac{\Gamma_0}{4\pi r_R}$, respectively. The directions of the induced velocities are as indicated in the figure. It is noted that the horizontal components of these induced velocities have a cancelling effects, as these are in opposite direction. However, the vertical components for these induced velocities add up and are given by

$$dw(y) = \frac{\Gamma_0}{4\pi}\left[\frac{\cos\theta_L}{r_L} + \frac{\cos\theta_R}{r_R}\right]. \tag{4.49}$$

In terms of the Cartesian coordinate system, the above incremental upwash is given by

$$dw(y) = \frac{\Gamma_0}{4\pi}\left[\frac{(s'+y)}{r_L^2} + \frac{(s'-y)}{r_R^2}\right]. \tag{4.50}$$

On a spanwise strip of width dy, the lift acting is given by

$$\text{Lift} = \rho\, U_\infty\, \Gamma_0\, dy$$

and the change in induced drag by the upwash in Equation (4.50) is given by

$$dD_v = \text{Lift}\,\frac{dw(y)}{U_\infty},$$

i.e.

$$dD_v = \rho\, dw(y)\, \Gamma_0\, dy.$$

Total change in induced drag is then obtained by integrating the above

$$\Delta D_v = -\int_{-s'}^{s'} \frac{\rho\,\Gamma_0^2}{4\pi}\left[\frac{(s'+y)}{r_L^2} + \frac{(s'-y)}{r_R^2}\right] dy.$$

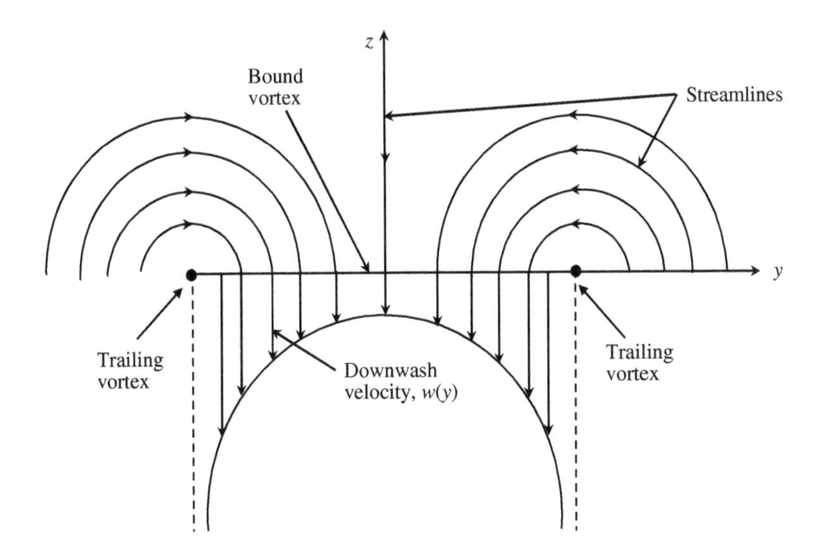

Figure 4.16 Induced velocity distribution due to a bound vortex

As the integrand is symmetric about $z = 0$, the net change in induced drag is given by

$$\Delta D_v = -2\int_0^{s'} \frac{\rho\,\Gamma_0^2}{4\pi}\left[\frac{(s'+y)}{r_L^2} + \frac{(s'-y)}{r_R^2}\right] dy. \tag{4.51}$$

Evaluating the integral one obtains

$$\Delta D_v = -\frac{\rho\,\Gamma_0^2}{2\pi}\left[\text{Ln}\frac{4H^2+(s'+y)^2}{4H^2+(s'-y)^2}\right|_0^{s'}.$$

Upon simplification, one gets

$$\Delta D_v = -\frac{\rho\,\Gamma_0^2}{4\pi}\text{Ln}\left[1+\frac{s'^2}{H^2}\right]. \tag{4.52}$$

Considering elliptic loading distribution over the wing, one substitutes $s' = \frac{\pi}{4}s$ in the above.

4.16 Prandtl's Lifting Line Equation or the Monoplane Equation

Consider a conventional unswept finite aspect ratio wing flying at incompressible flow speed. Using potential flow assumption and using Kutta-Jukowski theorem for the generated lift, we can express the lift coefficient of the wing as

$$C_L = \frac{\rho\,U_\infty\Gamma}{1/2\rho\,U_\infty^2\,c} = \frac{2\Gamma}{U_\infty\,c}. \tag{4.53}$$

With reference to Figure 4.10, we note that

$$C_L = a_\infty(\alpha_\infty - \alpha_0) = a(\alpha - \alpha_0) \text{ with } \alpha_\infty = \alpha - \epsilon.$$

Thus, we can express the lift coefficient as

$$C_L = \frac{2\Gamma}{U_\infty\, c} = a_\infty(\alpha - \alpha_0 - \epsilon).$$

Or

$$\frac{2\Gamma}{a_\infty\, c} = U_\infty(\alpha - \alpha_0) - U_\infty\, \epsilon. \tag{4.54}$$

Recall from Equation (4.14)

$$\epsilon = \frac{1}{4\pi U_\infty}\int \frac{d\Gamma/dy}{y - y_1} dy.$$

From Equation (4.54)

$$C_L\, c = \frac{2\Gamma}{U_\infty}$$

and therefore

$$\frac{d\Gamma}{dy} = \frac{U_\infty}{2}\frac{d\,(C_L c)}{dy}. \tag{4.55}$$

Thus, from Equations (4.53) and (4.54)

$$\frac{2\Gamma}{c\, a_\infty} = U_\infty\,(\alpha - \alpha_0) - \frac{1}{4\pi}\int \frac{d\Gamma/dy}{y - y_1}\, dy.$$

For a general load distribution given that

$$\Gamma(y_1) = 4s\, U_\infty \sum A_n \sin n\theta_1,$$

we obtain from above the following equation upon using Equation (4.35):

$$\frac{8s\, U_\infty}{a_\infty\, c}\sum A_n \sin n\theta_1 = U_\infty(\alpha - \alpha_0) - \frac{U_\infty}{\sin\theta_1}\sum n\, A_n \sin n\theta_1. \tag{4.56}$$

Introducing $\mu = \frac{a_\infty\, c}{8s}$, we rewrite Equation (4.56) as

$$\frac{1}{\mu}\sum A_n \sin n\theta_1 = (\alpha - \alpha_0) - \frac{\sum n\, A_n \sin n\theta_1}{\sin\theta_1}.$$

Alternatively

$$\mu(\alpha - \alpha_0) = \sum A_n \sin n\theta_1 \left[1 + \frac{\mu n}{\sin\theta_1}\right]. \tag{4.57}$$

This is known as the monoplane equation or the Prandtl's lifting line equation.
Note that the downwash integral has been simplified in terms of Glauert's integral by

$$\frac{1}{4\pi}\int \frac{d\Gamma/dy}{y - y_1}\,dy = \frac{1}{4\pi s}\int \frac{d\Gamma}{\cos\theta - \cos\theta_1}$$

$$= \frac{4s\,U_\infty}{4\pi s}\sum \int \frac{\cos n\theta}{\cos\theta - \cos\theta_1}\,d\theta$$

$$= U_\infty \sum \frac{n\,A_n \sin n\theta_1}{\sin\theta_1}.$$

Bibliography

Glauert, H. (1926) *The Elements of Aerofoil and Airscrew Theory.* Cambridge University Press, UK.
Lanchester, F.W. (1908) *Aerodonetics.* Constable, UK.
Lissaman, P.B. and Shollenberger, C.A. (1970) Formation flight of birds. *Science.* **168**(3934), 1003–1005.
Prandtl, L. (1921) *Application of Modern Hydrodynamics to Aeronautics. NACA TR* **116**.

5

Panel Methods

5.1 Introduction

If the flow is inviscid and irrotational, then the governing equation is

$$\nabla^2 \phi = 0. \tag{5.1}$$

This Laplace's equation was derived from the continuity equation, as given in Equation (2.31). We also noted in Chapter 3 that the governing equation for irrotational flow can also be written down with a Laplace's equation for the stream function given in Equation (3.4). For two-dimensional flows, ϕ and ψ being analytic function of (x, y), we can define a complex potential (W) as a function of $z\ (= x + iy)$. Subsequently, with the help of conformal mapping, we developed a general thin airfoil theory for two-dimensional flow past an airfoil. This is essentially valid for thin airfoil as the solution procedure required small perturbation assumption to hold, which is consistent with the governing linear equation. However in the process, the thickness of the airfoil was not incorporated and only the angle of attack and camber decided upon the lift generation process.

The two-dimensional potential flow model was refined for finite wing in Chapter 4, but once again the thickness of the geometry was not included. Both the thin airfoil theory and finite wing theory perform poorly near the stagnation point. Also these methods do not work for multi-element airfoils. All of these problem are eliminated within the irrotational flow assumption for the full aircraft, if we use panel method. This is the topic of discussion in this chapter.

However before we discuss about panel method, we need to comment about the property and solution of Equation (5.1). This is essentially a boundary value problem and belongs to elliptic partial differential equation class. Any elliptic PDE will have characteristics which occur in pair, each a complex conjugate of the other. Thus, this class of problems will have the highest derivative given by ∇^{2n}. For such an equation, one requires n boundary conditions. For Equation (5.1), we must have one boundary condition prescribed on all the segment of the boundary. A schematic of the problem is shown in Figure 5.1, for a 2D problem for the ease of understanding.

Theoretical and Computational Aerodynamics, First Edition. Tapan K. Sengupta.

Companion Website: www.wiley.com/go/sengupta

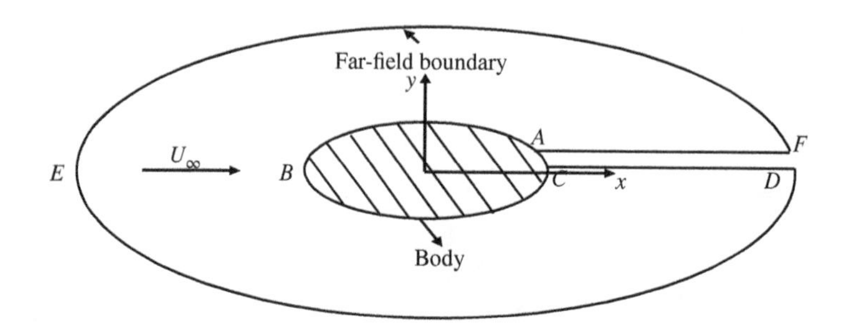

Figure 5.1 Schematic of the boundary value problem for Laplace's equation

In the figure, the body ABC is placed in a uniform flow U_∞ and let us say that we are interested in obtaining the solution in the domain inside the region marked as the far-field boundary. To make the flow domain simply-connected (i.e. to make the solution unique), we introduce a mathematical cut shown as CD and AF, which is nothing but the same line. The boundary for this case is the continuous line $ABCDEFA$. However for computational purposes, it is often considered to consist of ABC as the body, DEF as the far-field boundary and CD and AF as the periodic cut. This is due to the fact that physically the nature of the boundary conditions applied on each of the segments is qualitatively different.

For example, in solving Equation (5.1), we apply no-penetration condition for the normal velocity component on ABC; with vanishing perturbation at the far-field boundary, DEF and all variables are periodic between CD and AF.

To obtain the solution for general body shape, including the thickness of the body by panel method, we distribute elementary singularities studied in Chapter 3, which automatically satisfy Equation (5.1). Additionally, the solution for these elementary singularities also have the property of satisfying far-field boundary conditions required for the solution of Equation (5.1). Thus, one is left with the task of satisfying the body boundary conditions and the Kutta condition (for lifting surfaces) to fix the circulation only.

In panel methods, sources, doublets, vortices etc. are distributed either as basic units or in combination of these over the physical surface, which is segmented over small panels, which is the reason for the name as panel method. The boundary condition on the wing/body and the Kutta condition at the trailing edge are used to fix the strength of such distributions. It is noted that there is no unique distribution for such singularities and the accuracy of methods rest on using the requisite number of panels and their relative placement. The following issues are pertinent for the choice of singularities.

Sources and sinks can only represent symmetric flow, i.e. symmetric bodies at $\alpha = 0$, for nonlifting cases. From now on, we will not distinguish between source and sink, with the implicit understanding that the sink is a source with an opposite sign. For lifting cases, one must have doublet or vortices distributed over the panels. There are many methods currently in use. We will only discuss the method due to Hess and Smith (1967). But before doing this, we will discuss some issues related to singularity distribution.

5.2 Line Source Distribution

A string of sources are distributed along the x-axis for a two-dimensional body, as shown in Figure 5.2. Let m_i be the source strength located at x_i, then the stream function is given at a

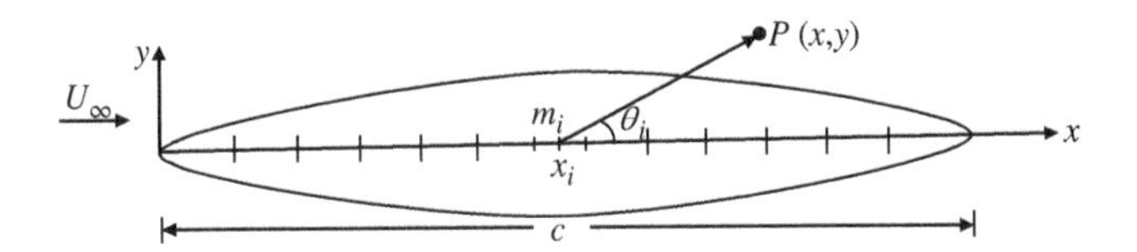

Figure 5.2 Source distribution for a slender body along the axes to solve Equation (5.1) by panel method

field point P with coordinate (x, y) by

$$\Psi = U_\infty\, y + \left(\sum_i \frac{m_i}{2\pi} \tan^{-1} \frac{y}{x - x_i} \right). \tag{5.2}$$

The second part of Equation (5.2), ψ_s is due to the effects of the elementary source distribution, as the first part is due to a uniform flow from left to right. Instead of a discrete distribution of sources, if the source was continuously distributed, then one can replace the summation by an integral for the stream function and velocity potential due to elementary source distribution as

$$\psi_s = \frac{1}{2\pi} \int m(t) \tan^{-1} \frac{y}{x - t} dt \tag{5.3}$$

$$\phi_s = \frac{1}{2\pi} \int m(t)\, \mathrm{Ln}\left[(x - t)^2 + y^2 \right] dt. \tag{5.4}$$

Let $Y(x)$ denote the closed body in Figure 5.2. Note that θ represents the angle subtended by the line joining the field point P to the source $m(t)$. The upstream stagnation point makes an angle π with the x-axis, while the downstream stagnation point makes $\theta = 0°$ with the x-axis. Both the stagnation points are located on the x-axis, i.e. for $y = 0$.

Thus, at the upstream stagnation point $\tan^{-1} \frac{y}{x-t} = \pi$ and at the downstream stagnation point $\tan^{-1} \frac{y}{x-t} = 0$.

From Equation (5.2), if the body is represented by the streamline $\Psi = 0$, then for the upstream and downstream stagnation points

$$\Psi = 0 = 0 - \int_0^c \frac{m(t)dt}{2\pi} \pi \tag{5.5}$$

and

$$\Psi = 0 = 0 - \int_0^c \frac{m(t)dt}{2\pi} 0. \tag{5.6}$$

While Equation (5.6) is trivially satisfied for a closed body, the source distribution must be such that is given by Equation (5.5) as

$$\int_0^c m(t)\, dt \equiv 0.$$

This is a necessary condition for source/sink distribution for any closed body, i.e. the sum of source strength must be equal to the sum of all the sinks needed to represent a closed body.

5.2.1 Perturbation Velocity Components due to Source Distribution

From Equations (5.3) and (5.4), one can calculate the perturbation components of the induced velocity due to source distribution of Figure 5.2 as

$$u_s(x, y) = \frac{1}{2\pi} \int_0^c m(t) \frac{x - t}{(x - t)^2 + y^2} dt \tag{5.7}$$

$$v_s(x, y) = \frac{1}{2\pi} \int_0^c m(t) \frac{y}{(x - t)^2 + y^2} dt. \tag{5.8}$$

The subscript on the left-hand side indicate the contributions due to the source distribution. As we are dealing with potential flow with linear governing equation, we can linearize the body boundary condition as

$$v_s \cong U_\infty \frac{dY}{dx}. \tag{5.9}$$

In Equation (5.8), the function $f(t) = \frac{y}{(x-t)^2+y^2}$ works like a sampling function or a distribution. This can be understood by considering a slender body, such that $y \to \epsilon$. Let us plot $f(t)$ against t, as shown in Figure 5.3. The function $f(t)$ acts like the Dirac's's comb, since f has a large peak only at $t = x$ and is small everywhere else. Then multiplying $m(t)$ with $f(t)$ and integrating will have dominant contribution coming only from $t = x$. Note when we use it for thin airfoil, i.e. for $(Y \to 0)$, then $f(t)$ acts exactly like a delta function. As a consequence, one can write Equation (5.8) as

$$\operatorname*{Lim}_{Y \to 0} v_s(x, Y) \cong \frac{m(x)}{2\pi} \int_{-\infty}^{\infty} \frac{Y}{(x - t)^2 + Y^2} dt. \tag{5.10}$$

In approximating Equation (5.10), we have taken $m(t)$ out of the integral and replaced it, by its value at $t = x$ and extended the limit from $-\infty$ to $+\infty$. The right-hand side of Equation (5.10) can be further simplified as

$$v_s \simeq \frac{m(x)}{2\pi} \tan^{-1} \frac{t - x}{Y} \Big|_{t=-\infty}^{\infty} = \frac{m(x)}{2\pi} \left[\frac{\pi}{2} - \left(-\frac{\pi}{2} \right) \right].$$

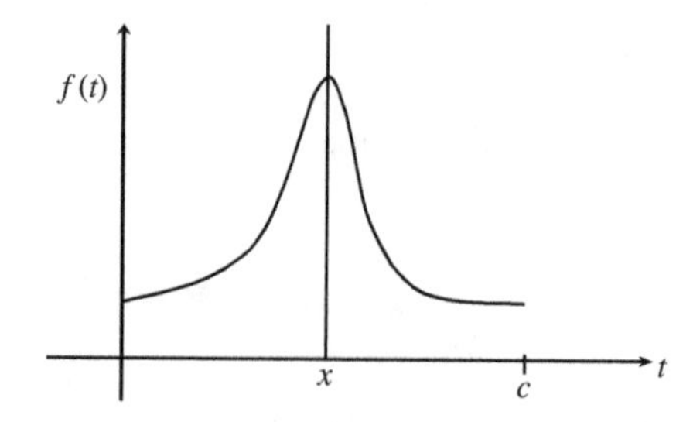

Figure 5.3 The variation of the function, f as a function of its argument, which appears in Equation (5.8)

Therefore, one gets

$$\operatorname*{Lim}_{Y\to 0} v_s = \frac{m(x)}{2}.$$

From Equation (5.9)

$$m(x) \simeq 2\,U_\infty \frac{dY}{dx}.$$

From Equation (5.7) we have

$$u_s = \frac{1}{2\pi}\int_0^c m(t)\frac{x-t}{(x-t)^2+Y^2}dt. \tag{5.10a}$$

Now for $Y \to 0$, we can simplify the above to get

$$u_s \cong \frac{1}{2\pi}\int_0^c m(t)\frac{dt}{(x-t)} = \frac{2U_\infty}{2\pi}\int \frac{Y'}{(x-t)}dt.$$

This approximation is good when t is not too close to x. On the slender body surface considered here, $Y \to 0$, and then

$$2\pi\, u_s(x, Y(x)) \simeq \int_0^{x-\epsilon} \frac{m(t)}{x-t}dt + \int_{x-\epsilon}^{x+\epsilon} \frac{(x-t)m(x)}{(x-t)^2+Y^2}dt + \int_{x+\epsilon}^{c} \frac{m(t)}{(x-t)}dt.$$

Let us perform a coordinate transformation $Z = (x - t)$, then the second integral in the above is

$$\simeq m(x)\int_{-\epsilon}^{\epsilon} \frac{-Z\,dZ}{Z^2+Y^2(x)}.$$

The integrand here is odd and limits are even and therefore the integral vanishes.

$$u_s(x, Y(x)) \simeq \frac{U_\infty}{\pi}\int_0^c \frac{Y'(t)dt}{(x-t)}. \tag{5.11}$$

and with the help of this perturbation streamwise velocity, using linearization in Equation (2.46), one can obtain the pressure coefficient as

$$C_p \simeq \frac{2u_s}{U_\infty}. \tag{5.12}$$

Note that in Equation (5.10a), if $m(t)$ is constant (M) then for $y \to \epsilon$

$$u_s = \frac{M}{2\pi}\int_0^c \frac{dt}{(x-t)} = \frac{M}{2\pi}\mathrm{Ln}\frac{x}{x-c}. \tag{5.10b}$$

From this expression, it is readily apparent that for constant strength source distribution, at $x = 0$ or c, the induced velocity displays a logarithmic singularity. This result in Equation (5.10b) will be used to explain Hess and Smith's method, discussed next.

5.3 Panel Method due to Hess and Smith

We have already noted that there are multitude of panel methods depending upon the type of singularities, variation of singularity strength and the distribution of these singularities. The first practical method was devised by Hess and Smith in mid-sixties and it is based on distribution of sources and vortices on the surface of the geometry in Hess and Smith (1964, 1967) and Katz and Plotkin (1992). Thus in this method, the velocity potential is given by

$$\phi = \phi_\infty + \phi_s + \phi_v, \tag{5.13}$$

where total velocity potential ϕ is composed of $\phi_\infty = U_\infty(x\cos\alpha + y\sin\alpha)$, due to the oncoming flow at an angle of attack, α; ϕ_s is due to the source distribution and ϕ_v is due to vorticity distribution. The last two contributions are due to varying source and vortex strengths, $q(s)$ and $\gamma(s)$ measured along the surface-conforming curve of each panels. Thus

$$\phi_s = \int \frac{q(s)}{2\pi}\text{Ln}\, r\, ds \quad \text{and} \quad \phi_v = -\int \frac{\gamma(s)}{2\pi}\theta\, ds. \tag{5.14}$$

Various geometric terms appearing in these expressions are defined in Figure 5.4. The values of $q(s)$ and $\gamma(s)$ are obtained from flow tangency body boundary condition, augmented by the Kutta condition. In describing thin airfoil theory, we noted that Kutta condition fixes the net circulation. We will use the same condition for different three-dimensional geometries, but with different criteria of zero loading at such edges. In Hess and Smith's method, Kutta condition is specified for airfoil by fixing constant strength vorticity over the whole surface, while flow tangency condition applied at as many points, as the number of unknown singularities to fix the source strength distribution. The source strength is considered fixed over each panel, but there is variation of strength of the sources from panel to panel. For such discrete number of panels, we can superpose the solution, as the governing equation is linear as

$$\phi = \phi_\infty + \sum_{j=1}^{N}\left[\frac{q_j(s)}{2\pi}\text{Ln}\, r_j - \frac{\gamma}{2\pi}\theta_j\right] ds. \tag{5.15}$$

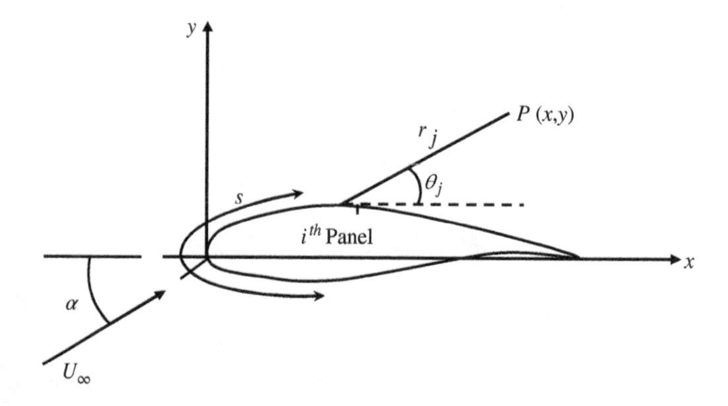

Figure 5.4 Schematic of panel method as applied for an airfoil for general lifting case

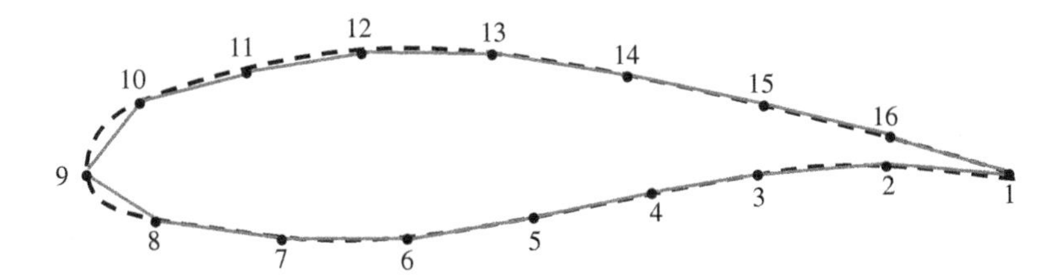

Figure 5.5 For the airfoil, straight panels approximate the curved airfoil surfaces, with the difference between the two minimized by taking more number of panels

In this method, the body surface is approximated by a collection of panels. In general, the integrals in Equation (5.14) are hard to evaluate, if the surfaces are curved. To avoid complications and reduce accumulation of errors, the panels on which the sources and vortices are distributed are taken as straight lines, as demonstrated in Figure 5.5.

Thus, we select a number of points (N) on the body contour, which are called the nodes and we connect such nodes by straight lines, which are the panels used in the method. Thus, the sources and vortices are distributed on straight line panels. Hess and Smith proposed that the source strength be constant on each panel and vorticity should be constant over all panels. In Equation (5.15), there are thus $(N + 1)$ unknowns: γ and $(q_j, j = 1, 2...N)$, which are solved by satisfying flow tangency condition at N control-points and satisfying the Kutta condition.

From the expressions in Equations (5.7) and (5.8) for $y = 0$ at $x \to 0$ or c; u_s and v_s are given by expressions which have singularities (logarithmic) when q is constant, as shown in Equation (5.10b). Thus the end of each panel, i.e. the nodes cannot be treated as control points. Generally the mid point of each panel (not the actual geometry) is used as control points. For Kutta condition, we equate the tangential velocities of the first and last panel at the respective mid-points.

We define the numerological sequence, as shown in Figure 5.5. Let the ith panel be defined by the ith and $(i + 1)^{th}$ nodes as shown in Figure 5.6. Let the inclination of the panel to the x-axis (of the global axis system) be θ_i, then

$$\sin\theta_i = \frac{y_{i+1} - y_i}{l_i} \quad \text{and} \quad \cos\theta_i = \frac{x_{i+1} - x_i}{l_i}.$$

with $l_i = [(x_{i+1} - x_i)^2 + (y_{i+1} - y_i)^2]^{1/2}$.

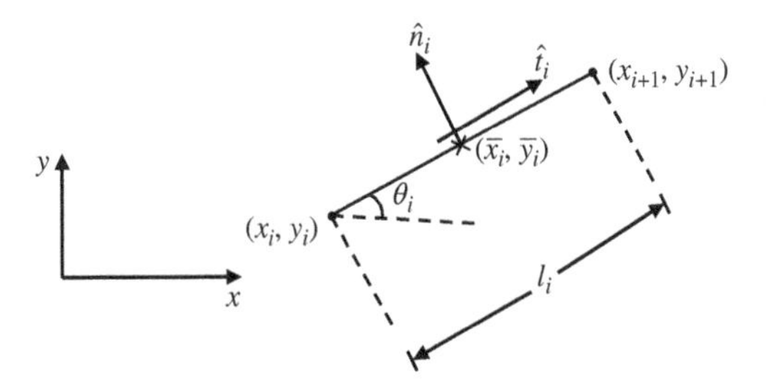

Figure 5.6 A sketch of a panel showing the notational convention in a global coordinate system

The unit normal and unit tangent of the ith panel are given by

$$\hat{n}_i = -\hat{i}\sin\theta_i + \hat{j}\cos\theta_i \tag{5.16}$$

$$\hat{t}_i = \hat{i}\cos\theta_i + \hat{j}\sin\theta_i. \tag{5.17}$$

The unit normal in Equation (5.16) is defined as positive, if it is directed outwards for the sequence of points adopted and the unit tangent in Equation (5.17) is from the ith to the $(i+1)^{th}$ point. These quantities are indicated in the sketch of a panel in Figure 5.6, once again in the global coordinate system.

The mid-point of the ith panel in Figure 5.6 is given by

$$\bar{x}_i = \frac{1}{2}(x_i + x_{i+1}) \text{ and } \bar{y}_i = \frac{1}{2}(y_i + y_{i+1}).$$

One denotes the velocity ($\vec{V}$) components at the mid-point of ith panel as

$$u_i \equiv u(\bar{x}_i, \bar{y}_i) \quad \text{and} \quad v_i \equiv v(\bar{x}_i, \bar{y}_i).$$

The flow tangency condition is applied at the panel mid-point as $\vec{V}_i \cdot \hat{n}_i \equiv 0$, i.e. using Equation (5.16) one can write this as

$$0 = -u_i \sin\theta_i + v_i \cos\theta_i. \tag{5.18}$$

The Kutta condition at the trailing edge of the airfoil is given by, using Equation (5.17) as

$$(\vec{V}_i \cdot \hat{t}_i)_{i=1} = -(\vec{V}_i \cdot \hat{t}_i)_{i=N}.$$

This is further simplified as

$$u_1\cos\theta_1 + v_1\sin\theta_1 = -(u_N\cos\theta_N + v_N\sin\theta_N). \tag{5.19}$$

A negative sign is used on the right-hand side of Equation (5.19), because the tangent directions are opposite for the first and last panel (N^{th} panel).

Because the induced velocities can be superposed (because we have a linear problem in hand), therefore we can write the velocity components at the middle of the ith panel as obtained by summing individual effects of sources and vortices at all the N panels as

$$u_i = U_\infty \cos\alpha + \sum_{j=1}^{N} q_j\, u_{ij}^{(s)} + \gamma \sum_{j=1}^{N} u_{ij}^{(v)} \tag{5.20}$$

$$v_i = U_\infty \sin\alpha + \sum_{j=1}^{N} q_j\, v_{ij}^{(s)} + \gamma \sum_{j=1}^{N} v_{ij}^{(v)}. \tag{5.21}$$

where $u_{ij}^{(s)}$, $u_{ij}^{(v)}$, $v_{ij}^{(s)}$ and $v_{ij}^{(v)}$ are called the influence coefficients. Physically, these represent respective velocity components due to unit source or vortex indicated by the superscript in parenthesis.

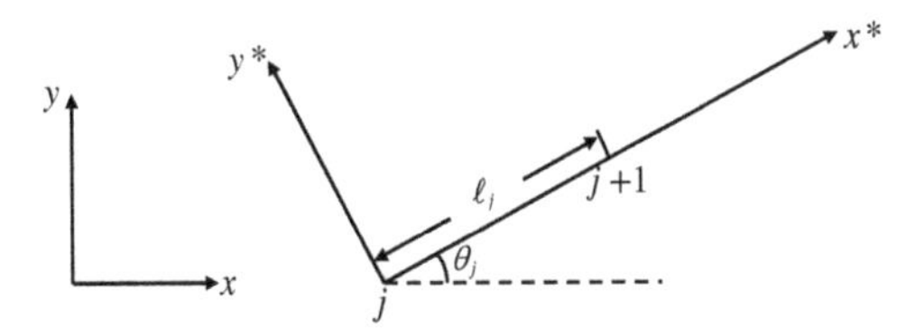

Figure 5.7 Local and global coordinate systems used in obtaining the influence coefficients

For example, $u_{ij}^{(s)}$ denotes x-component of velocity at the ith panel mid-point, due to a unit source distribution on the jth panel. To evaluate the influence coefficients, we work with a panel-fitted coordinate system, as shown in Figure 5.7, for the jth panel indicated by quantities with asterisk. Also, the global coordinate system is shown in this figure for reference. The x^*-axis of the jth panel makes an angle θ_j with the x-axis of the global coordinate system. Thus, the global and the local velocities are related by

$$u = u^* \cos\theta_j - v^* \sin\theta_j \tag{5.22}$$

$$v = u^* \sin\theta_j + v^* \cos\theta_j. \tag{5.23}$$

Let us next obtain the velocity components at the point $(x_i,\ y_i)$ by a unit-strength source distribution on the jth panel and is given by

$$u_{ij}^{*(s)} = \frac{1}{2\pi}\int_0^{l_j} \frac{x_i^* - t}{(x_i^* - t)^2 + y_i^{*2}}dt = -\frac{1}{2\pi}\text{Ln}\left[(x_i^* - t)^2 + y_i^{*2}\right]^{1/2}\Big|_0^{l_j}$$

$$v_{ij}^{*(s)} = \frac{1}{2\pi}\int_0^{l_j} \frac{y_i^*}{(x_i^* - t)^2 + y_i^{*2}}dt = \frac{1}{2\pi}\tan^{-1}\frac{y_i^*}{x_i^* - t}\Big|_0^{l_j}.$$

In these expressions, (x_i^*, y_i^*) is the local or panel-fitted coordinate corresponding to (x_i, y_i) in the global coordinate system, as shown in Figure 5.8.

The distance between the jth and i^{th} nodes is indicated by r_{ij} and these distances are not coordinate system dependent. Similarly, r_{ij+1} is defined in Figure 5.8. These two lines indicated by r_{ij} and r_{ij+1} makes angles γ_0 and γ_l with the x^*-axis of the local coordinate system. If the angle between r_{ij} and r_{ij+1} be defined as β_{ij}, then the influence coefficients due

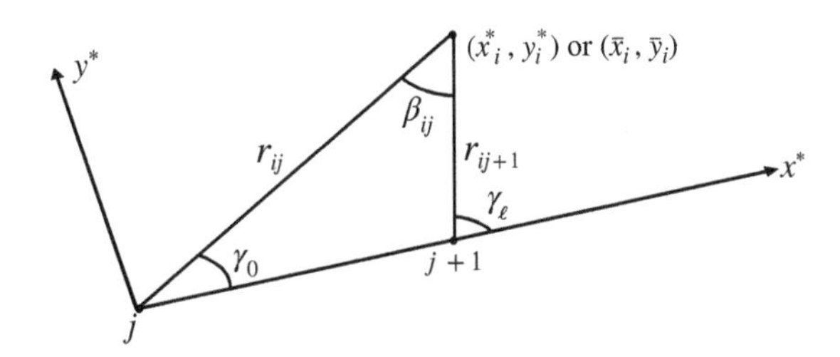

Figure 5.8 Local coordinate system indicating various distances and angles used in calculating influence coefficients

to the unit source are given by

$$u_{ij}^{*(s)} = -\frac{1}{2\pi}\,\text{Ln}\frac{r_{ij+1}}{r_{ij}} \tag{5.24}$$

$$v_{ij}^{*(s)} = \frac{1}{2\pi}(\gamma_l - \gamma_0) = \frac{\beta_{ij}}{2\pi}. \tag{5.25}$$

Thus, there is no need to revert back to global coordinates, as these expressions do not depend upon any coordinate system.

From Equations (5.24) and (5.25) one can calculate the self-induced effect due to the uniform source distribution as

$$u_{ii}^{*(s)} \equiv 0 \text{ and } v_{ii}^{*(s)} = \frac{1}{2},$$

if the point is approached from outside, because then $\beta_{ij} = \pi$. Instead if one approaches the panel from inside, then $\beta_{ij} = -\pi$ and therefore

$$v_{ii}^{*(s)} = -\frac{1}{2}.$$

However in aerodynamics, we will only be interested in the external flow problem.

5.3.1 *Calculation of Influence Coefficients*

With respect to Figure 5.8, if one works with the global coordinate system, then for the ith panel the influence of the jth panel is calculated by first noting the inclinations of r_{ij} and r_{ij+1} with respect to the jth panel as

$$\gamma_0 = \tan^{-1}\frac{\bar{y}_i - y_j}{\bar{x}_i - x_j} \text{ and } \gamma_l = \tan^{-1}\frac{\bar{y}_i - y_{j+1}}{\bar{x}_i - x_{j+1}}.$$

The included angle between r_{ij} and r_{ij+1} is obtained as

$$\beta_{ij} = \gamma_l - \gamma_0$$

$$\beta_{ij} = \tan^{-1}\frac{\bar{y}_i - y_{j+1}}{\bar{x}_i - x_{j+1}} - \tan^{-1}\frac{\bar{y}_i - y_j}{\bar{x}_i - x_j} = \tan^{-1}\frac{\tan\gamma_l - \tan\gamma_0}{1 + \tan\gamma_l\ \tan\gamma_0}$$

$$\beta_{ij} = \tan^{-1}\frac{(\bar{y}_i - y_{j+1})(\bar{x}_i - x_j) - (\bar{y}_i - y_j)(\bar{x}_i - x_{j+1})}{(\bar{x}_i - x_{j+1})(\bar{x}_i - x_j) + (\bar{y}_i - y_{j+1})(\bar{y}_i - y_j)}. \tag{5.26}$$

Thus using Equation (5.26) in Equation (5.25) and Equation (5.24), one obtains the influence coefficients at the ith panel control point due to the unit source distribution on the jth panel.

Having obtained the influence coefficients due to source, one can also calculate the influence coefficients due to vortex distribution as follows. Velocity component at the $(x_i^*,\ y_i^*)$ due to unit strength vortex on the jth panel is given with respect to the local coordinate system fixed with the jth panel by

$$u_{ij}^{*(v)} = -\frac{1}{2\pi}\int_0^{l_j} \frac{y_i^*}{(x_i^* - t)^2 + y_i^{*2}} dt = \frac{\beta_{ij}}{2\pi} = v_{ij}^{*(s)} \tag{5.27}$$

$$v_{ij}^{*(v)} = -\frac{1}{2\pi}\int_0^{l_j} \frac{x_i^* - t}{(x_i^* - t)^2 + y_i^{*2}} dt = \frac{1}{2\pi}\,\text{Ln}\frac{r_{ij+1}}{r_{ij}} = -u_{ij}^{*(s)} \tag{5.28}$$

Having obtained the various influence coefficients, one can use Equations (5.20) and (5.21) in the flow tangency condition given by Equation (5.18) as $-u_i \sin\theta_i + v_i \cos\theta_i = 0$, to get

$$-\sin\theta_i\left\{U_\infty \cos\alpha + \sum_{j=1}^{N} q_j\, u_{ij}^{(s)} + \gamma \sum_{j=1}^{N} u_{ij}^{(v)}\right\}$$

$$+\cos\theta_i\left\{U_\infty \sin\alpha + \sum_{j=1}^{N} q_j\, v_{ij}^{(s)} + \gamma \sum_{j=1}^{N} v_{ij}^{(v)}\right\} = 0.$$

Now using Equations (5.22) and (5.23) to replace the velocity components in global coordinate system by local coordinate system values, one gets

$$-\sin\theta_i\left\{\sum_{j=1}^{N} q_j\left(u_{ij}^{*(s)}\cos\theta_j - v_{ij}^{*(s)}\sin\theta_j\right) + \gamma\sum_{j=1}^{N}\left(u_{ij}^{*(v)}\cos\theta_j - v_{ij}^{*(v)}\sin\theta_j\right)\right\}$$

$$+\cos\theta_i\left\{\sum_{j=1}^{N} q_j\left(u_{ij}^{*(s)}\sin\theta_j + v_{ij}^{*(s)}\cos\theta_j\right) + \gamma\sum_{j=1}^{N}\left(u_{ij}^{*(v)}\sin\theta_j + v_{ij}^{*(v)}\cos\theta_j\right)\right\}$$

$$= U_\infty(\sin\theta_i \cos\alpha - \cos\theta_i \sin\alpha).$$

Now clubbing the terms with the various influence coefficients, one gets

$$\sum_{j=1}^{N} q_j\left\{u_{ij}^{*(s)}(-\sin\theta_i\cos\theta_j + \cos\theta_i\sin\theta_j) + v_{ij}^{*(s)}(\sin\theta_i\sin\theta_j + \cos\theta_i\cos\theta_j)\right\}$$

$$+\gamma\sum_{j=1}^{N}\left\{u_{ij}^{*(v)}(-\cos\theta_j\sin\theta_i + \cos\theta_i\sin\theta_j) + v_{ij}^{*(v)}(\sin\theta_i\sin\theta_j + \cos\theta_i\cos\theta_j)\right\}$$

$$= U_\infty \sin(\theta_i - \alpha).$$

Further simplification yields

$$\sum_{j=1}^{N} q_j\left\{u_{ij}^{*(s)}\sin(\theta_j - \theta_i) + v_{ij}^{*(s)}\cos(\theta_i - \theta_j)\right\}$$

$$+\gamma\sum_{j=1}^{N}\left\{u_{ij}^{*(v)}\sin(\theta_j - \theta_i) + v_{ij}^{*(v)}\cos(\theta_j - \theta_i)\right\} = U_\infty \sin(\theta_i - \alpha).$$

The above equation is written when body boundary condition of flow tangency is satisfied at the ith panel, due to source and vortex distribution at all other panels over the airfoil surface. This can be written as an algebraic equation given by

$$\sum_{j=1}^{N} q_j\, A_{ij} + \gamma\, A_{iN+1} = b_i, \tag{5.29}$$

where

$$A_{ij} = -u_{ij}^{*(s)} \sin(\theta_i - \theta_j) + v_{ij}^{*(s)} \cos(\theta_i - \theta_j),$$

$$A_{iN+1} = \sum_{J=1}^{N} -u_{ij}^{*(v)} \sin(\theta_i - \theta_j) + v_{ij}^{*(v)} \cos(\theta_i - \theta_j)$$

$$\text{and} \quad b_i = U_\infty \sin(\theta_i - \alpha).$$

The above equation has to be supplemented by the Kutta condition to close the system of unknowns. The Kutta condition is re-written for convenience as

$$u_1 \cos\theta_1 + v_1 \sin\theta_1 + u_N \cos\theta_N + v_N \sin\theta_N = 0,$$

which could be written with the help of Equations (5.21) and (5.22) as

$$\cos\theta_1 \left[U_\infty \cos\alpha + \sum_{j=1}^{N} (q_j\, u_{1j}^{(s)} + \gamma u_{1j}^{(v)})\right] + \sin\theta_1 \left[U_\infty \sin\alpha + \sum_{j=1}^{N} (q_j\, v_{1j}^{(s)} + \gamma v_{1j}^{(v)})\right]$$

$$+ \cos\theta_N \left[U_\infty \cos\alpha + \sum_{j=1}^{N} (q_j\, u_{Nj}^{(s)} + \gamma u_{Nj}^{(v)})\right] + \sin\theta_N \left[U_\infty \sin\alpha + \sum_{j=1}^{N} (q_j\, v_{Nj}^{(s)} + \gamma v_{Nj}^{(v)})\right] = 0.$$

This is written in short-form by introducing another summation below for which the index runs for $k = 1$ and N, while j takes value for all the panels

$$\sum_{k=1,N} \left\{ \cos\theta_k \sum_{j=1}^{N} q_j \left(u_{kj}^{*(s)} \cos\theta_j - v_{kj}^{*(s)} \sin\theta_j \right) + \sin\theta_k \sum_{j=1}^{N} q_j \left(u_{kj}^{*(s)} \sin\theta_j + v_{kj}^{*(s)} \cos\theta_j \right) \right.$$

$$\left. + \gamma \cos\theta_k \sum_{j=1}^{N} \left(u_{kj}^{*(v)} \cos\theta_j - v_{kj}^{*(v)} \sin\theta_j \right) + \gamma \sin\theta_k \sum_{j=1}^{N} \left(u_{kj}^{*(v)} \sin\theta_j + v_{kj}^{*(v)} \cos\theta_j \right) \right\} = \mathrm{R}_1,$$

where the right hand side term is written as R_1, whose actual form is given in the equation below

$$\sum_{k=1,N} \left[\sum_{j=1}^{N} q_j \left\{ u_{kj}^{*(s)} \cos(\theta_k - \theta_j) + v_{kj}^{*(s)} \sin(\theta_k - \theta_j) \right\} \right.$$

$$\left. + \gamma \sum_{j=1}^{N} \left\{ u_{kj}^{*(v)} \cos(\theta_k - \theta_j) + v_{kj}^{*(v)} \sin(\theta_k - \theta_j) \right\} \right] = -U_\infty \left[\cos(\theta_1 - \alpha) + \cos(\theta_N - \alpha) \right].$$

This also can be written in short hand form as

$$\sum_{k=1,N}\left[\sum_{j=1}^{N} q_j \bar{A}_{kj} + \gamma A_{kk}\right] = b_{N+1}$$

or

$$\sum_{j=1}^{N} A_{N+ij}\, q_j + \gamma A_{N+1,N+1} = b_{N+1}. \tag{5.30}$$

Thus, one assembles all the equations represented by Equations (5.29) and (5.30) for all the $(N+1)$ unknowns: q_j and γ. Once the singularity strengths are obtained, then one can evaluate the tangential velocity on the airfoil surface using Equation (5.17) as

$$v_{t,i} = u_i \cos\theta_i + v_i \sin\theta_i.$$

This helps us obtain the pressure coefficient as

$$C_p(\bar{x}_i, \bar{y}_i) = 1 - \frac{v_{ti}^2}{U_\infty^2}.$$

5.4 Some Typical Results

There are many versions of panel methods in use. In the following some results for typical aerofoils are shown here obtained by Hess and Smith's method. We have noted that panel methods allow usage of singularities on the equivalent straight panels joining the nodes and thus, automatically incorporate thickness effects. Another shortcoming of all irrotational flow solution methods is the poor accuracy of the solution near the stagnation points. This is usually reduced in panel methods by cosine distribution of nodes, which packs in panels of small length near the leading and trailing edges.

In Figure 5.9, three different airfoils are shown for which loads have been calculated by panel methods. The classical NACA0012 airfoil is shown which is a symmetric section and used in aircraft control surfaces and as rotor blades in helicopters. The SHM-1 airfoil has natural laminar flow (NLF) properties and is described in Chapter 10. This is designed for general aviation usage for a cruise Reynolds number of 11.7 million. Hence the panel method results would be appropriate for such high Reynolds number; yet the results are of less use as this airfoil is designed for delayed transition, which cannot be predicted from irrotational flow results directly. This could however be used for various semi-empirical methods for transition prediction along with panel method and boundary layer solutions. However, the panel method solution can also be used as the initial condition for solving Navier-Stokes equation for impulsive start problem. According to Fujino *et al.* (2003) and Fujino (2005) the chosen SHM-1 airfoil section has less nose-down pitching moment (no more negative than −0.04) as compared to other NLF sections. Also, the larger thickness of this airfoil (15%) allows this business jet to carry more fuel increasing its range. Thus this airfoil is aerodynamically efficient in terms of low pitching moment and and low drag for cruise performance at a Mach

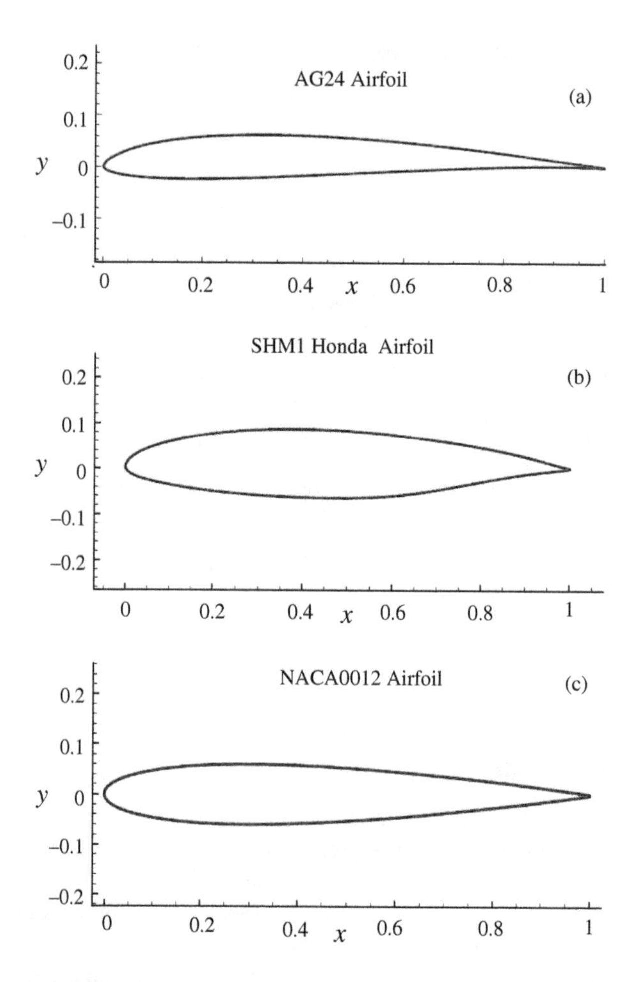

Figure 5.9 Three different aerofoils shown for different applications: (a) AG24 airfoil used for micro air vehicle applications; (b) SHM-1 airfoil used for general aviation purpose for its natural laminar flow (NLF) properties and (c) NACA 0012 airfoil used in aircrafts and rotorcrafts

number of 0.69. It also has docile stall characteristic and engines are mounted over the wing to derive added aerodynamic benefits, which will be described in later chapters. Lesser pitching moment implies lesser need for horizontal tail plane deflection, which otherwise contribute to trim drag.

The AG24 airfoil is designed for low Reynolds number applications, mostly for radio controlled model aircrafts and possibly for micro air vehicle. At the low Reynolds number of operation, the flow over the section will be highly dominated by viscous action and hence the panel method results would be of less relevance.

In Figure 5.10a, the pressure distributions over the NACA0012 airfoil are shown plotted. It is readily evident for this symmetric airfoil, lift is zero for zero angle of attack. As is the convention, negative pressure coefficient is plotted along the positive y-axis. The computed pressure

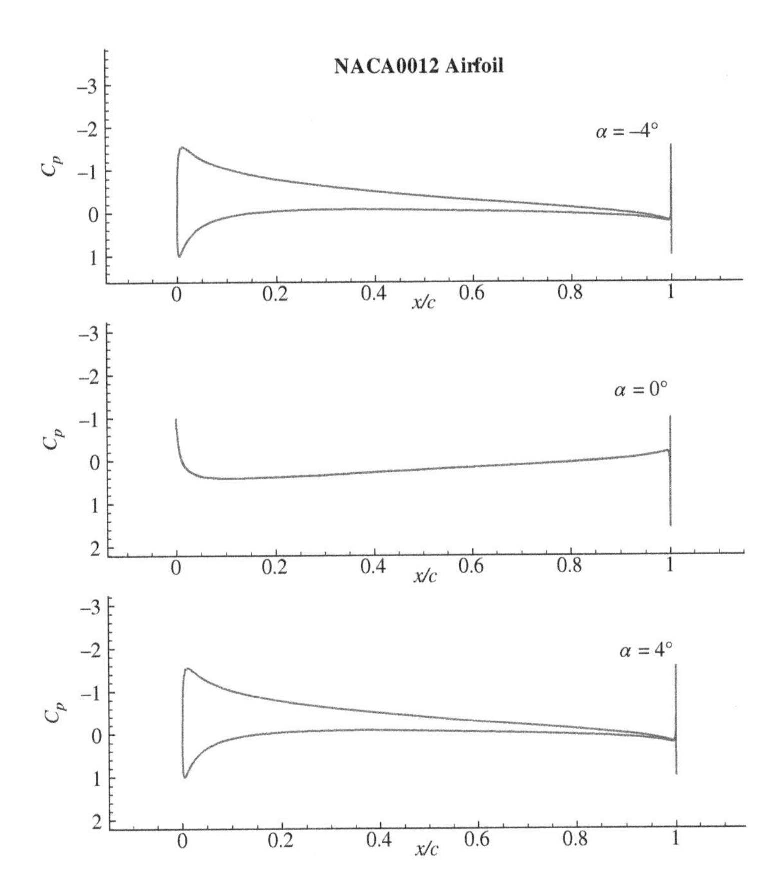

Figure 5.10(a) C_p distributions for the NACA0012 airfoil for the indicated angles of attack

also shows perfect symmetry for $\alpha = \pm 4°$. Experimental data and design methodologies of all NACA series of airfoils are given in Abbott and Doenhoff (1960).

In Figure 5.10b, pressure distribution over the SHM-1 airfoil is shown for four typical angles of attack. This is a thick and cambered airfoil and the special case of $\alpha = -1.4°$ is shown for which one gets the nonlifting solution. For zero angle of attack, the airfoil develops a net positive lift. This pressure distribution also shows a very positive feature of the airfoil, the loading shows a lower positive value near the trailing edge, which will mitigate the nose down tendency of airfoils of conventional design, as seen in Figure 5.10a. This is also true for the zero lift case, where the airfoil has comparably lesser stabilizing pitching moment, as has been stated in Fujino *et al.* (2003). The original method of designing this airfoil is by considering the upper and lower surface separately and using conformal mapping method, which was validated by subsequent panel method analysis, augmented by boundary layer calculation. For the cruise $C_l = 0.26$, the wing upper surface has favourable pressure gradient up to 42% of the chord length from the leading edge. The favourable pressure gradient on the lower surface runs longer up to 63% of the chord length. On the upper surface, the favourable pressure gradient is followed by mild concave pressure recovery. In contrast, on the lower surface the favourable pressure gradient is followed by steeper concave pressure recovery. According to

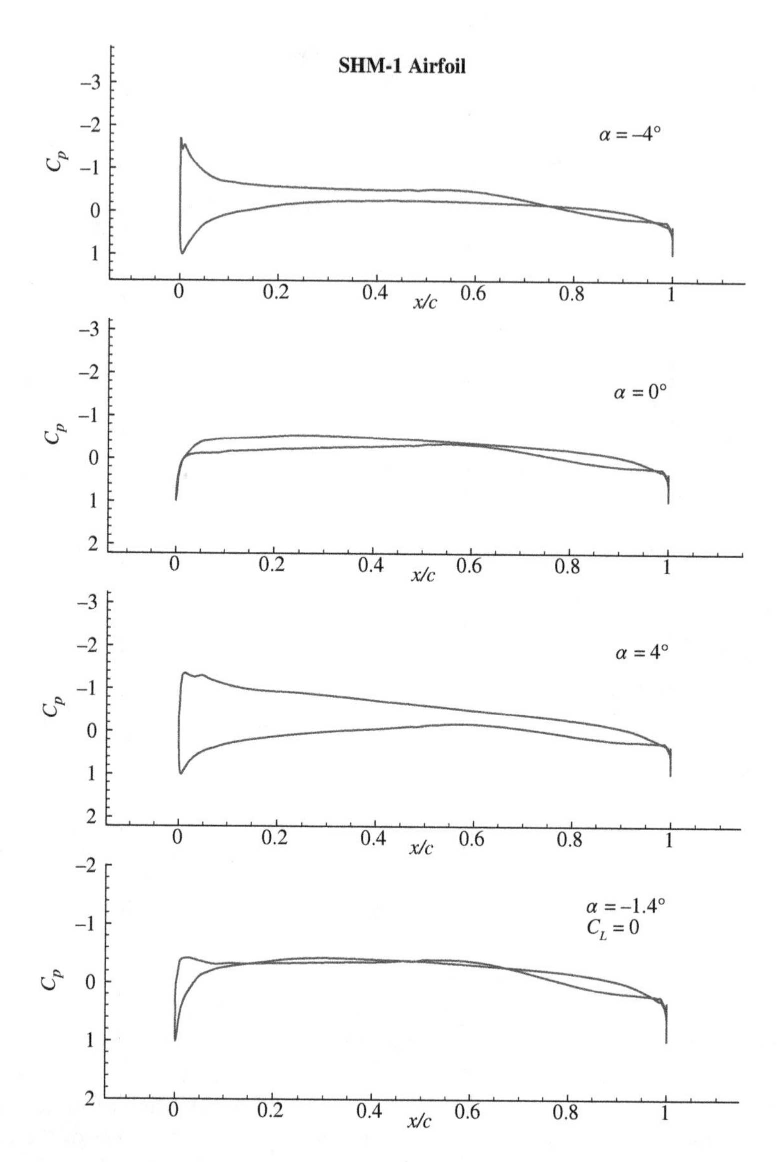

Figure 5.10(b) C_p distributions for the SHM-1 airfoil for the indicated angles of attack

Fujino *et al.* (2003), the upper surface trailing edge has been specifically designed with a very steep pressure gradient which confines the movement of separation point to produce high $C_{l,max}$ at low speeds. Such a procedure had been advocated earlier by Maughmer and Somers (1989).

In Figure 5.10c, pressure distribution is shown for AG24 airfoil, which is thinner compared to SHM-1 airfoil, but more cambered. As a result, the zero lift angle is lower ($\alpha_0 = -2.4°$). Because of the upward positive loading in aft part of the airfoil, the pitching moment is nearly zero. Thus, this airfoil also has better pitching moment property as compared to conventional

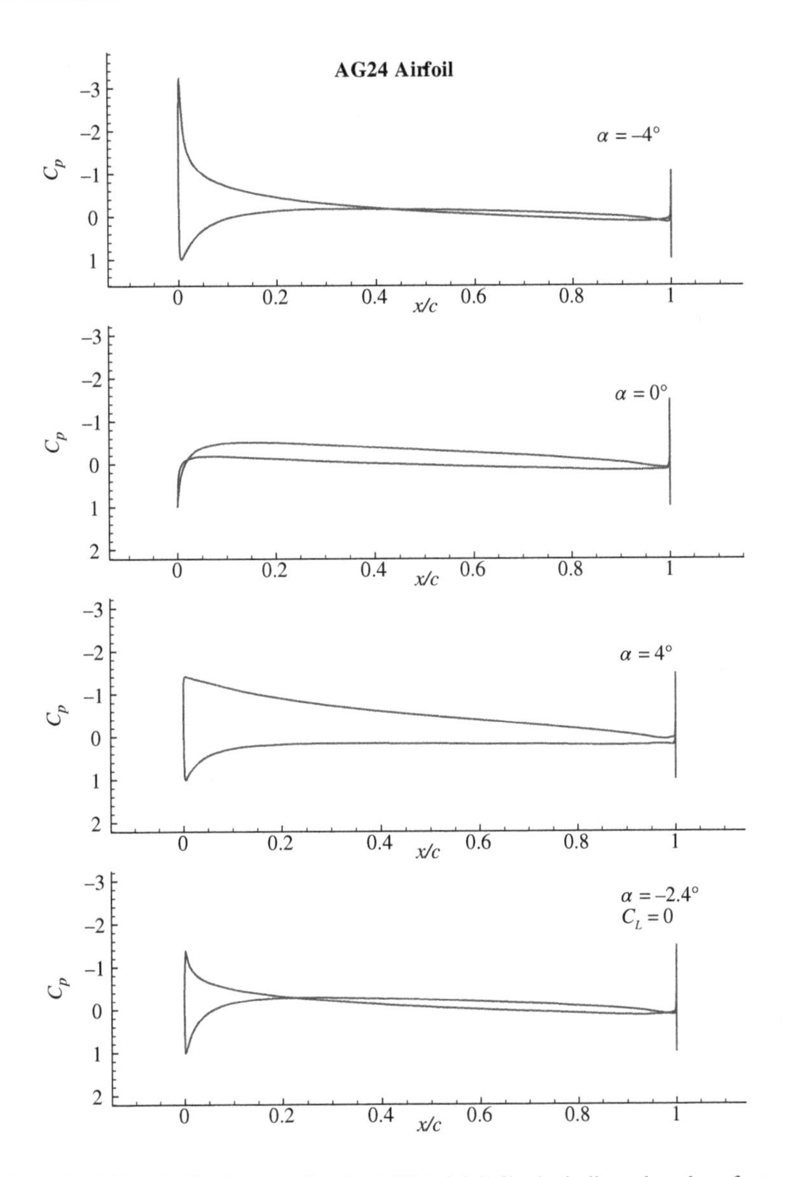

Figure 5.10(c) C_p distributions for the AG24 airfoil for the indicated angles of attack

NACA airfoils. However, it must be remembered that this airfoil is used for low Reynolds numbers and panel method results are of lesser relevance.

Although the airfoils shown here are not all developed for high Reynolds number applications, still the information from the pressure distribution shown in Figures 5.10, provides the impressed streamwise pressure gradient, which can be used as input for boundary layer analysis, as will be discussed in Chapter 7. The boundary layer solution is often used for predicting separation and transition apart from calculating skin friction drag. While for separation prediction, the criterion of vanishing shear stress at the wall can be used, for transition

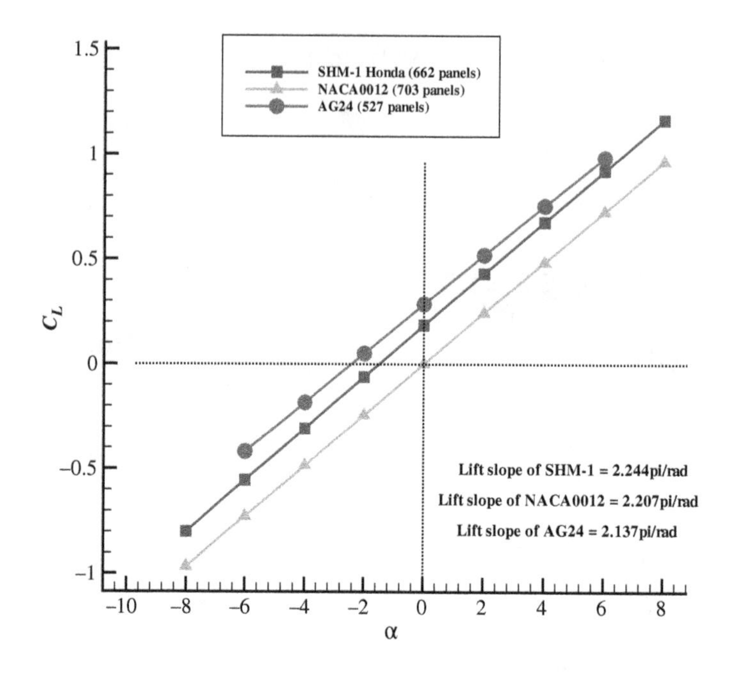

Figure 5.11 Lift curves for the airfoils of Figure 5.9

prediction the method of choice is essentially empirical, as discussed in Chapter 9. Transition prediction of flow over airfoil by solving Navier-Stokes equation is described in Chapter 10. In Figure 5.11, the lift curve of these airfoils are shown for different angles of attack. It is noted that all these airfoils have higher lift curve slope, as compared to the value predicted by thin airfoil theory.

Bibliography

Abbott. I.H. and von Doenhoff, A.E. (1960) *Theory of Wing Sections*. Dover Publications Inc., New York, USA.

Fujino, M., Yoshizaki, Y. and Kawamura, Y. (2003) Natural laminar flow airfoil development for a lightweight business jet. *J. Aircraft*, **40**(4), 609–615.

Fujino, M. (2005) Design and development of the HondaJet. *J. Aircraft*, **42**(3), 755–764.

Hess, J.L. and Smith, A.M.O. (1964) Calculation of nonlifting potential flow about arbitrary three-dimensional bodies. *J. Ship Research*, **8**, 22–44.

Hess, J.L. and Smith, A.M.O. (1967) Calculation of potential flow about arbitrary bodies. *Prog. Aerospace Sci.* **8**, 1–138.

Katz, J. and Plotkin, A. (1992) *Low Speed Aerodynamics: From Wing Theory to Panel Methods*. McGraw-Hill, New York, USA.

Maughmer, M.D. and Somers, D.M. (1989) Design and experimental results for a high-altitude long-endurance airfoil. *J. Aircraft*, **26**(2), 148–153.

6

Lifting Surface, Slender Wing and Low Aspect Ratio Wing Theories

6.1 Introduction

In Chapter 3, we have developed the possible chordwise loading on aircraft wing satisfying the thin airfoil assumption. Based on such chordwise loading, one can superpose a spanwise loading, as discussed in Chapter 4. In both these approaches, thickness of the wing is neglected. This problem can be circumvented by panel method described in Chapter 5, where effects of thickness of the wing and body can be incorporated directly. For low-aspect ratio straight wings, sweptback and delta wings, these approaches based on finite wing theory, lifting line equation and panel methods are not appropriate, due to the fact that strong rotationality becomes important. When such effects are not present, a more rational general approach is provided by lifting surface theory where simultaneously chordwise and spanwise loading distribution is obtained. One of the best-known methods of this kind is due to Falkner (1948), known as the vortex lattice method. In this method, wing surface is segmented into panels and on each panel a simple horse shoe vortex is placed, similar to that described in Chapter 4. The bound element of the horse shoe vortex remains on the wing planform and the trailing vortices extend to infinity in downstream direction. Thus, each panel carries one unknown to be evaluated, i.e. the singularity strength, and once again the unknown vortices' strength is determined by satisfying flow tangency condition at the collocation points simultaneously. In the limit of infinitely many such horse shoe vortices, one obtains a vortex sheet. The strength of this sheet per unit length will be denoted by $\gamma(x, y)$ and such distribution extends beyond the physical wing, as shown in Figure 6.1. As in Chapter 4, this is similar to the bound vortex, but is now a function of both x and y. Corresponding streamwise vortex distribution is denoted in Figure 6.1 as $\delta_w(y)$ which represents the wake of the wing. The lifting surface theory is a general one and simple in conception, which can be shown as equivalent to lifting line theory for moderate aspect ratio wing, and in the other extreme to slender wing theory applicable to low aspect ratio wings.

Theoretical and Computational Aerodynamics, First Edition. Tapan K. Sengupta.

Companion Website: www.wiley.com/go/sengupta

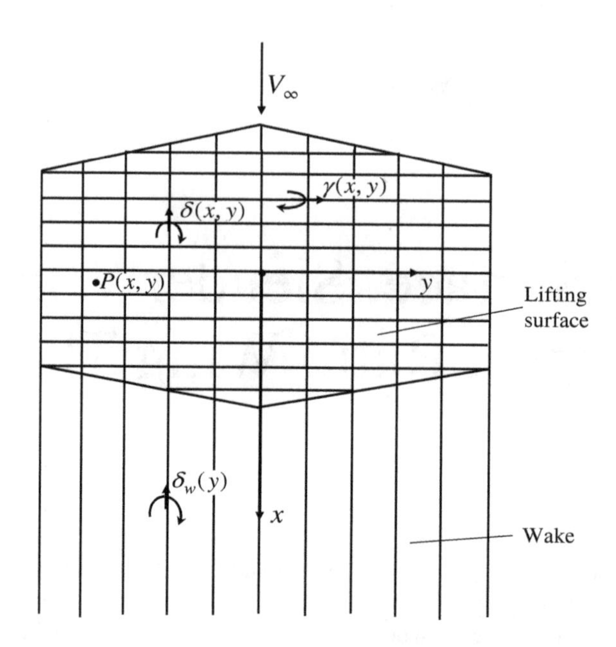

Figure 6.1 A schematic of a rectangular wing on which vortex lattice method is used. The bound circulation is given by γ and the vortex sheet strength representing the wake is given by δ

The panel method described in Chapter 5, uses elementary singularity distribution and is known for its general versatility, providing a tool which is used effectively for analysis and design at preliminary stages. In this chapter, we again move back to theoretical approximation, to what is known collectively as the lifting surface method. In describing the general lifting surface theory, we assign circulation distribution on the lifting body whose evaluation involves solving integral equations for the induced downwash. To explain aerodynamics of a lifting surface of area S moving in an unbounded expanse of fluid, we identify a finite region of outer surface area Σ, within which we solve the problem with Σ defining the far-field boundary. In this theory also, we distribute the circulation on the mean camber line, which is again projected on to the planform area of the lifting surface. A general development requires derivation of some theorems of vector calculus in obtaining the linearized aerodynamics problems.

6.2 Green's Theorems and Their Applications to Potential Flows

These theorems originate from Gauss' divergence theorem and are of general utility in potential flow theory. Let us develop these from the Gauss' divergence theorem for any vector, $\vec{V}$ given by

$$\oint\!\!\!\oint_{S+\Sigma} \vec{V} \cdot \hat{n}\, dS = -\iiint_{\mathcal{V}} \nabla \cdot \vec{V}\, d\mathcal{V}, \tag{6.1}$$

where the volume integral is related to the surface integral. Numerical utility of this theorem is readily noted from the right hand side, which requires evaluation of triple integral after calculating the divergence of $\vec{V}$. In contrast, on the left-hand side, one evaluates the double

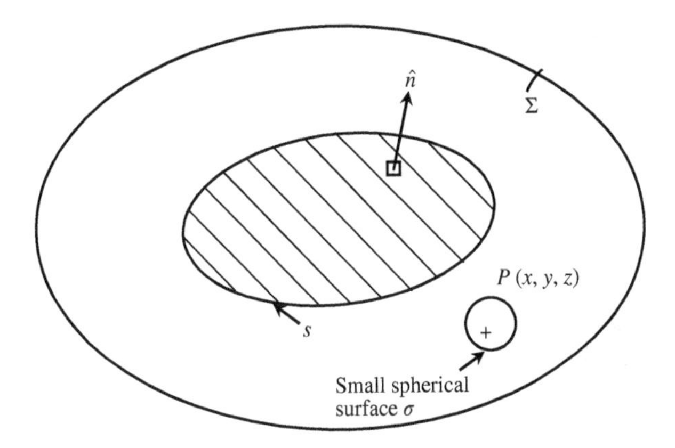

Figure 6.2 Applying reciprocal theorem for lifting surface theory to obtain ϕ at the point P exterior of a solid body indicated by surface area S, while it is within the far-field boundary denoted by Σ

integral of the flux of the vector out of the control volume defined with the help of unit normal ($\hat{n}$) to S and Σ, as shown in Figure 6.2.

Since $\nabla \cdot \vec{V}$ is the total flux per unit volume at a point, then $\nabla \cdot \vec{V}\ d\mathcal{V}$ is the total flux through the surface of the volume element $d\mathcal{V}$. Adding fluxes of adjoining such volume elements, the fluxes through the common surface elements balances out leaving no net contribution. As the total volume is assembly of such contiguous volume elements with common surface elements, the volume integral on the right-hand side of Equation (6.1) reduces to the outward flux through the surface areas of the control volume assembly, which have no cancelling surface elements other than $S + \Sigma$.

Let us define two functions ϕ and ϕ' possessing continuous, single-valued and finite first and second derivatives throughout $\mathcal{V}$ and let

$$\vec{V} = \phi\ \nabla\phi'. \tag{6.2}$$

Therefore

$$\vec{V} \cdot \hat{n} = \phi\left[\nabla\phi' \cdot \hat{n}\right] = \phi\frac{\partial\phi'}{\partial n} \tag{6.3}$$

and

$$\nabla \cdot \vec{V} = \nabla \cdot (\phi\nabla\phi') = \phi\nabla^2\phi' + \nabla\phi \cdot \nabla\phi'. \tag{6.4}$$

Substituting Equations (6.3) and (6.4) in Equation (6.1), one gets

$$\oiint_{S+\Sigma} \phi\frac{\partial\phi'}{\partial n} dS = -\iiint \left[\nabla\phi \cdot \nabla\phi' + \phi\nabla^2\phi'\right] d\mathcal{V}. \tag{6.5}$$

Again interchanging ϕ and ϕ', one gets from Equation (6.5) the following

$$\oiint_{S+\Sigma} \phi' \frac{\partial \phi}{\partial n} dS = -\iiint \left[\nabla\phi' \cdot \nabla\phi + \phi'\nabla^2\phi\right] d\mathcal{V}. \tag{6.6}$$

Equations (6.5) and (6.6) are known as Green's Theorem.

6.2.1 *Reciprocal Theorem*

Another important theorem is obtained if one considers ϕ and ϕ' to be harmonic functions, then on the right-hand side of Equations (6.4) to (6.6) one can substitute

$$\nabla^2\phi = \nabla^2\phi' \equiv 0.$$

Therefore from Equations (6.5) and (6.6) one gets the following identity

$$\oiint_{S+\Sigma} \phi \frac{\partial \phi'}{\partial n} dS = \oiint_{S+\Sigma} \phi' \frac{\partial \phi}{\partial n} dS. \tag{6.7}$$

This reciprocal theorem forms the basis of lifting surface theory, derived in the following.

6.3 Irrotational External Flow Field due to a Lifting Surface

Consider body with surface area S and any point P with coordinate (x, y, z) exterior of the body, which is within Σ and where we want to find the velocity potential ϕ. For the unbounded external flow, let us choose

$$\phi' = \frac{1}{r} = \frac{1}{\{(x-x_1)^2 + (y-y_1)^2 + (z-z_1)^2\}^{1/2}},$$

where (x_1, y_1, z_1) is any point on the lifting surface. So that $\nabla^2\phi' = \nabla^2\left(\frac{1}{r}\right) = 0$, everywhere except at $r = 0$, which is the singular point. To apply reciprocal theorem we exclude the point P by considering a sphere of dimension σ and exclude this sphere from a simply connected domain. Thus, one considers the domain which is exterior to S and interior to Σ while excluding σ. Thus, the application of reciprocal theorem to this domain yields

$$\oiint_{S+\Sigma} \phi \frac{\partial \phi'}{\partial n} dS + \oiint_{\sigma} \phi \frac{\partial \phi'}{\partial n} d\sigma = \oiint_{S+\Sigma} \phi' \frac{\partial \phi}{\partial n} dS + \oiint_{\sigma} \phi' \frac{\partial \phi}{\partial n} d\sigma.$$

Substituting $\phi' = \frac{1}{r}$ in the above and transposing terms one gets

$$\oiint_{\sigma} \phi \frac{\partial}{\partial n}\left(\frac{1}{r}\right) d\sigma = \oiint_{S+\Sigma} \frac{1}{r} \frac{\partial \phi}{\partial n} dS - \oiint_{S+\Sigma} \phi \frac{\partial}{\partial n}\left(\frac{1}{r}\right) dS + \oiint_{\sigma} \frac{\partial \phi}{\partial n} \frac{1}{r} d\sigma. \tag{6.8}$$

As σ is the small sphere around P, then one can express an infinitesimal area on the spherical surface, in term of a differential element of solid angle (Ω) measured from P as

$$d\sigma = r^2 d\Omega,$$

such that

$$\oiint_{\sigma} d\Omega = 4\pi.$$

Also, over the surface of σ one can express

$$\frac{\partial}{\partial n}\left(\frac{1}{r}\right) = \frac{\partial}{\partial r}\left(\frac{1}{r}\right) = -\frac{1}{r^2}.$$

Thus, the left hand side of Equation (6.8) is obtained by using the mean value theorem with respect to the point P

$$\operatorname*{Lim}_{\sigma\to 0} \oiint_{\sigma} \phi \frac{\partial}{\partial n}\left(\frac{1}{r}\right) d\sigma = -\phi_P \operatorname*{Lim}_{r\to 0} \oiint \frac{r^2 d\Omega}{r^2} = -4\pi\phi_P.$$

Similarly, the last term on the right hand side of Equation (6.8) is obtained as

$$\oiint_{\sigma} \frac{\partial \phi}{\partial n}\frac{1}{r} d\sigma = \left(\frac{\partial \phi}{\partial n}\right)_{\text{mean}} \operatorname*{Lim}_{r\to 0} \oiint \frac{r^2 d\Omega}{r} \equiv 0.$$

Therefore Equation (6.8) can be simplified as

$$\phi_P = -\oiint_{S+\Sigma} \frac{\partial \phi}{\partial n}\frac{1}{4\pi r} dS + \iint_{S+\Sigma} \phi \frac{\partial}{\partial n}\left(\frac{1}{4\pi r}\right) dS.$$

These integrals on the right-hand side evaluated over the far-field boundary, Σ yields zero value for the far-field boundary receding to infinite distance, where perturbations induced there is zero for ϕ and velocity components. Therefore, the value of velocity potential at the field point P is given by

$$\phi_P(x, y, z) = \oiint_{S} \phi \frac{\partial}{\partial n}\left(\frac{1}{4\pi r}\right) dS - \oiint_{S} \frac{\partial \phi}{\partial n}\frac{1}{4\pi r} dS. \tag{6.9}$$

For two-dimensional flow corresponding result is given as (left as an exercise).

$$\phi(x, y) = \frac{1}{2\pi} \oint_{c_B} \left[\frac{\partial \phi}{\partial n} \text{Ln}\, r - \phi \frac{\partial}{\partial n}(\text{Ln}\, r)\right] ds \tag{6.10}$$

where c_B indicates the closed line contour on the body. As $u = \phi_x$, so u also satisfies Equation (6.9), i.e.

$$u(x, y, z) = \oiint_{S'} \left[u\frac{\partial}{\partial n} - \frac{\partial u}{\partial n} \right] \frac{1}{4\pi r} dS. \tag{6.11}$$

Note the limit of the integral now includes both the wing and the wake behind in S' and both the upper and lower surfaces of the vortex sheet need to be considered. The induced velocity, u, is a measure of pressure coefficient, shown in Equation (2.46) as

$$C_p = 1 - \left[\frac{U_\infty + u}{U_\infty} \right]^2,$$

which for small perturbation u, reduces to

$$C_p = -\frac{2u}{U_\infty}.$$

Therefore, u is essentially the pressure coefficient. As the wake cannot sustain any pressure discontinuity, so the first term in Equation (6.11) is only nonzero over the physical wing surface.

We also note next that the second term for such a lifting surface in Equation (6.11) does not contribute from anywhere in S', for the following reason.

If (x_1, y_1, z_1) represents a point on the lifting surface, then for a general point in the field (x, y, z) requires

$$r = \sqrt{(x - x_1)^2 + (y - y_1)^2 + (z - z_1)^2}.$$

In Equation (6.11), for the second term in the bracket, we have the following on top of S'

$$\frac{\partial u}{\partial n} = \frac{\partial u}{\partial z_1}.$$

For irrotational flows, consequently from the zero spanwise vorticity condition we must have

$$\frac{\partial u}{\partial z_1} = \frac{\partial w}{\partial x_1}.$$

Similarly, on the bottom surface of S'

$$\frac{\partial u}{\partial n} = -\frac{\partial u}{\partial z_1} = -\frac{\partial w}{\partial x_1}.$$

Both w and its derivatives are continuous through all of S' and therefore based on the observations above, upper and lower surface contributions cancel each other, for the second bracketed terms of Equation (6.11).

Thus, Equation (6.11) simplifies to

$$u(x, y, z) = \iint\limits_S \left[u_u - u_l\right] \frac{\partial}{\partial n}\left(\frac{1}{4\pi r}\right) dS,$$

where S is now the wing planform projection of S', consisting of the physical wing.

We represent the lifting surface by rectangular infinitesimal elements given by $dx_1\, dy_1$. Also, we have noted earlier in Chapters 3 and 4 that the velocity discontinuity across the lifting surface represents the loading, which in this case is a two-dimensional function given by, $u_u - u_l = \gamma(x_1,\ y_1)$. Therefore

$$u(x, y, z) = \frac{1}{4\pi} \iint\limits_S \gamma(x_1, y_1) \frac{\partial}{\partial z_1}\left(\frac{1}{r}\right) dx_1\, dy_1. \tag{6.12}$$

Note that on the lifting surface, i.e. for $z_1 = 0$,

$$\frac{\partial}{\partial n}\left(\frac{1}{r}\right) = \pm \frac{\partial}{\partial z_1}\left(\frac{1}{r}\right)\Big|_{z_1=0}$$

and

$$\frac{\partial}{\partial z_1} = \frac{\partial}{\partial (z_1 - z)} = -\frac{\partial}{\partial z}.$$

Using these in Equation (6.12), one obtains

$$u(x, y, z) = -\frac{1}{4\pi} \iint\limits_S \gamma(x_1, y_1) \frac{\partial}{\partial z}\left(\frac{1}{r}\right) dx_1\, dy_1. \tag{6.12a}$$

Now for a lifting wing problem, $w(x, y, 0)$ is considered known over S. Also, as $u = \phi_x$ and $w = \phi_z$, so

$$w(x, y, z) = \frac{\partial}{\partial z} \int_{-\infty}^{x} u(x_0, y, z) dx_0, \tag{6.13}$$

where the vanishing perturbation condition at far upstream location is used, i.e. $\phi(-\infty, y, z) = 0$.

Using Equation (6.13) in (6.12a)

$$w(x, y, z) = -\frac{1}{4\pi} \iint \gamma(x_1, y_1) \times \left\{ \frac{\partial^2}{\partial z^2} \int_{-\infty}^{x} \frac{dx_0}{\sqrt{(x_0 - x_1)^2 + (y - y_1)^2 + z^2}} \right\} dx_1\, dy_1.$$

Now, after considerable algebra (as reported in Multhop (1950) and Thwaites (1960)) the above can be reduced for $z \to 0$, to one of the following forms

$$w(x, y, 0) = -\frac{1}{4\pi} \iint_S \frac{\gamma(x_1, y_1)}{(y - y_1)^2} \left[1 + \frac{x - x_1}{\sqrt{(x - x_1)^2 + (y - y_1)^2}}\right] dx_1\, dy_1 \tag{6.14}$$

or

$$w(x, y, 0) = -\frac{1}{4\pi} \iint_S \frac{\partial \gamma}{\partial y_1} \frac{1}{(y - y_1)} \left[1 + \frac{\sqrt{(x - x_1)^2 + (y - y_1)^2}}{x - x_1}\right] dx_1\, dy_1. \tag{6.15}$$

Thus the downwash given in Equation (6.14), can be written in nondimensional form as

$$\frac{w(x, y, 0)}{U_\infty} = -\frac{1}{8\pi} \iint \frac{2\gamma}{U_\infty} \frac{1}{(y - y_1)^2} \left[1 + \frac{x - x_1}{\sqrt{(x - x_1)^2 + (y - y_1)^2}}\right] dx_1\, dy_1.$$

Defining $l(x_1, y_1)$ as the load function or the pressure coefficient from

$$\left(\frac{2\gamma}{U_\infty}\right) = \frac{\Delta p}{\frac{1}{2}\rho U^2} = l(x_1, y_1),$$

the downwash at any such point on the surface of the wing ($z = 0$) is given by

$$\frac{w(x, y)}{U_\infty} = -\frac{1}{8\pi} \iint_S \frac{l(x_1, y_1)}{(y - y_1)^2} \left[1 + \frac{x - x_1}{\sqrt{(x - x_1)^2 + (y - y_1)^2}}\right] dx_1\, dy_1. \tag{6.16}$$

Equation (6.16) can also be written as

$$\frac{w(x, y)}{U_\infty} = \frac{1}{8\pi} \frac{\partial}{\partial y} \iint_S \frac{l(x_1, y_1)}{(y - y_1)} \left[1 + \frac{\sqrt{(x - x_1)^2 + (y - y_1)^2}}{(x - x_1)}\right] dx_1\, dy_1. \tag{6.17}$$

Equation (6.17) is preferable to Equation (6.16), since the integral involves evaluation of only a Cauchy principal value. Higher order singularity in Equation (6.16) is more difficult to handle though Multhop has defined a principal valuc to bc taken at $y = y_1$ as

$$\int_a^b \frac{f(y_1)}{(y - y_1)^2} dy_1 = \operatorname*{Lim}_{\epsilon \to 0} \left[\int_a^{y-\epsilon} \frac{f(y_1)\, dy_1}{(y - y_1)^2} + \int_{y+\epsilon}^b \frac{f(y_1)}{(y - y_1)^2} dy_1 - \frac{2f(y)}{\epsilon}\right].$$

Derivation of Equations (6.16) and (6.17) is based on elementary horse shoe vortex. It is not applicable to cases in which the bound vortex has a discontinuity in direction, such as in a swept wing. Despite this restriction, let us look at some cases arising out of this downwash integral equation.

6.3.1 Large Aspect Ratio Wings

Let us consider the class of wings which are unswept and of very large aspect ratio. For such a wing we may use the simplification

$$(x - x_1)^2 \ll (y - y_1)^2. \tag{6.18}$$

This clearly does not hold for the region around $y = y_1$. But, it can be shown that the error will not be serious provided

$$S\frac{\partial C_l}{\partial y} \ll 1.$$

The error is likely to be greatest near the wing tips, but again the fact that $C_l \to 0$ towards the tips has a mitigating effect.

With this assumption, Equation (6.17) can be written as

$$\frac{w}{U_\infty} = \frac{1}{8\pi}\frac{\partial}{\partial y}\iint_S \left[1 + \frac{|y - y_1|}{x - x_1}\right]\frac{l(x_1, y_1)}{y - y_1}dx_1\,dy_1.$$

Let us say that the leading and trailing edges of the wing be defined as $x_L(y)$ and $x_T(y)$, respectively, while the spanwise coordinate varies from $-s$ to $+s$. Then the above expression can be further simplified as

$$\frac{w}{U_\infty} = \frac{1}{8\pi}\frac{\partial}{\partial y}\int_{x_L}^{x_T}\int_{-s}^{s}\frac{l(x_1, y_1)}{y - y_1}\,dx_1\,dy_1 + \frac{1}{8\pi}\frac{\partial}{\partial y}\int_{x_L}^{x_T}\int_{-s}^{s}\frac{|y - y_1|}{y - y_1}\frac{l(x_1, y_1)}{x - x_1}dx_1\,dy_1.$$

The second expression (I_2) above can be further simplified by considering the following subranges: $y > y_1$ and $y < y_1$. In these subranges, the first factor takes the value ± 1, and hence it is simplified further as

$$I_2 = \frac{1}{8\pi}\frac{\partial}{\partial y}\int_{x_L}^{x_T}\int_{-s}^{y}\frac{l(x_1, y_1)}{x - x_1}dx_1\,dy_1 - \frac{1}{8\pi}\frac{\partial}{\partial y}\int_{x_L}^{x_T}\int_{y}^{s}\frac{l(x_1, y_1)}{x - x_1}dx_1\,dy_1.$$

These are differentials of integration, with ranges function of the unknown and hence can be simplified further and one notes that both these expressions contribute individually as

$$\frac{1}{8\pi}\int_{x_L}^{x_T}\frac{l(x_1, y)}{x - x_1}dx_1.$$

Thus, the downwash integral is

$$\frac{w}{U_\infty} = \frac{1}{8\pi}\frac{\partial}{\partial y}\iint \frac{l(x_1, y_1)}{y - y_1} dx_1\, dy_1 + \frac{1}{4\pi}\int_{x_L}^{x_T} \frac{l(x_1, y)}{x - x_1} dx_1. \tag{6.19}$$

Since first integral in Equation (6.19) is a function of y alone, the second integral also must be a function of y only, i.e.

$$\int_{x_L}^{x_T} \frac{l(x_1, y)}{x - x_1} dx_1 = F(y). \tag{6.20}$$

This is an integral equation from which the unknown l as a function of x can be determined without the knowledge of the spanwise loading.

The loading function $l(x, y)$ must satisfy the Kutta-Jukowski condition, i.e.

$$l(x_T, y) = 0. \tag{6.21}$$

Solution of Equation (6.20) subject to the above condition in Equation (6.21) gives

$$l(x, y) = \frac{F(y)}{\pi}\left[\frac{x_T(y) - x}{x - x_L(y)}\right]^{1/2}.$$

Since

$$C_l(y) = \frac{1}{c(y)}\int_{x_L(y)}^{x_T(y)} l(x, y) dx.$$

Therefore

$$F(y) = 2C_l(y).$$

Hence one obtains the chordwise loading as

$$l(x, y) = \frac{2C_l(y)}{\pi}\left[\frac{x_T(y) - x}{x - x_L(y)}\right]^{1/2}. \tag{6.22}$$

Substituting this in Equation (6.19) noting that $\alpha = w/U_\infty$, we get

$$\alpha(y) = \frac{C_l(y)}{2\pi} - \frac{1}{8\pi}\int_{-s}^{s} \frac{C_l(y_1)\, c(y_1)}{(y - y_1)^2} dy_1.$$

Integrating by parts and ensuring that $C_L c$ vanishes at wing tips, we find

$$\frac{C_l(y)}{2\pi} = \alpha - \frac{1}{8\pi}\int_{-s}^{s}\frac{d}{dy}(C_l\ c)\frac{dy}{y-y_1}. \tag{6.23}$$

This is Prandtl's classical lifting line equation, with the first factor of the integrand on the right-hand side represents the spanwise loading, which can be understood from

$$C_l = \frac{\rho\, U_\infty \Gamma}{1/2\ \rho\, U_\infty^2\ c} = \frac{2\Gamma}{U_\infty\ c}.$$

Thus,

$$C_l\ c = 2\Gamma/U_\infty.$$

Differentiating this with y one obtains

$$\frac{d}{dy}(C_l\ c) = \frac{2}{U_\infty}\frac{d\Gamma}{dy}.$$

This equivalence can be noted by rewriting Equation (4.54) as

$$\frac{2\Gamma}{a_\infty\ c} = U_\infty(\alpha - \alpha_0) - U_\infty\ \epsilon. \tag{4.54}$$

Recall from Equation (4.14)

$$\epsilon = \frac{1}{4\pi U_\infty}\int\frac{d\Gamma/dy}{y-y_1}dy.$$

One notes that Equation (6.23) and Equation (4.54) are identical, as $a_\infty = 2\pi$.

6.3.2 Wings of Small Aspect Ratio

For wings of small aspect ratio as depicted in Figure 6.3, one can use the following approximation

$$(y-y_1)^2 \ll (x-x_1)^2 \tag{6.24}$$

in Equation (6.17) for simplification. We note that this assumption fails at $x = x_1$. However, the assumption is reasonable in a overall sense, except near unswept leading edge. The downwash equation, Equation (6.17), now simplifies to

$$\frac{w(x,y)}{U_\infty} = \frac{1}{8\pi}\frac{\partial}{\partial y}\left[\iint_S l(x_1,y_1)\left\{1+\frac{|x-x_1|}{x-x_1}\right\}\frac{dx_1\ dy_1}{y-y_1}\right]. \tag{6.25}$$

The term in the second bracket has the value 2 for $x_1 < x$ and is zero for $x_1 > x$. This implies that the integration along x needs to be extended only from the leading edge $x_1 = x_L$ to the plane x = constant. Any parts of the wing behind $x_1 = x$, do not contribute to the downwash

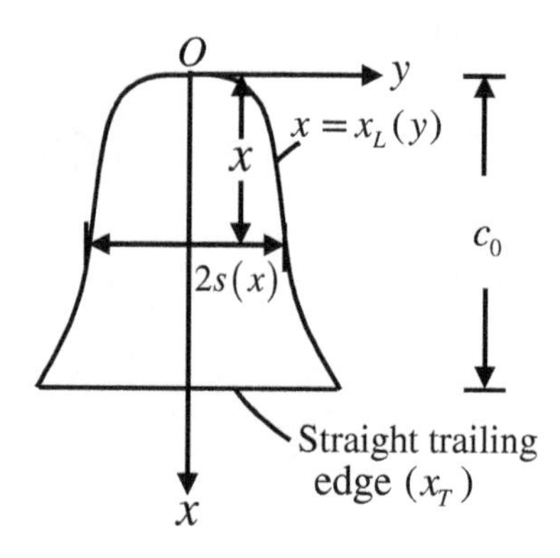

Figure 6.3 A schematic low aspect ratio wing with a nonrectangular section

at $x_1 = x$. This zone of influence is indicated in Figure 6.4, by the shaded region for points along $x_1 = x$. Thus the downwash integral simplifies to

$$\frac{w}{U_\infty} = \frac{1}{4\pi}\frac{\partial}{\partial y}\left[\int_{x_L(y)}^{x} dx_1 \int_{-s(x)}^{s(x)} \frac{l(x_1, y_1)}{y - y_1} dy_1\right], \tag{6.26}$$

where $s(x)$ is the local semi-span for $x_1 = x$. Performing the differentiation and changing the order of integrations one obtains

$$\frac{w}{U_\infty} = -\frac{1}{4\pi}\int_{-s(x)}^{s(x)} \frac{dy_1}{(y - y_1)^2}\left[\int_{x_L(y_1)}^{x(y_1)} l(x_1, y_1)\, dx_1\right]. \tag{6.27}$$

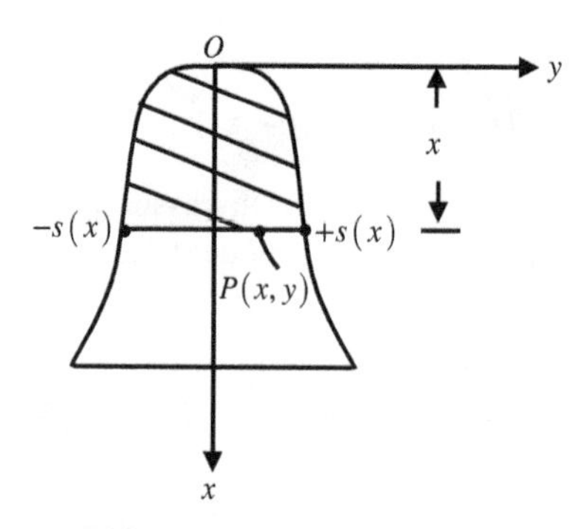

Figure 6.4 The domain of influence for a point at $x_1 = x$ on the lifting surface for the schematic of Figure 6.3, shown by shaded region

Now

$$\int_{x_L(y_1)}^{x} l(x_1, y_1)\, dx_1 = L(x,\ y_1) = \Big(x - x_L(y_1)\Big) C_l(y_1),$$

where $L(x, y_1)$ may be considered to be the load on the section at the spanwise section $y = y_1$ up to the plane $x = x_1$.

Since $L(x, y) \to 0$ as $y \to \mp\, s(x)$ and for $x = x_L(y)$, then from Equation (6.27) we have

$$\frac{w}{U_\infty} = -\frac{1}{4\pi}\int_{-s(x)}^{s(x)} L(x, y_1)\, d\left(\frac{1}{y - y_1}\right) = \left[-\frac{1}{4\pi}\frac{L(x, y_1)}{y - y_1}\right]_{-s(x)}^{s(x)} + \frac{1}{4\pi}\int_{-s(x)}^{s(x)} \frac{\partial L}{\partial y_1}\frac{dy_1}{y - y_1}.$$

The first term in this expression is zero because of tip conditions, and thus one gets

$$\frac{w}{U_\infty} = \frac{1}{4\pi}\int_{-s(x)}^{s(x)} \frac{\partial L}{\partial y_1}\frac{dy_1}{y - y_1}. \tag{6.28}$$

Comparing this result with the one for large aspect ratio wing given by

$$\frac{C_l}{2\pi} = \alpha - \frac{1}{8\pi}\int_{s(x)}^{s(x)} \frac{d}{dy}(C_l\, c)\frac{dy}{y - y_1}.$$

We find the downwash on this plane for the low aspect ratio wing equals twice the induced downwash on a large AR wing. Therefore, the induced drag experienced by a low aspect ratio wing will be more than that for the large aspect ratio wing.

6.4 Slender Wing Theory

We shall confine ourselves here to considering the aerodynamic characteristics of flat plate wing of slender platform. More general problems of wings and bodies of finite thickness distribution are treated in slender body theory.

The basic idea is that at small angle of attack α, flow pattern in any transverse plane approximates to that of two-dimensional incompressible potential flow, with oncoming velocity $U_\infty \alpha$ for the flow over the flat plate wing considered in Figure 6.5(a). This wing planform

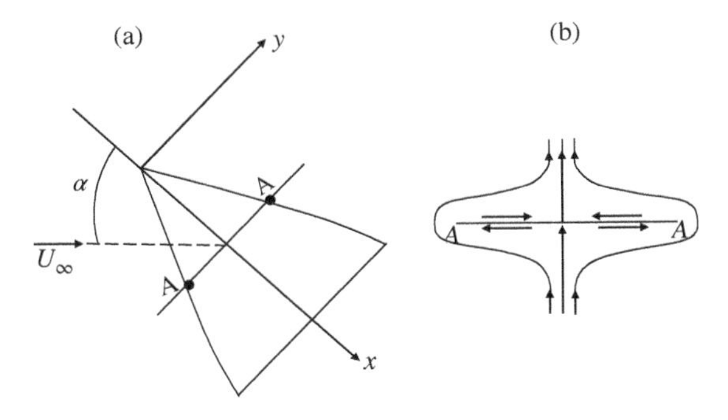

Figure 6.5 Irrotational potential flow model for low aspect ratio plane wing

resembles a delta wing with sharp leading edge and negligible thickness. Such a potential flow at any streamwise section represents a flow normal to a 2D flat plate, as shown in Figure 6.5(b). While the flow is readily expected to separate from sharp corners, an assumption to the contrary is made in slender wing theory due to the streamwise velocity being much larger in magnitude. This is approximately equal to U_∞, which circumvents the tendency of the flow to separate. However counter-intuitive it may appear, this is indeed true for very small of attacks. From a practical point of view though, such a geometry is used for large angles of attack operation of wings at super-critical speeds. Even when an aircraft with delta wing is in operation during landing and take-off, angles of attack are significantly higher than what is experienced by large aspect ratio wings. Hence the early development in this area should be read more to understand the qualitatively different nature of flow over low aspect ratio wings, as opposed to high aspect ratio wings. In the latter part of the chapter, we will focus upon the actual operation of low aspect ratio wings, which display the actual leading edge vortex which is commonly noticed at the leading edge of slender wings of delta or similar planforms.

Consider the planform shown in Figure 6.6, with the following geometry of the planform. Here the root chord is denoted by c_0 and the maximum span is defined as $b = 2S$. Such a wing is set at an angle of attack, α, with the free stream velocity in the plane of symmetry given by U. The y-axis is positive to the starboard, while z-axis is positive downwards. Also, the trailing edge is considered as straight and unswept. The theory then leads to the statement that near the wing, the flow in the transverse section at the station is that of a flow past flat plate of span $2s(x)$, in a uniform stream.

The velocity normal to the plate is $U\alpha$. The problem is solved easily by complex variable method by conformally mapping the cross flow plane ($Z = y + iz$) transformed to the Z^* ($= y^* + iz*$) plane given by

$$Z^{*2} = Z^2 - s^2(x). \tag{6.29}$$

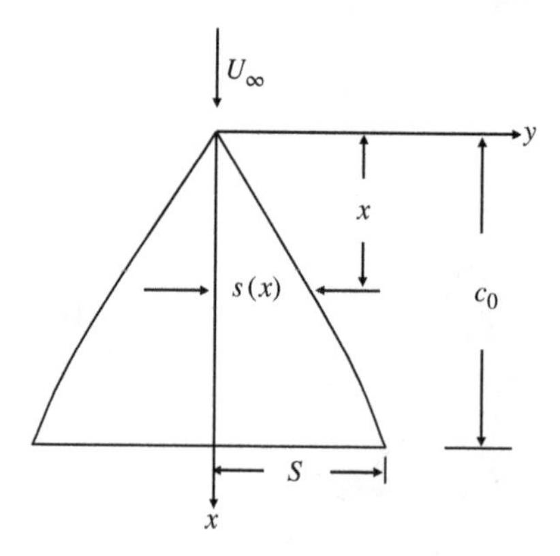

Figure 6.6 The geometry of a plane low aspect ratio wing modelled by slender wing theory

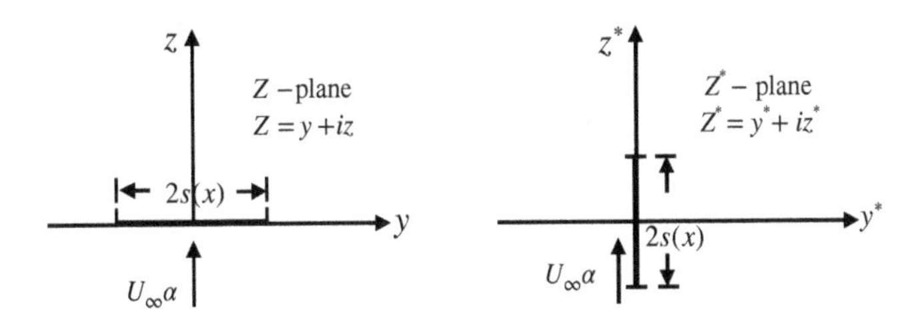

Figure 6.7 The flow in physical ($Z = y + iz$) and transformed plane ($Z^* = y^* + iz^*$) as modelled by slender wing theory

The flow in both the planes are depicted in Figure 6.7. The complex potential is constructed from the flow in Z^*-plane as

$$W = -i\,U_\infty \alpha\,Z^* \qquad (6.30)$$
$$= -i\,U_\infty \alpha \sqrt{Z^2 - s^2(x)}.$$

The quantity under the radical sign is negative for $z = 0$ and this fact is used in the following. We also note the fact that even for this three-dimensional flow, use of strip theory approach allows one to employ tools of studying two-dimensional flows, e.g. the use of complex variable method to define a complex potential

$$W = \phi + i\psi. \qquad (6.31)$$

Now on the upper surface of the wing for $z = 0^+$

$$\phi = U_\infty \alpha \left[s^2(x) - y^2\right]^{1/2}. \qquad (6.32)$$

Similarly for the lower surface, for $z = 0^-$

$$\phi = -U_\infty \alpha \left[s^2(x) - y^2\right]^{1/2}. \qquad (6.33)$$

From the definition of velocity component in the streamwise direction, one can use the above expressions to calculate the pressure differential across the plate as following

$$\Delta C_p = C_{p_l} - C_{p_u} = \frac{2u_u}{U_\infty} - \frac{2u_l}{U_\infty}.$$

Despite the fact that we are looking at the problem in (y, z)-plane, we can calculate u_u and u_l, because of the implicit dependence of ϕ on x through $s(x)$ and the pressure differential depends upon $\frac{ds}{dx} \neq 0$, so that

$$\Delta C_p = \frac{4\alpha\ s(x)}{[s^2 - y^2]^{1/2}} \frac{ds}{dx} \tag{6.34}$$

$$= \frac{4\alpha}{\sin\theta} \frac{ds}{dx},$$

where $y = s(x)\cos\theta$

Note that $\Delta C_p \to \infty$ when $y \to s(x)$, yielding the infinite leading edge suction.

Having obtained ΔC_p, one can integrate the same to obtain the lift acting on such a wing from

$$\delta L = \left[\frac{1}{2}\rho\, U_\infty^2 \int_{-s}^{s} \frac{4\alpha\ s(x)}{(s^2 - y^2)^{1/2}} \frac{ds}{dx} dy\right] \delta x.$$

Or

$$\frac{\delta L}{1/2\ \rho\ U_\infty^2} = 4\alpha\ s(x)\frac{ds(x)}{dx}\delta x \int_{-s}^{s} \frac{dy}{(s^2 - y^2)^{1/2}}.$$

As $y = s(x)\cos\theta$, therefore $dy = -s(x)\sin\theta\ d\theta$, and the integral in the above simplifies to

$$I = \int_{\pi}^{0} \frac{-s(x)\sin\theta\ d\theta}{s(x)\sin\theta} = \pi.$$

Therefore

$$\frac{\delta L}{1/2\ \rho\ U_\infty^2} = 4\pi\alpha\ s(x)\frac{ds(x)}{dx}\delta x.$$

If S_w is the planform area of the wing, then the lift coefficient is given by

$$C_L = \frac{\int \delta L}{1/2\ \rho U_\infty^2 S_w} = \frac{4\pi\alpha}{S_w} \int_0^c s(x)\frac{ds(x)}{dx}\, dx$$

$$= \frac{2\pi\alpha}{S_w} \int_0^c \frac{d}{dx}(s^2)\, dx = \frac{2\pi\alpha}{S_w} S^2 = \frac{\pi\alpha}{2}\left(\frac{4S^2}{S_w}\right).$$

Hence

$$C_L = \frac{\pi\alpha}{2} AR. \tag{6.35}$$

Thus the lift curve slope is obtained as

$$\frac{dC_L}{d\alpha} = \frac{\pi AR}{2}. \tag{6.36}$$

Theoretical C_{L_α} of a wing for the limiting case of $AR \rightarrow 0$ is found to be in good agreement with experiments, only for small angles of incidence.

6.5 Spanwise Loading

Consider the plane low aspect ratio wing shown in Figure 6.6, which has the straight trailing edge, as required in slender wing theory. The planform has a pointed apex and the straight trailing edge has the semi-span, S. To calculate the lift on a chordwise strip of width δy at a distance y from the wing centreline, we note that the leading edge of the strip is given by $x_L(y)$. At that chordwise location, the local span is given by $s(x)$. The lift acting on this strip in Figure 6.8 is obtained by using ΔC_p from Equation (6.34) as

$$\delta L = \left[\frac{1}{2}\rho U_\infty^2 \int_{x_L(y)}^{c_0} \Delta C_p \, dx\right]\delta y = \left[\frac{1}{2}\rho U_\infty^2 \int \frac{4\alpha \, s(x)}{[s^2 - y^2]^{1/2}} \frac{ds}{dx} dx\right]\delta y$$

$$\delta L = \frac{1}{2}\rho U_\infty^2 4\alpha \left[(S^2 - y^2)^{1/2} - \left(s^2(x) - y^2\right)^{1/2}\right]\delta y. \tag{6.37}$$

Hence the spanwise loading is elliptic.

To investigate the Kutta condition at the trailing edge, we note that for unswept trailing edge, this condition is satisfied only when $\frac{ds}{dx}$ at $x = c_0$ is zero.

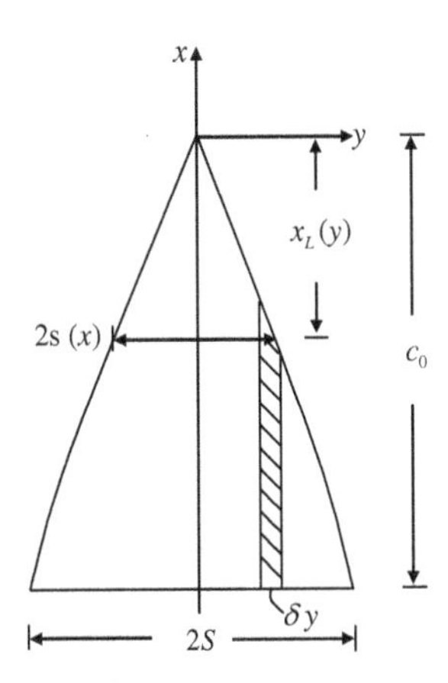

Figure 6.8 Spanwise loading of a low aspect ratio wing as obtained using slender wing theory

Since the spanwise loading is elliptic, one can immediately state the induced drag of the slender wing to be given by

$$C_{D_v} = \frac{C_L^2}{\pi AR} = C_L \frac{\alpha}{2}. \tag{6.38}$$

6.6 Lift on Delta or Triangular Wing

Consider the delta or the triangular planform wing shown in Figure 6.9, which is characterized by the semi-apex angle Λ. To calculate the potential lift acting on this planform, one uses from Equation (6.34) the differential pressure coefficient as

$$\Delta C_p = \frac{4\alpha\, s(x)}{(s^2 - y^2)^{1/2}} \frac{ds}{dx}.$$

From Figure 6.9 one notes that at any streamwise station x,

$$s(x) = x \tan \Lambda.$$

And thus

$$\frac{ds}{dx} = \tan \Lambda \ \ (= C_1).$$

Therefore

$$\Delta C_p = 4\alpha\, C_1 \frac{s(x)}{(s^2 - y^2)^{1/2}}.$$

Using the transformation $y = s(x) \cos \theta$ one further simplifies the above

$$\Delta C_p = 4\alpha\, C_1 / \sin \theta.$$

This expression clearly shows that the loading is constant along rays on which θ remains constant. From the pointed apex, these lines are the rays and this property of constant loading along rays constitute conical flows.

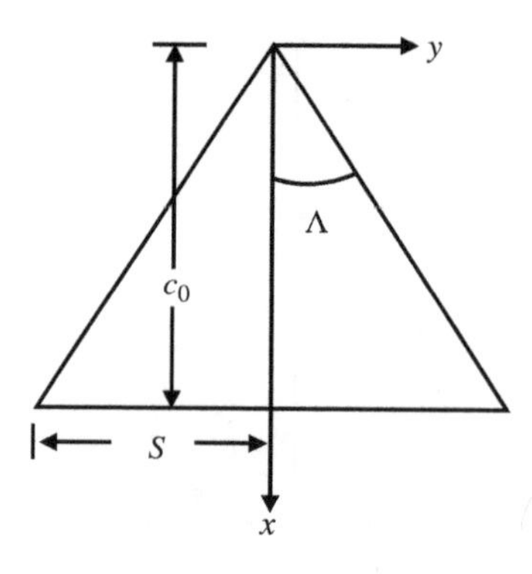

Figure 6.9 Delta wing or triangular wing characterized by the semi-apex angle Λ

One can also use Equation (6.34) to obtain the lift as

$$L = \frac{1}{2} \rho U_\infty^2 \int \frac{4\alpha\, s(x)}{\sqrt{s^2 - y^2}} \frac{ds}{dx} dy\, dx.$$

Or one can directly use the formula for the lift coefficient in Equation (6.35), and for the geometry of Figure 6.9 one gets

$$AR = \frac{4S^2}{S_w} \quad \text{and } S_w = \frac{1}{2}\, 2S\, c_0.$$

Since $S = c_0 \tan\Lambda$, one obtains

$$AR = \frac{4S^2}{S\, c_0} = \frac{4S}{c_0} = 4\tan\Lambda.$$

Thus, the lift coefficient can be obtained for the delta wing as

$$C_L = 2\pi\alpha \tan\Lambda. \tag{6.39}$$

Note that this lift is created by the pressure distribution, as obtained from ΔC_p. The other component of C_{L_p} shown in Figure 6.10, which would have caused a drag, similar to induced drag, is cancelled by the leading edge suction created in the slender wing theory. We will note that at higher angles of attack, a leading edge vortex is created, which actually turns this leading edge suction, in fact, creates additional lift, termed in the following as the vortex lift and is investigated next.

6.6.1 Low Aspect Ratio Wing Aerodynamics and Vortex Lift

We noted the lift experienced by the delta wing in Equation (6.39) is strictly due to pressure distribution, which acts normal to the surface of the wing. Thus, it is important to realize that this pressure induced lift acts perpendicular to the planform. Denoting the resultant lift acting upon the slanted surface shown in Figure 6.10 in the side-view by C_{L_p}, one can obtain the lift

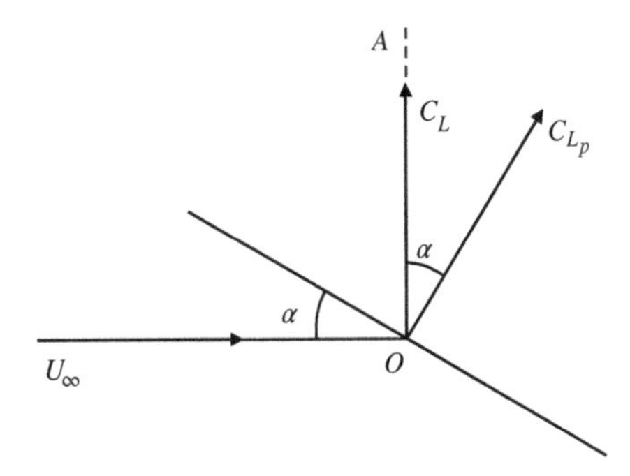

Figure 6.10 Potential flow lift acting on a low aspect ratio delta wing and its representation

acting perpendicular to the oncoming flow direction as

$$C_L(\text{along OA}) = C_{L_p} \cos \alpha,$$

where

$$C_{L_P} = 2\pi\alpha \tan \Lambda.$$

However, low aspect ratio delta wings flying in symmetric flight condition display a pair of vortices attached to the leading edge, as shown in Figure 6.11. Apart from the primary vortex shown attached to the leading edge of the delta wing, one also notices secondary weaker vortices attached inboard of the primary vortex. Thus over the wing, apart from the streamwise flow in the inboard, one also notices flow going to the outboard direction, beneath the primary vortex. Additional tip flow is noted in the near vicinity of secondary vortices. The presence of these vortex systems over the delta-wing makes such wing planform very attractive for practical usage.

Phenomenologically, at small angles of attack one notices good correlation between potential lift with the measured lift, where the potential lift indicated above by C_{L_p} is obtained based on attached flow model with Kutta condition applied at the trailing edge by the developed slender wing theory. However, with increase in angle of attack, one notices the formation of attached leading edge vortex which provides additional lift, termed as the vortex lift, with the major contribution coming from the primary vortex pair. The pair of vortex forms with its axis remaining aft of the leading edge and at the trailing edge it merges with the tip vortices.

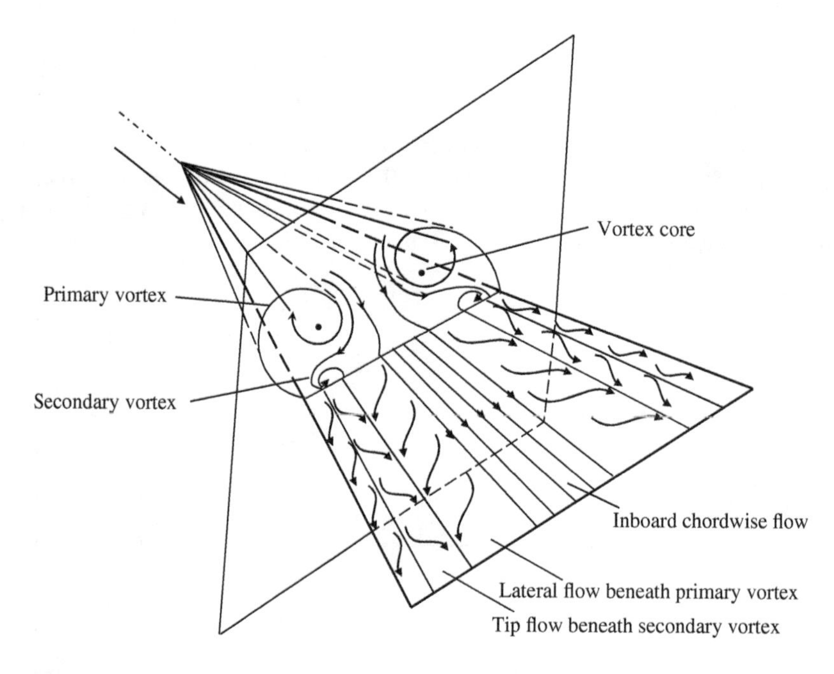

Figure 6.11 Flow over a delta wing in symmetric flight at high angle of incidence

Pohlhamus (1966, 1971a) hypothesized that the primary vortex pair causes the stagnation line to move on the top surface, rotating the suction force vector by ninety degrees to create a vortex lift. This turning of drag into lift was modelled empirically by Pohlhamus, who stated that the drag caused in the cross flow plane is essentially the additional lift created by the separated flow pattern.

In this model, for large angles of attack the potential flow lift is written as

$$C_{L_p} = \underbrace{2\pi \tan \Lambda}_{K_P} \, \sin\alpha \cos\alpha, \tag{6.40}$$

where we have replaced α by $\sin\alpha$ in Equation (6.39). To estimate the vortex lift, we consider a spanwise strip on the delta wing of width dx and length $2s(x)$, which experiences a cross flow velocity $U_\infty \sin\alpha$. If the drag coefficient experienced by this equivalent two-dimensional section as C_{D_p}, then the cross flow creates an equivalent *drag force* of magnitude

$$\frac{1}{2}\rho\, U_\infty^2 \sin^2\alpha \; 2s(x)\; C_{D_p}.$$

Usually, this equivalent flat plate drag coefficient is approximately equal to 1.95, as suggested in Houghton and Carpenter (1993). Since, this *drag* acts perpendicular to the planform, then the actual vortex lift is equivalent to this and is given by

$$L_v = \frac{1}{2}\rho\, U_\infty^2 \sin^2\alpha \; C_{D_p} \int_0^{c_0} 2s(x)\, dx.$$

This lift also acts perpendicular to the plane wing. As $s(x) = x\tan\Lambda$, with respect to the free stream the vortex lift coefficient is given by

$$C_{L_v} = C_{D_p} \sin^2\alpha \cos\alpha. \tag{6.41}$$

Thus, the total lift experienced by a delta wing is given by

$$C_L = K_p \; \sin\alpha \cos^2\alpha + K_v \sin^2\alpha \cos\alpha, \tag{6.42}$$

where K_v is the coefficient associated with the vortex lift and would be a function of wing planform shape and flow conditions. This has been generalized for arrow, delta and diamond planforms in the literature, including which are reported in Pohlhamus (1971). The definition sketch for the composite geometry which can accommodate all these planforms is given in Figure 6.12a. Corresponding lift coefficient functions K_p and K_v are shown in Figures 6.12b and 6.12c. Variation of K_p and K_v with AR for arrow, delta and diamond wings are shown in Figures 6.12b and 6.12c, as reported in Pohlhamus (1971). As this model of lift coefficient is based upon potential flow with leading edge suction analogy, the agreement between experiments and the empirical correlation breaks down even at moderate angles of attack.

The delta planform is favoured for its low speed performance due to the presence of leading edge vortex systems shown in Figure 6.11. This planform allows operation of aircrafts with such wing at very high angles of attack without flow separation, stall and abrupt loss of lift. In fact, the angle at which such wing planform can operate during landing and take-off has

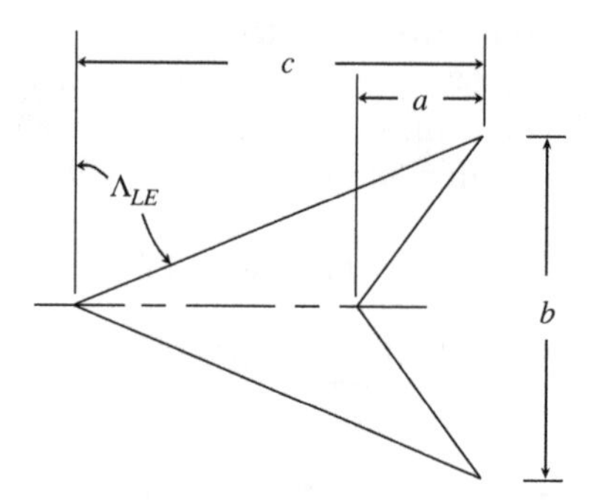

Figure 6.12(a) Arrow, delta and triangular wing characterized by the apex angle Λ

forced the designers of Concorde aircraft to provide a nose droop during landing for the pilot to have better visibility ahead.

The lift curve slope of delta type wing is given by

$$\frac{dc_L}{d\alpha} = K_P \cos 2\alpha + K_v(2 \sin \alpha \cos^2 \alpha - \sin^3 \alpha)$$

$$= K_P \cos 2\alpha + K_v(2 \sin \alpha - 3 \sin^3 \alpha).$$

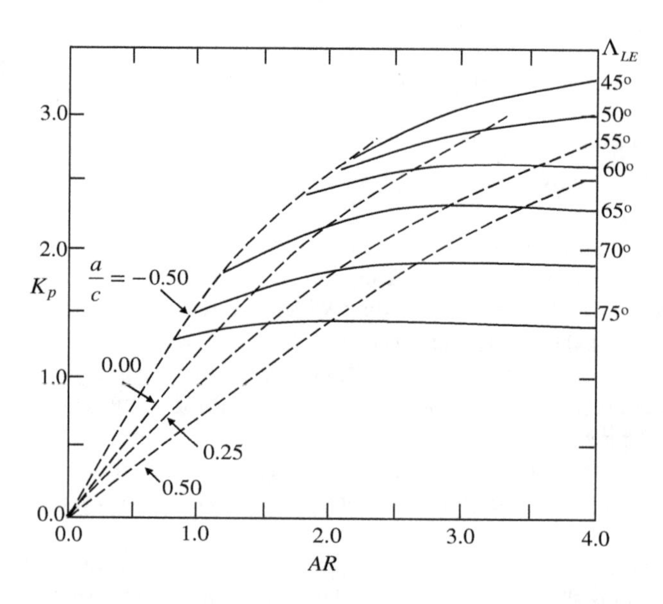

Figure 6.12(b) Variation of K_p with aspect ratio for different low aspect ratio triangular wing. The geometry is as given in Figure 6.12(a)

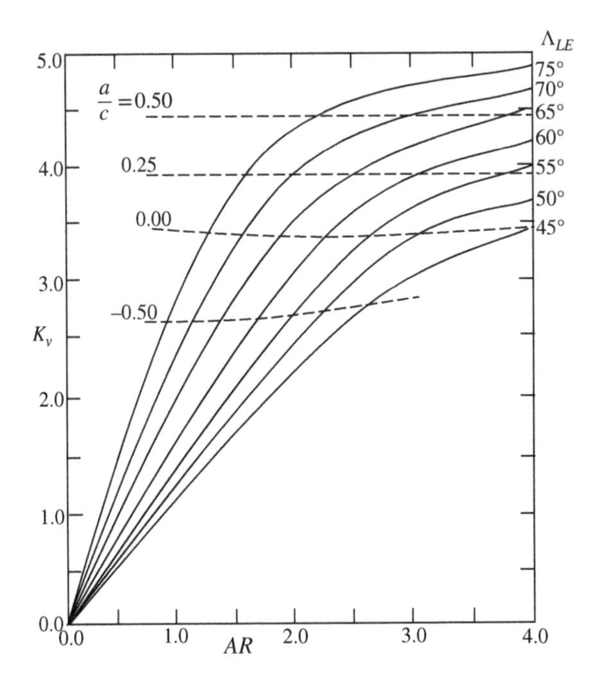

Figure 6.12(c) Variation of K_v with aspect ratio for different low aspect ratio triangular wing

In Figure 6.13, the potential and vortex lift experienced by a 70-degree delta wing is shown for variation with angle of attack, with the results taken from Pohlhamus (1971) and Heron (2007). It is clearly evident that with increase in angle of attack, the potential lift decreases due to breakdown of attached flow model used to evaluate C_{L_p}, while the vortex lift continues to increase, again due to the separation of flow from the leading edge forming the attached primary vortex system. In reality the total lift variation with angle of attack is different, as shown in Figure 6.13 with the experimental data taken from Heron (2007). It is interesting to note the experimental data points in Figure 6.13, shows saturation of total lift and subsequent fall of the same for $\alpha \geq 40°$.

The reason for the gentler drop of lift with high angles of attack of a delta wing is qualitatively different from that is noted for high aspect ratio wings. We have already noted that the vortex lift is due to the primary vortex attached to the leading edge of the delta wing, as shown in the sketch of Figure 6.11. This primary vortex is extremely stable due to its structure, which is explained by the feature of the axial and azimuthal (also called the swirl) velocity distribution in and around the primary core shown in Figures 6.14 and 6.15, respectively.

According to Lee and Ho (1990) and Mitchel and Molton (2002), the origin of vortex lift can be explained with the vorticity transport equation (VTE), which is nothing but the Navier–Stokes equation. VTE provides the balance between the generation of vortices due to no-slip condition and its convection and diffusion given by the various operators in the Navier–Stokes equation. While this is the general principle for all viscous flows, the delta wing at relatively higher angle of attack has the additional vorticity generated by the leading edge separation and subsequent reattachment of the primary bubble. The connector between the primary vortex core transports vorticity from the leading edge to the core and creates

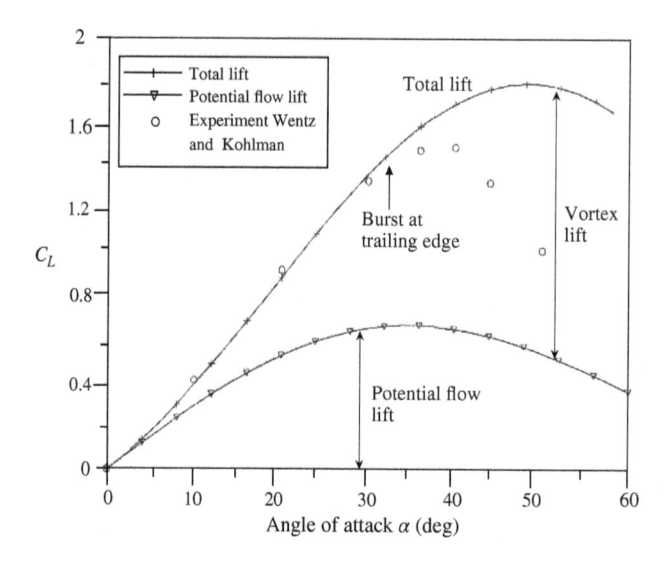

Figure 6.13 Comparison of lift data between empirical correlation provided in Equation (6.42) with experimental data in Heron (2007). Total lift is split into potential and vortex lift contributions

two additional thin shear layers. These shear layers also generate additional vorticity due to an instability mechanism of Kelvin-Helmholtz type (described in Chapter 9) which feeds the primary core. This is an additional mechanism of extracting energy from the free stream. This vorticity is convected by the primary free stream flow, with the velocity at the primary core seen to be very high in Figure 6.14, as compared to the free stream speed.

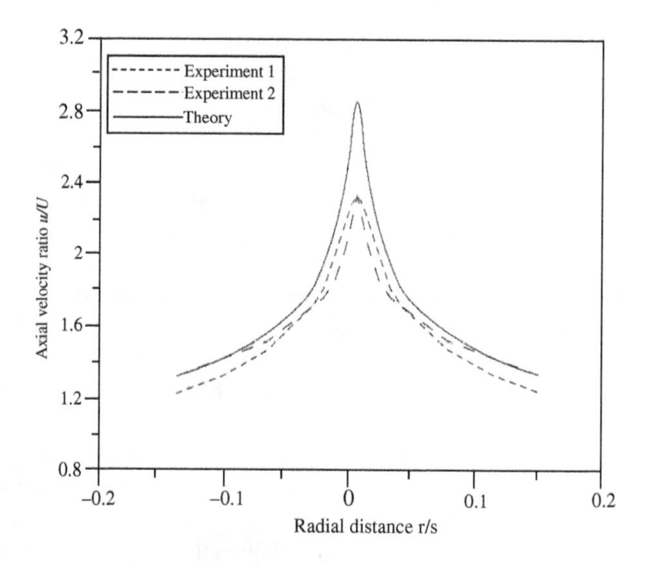

Figure 6.14 Axial velocity in the core of the primary vortex, as computed theoretically and compared with experimental data. *Source:* Heron (2007)

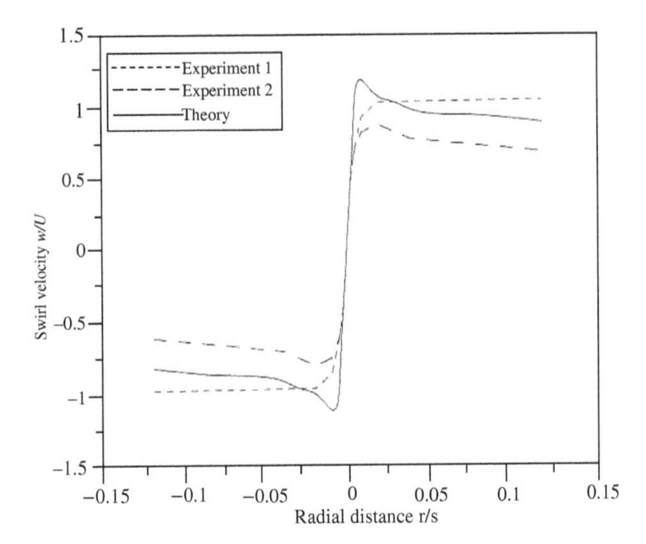

Figure 6.15 Azimuthal or swirl velocity in the core of the primary vortex, as computed theoretically and compared with experimental data. *Source:* Heron (2007)

Core of the primary vortex is nearly axi-symmetric and does not display discontinuity of vorticity distribution. The core's dimension increases with streamwise distance, with its inner structure showing viscous and inviscid subcores, with its size increasing almost up to 10% of wing-span. The axial velocity profile, shown in Figure 6.14, displays jet-like behaviour with very high peak velocity. This high velocity of the core is also responsible for strong entrainment of surrounding air which leads to strong stability of the primary vortex, even at very high angles of attack associated with adverse pressure gradient in the lee-side of the wing. Additionally the axial and swirl components of velocity is very tightly coupled, as both the components display the same primary vortex describing a helical path. Such high rotation of the core also stabilizes the primary vortex as in gyroscopic motion against ambient disturbances. Even when viewed separately, the swirl component also shows the presence of a viscous subcore, surrounded by a potential vortex. In Leibovich (1984), both these components are described empirically as

$$u(r) = U_1 + U_2 e^{-Cr^2}$$

and

$$w(r) = \frac{V_1}{r}\left[1 - e^{-Cr^2}\right].$$

The constants, U_1, U_2, V_1 and C can be found experimentally.

The vortex lift and its stability with respect to ambient disturbances deteriorate with further increase in angle of attack. The increased receptivity to disturbances at higher angles of attack is directly related to progressive increase in adverse pressure gradient on the lee-side of the wing as one approaches the trailing edge of the delta wing. Thus, the lift degradation occurs due to disruption in the primary vortices near the trailing edge. With further increase in angle of attack, adverse pressure gradient becomes more and more prominent at locations upstream. This is noted in the gentle loss of lift in Figure 6.13. This deterioration of lift associated with loss of coherence of primary vortex is termed as vortex breakdown and is the topic of discussion next. A schematic of vortex breakdown is shown in Figure 6.16, showing a dye visualization picture for flow over a delta wing at high angle of attack.

6.7 Vortex Breakdown

We note that the creation of lift by attached leading edge vortex is essentially a nonlinear phenomenon and the loss of lift is gradual in this flow field, due to vortex breakdown at angles beyond 35° to 40°, as shown in a typical plot in Figure 6.13. The primary vortex has a stable core and this vortex induces effects which make the wing withstand higher angles of attack. Despite the presence of vortex lift, the total lift sustained by a delta wing remains very low. As compared to any large aspect ratio wing, which has a theoretical estimate of lift curve slope of 2π per radian (or roughly about 0.1 per degree), a delta wing has a value which is order of magnitude smaller. For example, Anderson (2010) reports a 60° delta wing lift curve, which has $C_{L_\alpha} = 0.05$ per degree, although the wing sustains a $C_{L_{max}} = 1.3$ at 35°. One of the redeeming features of the delta wing is its stall characteristics, which is very gentle and is hence preferred for high maneuverability of military aircraft, which when slowing down from high supersonic to subsonic speeds during a manoeuvre can experience an angle of attack far in excess of that is common for transport aircrafts. These types of aircrafts most

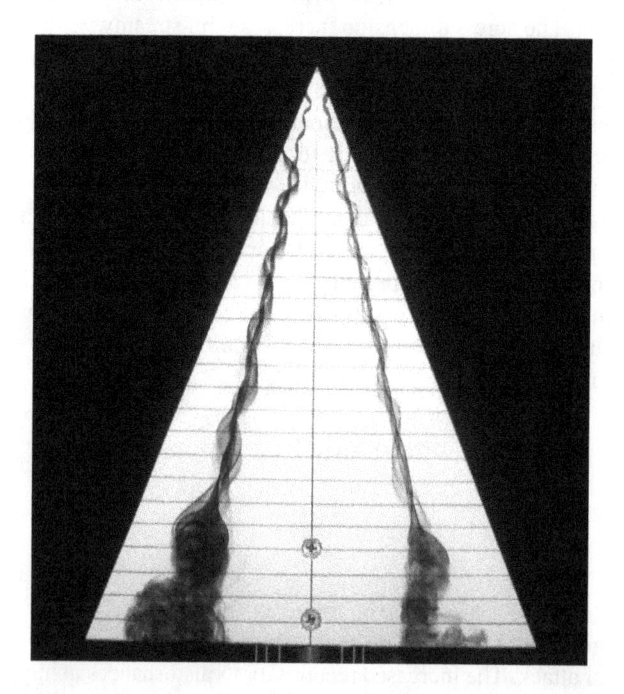

Figure 6.16 Vortex breakdown over delta wing at high angle of attack shown by flow visualization technique. *Source:* Courtesy of T.T. Lim, NUS Singapore. (See colour version of this figure in colour plate section)

often flying at supersonic speed have sharp leading edge, which serves two purposes. First, at supersonic speeds the wave drag of such wings with sharp leading edges are significantly lower as compared to wings with rounded leading edge. Secondly, a wing with sharp leading edge has a very well-defined point of separation and the load experienced remains quite steady. In contrast, any wing with rounded leading edge has the tendency to show unsteady movement of the point of separation, with the resultant wing loading unsteady. Although one is discussing irrotational flows in this chapter, it is worthwhile to remember that the drag experienced by a low aspect ratio wing aircraft will be significantly higher as compared to high aspect ratio wings and for reason of fuel economy alone, readers will not find transport or cargo aircrafts with low aspect ratio wing, with the sole exceptions of Concorde and TU-144 aircrafts, both of which are no more operational.

One of the notable exception to the above is the highly rounded shape of wing leading edge noted for the space shuttle, the aerodynamics of which is different as it operates at hypersonic speeds. The rounded leading edge of space shuttle is due to multiple reasons of (i) avoiding high heat transfer during re-entry, for which a wing with sharp leading edge will melt and (ii) the requirement of aerodynamic braking action of the shuttle due to large drag needed to slow down the vehicle to subsonic speed, where such rounded shapes are known also to provide higher aerodynamic efficiency in terms of lift to drag ratio.

The reason that the delta wing loses lift slowly, as compared to large aspect ratio wings, is the way unsteady separation spreads over these two classes of wings. For high aspect ratio wings, adverse pressure gradient is experienced near the leading edge of the wing, for increased angle of attack. Once the flow separates there, the flow degradation is global over the full chord length at any spanwise station. In contrast, for low aspect ratio wings the presence of primary and secondary leading edge vortices stabilize the flow over the wing at high angles of attack. Only when the angle of attack is increased greatly, then these primary vortices suffer instability starting from near the trailing edge of the wing. Such flow instability leads to *vortex breakdown* for the primary system, which accounts for the predominant fraction of vortex lift. At these relatively high angles of attack, the core of the primary vortices lose coherence at more and more upstream location with increasing α. Due to this progressive upstream movement of the location of *vortex breakdown*, the loss of lift is also similarly slow. Early experimental demonstration of vortex breakdown was shown in Lamboune and Bryer (1962). A recent rendition of the same phenomenon is shown in Figure 6.16, by flow visualization using dye injection from locations near the leading edge of the delta wing placed at very high angle of attack. One can note the visualized primary vortex made visible by the two different colours of the dye. The coherence of the primary vortex is kept intact up to long streamwise stretch, as evidenced by the tightly wrapping of the two coloured dyes. The loss of coherence due to instability of the primary vortex core is a distinctly visible diffusion of the dyes spread over a larger area of the planform.

The variation of flow features over delta wings of different semi-apex angle at different angles of attack is shown in Figure 6.17 taken from Heron (2007). For a fixed semi-apex angle of $\Lambda = 70°$, if the angle of attack is kept very small, then the vortex lift is absent and the lift is totally of potential flow origin, with practically nonexistent leading edge separation. With increase in angle of attack, flow starts separating from the leading edge giving rise to vortex lift, which increases with α till an angle of attack is reached when the vortex breakdown occurs at the trailing edge, as also indicated in Figure 6.13. In the region III of Figure 6.17, the progressive loss of lift is associated with point of vortex breakdown moving upstream. This is also referred

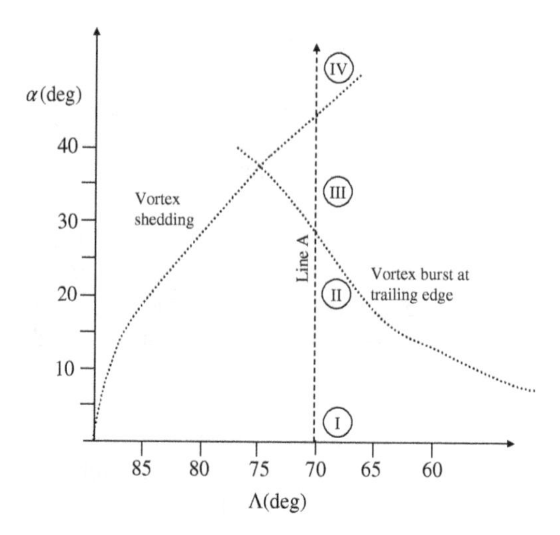

Figure 6.17 Flow field features of delta wing shown for different apex angles for different angles of attack

to as vortex bursting in Heron (2007), implying a sudden increase in the size of the core, as if it is decelerated substantially and diffusion of the core is visible, as noted in Figure 6.16. This diffusion is more than the molecular diffusion, due to rapid mixing arising out of turbulence being set up following the nonlinear stage of instability. When the angle of attack is increased further, then the vorticity production is more than what can be convected and diffused, leading to accumulation of vorticity in the core, which after some interval of time becomes untenable, leading to shedding of vortices. This process is inherently asymmetric and the shedding is alternating type. Such periodic asymmetric variation of lift also causes a rolling moment which is at the frequency of vortex shedding and this phenomenon is also called a wing rock. It is noted that the tendency of wing rock increases with larger semi-apex angle, as shown in Figure 6.17.

6.7.1 Types of Vortex Breakdown

Vortex breakdown is associated with the instability of primary vortex core which appears to be aligned as in a conical flow. As the core defines axial and azimuthal motion in a strongly coupled manner, the instability can occur in two principal way. First, it can occur in an axisymmetric manner with significant slowing down of flow in the core and attendant radial bulge is followed by breakdown, which is called the bubble-type breakdown. The other type of breakdown occurs in a nonaxisymmetric manner, where the core defines irregular helical path and is termed as the spiral-type breakdown. Spiral type breakdown is more common than the bubble-type breakdown in actual flow.

These two principal ways of *vortex breakdown* are shown in the flow visualization pictures in Figures 6.18a and 6.18b. The studies of such breakdown mechanisms are often conducted in a cylindrical vessel to create a predominantly axial motion. The azimuthal or swirl motion is created by imparting rotation to the flow, either physically rotating the vessel or by indirect means. This type of arrangement helps in creating vortex breakdown in a reproducible manner in the lab. In the lab scenario, the core of the vortex meanders in spiral breakdown, as shown

Figure 6.18(a) Spiral-type vortex breakdown created in a lab facility inside a cylindrical container. *Source:* Lim and Cui (2005). Reproduced with permission from the American Institute of Physics

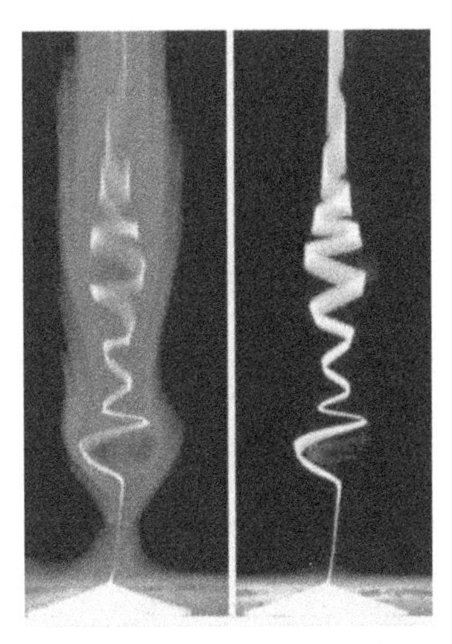

Figure 6.18(b) Bubble-type vortex breakdown, created and visualized in a lab environment. *Source:* Lim and Cui (2005). Reproduced with permission from the American Institute of Physics

in Figure 6.18a, which occurs over a stretch in the streamwise direction. This figure depicting the spiral-type vortex breakdown is from Lim and Cui (2005). Bubble type *vortex breakdown* is noted along with axisymmetric bulge of the vortex core in the second type, which leads to very rapid growth of the radius of the core and breakdown occurs suddenly, resulting in strong unsteadiness of the flow and buffeting of the wing. Note that Lamboune and Bryer (1962) experimentally created both type of breakdown in the same experiment on two sides of the same delta wing at high angle of attack.

Both types of vortex breakdown are due to growth of disturbances in and around the core of the primary vortices over the delta wing. At larger angles of attack, flow on the lee-side of the wing experiences adverse pressure gradient which destabilizes the flow in the core and the growing disturbances follow either of the routes described above. The phenomenon of eventual vortex breakdown relates to the secondary and tertiary stages of disturbance growth. The distinctive feature of vortex bursting is the sudden expansion in the radial size of the vortex, and the axial velocity, which increased linearly along the streamwise direction to almost three times the free stream speed, decreases rapidly within one to two core diameters to form a saddle point. Details of of the physical events leading up to vortex breakdown over a delta wing is documented by Wedemeyer (1982), Leibovich (1984) and Mitchell and Molton (2002). The life cycle of vortex breakdown in a controlled lab environment is shown graphically in Figure 6.19, which is produced by Cui, Lim and Lopez (2013) to visually demonstrate the growth of the disturbances following the stages of disturbance growth over delta wing.

The location where the vortex eventually bursts is not fixed, showing temporal fluctuation with respect to axial and cross flow directions. In contrast, for spiral type breakdown, the vortex core defines an expanding helical corkscrew, rotating in the same direction of the original leading edge vortex, as described in Wedemeyer (1982). For flow over delta wings occurrence of spiral breakdown is more common, as noted in Sarpakaya (1971). A more recent review of vortex breakdown is provided by Gurusal (2004).

6.8 Slender Body Theory

This is a generalization of slender wing theory to treat wing-body combinations, which is of interest for bodies of revolution fitted with small control surfaces. This is also based on irrotational flow model and the flow field and loads are estimated for such three-dimensional bodies by considering thin slices in cross flow planes and summing up the effects over the full body by superposition, as shown in Figure 6.20. The body is shown to be inclined at an angle of attack, α, to the oncoming flow with speed, U_∞. In the side view, only one control surface is visible along the x-axis. Looking at the cross-section at the marked station AA, shown on the right of Figure 6.20 as in the yz-plane, one notices the symmetrically placed pair of control surfaces. As we proceeded in slender wing theory, we will consider the flow in the cross-flow plane in isolation and consider the cross flow to show ideal potential flow features, which is the lift acting on the wing-body combination. Here, the body of revolution is characterized by the radius, R and wing by the semi-span, s.

Once again the complex potential for the cross flow is written for the complex variable, $Z = y + iz$, for the vertical velocity $U_\infty\, \alpha$, with the help of velocity potential and stream function by

$$W(Z) = \phi(y, z) + i\psi(y, z). \tag{6.43}$$

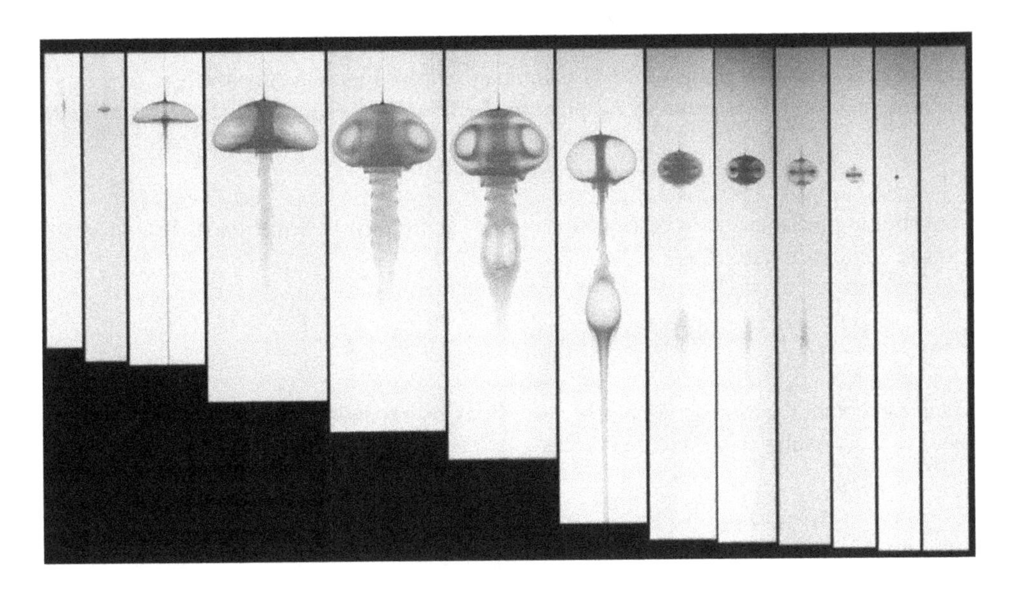

Figure 6.19 Life cycle of bubble type vortex breakdown obtained by dye visualization. The breakdown is created inside a confined cylindrical container and the phenomenon is to be tracked from right to left. *Source:* Cui (2009). (See colour version of this figure in colour plate section)

This is constructed in two steps via the following conformal mappings. The first Jukowski transformation takes one from the complex Z-plane to Z_1-plane via

$$Z_1 = Z + R^2/Z. \tag{6.44}$$

The transformed body stretches the semi-span to $s_1 = s + R^2/s$. This transformation takes the wing-body from Z-plane to a slit in the Z_1-plane. The flow is still normal to this slit. For the ease of constructing a complex potential, one transforms the horizontal slit in Z_1-plane to a vertical slit in Z_2-plane, by the application of the second Jukowski transformation:

$$Z_2 = (Z_1^2 - s_1^2)^{1/2}. \tag{6.45}$$

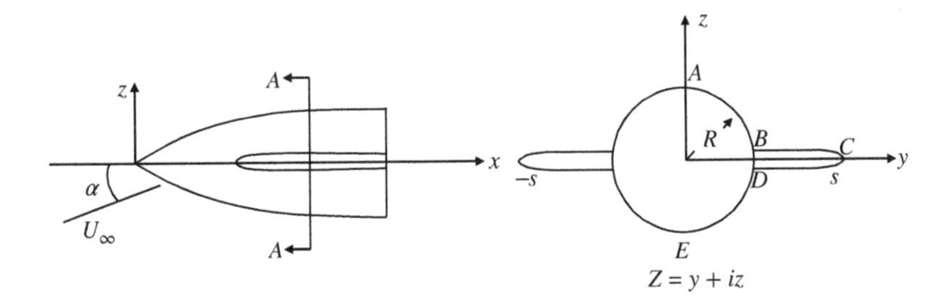

Figure 6.20 Flow past a wing-body combination analyzed by conformal mapping

The requisite Jukowski transforms are depicted further in Figure 6.21, which shows the last two transforms take the wing-body combination from Figure 6.20. Now it is very easy to construct the complex potential in Z_2-plane as the flow is now along the slit and is given by

$$W(Z_2) = i\, U_\infty\, \alpha Z_2. \tag{6.46}$$

Substituting the transforms of Equations (6.44) and (6.45) in Equation (6.46), we get the requisite complex potential as

$$W(Z) = -i\, U_\infty\, \alpha \left[(Z + R^2/Z)^2 - (s + R^2/s)^2 \right]^{1/2}. \tag{6.47}$$

One can obtain the load acting on the wing-body by expanding Equation (6.47) in inverse power of Z as Ashley and Landahl (1965):

$$W = -i\alpha U_\infty \left\{ Z - \frac{1}{2Z} \left[\left(s + \frac{R^2}{s} \right)^2 - 2R^2 \right] + \cdots \right\}. \tag{6.48}$$

One can use this in Blasius' Theorem to calculate the lift acting from the residue evaluated for the pole of Equation (6.48) at $Z = 0$ as in Ashley and Landahl (1965),

$$L = \pi \rho_\infty\, U_\infty^2\, \alpha \left(s^2 - R^2 + \frac{R^4}{s^2} \right)_B, \tag{6.49}$$

where the subscript B on the right hand side indicates the quantities to be evaluated at the base of the body. From the general result of Equation (6.49), one can evaluate the lift acting on the body alone by substituting $s_B = R_B$ as

$$L_B = \pi \rho_\infty\, U_\infty^2\, \alpha R_B^2.$$

Thus, the lift coefficient is obtained (based on the base area) as

$$C_{L_B} = 2\alpha.$$

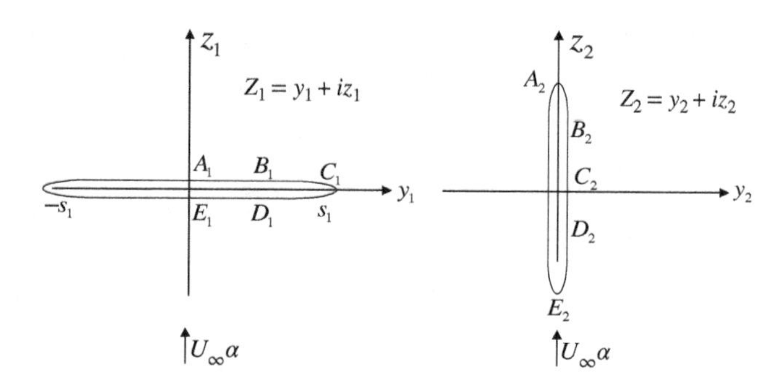

Figure 6.21 Jukowski transforms used to calculate potential lift for flow past a wing-body combination

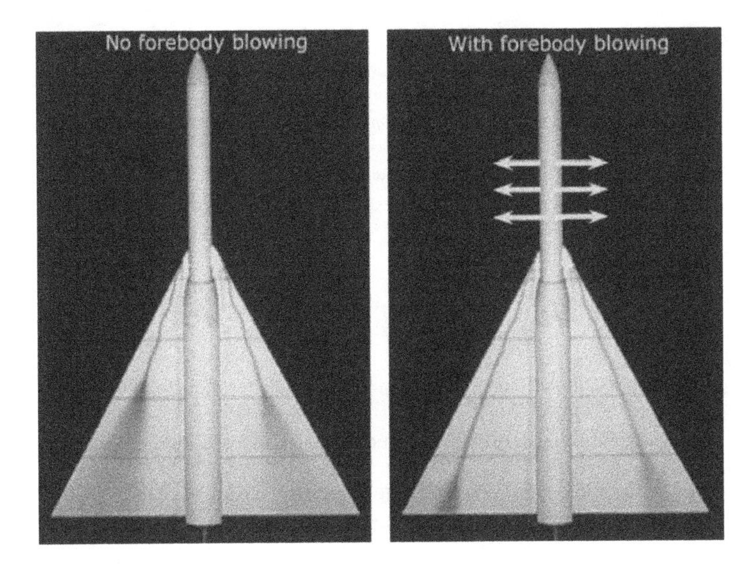

Figure 6.22 Control of vortex breakdown on delta wing by spanwise blowing from the forebody. On the left is the uncontrolled flow over the wing showing vortex breakdown by the radial diffusion of the injected dyes. Same flow is controlled by spanwise blowing shown on the right frame. *Source:* Courtesy of T.T. Lim

Similarly, the lift acting on the wing alone is obtained by substituting $R_B = 0$ in Equation (6.49) as

$$L_W = \pi \rho_\infty \, U_\infty^2 \, \alpha s_B^2.$$

Correspondingly, the lift coefficient is obtained as before as

$$C_{L_W} = \frac{\pi}{2} AR\alpha.$$

The body alone result is due to Munk (1924) and wing alone result is due to Jones (1946).

From Equation (6.49) it may appear that the effect of fuselage is to cause loss of lift only. However, sometimes it is possible to improve upon the lift created on the wing, by providing active control on the forebody. In Figure 6.22, we show this effect by spanwise blowing from the forebody, which delays vortex breakdown on the delta wing.

Bibliography

Anderson, J.D. (2010) *Fundamentals of Aerodynamics*. Fifth special Indian edn., Tata McGraw-Hill, New Delhi, India.

Ashley, H. and Landahl, M. (1965) *Aerodynamics of Wings and Bodies*. Dover Publ. Inc., New York, USA.

Cui, Y.D. (2009) Studies of vortex breakdown and its stability in a confined cylindrical container. Ph. D. thesis, National Univ. Singapore.

Falkner, V.M. (1948) The solution of lifting plane problem by vortex lattice method. *British ARC*, R & M **2591**.

Gurusal, I. (2004) Recent developments in delta wing aerodynamics. *The Aeronautical J.*, **108**, 1087, 437–452.

Heron, I. (2007) *Vortex burst behaviour of a dynamically pitched delta wing under the influence of a von Karman vortex street and unsteady freestream.* Ph. D. thesis submitted to Dept. of Aerospace Engineering, Wichita State Univ., USA.

Houghton, E.L. and Carpenter, P.W. (1993) *Aerodynamics for Engineering Students.* 4th edn., Edward Arnold Publishers Ltd., UK.

Jones, R.T. (1946) Properties of low-aspect ratio pointed wings at speeds below and above the speed of sound. *NACA Report* **835**.

Lamboune, N.C. and Bryer, D.W. (1962) The bursting of leading edge vortices: Some observation and discussion of the phenomenon. *British ARC* R & M **3282**.

Lee, M. and Ho, C.M. (1990) Lift force of delta wings. *ASME Applied Mechanics Review*, **43**(9), 209–221.

Leibovich, S. (1984) Vortex stability and breakdown: Survey and extension. *AIAA J.,* **22**(9), 1192–1206.

Lim, T.T. and Cui, Y.D. (2005) On the generation of a spiral-type vortex breakdown in an enclosed cylindrical container. *Physics Fluids*, **17**(4), 044105.

Mitchell, A.M. and Molton, P. (2002) Vortical substructures in the shear layers forming leading-edge vortices. *AIAA J.*, **40**(8), 1689–1692.

Multhop, H. (1950) Method of calculating the lift distribution of wings. (Subsonic lifting surface theory), *British ARC* R & M **2884**.

Munk, M.M. (1924) The aerodynamic forces acting on airship hulls. *NACA Report* 184.

Pohlhamus, E.C. (1966) A concept of the vortex lift of sharp-edge delta wing based on a leading edge suction analogy. *NASA TN D*-**3767**.

Pohlhamus, E.C. (1971) Charts for predicting the subsonic vortex-lift characteristic of arrow delta and diamond wings. *NASA TN D*-**6243**.

Pohlhamus, E.C. (1971a) Predictions of vortex-lift characteristics by a leading-edge suction analogy. *J. Aircraft*, **8**(4), 193–199.

Sarpakaya, T. (1971) Vortex breakdown in swirling conical flows. *AIAA J.*, **9**(9), 1792–1799.

Thwaites, B. (1960) *Incompressible Aerodynamics.* Oxford University Press, UK.

Wedemeyer, E.H. (1982) Vortex breakdown. AGARD LS-121; In *High Angle of Attack Aerodynamics.* Chapter 9, 1–17.

Wentz, W.H. and Kohlman, D.L. (1971) Vortex breakdown on slender sharp-edged wings. *J. Aircraft*, **8**(3), 156–161.

7

Boundary Layer Theory

7.1 Introduction

Fluid mechanics developed in two distinct tracks in early phase of its development, with one side interested in developing it as a branch of applied mathematics by neglecting the viscous effects, collectively these practitioners were called the hydrodynamicists. This dealt with ideal fluid flow with the primary intention to develop analytic solutions to problems. The major drawback of this approach is its inability to explain fluid resistance. This was of major concern to the practitioners of the other branch of fluid mechanics, who were involved in developing empirical correlations to account for fluid dynamic drag. This was the discipline of hydraulics. These branches developed in parallel, till the time of Prandtl (1904), who brilliantly showed the connection between these two disciplines. This is the subject matter of discussion in this chapter, which was briefly introduced in Section 2.14.

At the turn of the twentieth century, it was generally believed that for small coefficient of viscosity or equivalently for very large Reynolds number flows, one can 'switch off' the viscous terms in Navier–Stokes equations, giving one the so-called Euler equation. This process is fraught with difficulties, as this cannot account for drag components due to viscous actions, namely the skin friction drag and the pressure drag which arises due to additional phase shift due to viscous action. Note that in inviscid approaches, irrespective of the shape of body integrated pressure always produces zero drag – which is known as the D'Alembert's paradox. We would also like to note that the induced drag estimated in Chapter 4 for irrotational flow is also due to pressure distribution. However, this drag is estimated as a model arising due to tilting of the lift vector and this observation is actually empirical and not calculated from first principle on calculated pressure distribution. This is often a source of confusion and misconception among the beginners.

It is worth noting that the viscous terms are linear, while the retained convective acceleration terms are nonlinear. Thus, removal of viscous terms does not yield added advantage in solving the Euler's equation, instead of Navier–Stokes equation. With the added irrotational flow assumption, the governing equation becomes the linear Laplace's equation, which allows evaluation of lift force, but not the drag. For arbitrary body shapes, even this does not make

Theoretical and Computational Aerodynamics, First Edition. Tapan K. Sengupta.

Companion Website: www.wiley.com/go/sengupta

the lifting problem analytically tractable, except for some special two-dimensional flows, where one can use conformal mapping. Solution of Laplace's equation for a practical complex aerodynamic shape is feasible for lift evaluation with the advent of digital computer for incompressible flows. Some rudimentary aerodynamic shape for supersonic inviscid flow can also be analyzed. However, the most interesting and aerodynamically efficient transonic flow regime evaded analysis for decades. It was a major achievement of Murman and Cole (1971) to solve transonic flows with a simple and single approach for mixed characteristics domain. In some parts of the flow domain, flow is supersonic governed by hyperbolic PDE, while in the rest of the domain, the flow is subsonic governed by elliptic PDE. Murman and Cole showed us how to solve the complete domain by a single numerical procedure and heralded analyses of flight vehicles operating in transonic flow regime. One must concede that computing wave drag by transonic small perturbation (TSP) equation is conceptually difficult to justify for transonic flows, as in the procedure of Murman and Cole. Subsequent computations using Euler equation have partially deflected the original criticism against TSP and full potential equation (FPE) approaches. However, none of these advances answers how to compute viscous drag.

In the pre-computer era, a revolutionary advance was made by Ludwig Prandtl in 1904, when he clarified the essential influence of viscosity in fluid flows at high Reynolds numbers for unseparated flows and showed how Navier Stokes equations can be simplified to yield approximate solutions for these cases, yielding skin friction drag. Comparing Euler's equation with the Navier–Stokes equation, one notices that the discarded term in Euler's equation is the highest order derivatives (second order spatial derivatives). The inviscid approximation thus reduces the order of the differential equation. Naturally, one cannot satisfy all boundary conditions when solving the inviscid equations. This observation is used in the following in mathematically classifying any flows as perturbation problems.

7.2 Regular and Singular Perturbation Problems in Fluid Flows

The governing dimensional Navier–Stokes equation for incompressible flow can be written in vectorial notation as

$$\nabla \cdot \vec{V} = 0 \tag{7.1}$$

$$\frac{\partial \vec{V}}{\partial t} + \vec{V} \cdot \nabla \vec{V} = -\nabla p/\rho + \nu \nabla^2 \vec{V}. \tag{7.2}$$

When these equations are nondimensionalized with the help of characteristic length (L) and velocity scales (U_∞), Equation (7.1) remains the same with the spatial operator (∇) and the velocity vector, $\vec{V}$ are now with respect nondimensional independent variables. With the help of these two scales, one can construct a suitable time scale and when one uses it in Equation (7.2), one obtains the nondimensional Navier–Stokes equation as

$$\frac{\partial \vec{V}}{\partial t} + \vec{V} \cdot \nabla \vec{V} = -\nabla p + \frac{1}{Re} \nabla^2 \vec{V}, \tag{7.3}$$

where p has been nondimensionalized by ρU_∞^2 and $Re = \frac{U_\infty L}{\nu}$.

For Navier–Stokes equation, Re plays the role in defining a small parameter, $\epsilon = 1/Re$. If one simple-mindedly removes so-called *small terms* multiplied by this parameter, then one gets the reduced order Euler equation. In the perturbation method, the solution of this is known to constitute the outer solution. As a consequence of reducing order of the governing equation, one cannot satisfy all the boundary conditions. For viscous fluid flow, this leads to inability of satisfying the no-slip conditions at the wall. These have to be satisfied for Navier–Stokes equation, but for the difficulties of solving it, prevented early success for real fluid flow past bodies.

The statement that Euler equations applies at higher *Re* implies the region of validity of outer solution increases with *Re*. Yet very close to the body, we will have a region where viscous terms are significant and must be retained to satisfy the no-slip conditions at the wall. This also readily implies that besides the introduced integral length scale, L through *Re*, the flow introduces a smaller length scale which is inversely proportional to the Reynolds number of the problem, as described in Section 2.14. This defines the narrow region close to the boundary and is called the inner layer or boundary layer providing the inner solution, complementing the above-mentioned outer solution via smooth blending of these two solutions.

Prandtl identified the above structure of the flow field, with the 'outer solution' obtained from potential theory or Euler solution. Prandtl for the first time identified the 'inner solution', now very commonly known as thin shear layer or the boundary layer solution. Existence of such a layer does not only depend on the smallness of ϵ $(= 1/Re)$, it is intimately related to the structure of the solution. Whenever a thin shear layer (δ) exists, it introduces a new length scale much smaller than the integral dimension of the flow field. One can estimate the thickness of such a layer (δ) by performing an order of magnitude analysis of various terms in the Navier–Stokes equation as shown in Section 2.14. From the equivalence of viscous term with convective acceleration terms inside the inner layer, one estimates the thickness of the inner layer as

$$\frac{\delta}{L} \sim \left(\frac{UL}{\nu}\right)^{-1/2} = Re^{-1/2}.$$

7.3 Boundary Layer Equations

Physically the boundary layer assumption stems from the observation that in high *Re* flows, gradients of stress or heat fluxes are significant only in relatively thin layers. In the simplest case of a two-dimensional flow, the shear stress $\mu\frac{\partial u}{\partial y}$ is significant only in a limited range of y, inside the boundary layer close to the solid surface ($y = 0$). Outside this region, $\frac{\partial u}{\partial y}$ is of the same order as $\frac{\partial u}{\partial x}$ and both being smaller, but within the layer

$$\frac{\partial u}{\partial y} \gg \frac{\partial u}{\partial x}.$$

This is the thin shear layer approximation and the shear layer thickness δ satisfies $\frac{d\delta}{dx} \ll 1$, by definition of shear layer growing slowly in the x-direction.

Without referring to whether the flow is laminar or turbulent, we develop the boundary layer equation for steady or time-averaged two-dimensional motion. Suppose the boundary layer is forming on a flat wall, with x in the flow direction and y normal to the wall. A free stream velocity U_e, outside the boundary layer is prescribed as a function of x. Here, U_e plays the role of matching the inner and outer solutions. Let the associated pressure be p_e. As the predominant convection is in x-direction, the boundary layer has length scales L and δ, in x- and y-directions, respectively. Velocity components in these x- and y-directions scales by U and V, respectively.

The order of magnitude of the pressure differences across the boundary layer in the y-direction may not be the same as the order of magnitude of the imposed pressure differences outside the boundary layer in the x-direction; we denote the scale of the former variation by γ and the scale of latter variation by Π. We consider each equations in turn and label the terms with the orders of magnitude immediately below each terms of the equations in the following.

7.3.1 Conservation of Mass

The governing equation for mass conservation of two-dimensional fluid flow is given by

$$\frac{\partial u}{\partial x} + \frac{\partial v}{\partial y} = 0 \tag{7.4}$$

$$\frac{U}{L} \quad \frac{V}{\delta}.$$

For strict conservation of mass both these terms must be of the same order of magnitude, which implies $V \sim U\delta/L$.

Thus, one concludes the normal velocity to be small compared to the streamwise component, as we have already noted that $\delta/L \sim 0(\epsilon)$, where $\epsilon \sim Re^{-1/2}$. This order of magnitude analysis for velocity components will assist one to perform the same for momentum conservation equations. This can be performed for the momentum conservation equation components, one at a time, supplemented by the observation that the viscous actions are confined in a narrow region in a shear layer next to the wall. For the sake of simplification, we only consider the steady flow condition, realizing that similar analysis can be performed identically for unsteady flows also, provided the thin shear layer approximation holds.

7.3.2 The x-Momentum Equation

Here, we consider the steady flow equation given by

$$u\frac{\partial u}{\partial x} + v\frac{\partial u}{\partial y} = -\frac{1}{\rho}\frac{\partial p}{\partial x} + \nu\frac{\partial^2 u}{\partial x^2} + \nu\frac{\partial^2 u}{\partial y^2} \tag{7.5}$$

$$\frac{U^2}{L} \quad \frac{VU}{\delta} \sim \frac{U^2}{L} \quad \frac{\Pi}{\rho L} \quad \frac{\nu U}{L^2} \quad \frac{\nu U}{\delta^2}.$$

In the above orders of magnitude written for each terms below Equation (7.5), term by term, we note that all the convection terms are of 0(1), and so must be the pressure gradient term for a general flow case with arbitrary pressure gradient. However, as $\delta/L \ll 1$, the streamwise diffusion term is significantly smaller as compared to the wall-normal diffusion term, i.e.

$$\nu\frac{\partial^2 u}{\partial y^2} \gg \nu\frac{\partial^2 u}{\partial x^2}.$$

This is true, as for large Reynolds numbers

$$\frac{(\nu \frac{\partial^2 u}{\partial y^2})}{(\nu \frac{\partial^2 u}{\partial x^2})} \simeq \left(\frac{L}{\delta}\right)^2 = Re.$$

Also, the viscous diffusion terms can be written in terms of their order of magnitude as

$$\frac{\nu U}{L^2} = \left(\frac{\nu}{UL}\right)\frac{U^2}{L} = Re^{-1}\left(\frac{U^2}{L}\right).$$

And

$$\frac{\nu U}{\delta^2} = \frac{\nu}{LU}\left(\frac{U^2}{L}\right)\frac{L^2}{\delta^2} = Re^{-1}\left(\frac{U^2}{L}\right)\left(\frac{L}{\delta}\right)^2.$$

This establishes that in the x-momentum equation, one can neglect the streamwise diffusion term as compared to the wall-normal diffusion term. One also notices that the latter term is of same order as the convection terms inside the boundary layer. Thus, one obtains the simplified equation inside the boundary layer as

$$u\frac{\partial u}{\partial x} + v\frac{\partial u}{\partial y} = -\frac{1}{\rho}\frac{\partial p}{\partial x} + \nu\frac{\partial^2 u}{\partial y^2}. \quad (7.6)$$

This simplification may not appear as very significant, as we are able to neglect only one linear term, while the nonlinear convection terms remain as these are in the original Navier–Stokes equation. However, this seemingly simple modification has effects which are extraordinary in consequence. This allows one to convert the elliptic-in-space Navier–Stokes equation, to another partial differential equation which is parabolic in space. Thus, a mathematically (and computationally) more challenging boundary value problem has been converted into a simpler marching problem. For example, in many boundary layer problems, it may mean that we will be solving ordinary differential equation, instead of solving the original partial differential equation. This aspect is further explained in this chapter, with reference to numerical solution.

7.3.3 The y-Momentum Equation

One can perform a similar order of magnitude analysis for the y-momentum equation, as shown next. The full dimensional equation is given by

$$u\frac{\partial v}{\partial x} + v\frac{\partial v}{\partial y} = -\frac{1}{\rho}\frac{\partial p}{\partial y} + \nu\frac{\partial^2 v}{\partial x^2} + \nu\frac{\partial^2 v}{\partial y^2} \quad (7.7)$$

$$\frac{UV}{L} \qquad \frac{V^2}{\delta} \qquad \frac{\gamma}{\rho\delta} \qquad \frac{\nu V}{L^2} \qquad \frac{\nu V}{\delta^2}$$

$$\frac{U^2\delta}{L^2} \qquad \frac{U^2\delta}{L^2} \qquad \frac{\nu U\delta}{L^3} \qquad \frac{\nu U}{L\delta}.$$

In Equations (7.5) and (7.7), the pressure term will be of the same order of magnitude as the largest of other terms. Hence

$$(B): \frac{\Pi}{\rho L} \sim \frac{U^2}{L} \sim \frac{\nu U}{\delta^2}$$

$$(C): \frac{\gamma}{\rho \delta} \sim \frac{U^2 \delta}{L^2} \sim \frac{\nu U}{L\delta}.$$

Therefore, we can estimate the ratio of the pressure variation in the wall-normal direction to the variation in the streamwise direction given by

$$\frac{\gamma}{\Pi} \sim \frac{\delta^2}{L^2}.$$

This clearly indicates that the pressure difference across the boundary layer is much smaller than that is in the streamwise direction. We may also infer that $\frac{\partial p}{\partial y}$ is of the order δ. The pressure increase (γ), across the boundary layer, which would be obtained by integrating Equation (7.7), is of the order δ^2. Thus, the pressure in a direction normal to the boundary layer is practically constant. It may be assumed equal to that at the outer edge of the boundary layer, where its value is determined from the ideal flow equation, Equation (7.6). For this reason, it is said that the pressure is 'impressed' upon the boundary layer by the outer flow. This suggests a way of decoupling the viscous and inviscid part of an actual flow in the presence of a boundary layer.

One conceptual model would be to calculate the ideal fluid flow generating the pressure distribution. This pressure distribution can be used to calculate the boundary layer equations which can be written down as

$$\begin{aligned} \frac{\partial u}{\partial x} + \frac{\partial v}{\partial y} &= 0 \\ u\frac{\partial u}{\partial x} + v\frac{\partial u}{\partial y} &= -\frac{1}{\rho}\frac{\partial p}{\partial x} + \nu\frac{\partial^2 u}{\partial y^2} \\ \frac{\partial p}{\partial y} &= 0. \end{aligned} \tag{7.8}$$

Above three equations are valid approximation of the Navier–Stokes equation inside the boundary layer. In actual applications, the last equation is omitted and in the x-component equation the pressure gradient term is written as the ordinary derivative, $\frac{dp}{dx}$.

This form of boundary layer equation is valid for shear layer developing over a plane surface or over a surface with negligible curvature. For a body with radius of curvature, R, the y-momentum equation reduces to

$$\frac{\partial p}{\partial y} = \frac{\rho u^2}{R}.$$

In the outer layer, Equation (7.5) simplifies to

$$\frac{\partial U_e}{\partial t} + U_e \frac{\partial U_e}{\partial x} = -\frac{1}{\rho} \frac{dp}{dx}.$$

For steady flows, one can discard the first term on the left-hand side to relate the pressure gradient distribution along the edge of the boundary layer with the edge velocity gradient obtained from ideal flow analysis. Thus,

$$-\frac{1}{\rho} \frac{dp}{dx} = U_e \frac{dU_e}{dx},$$

which could be integrated to obtain the Bernoulli equation for steady flows.

7.3.4 Use of Boundary Layer Equations

Boundary layer equations are large Reynolds number approximation to Navier–Stokes equation in the narrow region close to the boundary. Existence of boundary layer makes the inviscid results meaningful, those related to pressure distribution. As explained above, within the boundary layer the Navier–Stokes equation is represented by boundary layer equation and outside it, the viscous terms can be dropped altogether to reduce the Navier–Stokes equation to the Euler equation.

Inside the boundary layer, velocity normal to the surface (v) is negligibly small and the following steps are followed in studying the flow field.

Step 1: Calculate the velocity and pressure field, as if the flow were inviscid, using flow tangency conditions at the body surface ($y = 0$) and the usual far-field boundary condition at $y \to \infty$ and call this the outer solution.

Step 2: Solve the boundary layer equation, using the above tangential flow at the boundary as the external boundary condition for the boundary layer equation, along with the no-slip condition at the real wall

$$u = v = 0 \quad \text{at} \quad y = 0$$

and

$$u \to U_e \quad \text{as} \quad y/\delta \to \infty.$$

The two solutions merge smoothly, i.e.

$$U_e = \lim_{y/\delta \to \infty} u = \lim_{y/L \to 0} u.$$

In the boundary layer equation, the pressure gradient term is obtained from the inviscid part of the flow, i.e. from the Bernoulli's equation.

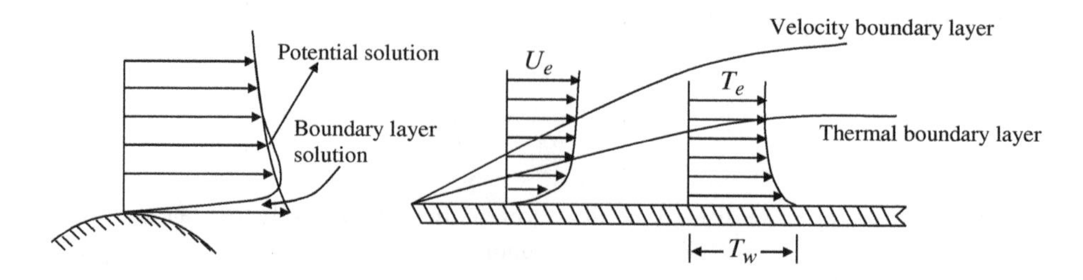

Figure 7.1 Boundary layers shown for (a) flow past a circular cylinder at the shoulder (left) and (b) a velocity and thermal boundary layer for flow past a heated flat plate

7.4 Boundary Layer Thicknesses

Application of no-slip boundary condition slows down the fluid close to the body, as compared to what is in the outer part of the shear layer, as shown schematically in Figure 7.1. On the left is the flow past a cylinder with the velocity profile shown at the shoulder. The ideal fluid flow analysis gives a value of $2U_\infty$ and this is compared with the boundary layer solution. On the right is the shear layer forming over a heated flat plate. For this flow on the right, apart from showing a velocity boundary layer, one also notices the presence of a thermal boundary layer for which the wall temperature is higher compared to the thermal boundary layer edge temperature.

To understand the role of the viscous effects inside a boundary layer, let us consider the control volume shown in Figure 7.2, with the far field being considered as one of the streamline for flow past a flat plate with a sharp leading edge. For the flow aligned with the plate (i.e. for zero angle of attack), the leading edge is the stagnation point and the inflow of the control volume is placed at the leading edge with a height H_i, so that the inflow velocity can be taken as U_e everywhere along the inflow. At the outflow, the velocity increases monotonically from zero value at the wall to its edge value, U_e at $y = Y$. The streamline at the far-field is so chosen that it remains outside the boundary layer everywhere. In general, boundary layer thickness (δ) is taken as that height where the streamwise velocity attains a value which is 99.9% of U_e. As the velocity near $y = \delta$ varies slowly, precise value of boundary layer thickness depends upon the chosen tolerance for the streamwise velocity. For example, depending upon the pressure

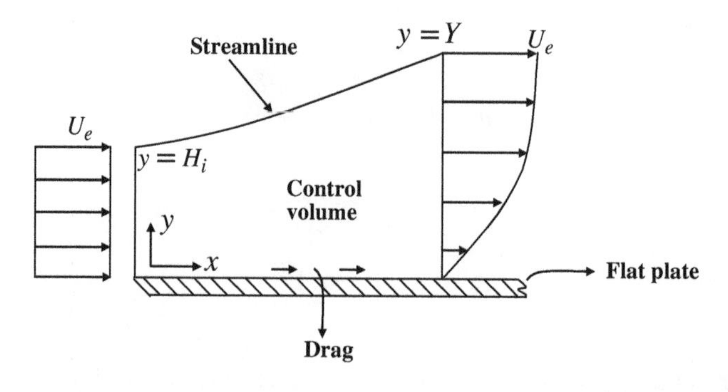

Figure 7.2 A control volume used to estimate various boundary layer thicknesses

gradient if we increase the value of tolerance from 99.9% to 99.99%, then the value of δ can increase by a factor of two or more, depending on the pressure gradient. Thus, there is a need to have a definitive quantitative measure of the boundary layer thickness, as discussed next.

7.4.1 Boundary Layer Displacement Thickness

The displayed control volume (with unit depth perpendicular to the plane of the paper) shown in Figure 7.2 is such that there is no mass source inside it and therefore for a steady incompressible flow mass conservation would require

$$\int\int_{CS} \rho\vec{V}\cdot dA = \int_0^Y \rho u\, dy - \int_0^{H_i} \rho U_e\, dy = 0.$$

where the left-hand side integral is performed over all the control surface elements defining the control volume. If the viscous action represents the lateral shifting of the streamline representing the upper lid of the control volume is given by the displacement thickness (δ^*), then $Y = H_i + \delta^*$.

Therefore the mass conservation equation can be written down as

$$U_e H_i = \int_0^Y u\, dy = U_e Y + \int_0^Y (u - U_e)\, dy$$

or

$$U_e(Y - H_i) = \int_0^Y (U_e - u)\, dy.$$

Note that $\delta^* = Y - H_i$ and the limit of the integral on the right-hand side can be extended to infinity without changing the value of the integral as the integrand is zero in the extended range. Thus

$$\delta^* = \int_0^\infty \left(1 - \frac{u}{U_e}\right) dy. \qquad (7.9)$$

This result is independent of the choice of H_i and provides the formal definition of boundary layer displacement thickness (δ^*). As this relates to the displacement of a streamline in the outer region, this provides an indication of mass defect due to viscous effect inside the boundary layer.

From the continuity equation

$$\begin{aligned} v(x,y) &= \int_0^y \frac{\partial v}{\partial y}(x,y^*)\, dy^* = -\int_0^y \frac{\partial u}{\partial x}(x,y^*)\, dy^* \\ &= -\frac{\partial}{\partial x}\int_0^y u\, dy^* \\ &= \frac{\partial}{\partial x}\int_0^\infty [U_e(x) - u(x,y^*)]\, dy^* - y\frac{dU_e}{dx}. \end{aligned}$$

When $y^* > \delta$: $u(x,y^*) \simeq U_e$.

Therefore for $y > \delta$: $v(x, y) = \frac{\partial}{\partial x} \int_0^y [U_e(x) - u(x, y^*)]\, dy^* - y\frac{dU_e}{dx}$.

As $\delta^*(x) = \int_0^\infty [1 - \frac{u}{U_e}]\, dy$ is the displacement thickness, then we obtain the wall-normal velocity as

$$v(x, y) = \frac{d}{dx}(U_e\, \delta^*) - y\frac{dU_e}{dx}. \tag{7.9a}$$

Thus, δ^* is the distance by which the external potential field of flow is equivalently displaced outwards as a consequence of the decrease in velocity in the boundary layer.

7.4.2 Boundary Layer Momentum Thickness

In the previous subsection, mass defect is quantified in terms of displacement thickness. Similarly, we can discuss about the momentum loss as a consequence of viscous effects in a boundary layer with reference to Figure 7.2, as manifested in terms of drag, i.e.

$$Drag = D = \int_0^Y u(\rho u)\, dy - \int_0^{H_i} U_e(\rho U_e)\, dy.$$

We note that there is no momentum transfer at the solid wall of the control surface due to no-slip boundary condition. In the same way, no momentum is lost through the top of the control volume, as the streamline outside the shear layer has no momentum crossing it. For incompressible flows, we have shown above that $H_i = \int_0^Y u/U_e\, dy$ and thus, the drag experienced by the wetted surface is given by

$$D = \rho \int_0^\infty u(U_e - u)\, dy.$$

Once again, we have extended the range of integration from Y to ∞, as the integrand is identically equal to zero in the extended range of ordinate. Thus

$$\frac{D}{\rho U_e^2} = \int_0^\infty \frac{u}{U_e}\left(1 - \frac{u}{U_e}\right) dy. \tag{7.10}$$

The quantity on either side of the above equation has a dimension of length and it is known as the momentum thickness, indicated by θ and given by

$$\theta = \int_0^\infty \frac{u}{U_e}\left(1 - \frac{u}{U_e}\right) dy. \tag{7.11}$$

Note that at any streamwise station, δ^* and θ are obtained by integrating the velocity distribution across all y and thus, these quantities are solely functions of x only. Also, from the expressions of δ^* and θ it is apparent that $\delta^* > \theta$, always. The ratio $H = \delta^*/\theta$ plays a very major role in understanding the property of boundary layers and is known as the shape factor. It is noted that the value of H is significantly different for laminar and turbulent flows. In fact, at transition H reduces abruptly and is often used to detect transition experimentally, by obtaining it from the measured velocity profile.

For compressible flows, one can perform similar analyses to obtain expressions for displacement and momentum thicknesses as

$$\delta^* = \int_0^\infty \left(1 - \frac{\rho u}{\rho_e U_e}\right) dy \tag{7.12}$$

$$\theta = \int_0^\infty \frac{\rho u}{(\rho U)_e}\left(1 - \frac{u}{U_e}\right) dy. \tag{7.13}$$

For shear layers one can also define an energy thickness.

If the expression of drag in Equation (7.10) is solely due to skin friction–as in case of a flat plate of length L, then this can be also expressed in term of the wall shear stress as

$$D = \int_0^L \tau_w(x)\, dx.$$

Wall shear stress and drag can be also expressed in nondimensional form by

$$C_f(x) = \frac{\tau_w(x)}{\rho U_e^2/2} \quad \text{and} \quad C_d = \frac{D}{\rho U_e^2 L/2}.$$

7.5 Momentum Integral Equation

After the differential form of boundary layer equation was described and physical understanding provided through the works of researchers at Goettingen, von Kármán (von Kármán 1921) came out with an integral representation of boundary layers. This greatly simplifies the governing partial differential equation to an ordinary differential equation. Here, we develop the boundary layer equation in integral form for two-dimensional steady compressible flows starting from

$$\frac{\partial(\rho u)}{\partial x} + \frac{\partial(\rho v)}{\partial y} = 0 \tag{7.14}$$

$$\rho u\frac{\partial u}{\partial x} + \rho v\frac{\partial u}{\partial y} = -\frac{dp}{dx} + \frac{\partial \tau}{\partial y}, \tag{7.15}$$

where $\tau = \mu\frac{\partial u}{\partial y}$ is written in the simplified form obtained using boundary layer assumption. Here, u and v are velocity components in x- and y-directions, respectively.

Multiplying Equation (7.14) by u and adding it to Equation (7.15) one gets

$$\frac{\partial}{\partial x}(\rho u^2) + \frac{\partial}{\partial y}(\rho u v) = \rho_e U_e\frac{dU_e}{dx} + \frac{\partial \tau}{\partial y}. \tag{7.16}$$

Note that the pressure gradient is related to boundary layer edge velocity via the Bernoulli's equation $\frac{dp}{dx} = -\rho_e U_e\frac{dU_e}{dx}$. Integrating Equation (7.16) with respect to y from 0 to h, with h a constant, independent of x and sufficiently greater than the shear layer thickness δ, so that all quantities reach their external flow values much before $y = h$. Also, we note that in the free stream $\tau \to 0$, so integration of Equation (7.16) yields

$$\int_0^h \frac{\partial}{\partial x}(\rho u^2)\, dy + \rho_h\, v_h\, U_e = \int_0^h \rho_e U_e \frac{dU_e}{dx}\, dy - \tau_w, \tag{7.17}$$

where quantities with subscript h indicate that the corresponding quantities have been evaluated at h. Integration of Equation (7.14) with respect to y gives

$$\rho_h v_h = -\int_0^h \frac{\partial}{\partial x}(\rho u)\, dy.$$

Thus, Equation (7.17) simplifies to

$$\int_0^h \left[\frac{\partial}{\partial x}(\rho u^2) - U_e \frac{\partial}{\partial x}(\rho u) - \rho_e U_e \frac{dU_e}{dx}\right] dy = -\tau_w.$$

This equation can be rearranged to get

$$\int_0^h \left[-\frac{\partial}{\partial x}[\rho u(U_e - u)] - \frac{dU_e}{dx}(\rho_e U_e - \rho u)\right] dy = -\tau_w.$$

Since $U_e - u \equiv 0$ for $y \geq h$, so we can extend the upper limit to any height above h.

We have chosen h to be independent of x, we can also change the order of differentiation and integration, i.e. the first part of the integral can be written as

$$-\frac{d}{dx}\left[\int_0^h \rho u(U_e - u)\, dy\right].$$

This is treated as ordinary derivative, because the argument in the parenthesis is independent of y. One can also replace h by ∞, for the limit of the integral and rearrange to get

$$\frac{d}{dx}\left[\rho_e U_e^2 \int_0^\infty \frac{\rho u}{\rho_e U_e}\left(1 - \frac{u}{U_e}\right) dy\right] + \rho_e U_e \frac{dU_e}{dx} \int_0^\infty \left(1 - \frac{\rho u}{\rho_e U_e}\right) dy = \tau_w. \tag{7.18}$$

With the definition of displacement and momentum thicknesses in Equations (7.12) and (7.13) for compressible flows when used in Equation (7.18), one obtains

$$\frac{d}{dx}(\rho_e U_e^2 \theta) + \rho_e U_e \frac{dU_e}{dx}\delta^* = \tau_w. \tag{7.19}$$

Writing this equation in terms of pressure gradient, one obtains

$$\frac{d}{dx}(\rho_e U_e^2 \theta) = \tau_w + \delta^* \frac{dp}{dx}. \tag{7.20}$$

For compressible flows, using the isentropic relation

$$\frac{1}{\rho_e}\frac{d\rho_e}{dx} = -\frac{M_e^2}{U_e}\frac{dU_e}{dx}$$

and the definitions for $C_f = \tau_w/(\rho_e U_e^2/2)$ and $H = \delta^*/\theta$, Equation (7.19) simplifies to

$$\frac{d\theta}{dx} + \frac{\theta}{U_e}\frac{dU_e}{dx}(H + 2 - M_e^2) = \frac{C_f}{2}. \tag{7.21}$$

For a constant density incompressible flow, one can set $M_e = 0$ to obtain the corresponding momentum integral equation as

$$\frac{d\theta}{dx} + \frac{\theta}{U_e}\frac{dU_e}{dx}(H + 2) = \frac{C_f}{2}. \tag{7.22}$$

7.6 Validity of Boundary Layer Equation and Separation

The boundary layer assumption helps in reducing an initial-boundary value problem (Navier–Stokes equation) to a marching problem in space. Inside the boundary layer, information propagates downstream, as a consequence. Also the pressure gradient is impressed through the boundary layer unaltered from its edge value. However, the no-slip condition at the wall implies the slowing down of the convection velocity, as one moves toward the wall. Thus, in the presence of adverse pressure gradient ($\frac{dp}{dx} > 0$), convection becomes weaker as one approaches the wall, while the adverse pressure gradient has undiminished effects at those heights. This behaviour can violate the condition of information propagating downstream, as the pressure gradient becomes progressively more adverse in comparison to convective acceleration, close to the wall.

Let us determine circumstances under which some of the retarded fluid in the boundary layer can be transported into the main stream, i.e. to find when separation of the flow from the wall may occur. Retarded fluid cannot proceed in local streamwise direction, in the presence of sufficiently adverse pressure gradient, owing to its small kinetic energy. In the first place, the packet may be retarded due to the pressure gradient itself, apart from the effect of shear. In such a situation the shear layer next to the wall is deflected sideways (in the wall-normal direction). When this happens, we say the flow has separated.

The point of separation is defined as the limit between forward and reverse flow inside the shear layer, in the immediate neighbourhood of the wall, i.e. when

$$\left.\frac{\partial u}{\partial y}\right|_w = 0.$$

The close-up view of streamlines in the neighbourhood of point of separation is shown in the sketch of Figure 7.3. Mathematically, the wall streamline bifurcates at the point of separation, where the flow field is singular and the location is the half-saddle point. Also, in the presence of nonnegligible wall-normal component of velocity, the shear layer thickens rapidly. Generally speaking, boundary layer equation is valid only up to the point of separation. This is not valid due to

(i) boundary layer thickening, which is a consequence of v not being small any more;
(ii) the parabolic nature of the governing equation being lost due to the presence of reverse flow.

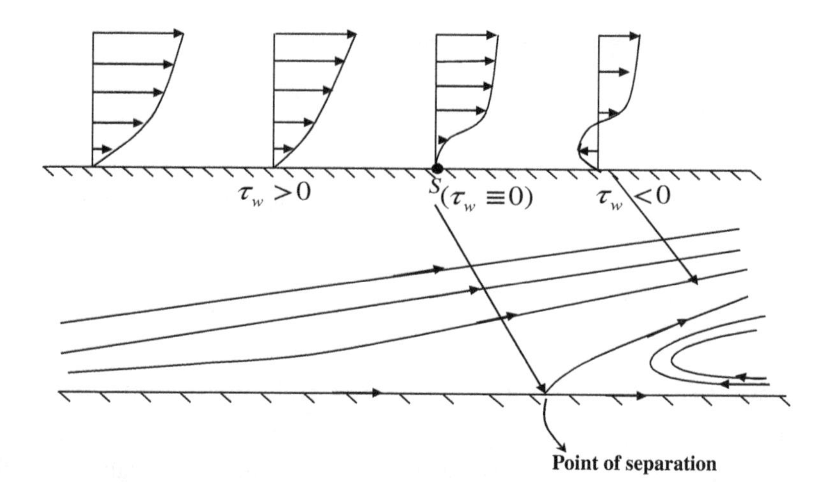

Figure 7.3 Boundary layer evolution in the presence of adverse pressure gradient. The point of separation is denoted by vanishing wall shear stress

Obviously when flow separates, then neither the boundary layer equation nor the inviscid flow equations are valid limiting solutions of Navier–Stokes equation and hence both of these equations fail together. The boundary layer equation fails due to mathematical singularity, while the inviscid solution loses its physical meaning. One is then left with no other choice but to solve the full Navier–Stokes equation.

The fact that steady separation occurs only in decelerated flows with $(dp/dx > 0)$ can be inferred from the relation between the pressure gradient and the velocity distribution with the aid of boundary layer equation. At the wall $(y = 0)$,

$$u,\ v = 0 \tag{7.23}$$

and

$$\mu \left.\frac{\partial^2 u}{\partial y^2}\right|_w = \frac{dp}{dx}. \tag{7.24}$$

Thus at the wall, curvature of the velocity profile u, depends only on the streamwise pressure gradient. For accelerated flows, $\frac{dp}{dx} < 0$ and then $\left.\frac{\partial^2 u}{\partial y^2}\right|_w = (u_{yy})_w < 0$. Note that the shear is zero in the free stream and is a decaying function of y. This implies that u_{yy} is negative in the far-stream and at the wall. Therefore this must be the case $(u_{yy} < 0)$ for all y. This case is demonstrated in the sketch of Figure 7.4. One notices that the shear (u_y) remains positive at all y, as u_{yy} remains negative at all heights.

For flows experiencing adverse pressure gradient, $\frac{dp}{dx} > 0$ and therefore the wall value satisfies $(u_{yy})_w > 0$ following Equation (7.24). In Figure 7.5, we show the evolution of the velocity profile with streamwise distance under the joint action of adverse pressure gradient and shearing action of viscosity. As discussed above, u_{yy} at the far-stream is once again negative. Thus, the passage of u_{yy} from a positive value at the wall to a negative value in the free stream can only occur when u_{yy} must be equal to zero at an intermediate height. So u_y

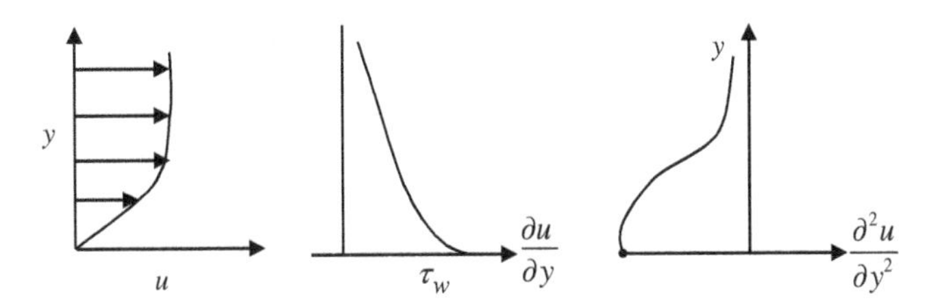

Figure 7.4 Favourable or zero pressure gradient boundary layer flow over a wedge shown by the velocity profile and its shear derivatives

must be an increasing function of y in the near vicinity of the wall as the following Taylor series expansion indicates

$$\frac{\partial u}{\partial y}(y) \sim \left(\frac{\partial u}{\partial y}\right)_{wall} + y\left(\frac{\partial^2 u}{\partial y^2}\right)_{wall}. \tag{7.25}$$

The consequence of this is sketched in Figure 7.5, for the velocity profile of an adverse pressure gradient flow. This helps us define a point of inflection for separated velocity profile. For such a case, τ increases first from its wall value and reaches a maximum at an intermediate height and decreases thereafter monotonically, as shown in Figure 7.5. At the height where τ attains the maximum value is also the location where $u_{yy} = 0$. This is the point of inflection. We note that the presence of point of inflection at an intermediate height is indicative of probable temporal instability by an inviscid mechanism, as can be shown by applying Rayleigh and Fjϕrtoft's theorem (see e.g. Sengupta (2012) for details).

Note that for zero pressure gradient boundary layer, the so-called point of inflexion is at the wall itself following Equation (7.24) and does not indicate flow instability by inviscid mechanism.

7.7 Solution of Boundary Layer Equation

Before one tries to solve the boundary layer equation (a nonlinear partial differential equation), it is worthwhile exploring certain cases, where by additional transformations, one can reduce the governing equations to a system of ODE, which are easier to solve numerically. This makes use of similarity of velocity profiles in the transformed coordinates. The similarity solutions are useful and are nontrivial examples of the underlying physical processes. Furthermore

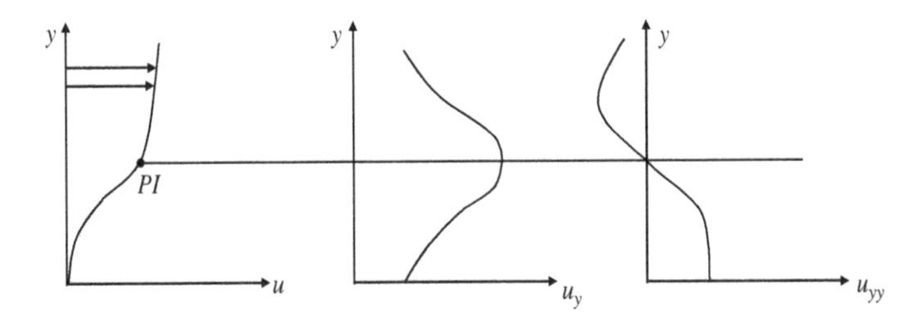

Figure 7.5 Adverse pressure gradient boundary layer flow over a wedge shown by the velocity profile and its shear derivatives

similarity transformations help in numerical solutions even for non-similar flows. So, first of all we will introduce similarity analysis.

7.8 Similarity Analysis

Consider two-dimensional 'uncoupled' laminar flow with negligible body force. The governing equations are as given in Equation (7.8). For external flows one can replace $\frac{1}{\rho}\frac{dp}{dx}$ by $-U_e\frac{dU_e}{dx}$. One can write the solution of these equations in the form

$$\frac{u}{U_e} = \phi_1(x, y), \tag{7.26}$$

subject to boundary conditions

$$\text{at } y = 0;\ u = v = 0$$

$$\text{and at } y = \delta;\ u = U_e.$$

There are special cases where Equation (7.26) can be written as

$$\frac{u}{U_e} = \phi_1(\eta), \tag{7.27}$$

where η is called the similarity variable and is a special function of x and y. We define here 'similar' solutions as those for which u has the property that two velocity profiles $u(x, y)$ located at different streamwise coordinates x are similar and any one is related to the others via a similarity transformation, i.e. velocity profiles at all x can be made congruent, when these are plotted in terms of a 'similarity variable'. With the existence of η as a single independent variable, instead of two variables x and y, makes it apparent the role of the transformation for the governing equations, Equation (7.8) to convert a set of partial differential equations to ordinary differential equation(s).

A detailed group theoretic approach to similarity transformation is given in Cebeci and Bradshaw (1977) and interested readers should consult the same. We provide a very brief description of the same. Here, instead of working with two equations in Equation (7.8), we work with one equation by introducing stream function ψ and we rewrite these equations as

$$\frac{\partial\psi}{\partial y}\frac{\partial^2\psi}{\partial x\,\partial y} - \frac{\partial\psi}{\partial x}\frac{\partial^2\psi}{\partial y^2} = U_e\frac{dU_e}{dx} + \nu\frac{\partial^3\psi}{\partial y^3}. \tag{7.28}$$

Note that the existence of ψ automatically satisfies the continuity equation. We consider special cases of flows for which the boundary layer edge velocity can be represented as

$$U_e = cx^m. \tag{7.29}$$

Although this might appear as a serious restriction on the flow type considered, we will shortly show how general flows are often treated for engineering calculations of flow past airfoils and wings by representing the edge velocity by an equivalent local representation, with m as a function of x.

The similarity coordinate is defined as

$$\eta = \frac{y}{x^{(1-m)/2}} = \frac{y}{x^{1/2}}\, x^{m/2} = \left(\frac{U_e}{cx}\right)^{1/2} y \tag{7.30a}$$

and the stream function is defined as

$$\psi = \left(\frac{U_e x}{c_1}\right)^{1/2} f(\eta). \tag{7.30b}$$

One notes that ψ has a dimension of L^2T^{-1} and η is dimensionless. This is possible by choosing $c \equiv \nu$ in the definition of η and $c_1 = 1/\nu$ for ψ, so that

$$\eta = \left(\frac{U_e}{\nu x}\right)^{1/2} y \;\text{ and }\; \psi = (U_e\, \nu x)^{1/2} f(\eta). \tag{7.31}$$

This is called the Falkner-Skan transformation. The essential idea is to introduce f, the nondimensional stream function. We will use Falkner-Skan transformation in boundary layer equation to render it to an ODE for similar flows. Falkner-Skan transformation is also used for nonsimilar flows.

For nonsimilar flows, the dependence of solution on x remains and this is reflected in the choice of ψ as follows

$$\psi = (U_e\, \nu x)^{1/2} f(x, \eta). \tag{7.32}$$

We carry out the transformations for Equation (7.24); first by moving over from (x, y)-plane to (x, η)-plane. The derivatives in the two planes are related by

$$\left(\frac{\partial}{\partial x}\right)_y = \left(\frac{\partial}{\partial x}\right)_\eta + \left(\frac{\partial}{\partial \eta}\right)_x \frac{\partial \eta}{\partial x} \tag{7.33}$$

and

$$\left(\frac{\partial}{\partial y}\right)_x = \left(\frac{\partial}{\partial \eta}\right)_x \frac{\partial \eta}{\partial y}. \tag{7.34}$$

So that $$-v = \frac{\partial \psi}{\partial x} = \left[f \frac{d}{dx}(U_e\, \nu x)^{1/2} + (U_e\, \nu x)^{1/2} \frac{\partial f}{\partial x}\right] + (U_e\, \nu x)^{1/2} f' \frac{\partial \eta}{\partial x}$$

and $$u = \frac{\partial \psi}{\partial y} = (U_e\, \nu x)^{1/2} f' \left(\frac{U_e}{\nu x}\right)^{1/2} = U_e f'.$$

Also $$\frac{\partial^2 \psi}{\partial y^2} = U_e \left(\frac{U_e}{\nu x}\right)^{1/2} f'' \;\text{ and }\; \frac{\partial^3 \psi}{\partial y^3} = U_e \left(\frac{U_e}{\nu x}\right) f'''.$$

As $$\frac{\partial \eta}{\partial x} = -\frac{\eta}{2x}, \;\text{ therefore, }\; v = -\frac{\partial}{\partial x}[(U_e\, \nu x)^{1/2} f] + \frac{\eta}{2}\left(\frac{U_e\, \nu}{x}\right)^{1/2} f'$$

and $$\frac{\partial u}{\partial x}=\left[\left(\frac{\partial}{\partial x}\right)_\eta-\frac{\eta}{2x}\left(\frac{\partial}{\partial \eta}\right)_x\right](U_e f')=U_e\frac{\partial f'}{\partial x}+f'\,\frac{dU_e}{dx}-\frac{\eta\, U_e}{2x}\,f''.$$

Thus, $$u\frac{\partial u}{\partial x}=U_e^2\,f'\,\frac{\partial f'}{\partial x}+U_e\,f'^2\,\frac{dU_e}{dx}-\frac{\eta\,U_e^2}{2x}f'\,f''.$$

Also, $$\frac{\partial u}{\partial y}=\left(\frac{U_e}{\nu x}\right)^{1/2}U_e\,f''.$$

Therefore, $$v\frac{\partial u}{\partial y}=\left\{\left(\frac{U_e}{\nu x}\right)^{1/2}U_e\,f''\right\}\left[\frac{\eta}{2}\left(\frac{U_e\,\nu}{x}\right)^{1/2}f'-\frac{\partial}{\partial x}\left\{(U_e\,\nu x)^{1/2}f\right\}\right]$$

$$=\frac{\eta U_e^2}{2x}f'\,f''-\left(\frac{U_e}{\nu x}\right)^{1/2}U_e f''\left[(U_e\nu x)^{1/2}\frac{\partial f}{\partial x}-(U_e\,\nu)^{1/2}\frac{f x^{-1/2}}{2}+(\nu x)^{1/2}f\frac{dU_e}{dx}\right]\frac{U_e^{-1/2}}{2}$$

$$=\frac{U_e^2\,\eta}{2x}f'\,f''-U_e^2 f''\,\frac{\partial f}{\partial x}+\frac{U_e}{2x}\,U_e\,f\,f''+\frac{U_e}{2}\,\frac{dU_e}{dx}\,f\,f''.$$

And finally

$$\nu\frac{\partial^2 u}{\partial y^2}=\frac{U_e^2}{x}f'''.$$

So for Equation (7.24), the left-hand side simplifies to

$$=\left[U_e^2 f'\frac{\partial f'}{\partial x}+U_e\,f'^2\frac{dU_e}{dx}-\frac{\eta\,U_e^2}{2x}f'\,f''\right]+U_e^2\left[\frac{\eta}{2x}f'\,f''-f''\,\frac{\partial f}{\partial x}+\frac{f\,f''}{2x}\right]+\frac{U_e}{2}\frac{dU_e}{dx}f\,f''$$

$$=U_e^2\,f'\,\frac{\partial f'}{\partial x}+U_e\,\frac{dU_e}{dx}f'^2-U_e^2\,f''\,\frac{\partial f}{\partial x}+\frac{U_e^2}{2x}\,f\,f''+\frac{U_e}{2}\,\frac{dU_e}{dx}\,f\,f''.$$

Similarly the right hand side of Equation (7.24) is given by

$$=U_e\,\frac{dU_e}{dx}+\frac{U_e^2}{x}f'''.$$

Thus the boundary layer equation simplifies to

$$U_e^2 f'\frac{\partial f'}{\partial x}+U_e\frac{dU_e}{dx}f'^2-U_e^2 f''\frac{\partial f}{\partial x}+\left(\frac{U_e^2}{2x}+\frac{U_e}{2}\frac{dU_e}{dx}\right)f\,f''=U_e\frac{dU_e}{dx}+\frac{U_e^2}{x}f'''. \quad (7.35)$$

Now multiplying both side by $\frac{x}{U_e^2}$ one gets

$$x\,f'\frac{\partial f'}{\partial x}+\frac{x}{U_e}\frac{dU_e}{dx}f'^2-xf''\frac{\partial f}{\partial x}+\left(\frac{1}{2}+\frac{x}{2U_e}\frac{dU_e}{dx}\right)f\,f''=\frac{x}{U_e}\frac{dU_e}{dx}+f'''. \quad (7.36)$$

Now from Equation (7.31a) $U_e = Cx^m$ and thus one obtains

$$\frac{dU_e}{dx} = mC\, x^{m-1} = \frac{mC}{x}\, x^m = \frac{m}{x}\, U_e,$$

which can be reorganized as

$$\frac{x}{U_e}\frac{dU_e}{dx} = m.$$

Thus, one rewrites Equation (7.36) as

$$x\, f'\, \frac{\partial f'}{\partial x} + m\, f'^2 - x\, f''\, \frac{\partial f'}{\partial x} + \left(\frac{m}{2} + \frac{1}{2}\right) f f'' = m + \; f''' \tag{7.37}$$

or

$$f''' + m(1 - f'^2) + \frac{m+1}{2}\, f f'' = x\left(f' \frac{\partial f'}{\partial x} - f''\, \frac{\partial f}{\partial x}\right). \tag{7.38}$$

This nonsimilar boundary layer equation can be shown as parabolic and can be equivalently solved by marching in x direction. Thus, one requires to solve the above equation subject to the wall boundary condition ($y = 0$),

$$\left.\begin{aligned} u &= 0 \\ v &= v_w(x) \end{aligned}\right\}. \tag{7.39a}$$

As $y = 0$ implies $\eta = 0$, so the boundary condition given by Equation (7.39a) translates to $f' = 0$ and

$$\frac{\partial \psi}{\partial x} = -v_w(x).$$

The latter can be written in integral form as

$$\psi = -\int_0^x v_w\, dx, \;\; \text{i.e.,} \;\; f_w = -\frac{1}{(U_e \nu x)^{1/2}} \int_0^x \; v_w\, dx$$

and the other boundary condition applies at the edge of the boundary layer ($y = Y_e$) is given for $y \rightarrow Y_e$: as $u \rightarrow U_e$ This is equivalently expressed as

$$\text{for } \eta = \eta_e : \;\; f' = 1, \tag{7.39b}$$

where a finite but large value of η_e corresponding to Y_e is taken as the far-field, where the other boundary condition is placed. It is customary to use for laminar boundary layer $\eta_e \sim 8$, which is independent of x.

Having developed the equations for nonsimilar flows, it is easy to see as to what happens for similar flows. Then obviously f will not be a function of x and the governing boundary

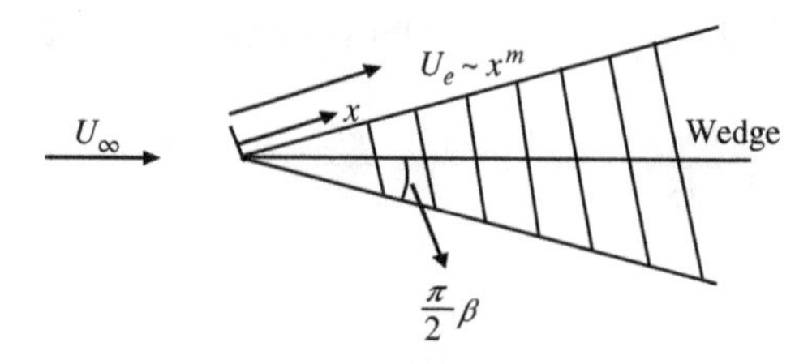

Figure 7.6 Flow over a wedge, with semi-wedge angle defined in the figure

layer equation simplifies to

$$f''' + \frac{m+1}{2} f f'' + m[1 - f'^2] = 0. \tag{7.40}$$

In such similar flows, the boundary conditions cannot depend on x and m also should be a constant. Similarly, f_w must also be a constant. The equation given by Equation (7.40) is also called the Falkner-Skan equation for flow over a wedge, as the flow is deflected through an angle of $\pi\ \beta/2$, where $\beta = \frac{2m}{m+1}$. A schematic of the flow past the wedge is shown in Figure 7.6. For similarity, m is constant, and as

$$m = \frac{x}{U_e}\frac{dU_e}{dx}.$$

So it is easy to see that m depends on whether the flow is accelerating or decelerating. For decelerating flows m takes negative values and it can take value up to $m = -0.0904$. Below this value of Falkner-Skan parameter, flow separates and which is indicated by $f_w'' = 0$ from the solution of Equations (7.38) and (7.40).

One can show that the solution of Falkner-Skan equation given by Equation (7.38) can be used to determine the following physical properties of the shear layer. For example, the wall friction coefficient can be obtained by using the transformed coordinate as

$$C_f = \frac{2f_w''}{\sqrt{R_x}}. \tag{7.41a}$$

Similarly the displacement and momentum thickness coordinate is given by

$$\frac{\delta^*}{x} = \frac{1}{\sqrt{R_x}}\int_0^{\eta_e}(1 - f')\, d\eta = \frac{\eta_e - f_e + f_w}{\sqrt{R_x}} = \frac{\delta_1^*}{\sqrt{R_x}} \tag{7.41b}$$

$$\frac{\theta}{x} = \frac{1}{\sqrt{Rx}}\int_0 f'(1 - f')\, d\eta = \frac{\theta_1}{\sqrt{R_x}}, \tag{7.41c}$$

where the subscripts 'w'and 'e' refer to wall and shear layer edge values, respectively.

The velocity profiles for wedge flows are characterized by β (also called the Hartree parameter) and the shape factor $H = \frac{\delta^*}{\theta}$. Typical nondimensional flow profiles of this family are plotted against nondimensional wall-normal coordinates in Figure 7.7. The wall-normal distance is normalized by the boundary layer thickness of the parallel shear layer.

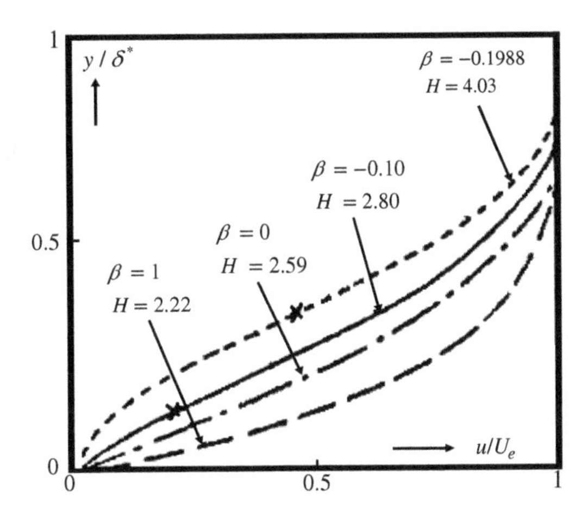

Figure 7.7 Typical mean velocity profiles of some thin shear layers. *Source:* Arnal (1984). Copyright NATO-CSO

The flow profiles with $H > 2.591$ correspond to velocity distributions with inflection point and these are flows with adverse pressure gradient. On the contrary, the flow profiles with $H < 2.591$ correspond to $\frac{dp}{dx} < 0$ (accelerated flows). The figure with $\beta = 0$ and $H = 2.591$ corresponds to the Blasius profile. The profile with $\beta = 1$ and $H = 2.22$ corresponds to stagnation point flow. The other two profiles in Figure 7.7, are for flows with adverse pressure gradients and the crosses on the profile indicate the locations of the inflection points. The profile for $\beta = -0.1988$ ($H = 4.032$) corresponds to the case of incipient separation of the boundary layer.

7.8.1 Zero Pressure Gradient Boundary Layer or Blasius Profile

This is a case of flow over a flat plate at zero incidence. The external potential flow, U_e is independent of x for such a case and $m = 0$, as $U_e = U_\infty$. The governing equation is then given by

$$f''' + ff'' = 0. \tag{7.42}$$

This equation can be numerically solved by using the boundary conditions given by Equations (7.39a) and (7.39b). Having solved the boundary value problem, one obtains the boundary layer thickness as, $\delta = 5\sqrt{\frac{\nu x}{U_\infty}}$. In in the top frame of Figure 7.8, the velocity profile (f') is shown as function of the similarity coordinate, η, with the shear (f'') shown in the middle frame and the curvature of the velocity profile (f''') is shown in the bottom frame. Also, one can use Equations (7.9) and (7.11) to obtain the displacement and momentum thicknesses of the Blasius boundary layer as

$$\delta^* = 1.7208\sqrt{\frac{\nu x}{U_\infty}} \tag{7.43a}$$

$$\theta = 0.664\sqrt{\frac{\nu x}{U_\infty}}. \tag{7.43b}$$

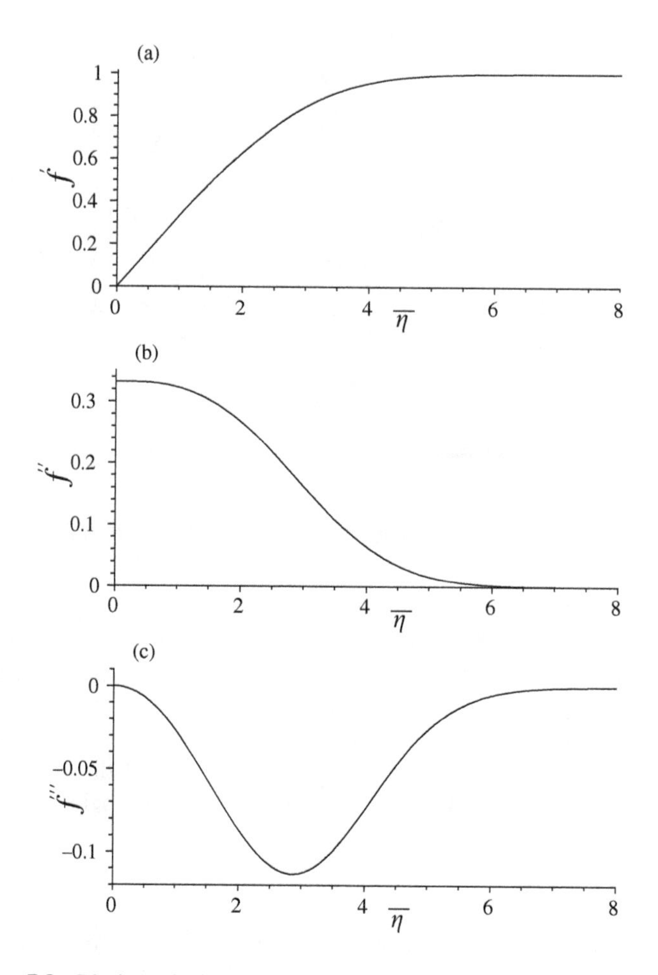

Figure 7.8 Blasius velocity profile for a zero pressure gradient boundary layer

Thus, the shape factor for the Blasius boundary layer is given by

$$H = \frac{\delta^*}{\theta} = 2.5911.$$

7.8.2 *Stagnation Point or the Hiemenz Flow*

This flow is valid for the shear layer in the immediate vicinity of the stagnation point. Such a flow turns by an angle of $\pi/2$, upon impinging on the stagnation point and from Figure 7.6, one must have $\beta = 1$. As $\beta = \frac{2m}{m+1}$, therefore one must have $m = 1$ for the stagnation point flow. The governing equation is obtained from Equation (7.38) by substituting $m = 1$ in it to obtain

$$f''' + ff'' - f'^2 + 1 = 0. \tag{7.44}$$

One notices that the stagnation point flow corresponds to favourable pressure gradient. Such pressure gradient is known to stabilize equilibrium flows without heat transfer (Sengupta 2012). This can be verified by stability calculations and would be explained in Chapter 9, again. In contrast, adverse pressure gradient destabilizes laminar flow, leading to flow transition. We have noted as examples of the role of adverse pressure gradient in destabilizing flows in the core of primary vortices forming over delta wing at high angles of attack in Chapter 6. The properties of the basic shear layers affect many aerodynamic flows and computing the shear layer accurately assists in predicting the stability of such flows.

Similarly, the heuristic reason for the effects of blowing and suction on flow instability can be provided by noting that wall normal blowing introduces mass of new fluid with zero stream-wise momentum into the boundary layer, this is akin to the flow due to adverse pressure gradient, which also results in reduced streamwise momentum of flow and at separation this value of streamwise momentum is zero.

Above two are examples of boundary layers associated with the shearing action of a bounding wall. In the following, we discuss about two examples of thin shear layers in the absence of any wall. Such free shear layers share common properties and their observability in nature depends upon the stability of these thin shear layers.

7.8.3 *Flat Plate Wake at Zero Angle of Attack*

This is an example of a free shear layer, where the shear layer thickness is smaller compared to the integral length scale. Also, the wake is considered to be developing without the presence of any applied pressure gradient. With reference to Figure 7.9, one can define a velocity defect function as the difference between the free stream speed (U_∞) and the local speed ($u(x, y)$) given by

$$u_1(x, y) = U_\infty - u(x, y).$$

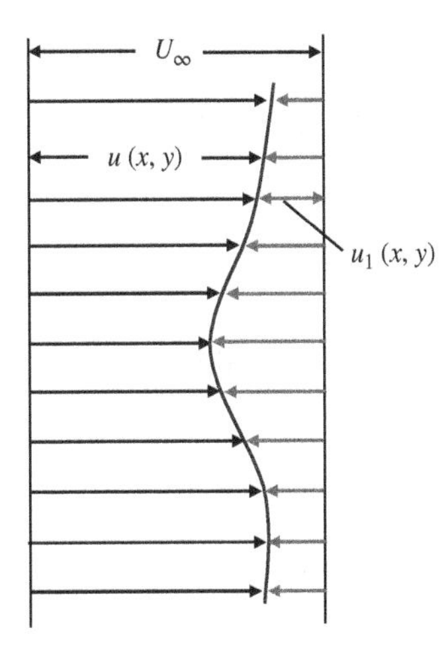

Figure 7.9 Flat plate wake at zero angle of attack

One considers the case where the defect $u_1(x, y)$ is small so that perturbation analysis remains a valid option. If one considers the steady flow case, then the thin shear layer equation is obtained from the x-momentum equation given by

$$U_\infty \frac{\partial u_1}{\partial x} = \nu \frac{\partial^2 u_1}{\partial y^2}, \tag{7.45}$$

where quadratic terms in u_1 and v_1 have been omitted, due to the small perturbation assumption. The required boundary conditions to solve this equation are
at $y = 0$: $\frac{\partial u_1}{\partial y} = 0$ and for $y \to \infty$: $u_1 \to 0$.
Solution of Equation (7.45) can be obtained by use of a similarity transformation

$$\eta = y\sqrt{\frac{U_\infty}{\nu x}} \text{ and } u_1 = U_\infty\, C\left(\frac{x}{l}\right)^{-1/2} g(\eta), \tag{7.46}$$

where l is the length of the plate. The exponent $-1/2$ for the dependent variable is taken on the ground that the momentum integral or the drag must be independent of x, since

$$2D = b\rho\, U_\infty \int_{-\infty}^{\infty} u_1\, dy = b\rho\, U_\infty^2\, C\sqrt{\frac{\nu l}{U_\infty}} \int_{-\infty}^{\infty} g(\eta) d\eta. \tag{7.47}$$

If we introduce Equation (7.46) into Equation (7.45), then we get the governing equation for the similarity function as

$$g'' + \frac{\eta}{2} g' + \frac{1}{2} g = 0. \tag{7.48}$$

This equation has to be solved subject to the boundary conditions: $g'(0) = 0$ and $g(\infty) \to 0$
One can integrate Equation (7.48) once to get

$$g' + \frac{1}{2}\eta g = 0.$$

where the first boundary condition has been used. Further integration would yield

$$g = \exp\left(-\frac{\eta^2}{4}\right).$$

One notes that this solution is the familiar error function, whose integral yields

$$\int_{-\infty}^{\infty} g(\eta)\, d\eta = 2\sqrt{\pi}.$$

So Equation (7.47) gives, $2D = 2\sqrt{\pi}\, C\, b\, \rho\, U_\infty^2 \sqrt{\frac{\nu l}{U_\infty}}$.
Equating this with the drag of a zero pressure gradient flow over flat plate wetted on both sides as given in Equation (7.43b), for the momentum thickness

$$2D = 1.328\, b\, \rho\, U_\infty^2 \sqrt{\frac{\nu l}{U_\infty}},$$

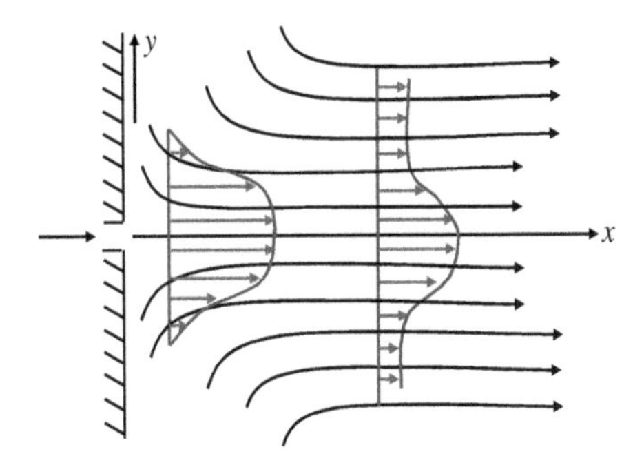

Figure 7.10 Two-dimensional laminar jet, as a similar flow

one gets

$$C = \frac{0.664}{\sqrt{\pi}}.$$

Thus the wake defect is given by

$$\frac{u_1}{U_\infty} = \frac{0.664}{\sqrt{\pi}}\left(\frac{x}{l}\right)^{1/2} \exp\left\{-\frac{y^2\,U_\infty}{4x\,\nu}\right\}. \tag{7.49}$$

This type of velocity profile is noted for $x > 3l$, for very low Reynolds numbers. However, the velocity distribution has point of inflection and hence is always unstable. In actual flows, there are no laminar wakes, all turn turbulent due to instabilities.

7.8.4 Two-dimensional Laminar Jet

This is also a typical example of a thin shear layer forming without any boundary present, apart from the jet nozzle or slit. The jet emerges from a long, narrow slit and mixes with the surrounding fluid. In practice, this flow also becomes easily turbulent.

The emerging jet entrains the surrounding fluid which is at rest. The entrainment is due to the action of shear stress in a real fluid. The jet spreads outwards in the x-direction. For conservation of momentum, i.e. to account for the momentum loss due to drag is manifested by a reduction of the maximum velocity of the jet as it moves downstream with lateral spreading.

For the analysis of 2D laminar jet, we make the following assumptions:

(i) The slit is infinitesimally small.
(ii) Flow velocity at slit is infinitely large, as the mass ejected must be finite.

A schematic of a laminar jet emerging out of a narrow slit is shown as sketched in Figure 7.10. Because the ambient fluid is at rest, so the static pressure (p) experienced by the emerging jet is the same for every x-station. By thin shear layer assumption, we have already seen that p is invariant with y. Also, for this thin shear layer assumption $\frac{dp}{dx} \equiv 0$.

Total momentum in x-direction, denoted by J, must remain constant and be independent of the distance x from the orifice. Hence

$$J = \rho \int_{-\infty}^{\infty} u^2 dy = \text{constant}. \tag{7.50}$$

The governing equation for the 2D laminar jet can be written with the assumption of a thin shear layer from the mass and x-momentum equation as

$$\frac{\partial u}{\partial x} + \frac{\partial v}{\partial y} = 0 \tag{7.51}$$

$$u\frac{\partial u}{\partial x} + v\frac{\partial u}{\partial y} = \frac{1}{\rho}\frac{\partial \tau}{\partial y}, \tag{7.52}$$

where for the thin shear layer

$$\tau = \mu \frac{\partial u}{\partial y}. \tag{7.53}$$

These equations have to be solved subject to the symmetry and other boundary conditions given by

$$\text{at } \ y = 0 : v = 0 \ \text{ and } \ \frac{\partial u}{\partial y} = 0$$

and far away from the centre-line, i.e.

$$\text{at } \ y \to \infty : \quad u = 0.$$

In trying to obtain the similarity solution, we represent

$$f'(\eta) = \frac{u(x, y)}{u_c(x)},$$

where $u_c(x)$ is the jet centre-line velocity. If we denote the shear layer thickness by $\delta(x)$, then one defines the nondimensional independent variable as $\eta = y/\delta(x)$. The similarity solution for the dependent variable is written for the stream function as following

$$\psi(x, y) = u_c(x)\ \delta(x)\ f(\eta). \tag{7.54}$$

Thus the streamwise velocity is defined in terms of the nondimensional stream function, f by, $u = u_c f'$.

The momentum in the x-direction is then obtained as

$$J = 2\rho M \int_0^{\infty} f'^2\, d\eta, \tag{7.55}$$

where $M = u_c^2\, \delta$. Using this information in Equation (7.52), we get

$$\frac{u_c^2}{2}\frac{d\delta}{dx}[f'^2 + ff''] = -\frac{\tau'}{\rho} = -\frac{\mu}{\rho}\frac{\partial^2 u}{\partial y^2} = -\frac{\nu u_c}{\delta}\, f''',$$

or

$$\frac{u_c\delta}{2\nu}\frac{d\delta}{dx}[f'^2 + f f''] = -f'''. \tag{7.56}$$

Note that the existence of ψ automatically satisfies the mass conservation equation, namely Equation (7.51). The above equation is to be solved subject to the boundary conditions:

$$\text{at } \eta = 0 : \ f = f'' = 0$$

$$\text{and at } \quad \eta = \eta_e : \ f' = 0. \tag{7.57}$$

Similarity this flow is possible, if in Equation (7.56) one has

$$\frac{u_c\delta}{2\nu}\frac{d\delta}{dx} = C_1(\text{constant}). \tag{7.58}$$

Additionally, if one puts arbitrarily $C_1 = 1$, then the governing equation becomes

$$f''' + (f')^2 + f f'' = 0. \tag{7.59}$$

From Equation (7.58), using the constant as unity, one gets $\frac{u_c\delta}{2\nu}\frac{d\delta}{dx} = 1$. Also as, $M = u_c^2\delta$, then $u_c = \sqrt{\frac{M}{\delta}}$. Therefore

$$\sqrt{\frac{\delta M}{2\nu}}\frac{d\delta}{dx} = 1.$$

Now as J is constant, therefore M must also be a constant and hence one can write as an order of magnitude analysis

$$\sqrt{\delta}\,\frac{d\delta}{dx} \sim 1,$$

which can be alternately also written as

$$\delta^{1/2}d\delta = \frac{2\nu}{\sqrt{M}}\,dx.$$

Upon integration one finds

$$\frac{2}{3}\,\delta^{3/2} = \frac{2\nu}{\sqrt{M}}\,x \text{ or } \delta^{3/2} = \frac{3\nu}{\sqrt{M}}\,x$$

This can be simplified to establish the dependence of the shear layer thickness on the stream-wise coordinate as

$$\delta = \left(\frac{9\nu^2}{M}\,x^2\right)^{1/3} \sim x^{2/3}. \tag{7.60}$$

and similarly the jet centre-line velocity scaling with x is obtained as

$$u_c = \sqrt{\frac{M}{\delta}} = M^{1/2} \bigg/ \left(\frac{9\nu^2 x^2}{M}\right)^{1/6} = \frac{M^{2/3}}{(3\nu x)^{1/3}} \sim x^{-1/3}. \tag{7.61}$$

To find solution of Equation (7.59) subject to boundary conditions in Equation (7.57), one can integrate Equation (7.59) once to obtain

$$f'' + ff' = \text{constant}. \tag{7.62}$$

Using the boundary condition at $\eta = 0$, this constant on the right-hand side turns out to be zero.

Integrating Equation (7.62) once again, we note that

$$f = \sqrt{2}\tanh\frac{\eta}{\sqrt{2}}. \tag{7.63}$$

Note that in arriving at this expression, we have used the boundary condition at $\eta = \eta_e$. From the above, the velocity profile is obtained as

$$f' = \text{sech}^2\frac{\eta}{\sqrt{2}}. \tag{7.64}$$

Inserting Equation (7.64) in Equation (7.55) one gets

$$M = \frac{3}{4\sqrt{2}}(J/\rho).$$

Using this in Equations (7.58) and (7.59), one gets

$$\delta = \left(\frac{12\sqrt{2}\,\nu^2 x^2}{J/\rho}\right)^{1/3}$$

$$u_c = \left[\frac{3}{32}\left(\frac{J}{\rho}\right)^2 \frac{1}{\nu x}\right]^{1/3}.$$

The jet mass flow rate can be obtained from the following integral as

$$\dot{m} = 2\rho\int_0^\infty u\,dy = 3.302\rho\left(\frac{J}{\rho}\right)^{1/3}(\nu x)^{1/3}.$$

7.8.5 *Laminar Mixing Layer*

Here for the mixing of two uniform streams, as shown in Figure 7.11, a similarity solution is shown to exist with certain restrictions. As before, starting with the conservation of mass and

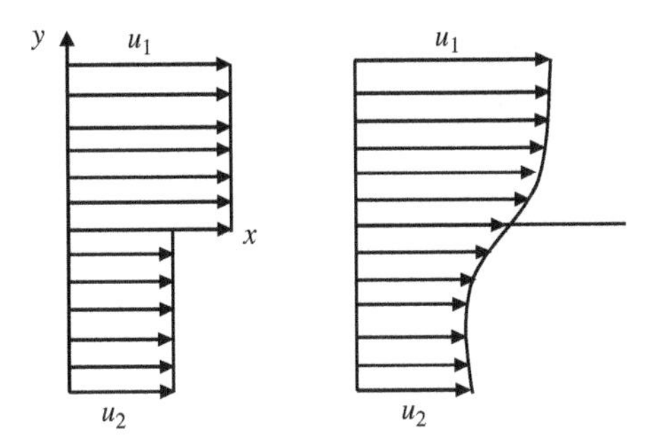

Figure 7.11 Two-dimensional laminar mixing layer, as an example of similar profile

x-momentum equation with thin shear layer approximation, the governing equations are

$$\frac{\partial u}{\partial x} + \frac{\partial v}{\partial y} = 0 \tag{7.65}$$

$$u\frac{\partial u}{\partial x} + v\frac{\partial u}{\partial y} = \frac{1}{\rho}\frac{\partial}{\partial y}\left(u\frac{\partial u}{\partial y}\right). \tag{7.66}$$

These have to be solved subject to the following boundary conditions:

$$y \to \infty: \quad u \to u_1$$

$$y \to -\infty: \quad u \to u_2. \tag{7.67}$$

As before we introduce the stream function so that Equation (7.65) is automatically satisfied. We define the nondimensional stream function $f(\eta)$ by

$$\psi(x, y) = u_1\delta(x)\, f(\eta), \tag{7.68}$$

where the independent similarity variable is given by $\eta = y/\delta(x)$, with $\delta(x)$ is the mixing shear layer thickness, which is considered thin via the requirement, $\frac{d\delta}{dx} \ll 1$. For this definition of similarity variable one has

$$\frac{\partial \eta}{\partial x} = -\frac{y}{\delta^2}\frac{d\delta}{dx} = -\frac{\eta}{\delta}\frac{d\delta}{dx}.$$

Thus, the velocity components are given by

$$u = u_1 f' \text{ and } v = u_1\frac{d\delta}{dx}\,[f'\eta - f]. \tag{7.69}$$

One notes that $y = 0$ is not always the line of symmetry. To obtain the thin shear layer equation in similarity variable, one simplifies to obtain

$$\frac{\partial u}{\partial x} = u_1 f''\frac{\partial \eta}{\partial x} = -\frac{u_1}{\delta(x)}\,\eta\frac{d\delta}{dx}f''$$

and

$$\frac{\partial u}{\partial y} = u_1 f'' \frac{\partial \eta}{\partial y} = \frac{u_1}{\delta(x)} f'' \text{ and } \frac{\partial^2 u}{\partial y^2} = \frac{u_1}{\delta^2} f'''.$$

Using these in x-momentum equation, one gets

$$-u_1 f' \frac{u_1}{\delta} \frac{d\delta}{dx} \eta f'' + u_1 \frac{d\delta}{dx} [f'\eta - f] \frac{u_1}{\delta} f'' = \frac{\nu\, u_1}{\delta^2} f'''$$

or

$$-\frac{u_1^2}{\delta} \frac{d\delta}{dx} \eta\, f'\, f'' + \frac{u_1^2}{\delta} \frac{d\delta}{dx} \eta f'\, f'' - \frac{u_1^2}{\delta} \frac{d\delta}{dx} f f'' = \frac{\nu\, u_1}{\delta^2} f'''$$

or

$$f''' + \frac{u_1 \delta}{\nu} \frac{d\delta}{dx} f\, f'' = 0. \tag{7.70}$$

For similarity solution, we demand that $\frac{u_1 \delta}{\nu} \frac{d\delta}{dx}$ = constant.
Taking this constant equal to 1/2 so as to define δ, we see that

$$\frac{\delta^2}{2} = c\left(\frac{\nu x}{u_1}\right).$$

If we arbitrarily choose $c = 1/2$, so that $\delta = \left(\frac{\nu x}{u_1}\right)^{1/2}$ and then Equation (7.70) becomes

$$f''' + 1/2\, f f'' = 0. \tag{7.71}$$

Equation (7.71) is similar to the Blasius profile, which needs to be solved subject to the boundary conditions at the far-fields as:

$\eta = \eta_e : f' = 1$

$\eta = -\eta_e : f' = \frac{u_2}{u_1} = \lambda$ (a parameter of the problem).

Or the same equation can be solved for the half-range with the boundary conditions

$$\eta = 0 : f = 0 \text{ or } f' = \frac{1}{2}(1 + \lambda).$$

7.9 Use of Boundary Layer Equation in Aerodynamics

We have already noted the existence of differential and integral form of boundary layer equations, both of which can be used to incorporate viscous effects to the results obtained by inviscid analysis. Actual usage of the boundary layer dictates which form will be more desirable. If one wants only the mean quantities for the boundary layer, then the integral formulation will be more expedient. However, if one is interested in the instability properties of the boundary layer, then the differential formulation of the boundary layer equation will be necessary. We note in passing that with the present state of development of scientific computing and the hardware at the desktop level, direct numerical simulation (DNS) of flow past airfoil is possible. Such results are provided in Chapter 10. However, we first describe the methodologies by which boundary layer information obtained from the differential formulation of boundary layer is used.

7.9.1 Differential Formulation of Boundary Layer Equation

Instead of using the differential form of boundary layer equation given by Equation (7.8), one gets more accurate results by solving Equation (7.38). This is the nonsimilar formulation of the corresponding marching problem defined by this equation.

$$f''' + m(1 - f'^2) + \frac{m+1}{2} f f'' = x\left(f'\frac{\partial f'}{\partial x} - f''\frac{\partial f}{\partial x}\right). \tag{7.38}$$

In the case of incompressible flows, first the inviscid equation is solved, as described by the panel method in Chapter 5. The slip velocity obtained on the surface of the airfoil by the inviscid equation, is used as the boundary layer edge velocity, $U_e(x)$. The boundary layer calculation is started from the stagnation point, where the solution is obtained by solving the Hiemenz flow equation

$$f''' + f f'' - f'^2 + 1 = 0. \tag{7.44}$$

As the governing equation is parabolized by boundary layer assumption, one can march from the stagnation point in two sweeps, along the upper and lower surfaces of the airfoil independently. However, one notes that the Falkner-Skan parameter (m) is calculated numerically from the $U_e(x)$ distribution obtained as the inviscid solution.

As one is marching downstream, from say a station ‘1’ to ‘2’, then one can calculate the right-hand side terms of Equation (7.38) in the following manner for any quantity, q

$$\frac{\partial q}{\partial x} \simeq \frac{q_2 - q_1}{\Delta x}.$$

One usually follows the robust and accurate Keller box scheme, described in Cebeci and Bradshaw (1977), in solving the resultant discretized equation. It is possible to incorporate boundary layer stability calculations (as is described in Chapter 9), to empirically predict transition. If the transition is enforced, as in many experimental cases, then corresponding transition and turbulence models can also be incorporated in this approach. Details of various possible methods can be obtained from Cebeci and Cousteix (2005) for the interested readers.

Apart from the use of tangential velocity, one can also use Equation (7.9) to incorporate the wall-normal component of boundary layer edge velocity in an iterative manner:

$$v(x, y) = \frac{d}{dx}(U_e\, \delta^*) - y\frac{dU_e}{dx}.$$

This wall-normal velocity obtained from the boundary layer solution is to be used in the panel code. Thus, this is done to incorporate viscous-inviscid interaction iteratively in many analysis and design methods.

7.9.2 Use of Momentum Integral Equation

The momentum integral equation has been obtained by von Kármán as

$$\frac{d\theta}{dx} + \frac{\theta}{U_e}\frac{dU_e}{dx}(H+2) = \frac{C_f}{2}. \tag{7.22}$$

This relates the boundary layer integral properties, but which requires the information about the velocity profile. As no specific assumption is made regarding the velocity profile, Equation (7.22) is equally applicable for laminar and turbulent flows. With appropriate velocity profile chosen, Equation (7.22) can be integrated to obtain the various integral properties of the boundary layer.

Among many variants of momentum integral equation solution methods available, we discuss here Pohlhausen's method in the following.

7.9.3 Pohlhausen's Method

To initiate the solution procedure of Equation (7.22), approximate velocity profiles in the form of polynomial expression are proposed in this method. If one decides to choose a cubic polynomial, then it is represented as

$$\bar{u} = \frac{u}{U_e} = p_0 + p_1\eta + p_2\eta^2 + p_3\eta^3, \tag{7.72}$$

where $\eta = y/\delta$ represents the nondimensional wall coordinate with respect to the shear layer thickness. The coefficients of the polynomial in Equation (7.72) are obtained by satisfying the boundary conditions at the wall ($\eta = 0$) and at the edge of the boundary layer ($\eta = 1$) as

$$\bar{u} = 0 \text{ at } \eta = 0 \quad \text{and} \quad \bar{u} = 1 \text{ at } \eta = 1. \tag{7.73a}$$

To fit a cubic polynomial of Equation (7.72), we must satisfy two additional conditions for the profile, one of which is that the approach of the velocity profile to the edge velocity (U_e) with η must be smooth without slope discontinuity there. This is equivalent to satisfying the following condition

$$\frac{\partial \bar{u}}{\partial \eta} = 0 \text{ at } \eta = 1. \tag{7.73b}$$

Having noted that the momentum integral relation provides a statement of variation of the boundary layer along the streamwise direction, its variation in the wall-normal direction can be embedded by using the boundary layer equation, Equation (7.16), at the wall as

$$\frac{\partial^2 \bar{u}}{\partial \eta^2} = \frac{\delta^2}{\nu}\frac{dU_e}{dx} \text{ at } \eta = 0. \tag{7.73c}$$

Use of the boundary conditions in Equations (7.73a) to (7.73c) in Equation (7.72) yields

$$p_0 = 0 \tag{7.74a}$$

$$p_1 + p_2 + p_3 = 1 \tag{7.74b}$$

$$p_1 + 2p_2 + 3p_3 = 0 \tag{7.74c}$$

$$2p_2 = -\Lambda, \tag{7.74d}$$

where $\Lambda = \frac{\delta^2}{\nu}\frac{dU_e}{dx}$ is the dimensionless weightage of pressure force with respect to the viscous force and is called the Pohlhausen parameter. Thus, the velocity profile can be represented as

$$\bar{u} = \frac{3}{2}\eta - \frac{1}{2}\eta^3 + \frac{\Lambda}{4}\left(\eta - 2\eta^2 + \eta^3\right). \tag{7.75}$$

It is evident that $\Lambda = 0$ represents a zero pressure gradient flow, while negative value of Λ represents a decelerating flow and positive value corresponds to accelerating or favourable pressure gradient flow. From Equation (7.75) one can obtain the zero wall shear condition from

$$\frac{\partial \bar{u}}{\partial \eta} = 0 \text{ at } \eta = 0,$$

where the flow will separate. This fixes the value of $\Lambda_{sep} = -6$ for separation. How accurate is this estimate? To answer this question, let us obtain a velocity profile which is a quartic, as compared to the cubic in Equation (7.72) as

$$\bar{u} = \frac{u}{U_e} = p_0 + p_1\eta + p_2\eta^2 + p_3\eta^3 + p_4\eta^4. \tag{7.76}$$

Now to fix five constant p_i's, we need to satisfy an additional boundary condition for the profile, in addition to those given in Equation (7.74). This is provided as the following condition as $\frac{\partial^2 \bar{u}}{\partial \eta^2} = 0$ at $\eta = 1$. Proceeding as before, we obtain the following expression for the velocity profile

$$\bar{u} = 2\eta - 2\eta^3 + \eta^4 + \frac{\Lambda}{6}\left(\eta - 3\eta^2 + 3\eta^3 - \eta^4\right). \tag{7.77}$$

As before, one can obtain the criterion for flow separation from

$$\frac{\partial \bar{u}}{\partial \eta} = 0 \text{ at } \eta = 0,$$

to obtain the separation criterion as $\Lambda_{sep} = -12$, a value which is double in magnitude, as compared to that is obtained from the cubic velocity profile. This is one of the clear indications that integral form of boundary layer equation is always less reliable, as compared to differential form. Comparing Equations (7.75) and (7.77), it is obvious that the quartic is more reliable, as this satisfies the second derivative at the edge of the shear layer in a definitive manner. Second derivatives of velocity profiles will be shown to strongly affect the flow instability in Chapter 9. This is also the main reason for preferring quartic velocity profile over cubic velocity profile.

If we denote the nondimensional displacement and momentum thicknesses with respect to boundary layer thickness (δ) by $d_{\delta*}$ and d_θ as

$$d_{\delta*} = \frac{\delta*}{\delta} = \int_0^1 (1-\bar{u})d\eta \tag{7.78}$$

$$d_\theta = \frac{\theta}{\delta} = \int_0^1 \bar{u}(1-\bar{u})d\eta. \tag{7.79}$$

It can be shown after some algebraic manipulation, the following estimates for the above ratios and the wall skin friction for the cubic and quartic velocity profiles as

$$[d_{\delta*}]_{\text{cubic}} = \frac{3}{8} - \frac{\Lambda}{48}$$

$$[d_\theta]_{\text{cubic}} = \frac{1}{280}\left(39 - \frac{\Lambda}{2} - \frac{\Lambda^2}{6}\right) \tag{7.80}$$

$$\text{and} \quad [C_f]_{\text{cubic}} = \frac{\mu}{\rho\, U_e \delta}\left(3 + \frac{\Lambda}{2}\right)$$

$$[d_{\delta*}]_{\text{quartic}} = \frac{3}{10} - \frac{\Lambda}{120}$$

$$[d_\theta]_{\text{quartic}} = \frac{1}{63}\left(\frac{37}{5} - \frac{\Lambda}{15} - \frac{\Lambda^2}{144}\right) \tag{7.81}$$

$$\text{and} \quad [C_f]_{\text{quartic}} = \frac{\mu}{\rho\, U_e \delta}\left(4 + \frac{\Lambda}{3}\right),$$

It has been stated in Cebeci and Bradshaw (1977) that Λ cannot take value greater than 12, otherwise the velocity profile will display an overshoot. Thus, it has been argued that Λ should always be restricted between -12 and $+12$. What are the corresponding ranges for the cubic velocity profile? This is left as an exercise for the reader.

The above condition of excluding velocity overshoot above the unity value, may appear as logical. Yet, in many flows with adverse pressure gradient and in flows with heat transfer, it can be shown (as in Sengupta, 2012) that the velocity profile actually has overshoot. The relationship between the critical values of Λ for which overshoot exists in $0 \le \eta \le 1$ with the critical height can be obtained from

$$\frac{d\bar{u}}{d\eta} = 0 \ \text{ and } \ \frac{d^2\bar{u}}{d\eta^2} < 0.$$

This is left as an exercise for the readers.

7.9.4 Thwaite's Method

Thwaite devised a method in Thwaite (1949), where H and C_f is represented as function of boundary layer edge velocity (U_e) and momentum thickness (θ), so that the momentum integral equation, Equation (7.22) can be numerically integrated. He suggested the following representations:

$$\left.\frac{\partial^2 u}{\partial y^2}\right|_w = -\frac{U_e}{\theta^2}\lambda \text{ and } \left.\frac{\partial u}{\partial y}\right|_w = \frac{U_e}{\theta}l. \tag{7.82}$$

With these equations defining the newly introduced parameters λ and l, so that the shape factor, H and l can be expressed to depend upon λ from the existing database. This was suggested and empirical correlations were provided by Thwaite. Using Equations (7.82) and (7.73c), it can be shown that

$$\lambda = \frac{\theta^2}{\nu}\frac{dU_e}{dx} \text{ and } \frac{C_f}{2} = \frac{\nu}{U_e^2}\left(\frac{\partial u}{\partial y}\right)_w = \frac{\nu l(\lambda)}{U_e\theta}. \tag{7.83}$$

Using Equation (7.83) in Equation (7.22) one gets

$$\frac{U_e}{\nu}\frac{d\theta^2}{dx} = 2\lambda(2-H) + 2l. \tag{7.84}$$

Using differential formulation results, Thwaite provided the following data fit for the right hand side of Equation (7.84) $[G(\lambda)]$ as

$$G(\lambda) = 0.45 - 6\lambda. \tag{7.85}$$

Using this expression for $G(\lambda)$ on the right-hand side of Equation (7.84) one gets

$$\frac{U_e}{\nu}\frac{d\theta^2}{dx} = 0.45 - 6\frac{\theta^2}{\nu}\frac{dU_e}{dx}.$$

Multiplying both sides by U_e^5 and rearranging, one obtains

$$\frac{1}{\nu}\frac{d}{dx}(\theta^2 U_e^6) = 0.45U_e^5. \tag{7.86}$$

Integrating this equation from $x = 0$ to any arbitrary station, one gets

$$\frac{\theta^2 U_e^6}{\nu} = 0.45\int_0^x U_e^5 dx + \left(\frac{\theta^2 U_e^6}{\nu}\right)_{x=0}. \tag{7.87}$$

In Thwaite's method, one obtains the distribution of θ with x from the solution of the above equation. In this method, H and C_f are obtained from the following empirical correlation given in Cebeci and Bradshaw (1977) as

For $0 \leq \lambda \leq 0.1$:

$$H = 2.61 - 3.75\lambda + 5.24\lambda^2$$

$$l = 0.22 - 3.75\lambda - 1.8\lambda^2.$$

And for $-0.1 \leq \lambda \leq 0$:

$$H = \frac{0.0731}{0.14 + \lambda} + 2.088$$

$$l = 0.22 + 1.402\lambda + \frac{0.018\lambda}{0.107 + \lambda}.$$

Bibliography

Arnal, D. (1984) Description and prediction of transition in two-dimensional incompressible flow. *AGARD Report* **709**, 2.1–2.71.

Cebeci, T. and Bradshaw, P. (1977) *Momentum Transfer in Boundary Layers.* Hemisphere Publishing Crop., Washington DC, USA.

Cebeci, T. and Cousteix, J. (2005) *Modeling and Computation of Boundary-Layer Flows.* Horizon Publishing, Long Beach, California, USA.

Murman, E.M. and Cole, J.D. (1971) Calculation of plane steady transonic flows. *AIAA J.*, **9**, 114–121.

Prandtl, L. (1904) Uber flussigkeitsbewegung bei sehr kleiner reibung. Proc. of *Third Int. Math. Cong., Heidelberg.* (Also in English as *NACA TM* **452**.)

Sengupta, T.K. (2012) *Instabilities of Flows and Transition to Turbulence.* CRC Press, Florida, USA.

Th. von Kármán. (1921) Über laminare und turbulente reibung. *Z. Angew. Math. Mech.,* **1**, 233–252.

8

Computational Aerodynamics

8.1 Introduction

Computational power currently available at the desktop enables one to probe computationally many detailed aerodynamic properties of flying vehicles. Aerodynamics, apart from numerical weather prediction, has all along provided impetus for many developments in computing. In this chapter, we dwell upon some of the main issues relevant to computational aerodynamics as practised today. These are related to fundamental aerodynamic concepts and not for computing flows past complete complex geometries often employed for flying vehicles in industry or for preliminary design of aerospace vehicles. The main emphasis here is on the fundamentals of aerodynamics, which even in the simplest symmetric steady flight of aircraft in longitudinal plane is concerned with lift, drag and pitching moment acting on the vehicle. While lift is inherently related to the ability of a heavier than air vehicle to lift not only its own weight in steady and level flight, but the lifting surface should be able to generate more lift to perform steady and unsteady manoeuvres. While any airborne object would experience drag due to pressure distribution and shear stresses, lifting surfaces experience additional drag due to lift generation, which has been explained theoretically in Chapter 4.

In Chapter 2, we have indicated hierarchies of various models employed in calculating forces and moments acting upon aerodynamic bodies and stated that the Navier-Stokes equation is adequate in defining all forces and moments acting due to active stresses on various surfaces. Thus, the problem of external aerodynamics involves evaluation of lift, drag and pitching moment, as the lowest common denominator. It is for this reason that airfoil as the quintessential element of the lifting surface is studied in such great detail, both theoretically and experimentally, providing these three aerodynamic properties.

In theoretical studies, one can start from the most rudimentary potential flow models characterized by panel methods described in Chapter 5, which allow one to obtain the lift and pitching moment acting on an aerodynamic surface from inviscid analysis for pressure distribution. As viscous effects are not considered in potential flows, one cannot obtain shear stresses and hence the skin friction. In the previous chapter we have explained how this can be circumvented by solving the boundary layer equation in obtaining the skin friction, when

Theoretical and Computational Aerodynamics, First Edition. Tapan K. Sengupta.

Companion Website: www.wiley.com/go/sengupta

the flow remains attached to the surface. Also it was explained in the previous chapter, how one can hierarchically first solve the potential flow to obtain inviscid pressure distribution on the surface. This is aided by the boundary layer assumption, which states that the impressed pressure on the boundary layer is transmitted unaltered through the shear layer. Having obtained the external pressure distribution, one can solve the boundary layer equation using a nonsimilar formulation – an extension of the Falkner-Skan transformation for the thin shear layer. For compressible flows, one can solve the Euler equation to obtain the inviscid pressure distribution and which can be used along with compressible boundary layer equation. This is called as the weak coupling between viscous and inviscid effects on lifting surfaces. Such an analysis becomes untenable for wings at larger angles of attack, due to flow separation and for other nonlifting aerodynamic surfaces. Additional difficulties of coupling the viscous and inviscid dynamics are also experienced in explaining the phenomena of instability and transition, which are often attempted in a semi-empirical manner – a topic dealt with in the following chapter.

All the above-mentioned restrictions of studying flow past aerodynamic surfaces at off-design conditions can be removed by solving the Navier-Stokes equation. In this chapter, all the subtle issues of scientific computing are described in brief. Interested readers are advised to read books, monograph and treatises on scientific computing, which are distinctly different from books on CFD. A carefully obtained methodology works across the speed regime and this is advocated in this book, i.e. no special efforts are needed for incompressible and compressible flows, both are computed by same method. We will present methods and results for incompressible and compressible flows past airfoils, specifically for the natural laminar flow (NLF) airfoils in Chapter 10. The methodology for solving incompressible flows past airfoils can be gainfully employed in studying time accurate transonic and supersonic flow past airfoils and this is shown in Chapter 11. Thus, it is apparent that the methodology for solving Navier-Stokes equation for flow past airfoils have to be performed very accurately and here an approach will be followed which uses structured grid with finite difference methods. For this purpose we need to develop a good-quality grid around airfoils. Here, we also emphasize the use of orthogonal grids for airfoils of different thickness and camber and describe a method to do so. Generated orthogonal grid with highly desirable properties will help one to perform direct simulation of flow past aerofoils.

8.2 A Model Dynamical Equation

Basic equations in aerodynamics are also the statements of conservation principles for mass, momentum and energy, which lead to space-time dependent quasi-linear PDEs. One chooses the physical principles to be satisfied as the basis for deriving the governing equations. For example, if we are only interested on incompressible flows without heat transfer, then one would just be concerned with mass and momentum conservation. For typical aerodynamic applications one is interested in high Reynolds number flows, which are convection dominated and can support discontinuities, such as shocks in the flow. Thus, it is imperative that readers appreciate the nuances of solution techniques of various space-time dependent equations. To develop numerical methods capable of simulating such flows, one obtains an appropriate model equation to capture the essence of such simulations. Also, notice that unlike solid mechanics problems, in fluid mechanics we are concerned with problems having a relatively larger number of degrees of freedom, as the flow is associated with significantly lower physical diffusion. This cannot be completely ignored from the whole flow domain, as noted in the previous chapter via the existence of a thin shear layer. This physical process of diffusion is very important at high wavenumbers. Thus, simulating flows one would have to be careful in resolving higher wavenumbers.

Computing governing fluid dynamical equations accurately, one must consider space-time dependence of the problem together. Noticing that the placement of a body in a fluid is equivalent to creating disturbances which propagate in space and time. It is therefore clear that the disturbance field can be considered as a agglomeration of the building blocks of the disturbance field, namely the waves. Such disturbance propagation can be described by the one-dimensional convection equation

$$\frac{\partial u}{\partial t} + c\,\frac{\partial u}{\partial t} = 0; \qquad c > 0. \tag{8.1}$$

The building blocks of any arbitrary aggregation of plane waves can be understood by defining certain wave parameters for a single-periodic function

$$u(x,t) = a \sin\left[\frac{2\pi}{\bar{\lambda}}(x - ct)\right]. \tag{8.2}$$

One can identify this as a specific solution for Equation (8.1) which has a nonzero right-running wave solution. In the above, the quantity in the square brackets represents the phase of the wave and a represents the amplitude of the wave. The quantity $\bar{\lambda}$ is the wavelength, since u does not change when x is changed by $\bar{\lambda}$, with t held fixed. One defines wavenumber k $(= \frac{2\pi}{\bar{\lambda}})$, which provides the number of full waves in a length of 2π. Thus, the representation of Equation (8.2) can be alternately written as

$$u(x,t) = a \sin\,[k(x - ct)]. \tag{8.3}$$

Keeping one's gaze fixed at a single point, the least time after which $u(x,t)$ retains the same value determines the time period T, and this is also the time required for the wave to travel one wavelength: $T = \frac{\bar{\lambda}}{c}$. The number of oscillations at a point per unit time is the frequency given by $f_0 = \frac{1}{T}$. One can define a circular frequency $\bar{\omega}$ by noting

$$\bar{\omega} = kc. \tag{8.4}$$

Thus, c $(= \bar{\omega}/k)$ has a dimension of speed and is appropriately called the phase speed, the rate at which the phase of the wave propagates. Such movement is not always physical and most often illusory. Equation (8.4) is known as the physical dispersion relation for obvious reasons.

If we look at the disturbance energy spectrum $(E(k))$ of many fluid dynamical systems, such as the one sketched in Figure 8.1 for flow over a flat plate, then it is apparent that a continuum of wavenumbers is present in the system. For such a space-time dependent system, it is important to know how the disturbance energy is transmitted or if the propagation of disturbances and the propagation of associated energy are different. These queries can be satisfactorily answered if we can work out the details of how neighbouring elements in the spectrum relate and interact with each other. Let us say that the spectrum and the dispersion relation $\bar{\omega} = \bar{\omega}(k)$ are continuous functions of their argument.

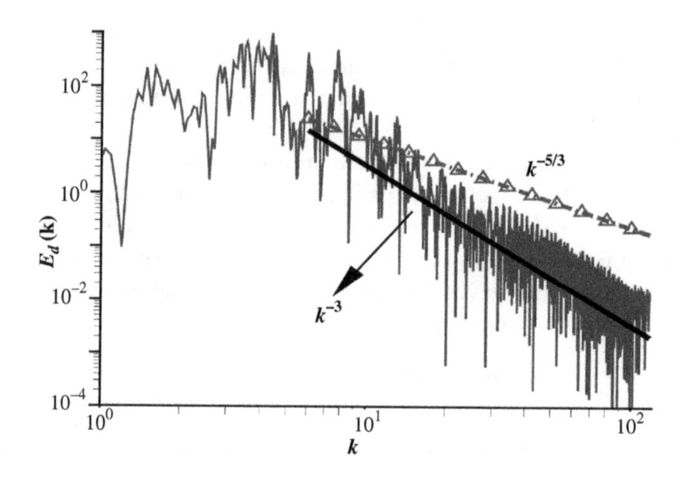

Figure 8.1 A typical energy spectrum of a two-dimensional flow over a flat plate. Note that the spectrum varies as k^{-3} for an intermediate wavenumber range. *Source:* Sengupta and Bhaumik (2011). Reproduced with permission from American Physical Society

Let us now track two such closely spaced neighbouring wavenumbers k_1 and k_2 ($= k_1 + dk$). Corresponding circular frequencies are also closely spaced with values $\bar{\omega}_1$ and $\bar{\omega}_2$ ($= \bar{\omega}_1 + d\bar{\omega}$). Also, the wavenumbers are so close that the amplitudes of the individual harmonic components are taken as same and their superposition gives rise to the wave-form

$$\begin{aligned} Y &= a\cos(k_1 x - \bar{\omega}_1 t) + a\cos(k_2 x - \bar{\omega}_2 t) \\ &= \left[2a\cos\left(\frac{dk}{2}x - \frac{d\bar{\omega}}{2}t\right)\right]\cos\left[\left(k_1 + \frac{dk}{2}\right)x - \left(\bar{\omega}_1 + \frac{d\bar{\omega}}{2}\right)t\right]. \end{aligned} \tag{8.5}$$

While the phase of the second factor resembles the phase of the original harmonic elements, it is the first factor that is of significant interest, which represents an amplitude varying slowly in space (with wavelength $4\pi/dk$) and time (with time period $4\pi/d\bar{\omega}$).

This modulation of the resultant amplitude occurs via its phase variation and the corresponding $x/t =$ constant line moves with the speed

$$V_g = \frac{d\bar{\omega}}{dk}, \tag{8.6}$$

which is defined as the group velocity. This is nothing but the slope of the dispersion relation given by Equation (8.4) for the governing equation given in Equation (8.1). A typical sketch for the disturbance variation of this group, composed of k_1 and k_2 is as shown in Figure 8.2.

As the energy of the system is proportional to the square of the resultant amplitude of the interacting waves, the energy of the composite system propagates at this speed, V_g, the speed at which the amplitude travels. Heuristically, note the location of the nodes where the resultant amplitude is zero. Thus, we note the physical implications of V_g as the speed with which energy travels in a system displaying a wide-band spectrum. One also notes that the group velocity is a consequence of the dispersion property, which indicates the circular frequency as a function of wavenumber. The carrier waves use the phase speed for phase variation, while the group velocity is associated with the propagation of amplitude.

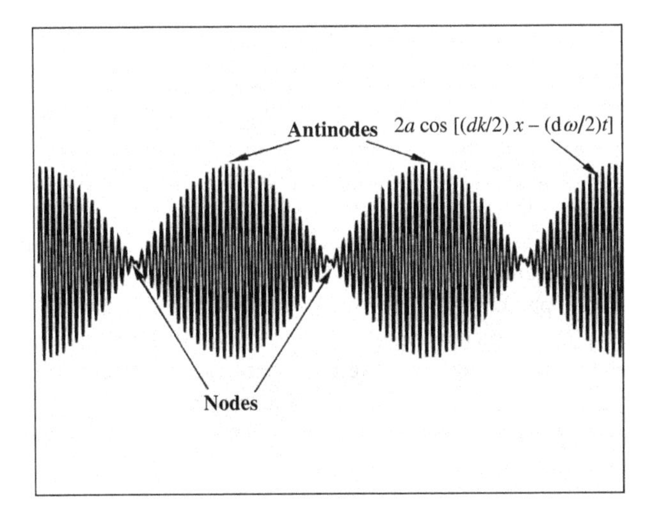

Figure 8.2 Phenomenon of modulation in a group of waves

Consider the propagation of waves governed by the 1D convection equation given by Equation (1), for which one can substitute the trial solution $u = ae^{i(kx-\bar{\omega}t)}$ to obtain the dispersion relation as given by Equation (8.4).

One readily notes that

$$\frac{d\bar{\omega}}{dk} = V_g = c. \tag{8.7}$$

This signifies that the phase speed and group velocity are indistinguishable for non-dispersive systems like the one given in Equation (8.1).

For a general system the dispersion relation can be reduced to

$$V_g = c + k\frac{\partial c}{\partial k}. \tag{8.8}$$

8.3 Space–Time Resolution of Flows

Most practical aerodynamic flows are essentially at high Reynolds numbers for which laminar flows become unstable at Reynolds numbers above critical values, causing the resultant flows to be either transitional or turbulent. Laminar flows can be either steady or unsteady, but transitional and turbulent flows are inherently unsteady. These flows exhibit space and time scales with broad spectra. Resolving these spectra is the main challenge in computing these flows. Thus, the computing tools employed in solving these problems must be *spectrally accurate* to resolve these scales. Not only must the spatial and temporal scales be resolved together, their relationship as given by the dispersion relation must also be obeyed. This indicates that the length and time scales are dependent via the dispersion relation.

8.3.1 Spatial Scales in Turbulent Flows and Direct Numerical Simulation

It is customary to treat the scales of turbulent flows to be related to equivalent eddy sizes in the evolving flow field. It is not necessary that such eddies exist in reality, and this is done for the ease of understanding and analysis. The largest scale is associated with the integral dimension of the fluid dynamical system denoted by l, at which the flow is fed with energy. For the flow over an aircraft wing, this can be visualized to be equivalent to the chord of the wing itself, for the analysis of steady and level flight. Without going into further details, one can state that the supplied energy at large length scale is redistributed to smaller length scales due to multiple physical processes caused by nonlinearity and dispersion.

In general for turbulent flows, one represents kinetic energy density (kinetic energy per unit volume) in the spectral plane, which is preferred due to easier interpretation and existence of a developed theory for homogeneous turbulence in wavenumber space. A dynamical balance between supplied energy and its dissipation is not easily understood for incompressible flows by focusing attention on momentum conservation equation alone. A clearer picture emerges for incompressible flows, if one looks at the distribution of total mechanical energy of the flow, as described in Sengupta *et al.* (2003).

For homogeneous flows one can relate this dissipation of energy directly to the diffusion terms appearing in the Navier-Stokes equation. For such flows, dissipation is given by $\nu||\nabla u||_2^2$. Extending this observation for general inhomogeneous flows, the energy budget of the disturbance field is in general considered to depend upon the wavenumber (k), dissipation (ϵ) and kinematic viscosity (ν). Consequently it is stated that for both internal and external flows, the dissipation peak is located at a higher wavenumber *as compared to the peak of energy spectrum*. If we define u as the velocity (which represents kinetic energy per unit mass) in the large scale, then we can define a Reynolds number given by

$$Re = \frac{ul}{\nu}.$$

It has been shown by Kolmogorov (see Tennekes and Lumley 1971) that the smallest excited length scale – known as the Kolmogorov scale – is given by

$$\eta_K = (\nu^3/\epsilon)^{\frac{1}{4}}. \tag{8.9}$$

Thus, the largest and the smallest length scales are related by

$$\frac{l}{\eta_K} = (Re)^{\frac{3}{4}}. \tag{8.10}$$

In turbulent boundary layers, there is a region very close to the wall where the velocity varies linearly with the distance from the wall. This is called the *viscous sublayer* and its thickness ξ is related to l by

$$\frac{l}{\xi} = Re. \tag{8.11}$$

For flow computations at high Reynolds numbers via DNS these scales are resolved. If the cut-off wavenumber is represented by k_c (related to η_K), then Equation (8.10) can also be written as

$$k_c l \sim O(Re^{3/4}). \tag{8.12}$$

This equation is often used to state grid requirements for DNS, as a ball-park number. Extending the logic for three-dimensional flows, this estimate shows that the resolution requirement scales as $(Re^{3/4})^3$ or roughly about Re^2. There is nothing rigorous about this oft-quoted relation.

Kolmogorov's scaling theory states that there exist length scales shorter than those are directly excited (l), but larger than the Kolmogorov scale (η_K), for which the energy spectrum will be independent of eventual dissipation by viscous action, if there exists a statistical steady state. At these intermediate scales – the *inertial subrange* – the dependence of $E(k)$ is solely determined by nonlinear energy transfer via vortex stretching by a cascade process and the overall energy flux is shown to depend as

$$E(k, \varepsilon) = C_k k^{-\frac{5}{3}} \varepsilon^{\frac{2}{3}}. \tag{8.13}$$

It has been shown in Davidson (2004), Doering and Gibbon (1995) and Sengupta and Bhaumik (2011) that there are many important two-dimensional flows which display turbulence spectrum totally different from that is given by Equation (8.13) for three-dimensional turbulent flows. For 2D viscous flows, enstrophy increases due to continual vorticity generation at the wall (which may not be compensated by dissipation). Enstrophy can be created or destroyed, when large eddies break by Kelvin–Helmholtz instabilities and/or complex vortex interactions and/or coalescence. It has been reasoned (Batchelor 1969, Kraichnan and Montgomery 1980) that enstrophy plays the same role in two-dimensional and three-dimensional flows for generating rotationality as the energy plays in three-dimensional flows in creating translational kinetic energy. Applying scaling arguments, one can formulate a dissipation scaling for two-dimensional turbulent flow. The details in Kraichnan and Montgomery (1980) and Doering and Gibbon (1995) show the high wavenumber energy spectrum to be given by

$$E(k, \chi) = \chi^{2/3} k^{-3}. \tag{8.14}$$

where χ is the enstrophy dissipation rate, given by $||\nabla\omega||_2^2$ for homogeneous flows (see Doering and Gibbon 1995, for details). In Sengupta *et al.* (2013) an alternate viewpoint is presented for inhomogeneous flows by noting that in all turbulent flows rotationality is very important. Hence enstrophy must be also considered to account for rotational motion, apart from $E(k)$ for translational motion. In this work, a transport equation is derived to define a cascade process, which is independent of dimensionality of the flow.

8.3.2 Computing Unsteady Flows: Dispersion Relation Preserving (DRP) Methods

It has been noted that Equation (8.1) as a model equation performs the role to check accuracy of any numerical method for space-time discretizations. This is a nondissipative and nondispersive equation which convects the initial solution to the right at the speed c, representing both phase speed and group velocity. As this equation is nondissipative and nondispersive, it provides a simple equation, yet a tough test for any combined space-time discretization method's properties. Explanations provided here are based on spectral analysis of numerical schemes, developed for full-domain analysis of computational methods (Sengupta *et al.* 2003a). In this analysis method, a dependent variable is represented by a hybrid space-time expansion

$$u(x_m, t^n) = \int U(kh, t^n)\, e^{ikx_m}\, dk. \tag{8.15}$$

8.3.3 Spectral or Numerical Amplification Factor

The spectral amplification factor or the gain of a numerical method for a governing equation involving $u(x, t)$ is defined by

$$G(\Delta t, kh) = \frac{U(kh, t^n + \Delta t)}{U(kh, t^n)}. \tag{8.16}$$

Note that G is a complex quantity, as the Fourier–Laplace amplitude $U(kh, t)$ is complex. In the continuum limit of $\Delta t \to 0$, one must have

$$|G| \equiv 1 \tag{8.17}$$

for the solution of Equation (8.1). However, real and imaginary parts of G depend on the differential equation solved and the discretization methods adopted for spatial and temporal derivatives considered together in the discrete equation. Analyses of numerical methods and associated error propagation have been performed following different routes by many researchers. In the classical approach attributed to von Neumann, the evolving error of the discretized differential equation with linear constant coefficients is assumed to follow an identical discrete equation. In this approach due to von Neumann, the difference between exact and computed solutions arises due to round-off error and error in the initial data. For periodic problems, error furthermore consists of the Fourier series and the individual normal modes are investigated. The main assumption for linear problems that the error and the signal follow the same dynamics in the von Neumann approach appears intuitively correct. However, it has been unambiguously shown in Sengupta *et al.* (2007) that this assumption is flawed for any discrete computing and the difference is due to dispersion and phase errors and when the numerical method is not strictly neutrally stable.

The exact spatial derivative of $u(x_j)$ at the same node is given by

$$\left[\frac{\partial u}{\partial x}\right]_{exact} = \int ikU\, e^{ikx_j}\, dk. \tag{8.18}$$

While solving Equation (8.1) by discrete methods, the spatial derivative u'_j (denoted by a prime) has been shown (Sengupta *et al.* 2003a, Sengupta and Dipankar 2004) to be equivalent to

$$\left[u'_j\right]_{numerical} = \int ik_{eq} U\, e^{ikx_j} dk. \tag{8.19}$$

Numerically the same derivative can also be estimated from the linear algebraic equation in the physical plane by $\{u'\} = \frac{1}{h}[C]\{u\}$. One obtains appropriate $[C]$ matrices for finite-domain nonperiodic discretized problems for different methods, which have been provided in the literature (Sengupta *et al.* 2003a, 2012a). The dimension of $[C]$ matrix corresponds to the

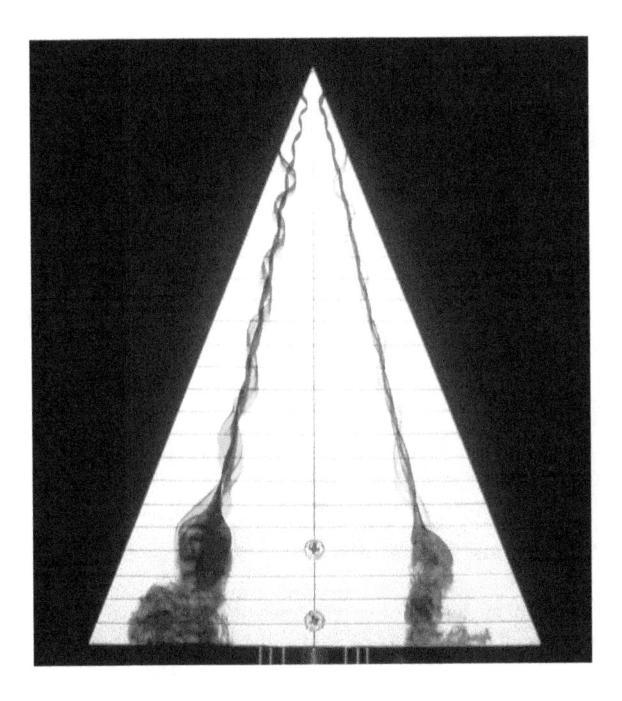

Figure 6.16 Vortex breakdown over delta wing at high angle of attack shown by flow visualization technique. *Source:* Courtesy of T.T. Lim, NUS Singapore

Theoretical and Computational Aerodynamics, First Edition. Tapan K. Sengupta.

Companion Website: www.wiley.com/go/sengupta

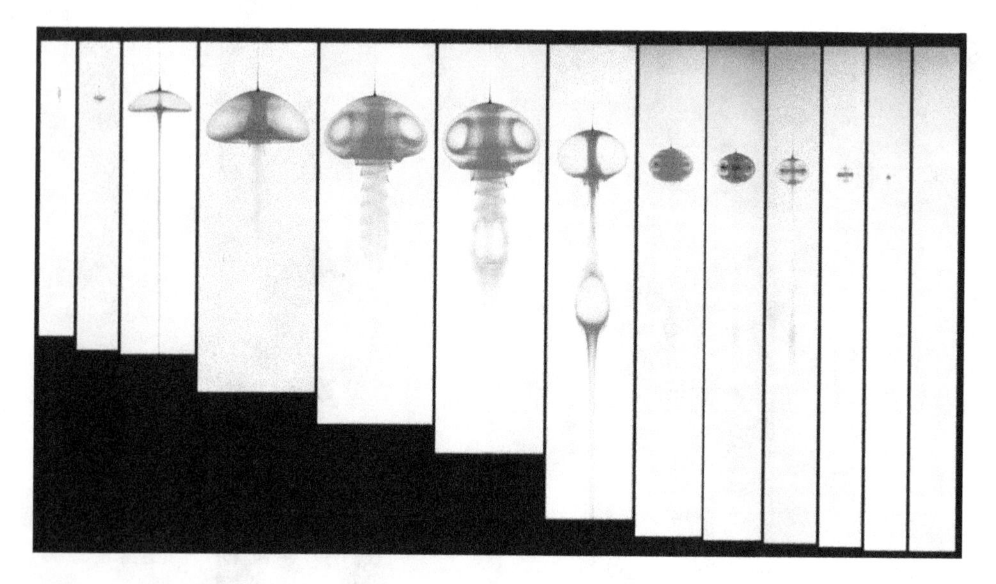

Figure 6.19 Life cycle of bubble type vortex breakdown obtained by dye visualization. The breakdown is created inside a confined cylindrical container and the phenomenon is to be tracked from right to left. *Source:* Cui (2009)

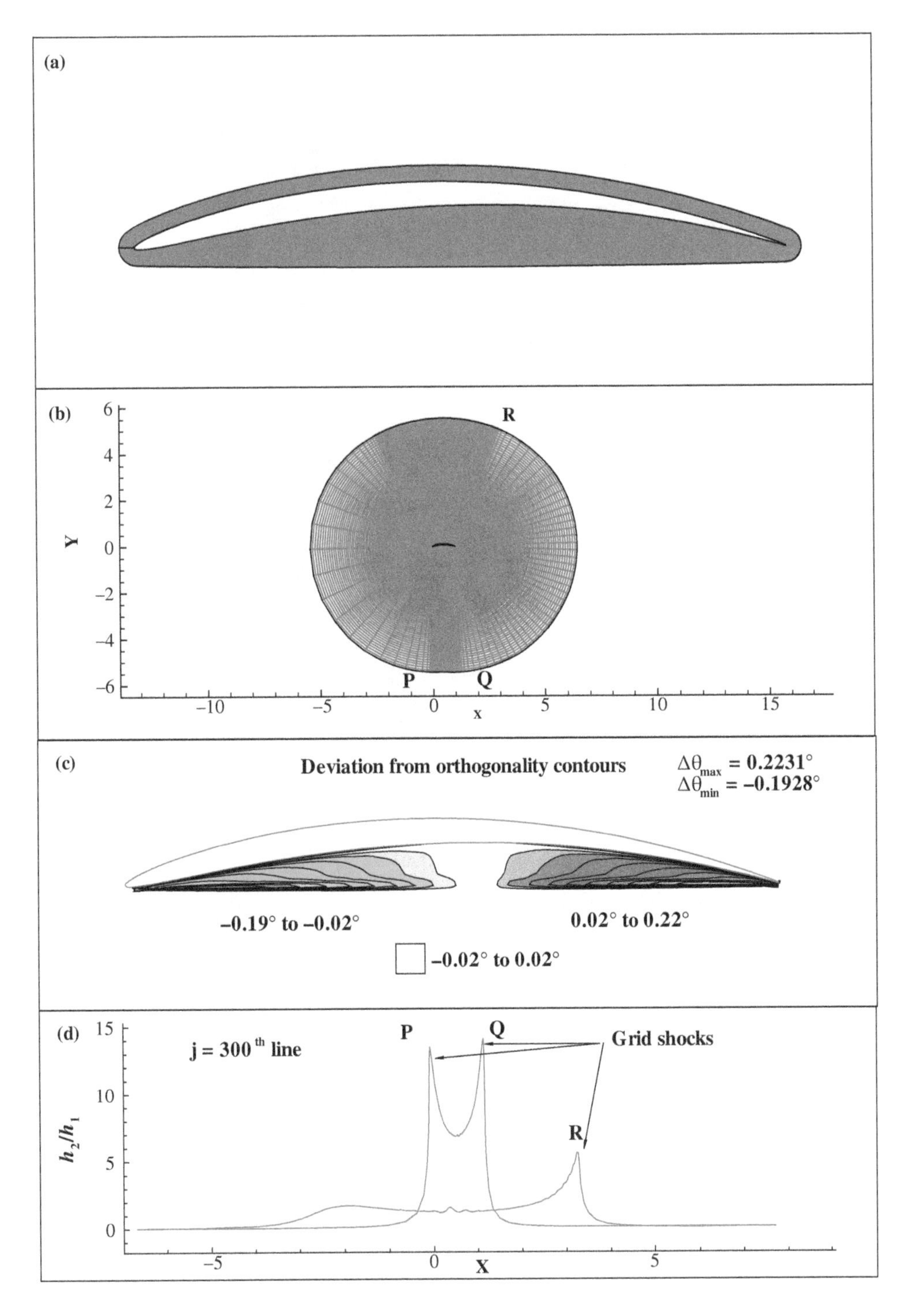

Figure 8.10 (a) Zoomed view of the orthogonal grid generated around NACA 8504 aerofoil; (b) complete grid; (c) deviation from orthogonality contours for the grid shown in Figure 8.10(b); (d) typical variation of grid scale parameter combination h_2/h_1 on the last azimuthal line. Note grid shocks at P, Q and R marked in frame (b)

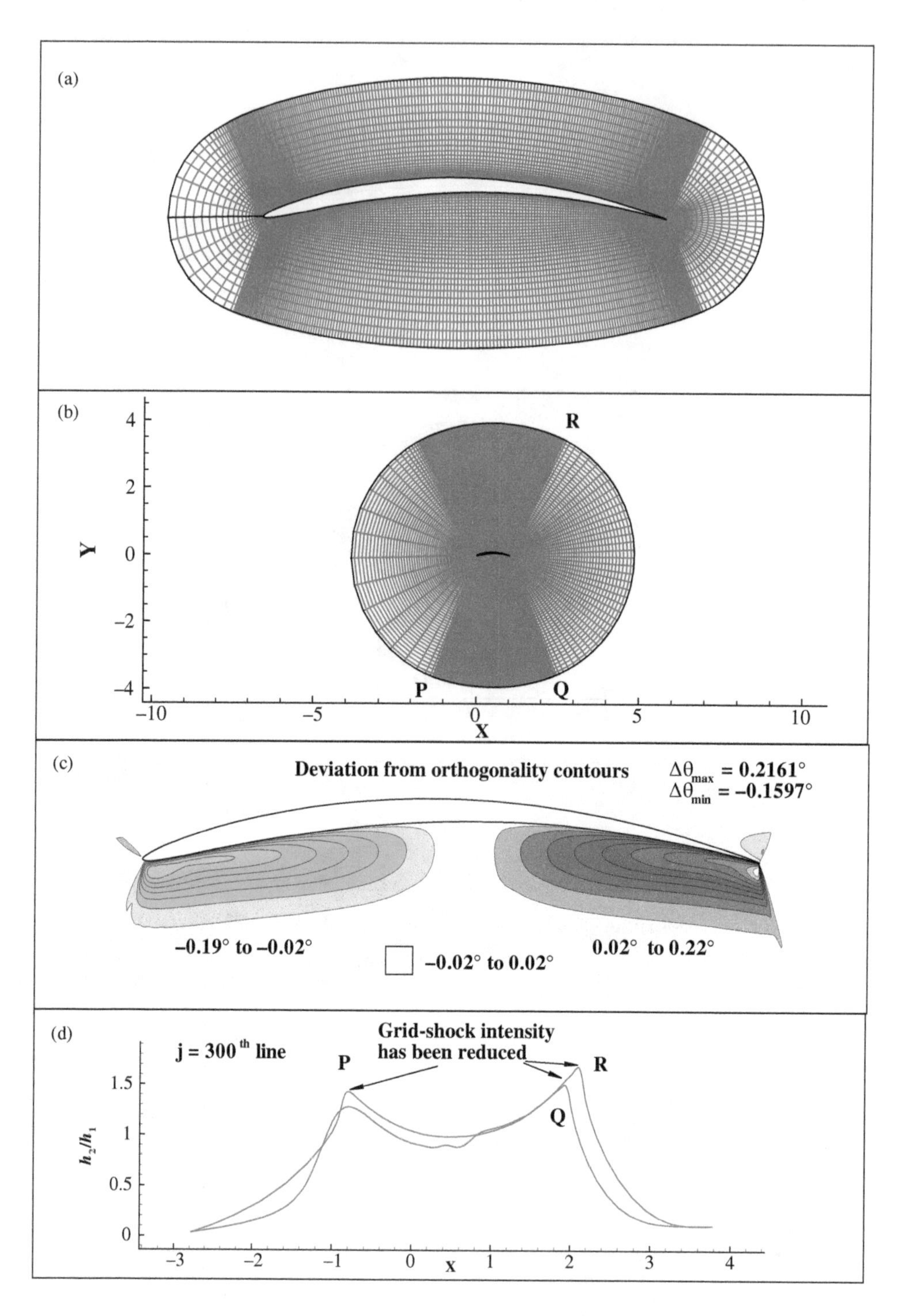

Figure 8.11 (a) Zoomed view of the orthogonal grid generated around NACA 8504 aerofoil; (b) complete grid; (c) deviation from orthogonality contours for the grid shown in Figure 8.11(b); (d) typical variation of grid scale parameter combination h_2/h_1 on the last azimuthal line. Note, intensity of grid-shocks at P, Q and R has been reduced

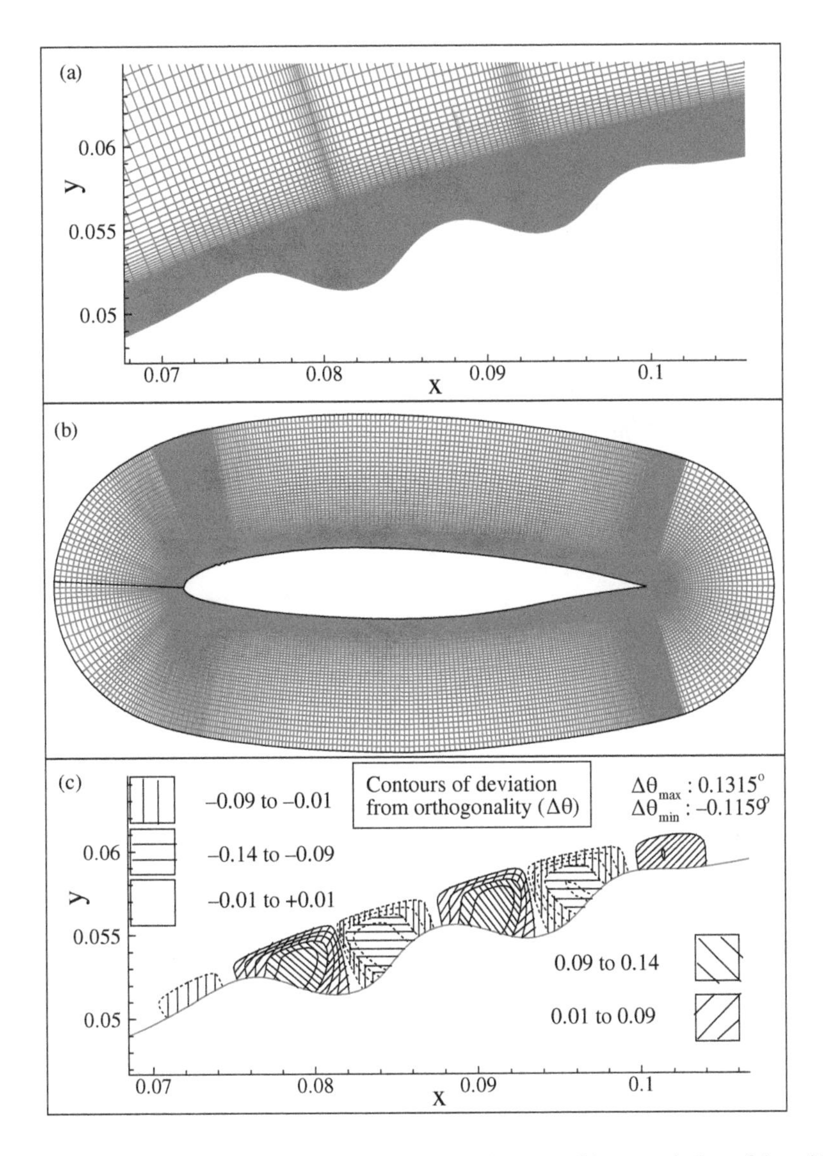

Figure 8.13 (a) Zoomed view of the grid near roughness elements; (b) zoomed view of the grid around complete aerofoil and (c) deviation from orthogonality contours for the generated grid in (b)

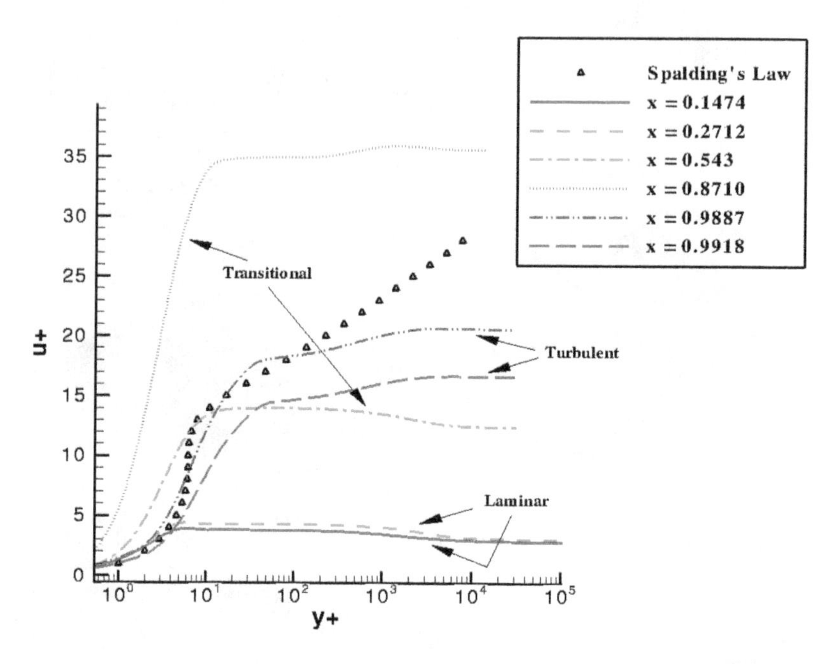

Figure 11.5 Distribution of mean streamwise velocity at various streamwise locations in wall coordinates, for flow around NACA 0012 airfoil with $Re_\infty = 3 \times 10^6$ at $\alpha = -0.14°$ and $M_\infty = 0.779$

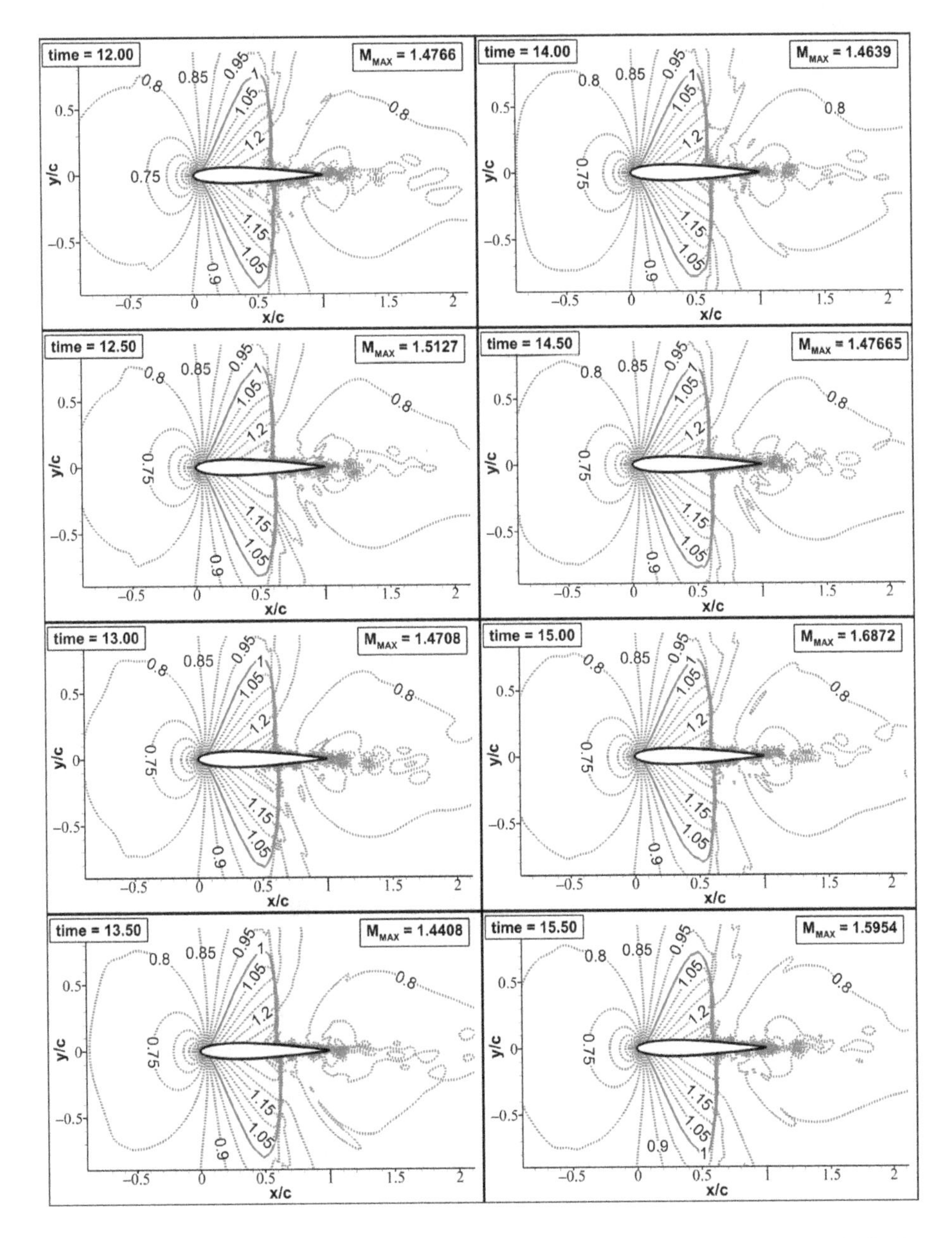

Figure 11.16 Computed M contours for flow around NACA 0012 airfoil with $M_\infty = 0.82$, $Re_\infty = 3 \times 10^6$ at $\alpha = -0.14°$

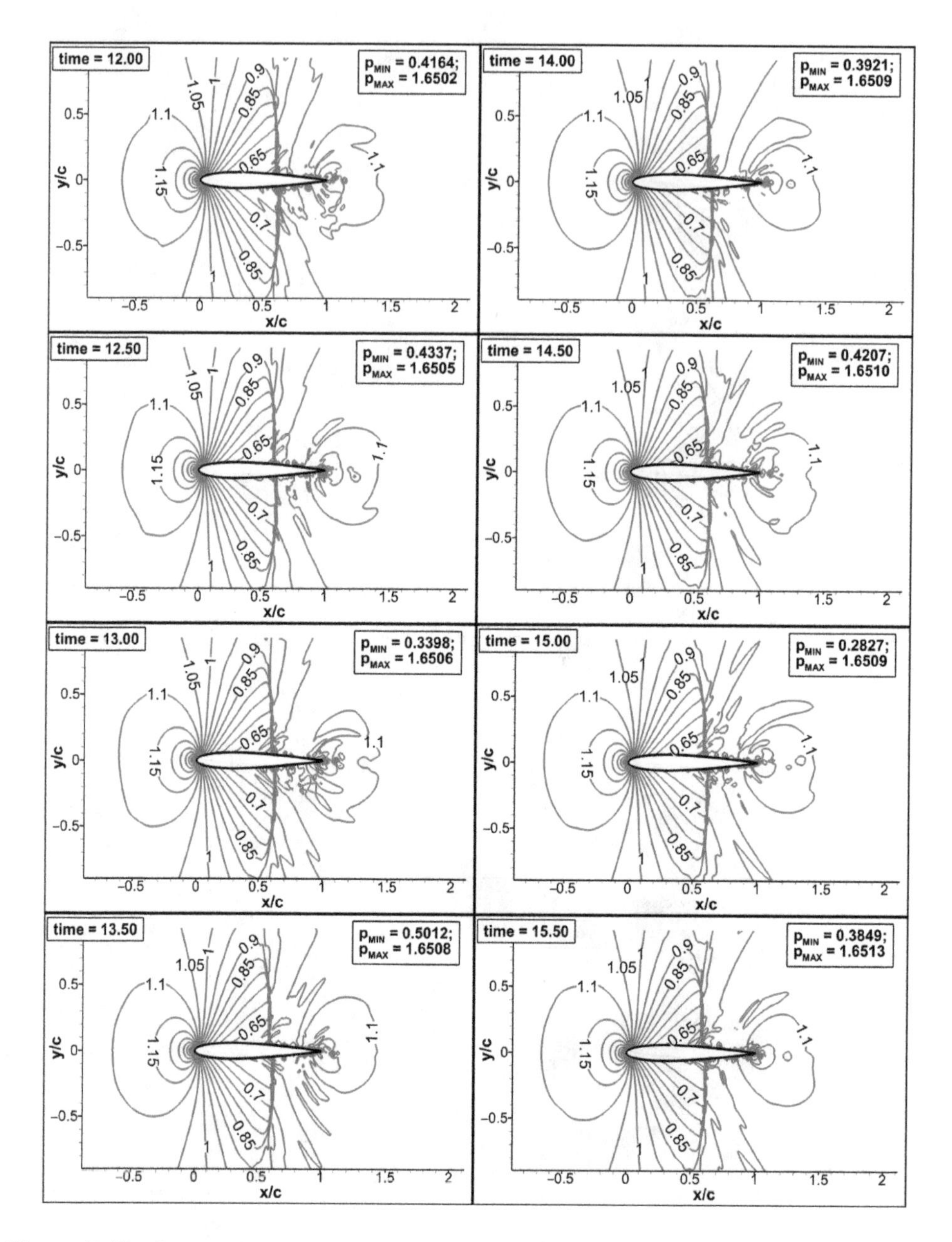

Figure 11.17 Computed p contours for flow around NACA 0012 airfoil with $M_\infty = 0.82$, $Re_\infty = 3 \times 10^6$ at $\alpha = -0.14°$

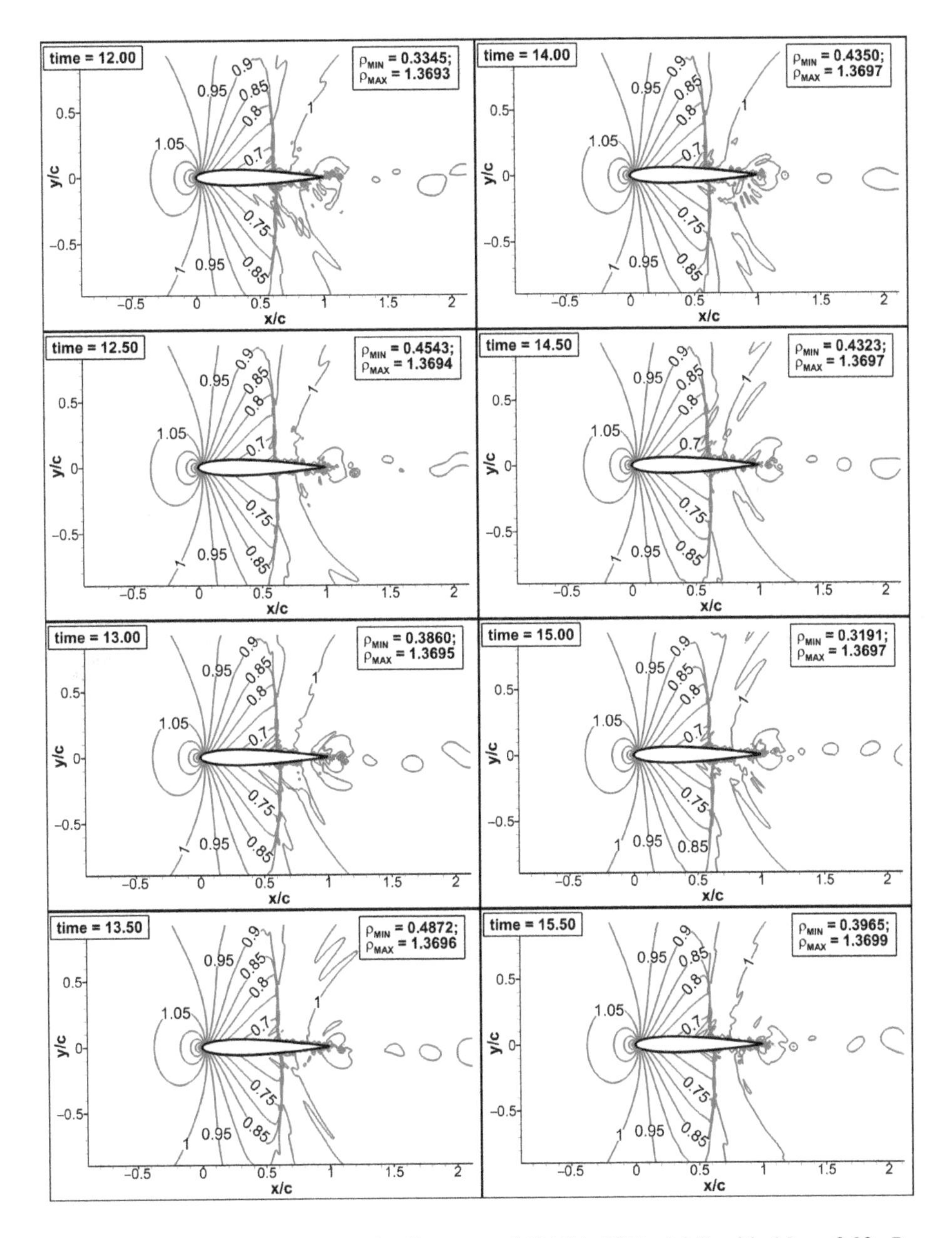

Figure 11.18 Computed ρ contours for flow around NACA 0012 airfoil with $M_\infty = 0.82$, $Re_\infty = 3 \times 10^6$ at $\alpha = -0.14°$

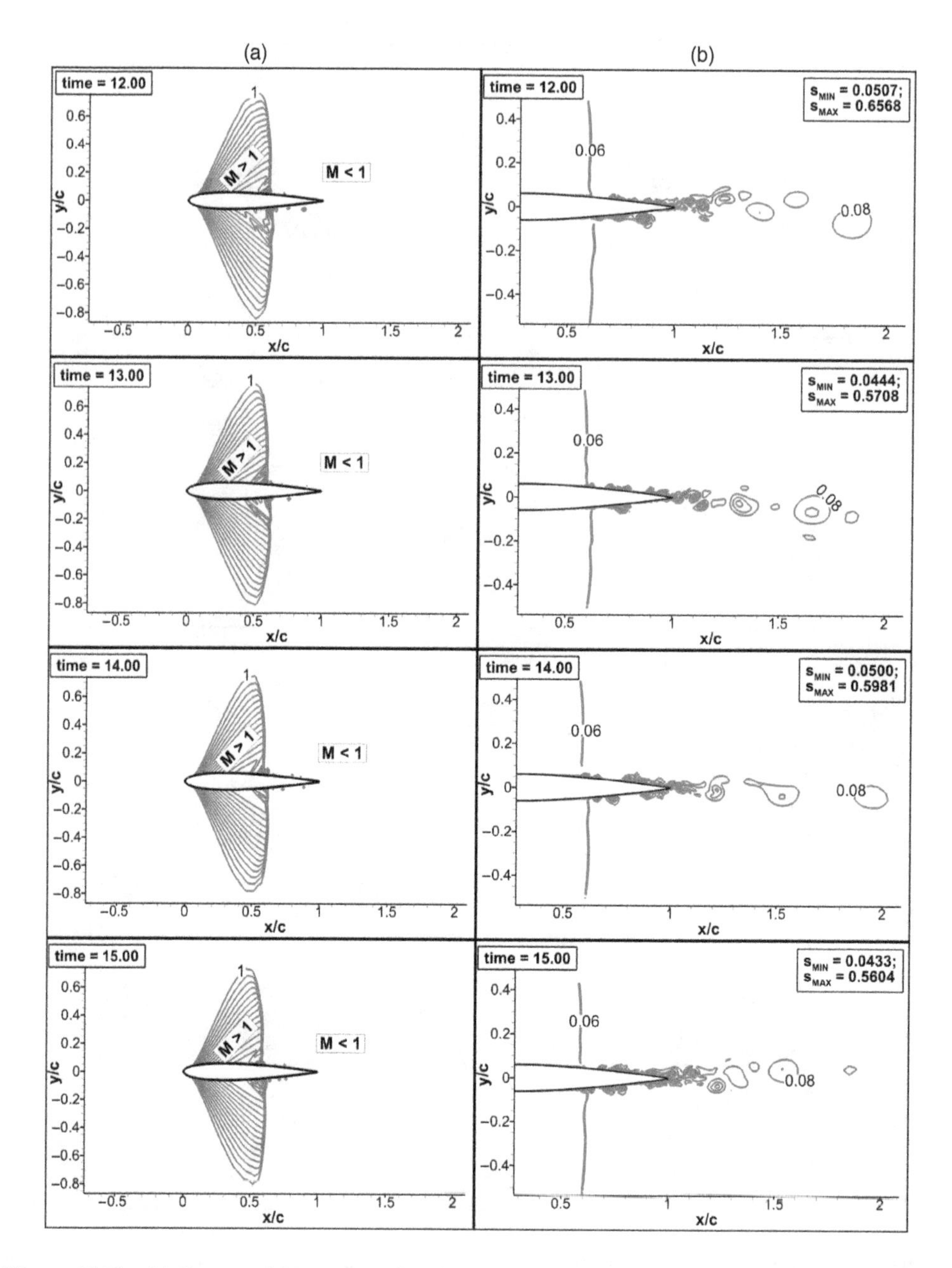

Figure 11.19 (a) Computed M contours showing subsonic, sonic and supersonic region (b) Computed s contours showing entropy generation in the region of the shock for $M_\infty = 0.82$ and $Re_\infty = 3 \times 10^6$ at $\alpha = -0.14°$

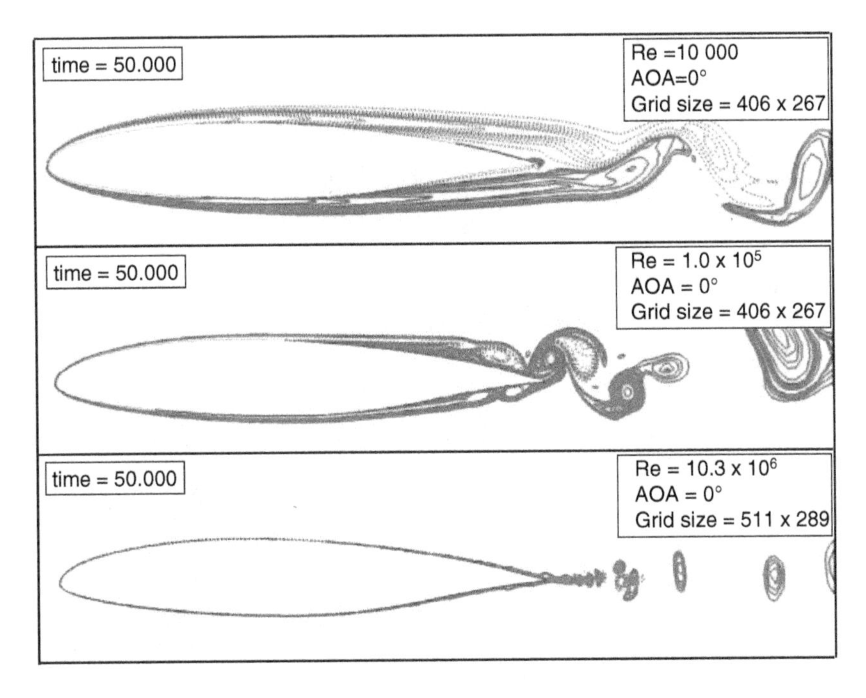

Figure 12.1 Flow past SHM-1 airfoil at the indicated Reynolds numbers, with the section placed at zero angle of attack

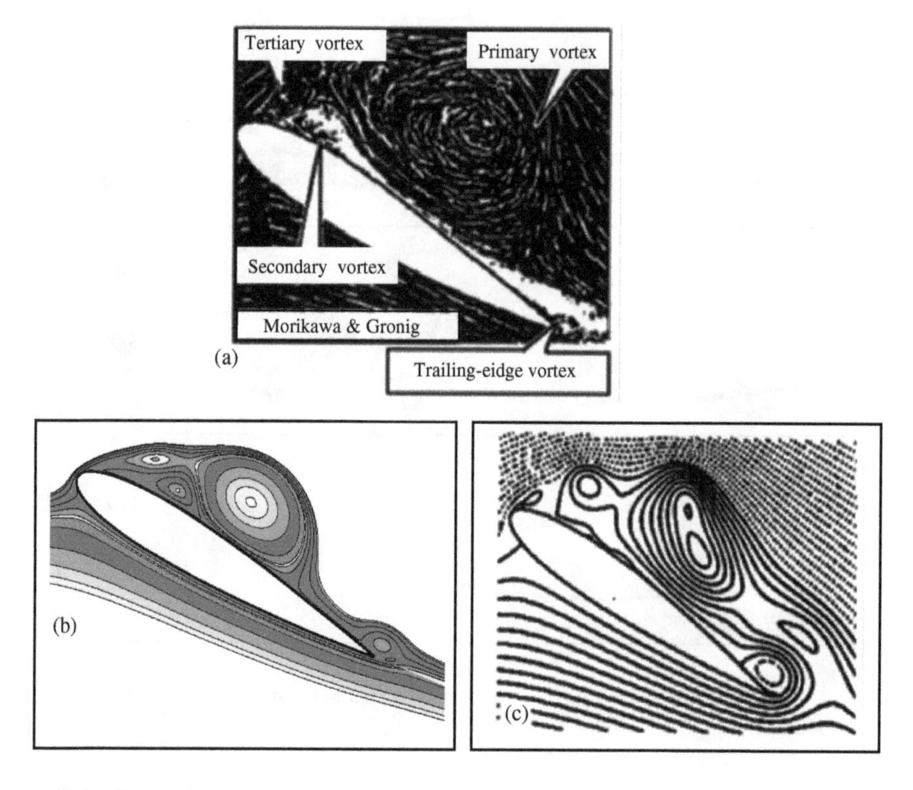

Figure 12.4 Comparison of (a) experimental visualization of Morikawa and Gröenig (1995) with the computations reported in Sengupta, Lim, Sajjan, Ganesh and Soria (2007a). Reproduced with permission from Cambridge University Press for the cases of (b) tangent hyperbolic start-up given by Equation (12.27) and (c) ramp start given by Equation (12.26). All the results are shown at $t = 2.903$

number of nodes or elements employed in the spatial discretization process. This implies that the derivative at the jth node/ element is evaluated as

$$u'_j = \frac{1}{h} \sum_{l=1}^{N} C_{jl} \, u_l,$$

where $u_l = u(x_l, t) = \int U(kh, t) \, e^{ikx_l} \, dk$ is the function value at the lth node and N is the number of nodes/ elements. Using the hybrid spectral representation, one can alternately write the numerical derivative as

$$u'_j = \int \frac{1}{h} \sum C_{jl} \, U(kh, t) \, e^{ik(x_l - x_j)} \, e^{ikx_j} \, dk. \tag{8.20}$$

Comparing Equations (8.19) with (8.20), one obtains

$$[ik_{eq}]_j = \frac{1}{h} \sum_{l=1}^{N} C_{jl} \, e^{ik(x_l - x_j)}. \tag{8.21}$$

Although in physical plane computations C_{jl}'s are real, $[k_{eq}]_j$ is in general complex, with real and imaginary parts representing numerical phase term and added numerical diffusion, respectively. It is determined by the spatial discretization method fixing the entries of $[C]$ matrix, incorporating different methods employed in the interior and near the boundaries. A logical representation of the discretization process is provided by the quotient k_{eq}/k. In Figure 8.3(a), real part of k_{eq}/k is plotted as a function of nondimensional wavenumber (kh) up to the Nyquist limit of π, for interior points only – for second order central finite differencing (CD_2) method and the symmetrized version of an optimized upwind compact scheme (SOUCS3) originally given in Dipankar and Sengupta (2006). Details of SOUCS3 scheme are given in Equations (8.31)–(8.35). Ideally, the ratio k_{eq}/k should be real and equal to one, as in Fourier spectral method. Departure from this ideal condition, amounts to phase mismatch and relates to filtering the unknown in obtaining the derivative numerically. Note that CD_2 method produces higher filtering at all wavenumbers, while SOUCS3 scheme has near-spectral accuracy over a significantly larger range of k. Higher resolution of SOUCS3 scheme makes it more accurate for numerical computations and from Figure 8.3(a), SOUCS3 is seen to resolve derivatives over a wavenumber range that is almost ten times, as compared to what is provided by the CD_2 scheme in each direction. For multi-dimensional problems, such disparities show that compact schemes are not only very economic, but are also highly accurate.

Unfortunately, above analysis of spatial discretization was considered all encompassing for some time and it was presumed that the Fourier spectral method is the best, irrespective of the time discretization strategy used for studying unsteady problems. In the following, we establish that any method's worth should be judged by space and time discretization considered together. While such a study is desirable for each and every equation that is to be solved, here we proceed with the modest goal of calibrating different methods, with respect to a model equation. Thus, other essential important numerical properties are obtained here via the spectral representation in Equation (8.1) which gives

$$\int \left[\frac{dU}{dt} + \frac{c}{h} \sum U C_{jl} \, e^{ik(x_l - x_j)} \right] e^{ikx_j} dk = 0. \tag{8.22}$$

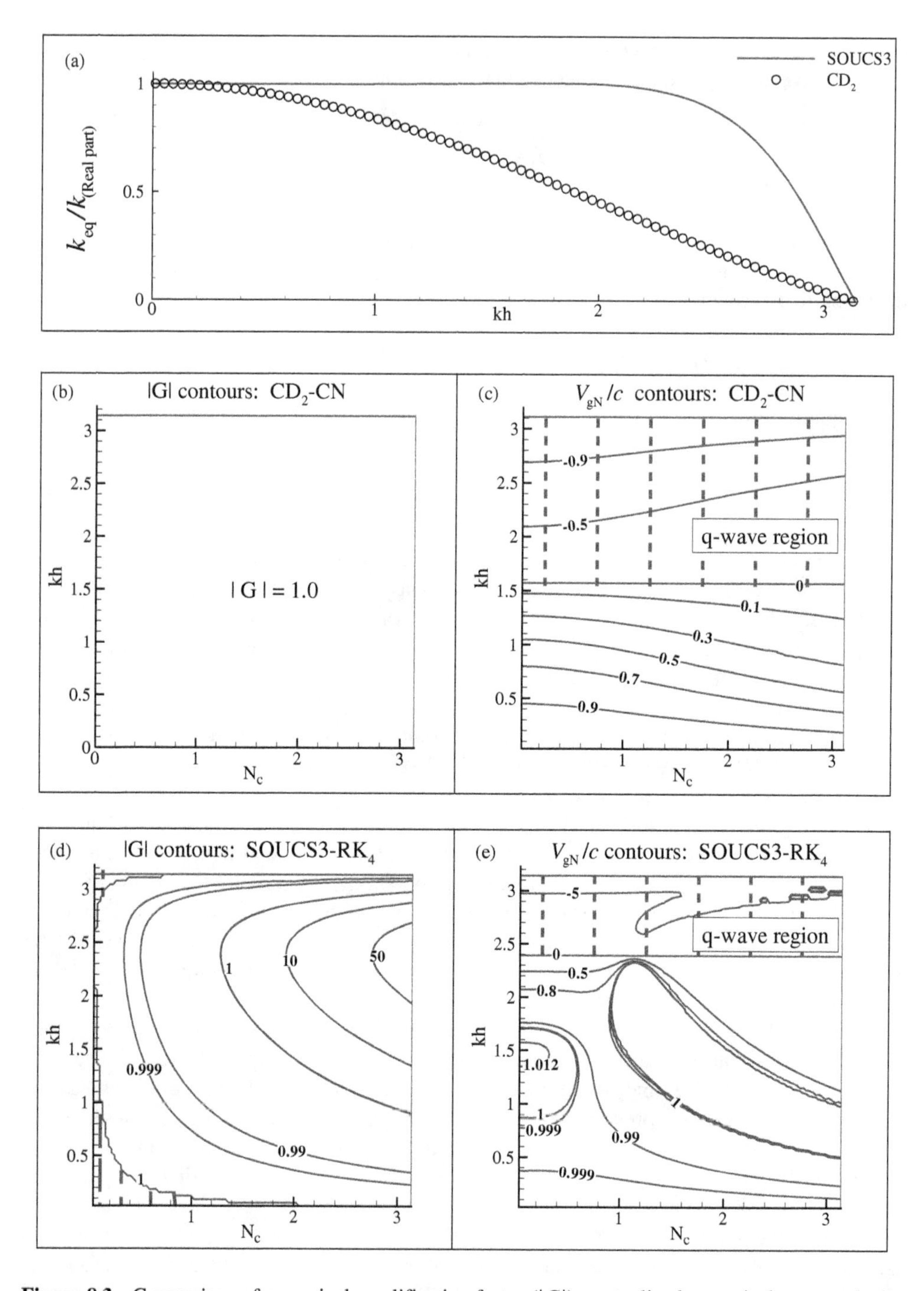

Figure 8.3 Comparison of numerical amplification factor ($|G|$), normalized numerical group velocity (V_{gN}/c) and phase error ($1 - c_N/c$) contours for the near-boundary node $j = 2$ (left column) and central node $j = 51$ (right column) for the 1D convection equation solved by the OUCS3-RK_4 method with 101 points

Two-time level, multi-stage higher order methods are needed ideally for accuracy and stability, as explained with respect to the four-stage, fourth-order Runge–Kutta (RK_4) method. If one denotes the right-hand side of Equation (8.1) by $L(u) = -c\frac{\partial u}{\partial x}$, then the steps used in (RK_4) are given by

$$\text{Step 1}: \quad u^{(1)} = u^{(n)} + \frac{\Delta t}{2} L[u^{(n)}]$$

$$\text{Step 2}: \quad u^{(2)} = u^{(n)} + \frac{\Delta t}{2} L[u^{(1)}]$$

$$\text{Step 3}: \quad u^{(3)} = u^{(n)} + \Delta t L[u^{(2)}]$$

$$\text{Step 4}: \quad u^{(n+1)} = u^{(n)} + \frac{\Delta t}{6}\{L[u^{(n)}] + 2L[u^{(1)}] + 2L[u^{(2)}] + L[u^{(3)}]\}.$$

For the RK_4 time integration scheme used with any spatial discretization scheme, G_j is obtained as Sengupta *et al.* (2007)

$$G_j = 1 - A_j + \frac{A_j^2}{2} - \frac{A_j^3}{6} + \frac{A_j^4}{24}, \tag{8.23}$$

where spatial discretization is represented generally as in Equations (8.20) and (8.21), so that

$$A_j = N_c \sum_{l=1}^{N} C_{jl}\, e^{ik(x_l - x_j)}.$$

where $N_c = \frac{c\Delta t}{h}$ is the Courant-Friedrich-Lewy (CFL) number.

It is also possible to use an implicit method like the Crank-Nicolson method with CD_2 discretization (Sengupta *et al.* 2012) to obtain the numerical properties. For CD_2-CN discretization scheme, one obtains G_j for the solution of Equation (8.1) as

$$G_j = \frac{1 - \frac{N_c^2}{4}\sin^2 kh}{1 + \frac{N_c^2}{4}\sin^2 kh} + i\frac{-N_c \sin kh}{1 + \frac{N_c^2}{4}\sin^2 kh}. \tag{8.24}$$

While $|G_j| \neq 1$ is a source of error itself, additional error arises due to dispersion, whose effects are subtle and often misunderstood. This is explained briefly next. If we represent the initial condition for Equation (8.1) as

$$u(x_j, t = 0) = u_j^0 = \int A_0(k)\, e^{ikx_j}\, dk, \tag{8.25}$$

then the general solution at any arbitrary time can be obtained as

$$u_j^n = \int A_0(k)\, [|G_j|]^n\, e^{i(kx_j - n\beta_j)}\, dk, \tag{8.26}$$

where $|G_j| = (G_{rj}^2 + G_{ij}^2)^{1/2}$ and $\tan(\beta_j) = -\frac{G_{ij}}{G_{rj}}$, with G_{rj} and G_{ij} as the real and imaginary parts of G_j, respectively. Thus, the phase of the solution at the jth node is determined by $n\beta_j = kc_N t$, where c_N is the numerical phase speed. Although the physical phase speed is a constant for all k for the nondispersive system, this analysis shows that c_N is in general k-dependent, i.e. the numerical solution is dispersive, in contrast to the nondispersive nature of Equation (8.1). The implications of this simple difference are profound, as demonstrated below.

The general numerical solution of Equation (8.1) is denoted as

$$\bar{u}_N = \int A_0\, [|G|]^{t/\Delta t}\, e^{ik(x-c_N t)}\, dk. \tag{8.27}$$

The numerical dispersion relation is now given as $\bar{\omega}_N = c_N k$, instead of $\bar{\omega} = ck$. Nondimensional phase speed and group velocity (from the general definition given in Equations (8.7) and (8.8)) at the jth node are expressed as

$$\left[\frac{c_N}{c}\right]_j = \frac{\beta_j}{\bar{\omega}\Delta t} \tag{8.28}$$

$$\left[\frac{V_{gN}}{c}\right]_j = \frac{1}{hN_c}\frac{d\beta_j}{dk}. \tag{8.29}$$

If the computational error is defined as $e(x,t) = u(x,t) - \bar{u}_N$, then one obtains the governing equation for its dynamics as given by Sengupta *et al.* (2007)

$$\frac{\partial e}{\partial t} + c\frac{\partial e}{\partial x} = -c\left[1 - \frac{c_N}{c}\right]\frac{\partial \bar{u}_N}{\partial x} - \int \frac{dc_N}{dk}\left[\int ik' A_0[|G|]^{t/\Delta t} e^{ik'(x-c_N t)} dk'\right] dk$$
$$- \int \frac{Ln\,|G|}{\Delta t} A_0\, [|G|]^{t/\Delta t}\, e^{ik(x-c_N t)}\, dk. \tag{8.30}$$

This is the correct error propagation equation, as opposed to that obtained using the assumption made in von Neumann analysis, where the right-hand side of Equation (8.30) is assumed to be identically equal to zero. This is on the premise that $c_N \cong c$, i.e. there is no dispersion error and furthermore the numerical method is perfectly neutral, so that the last term on the right-hand side of Equation (8.30) is also identically zero. Error can grow very fast when the numerical solution displays sharp spatial variation, due to the first term on the right-hand side of Equation (8.30). This happens for flows with shocks.

To understand the ramifications of Equation (8.30) for the model equation, one needs to look at the numerical properties of a specific combination of spatial and temporal discretization methods. For this purpose, we show the properties of the compact scheme OUCS3 introduced in Sengupta *et al.* (2003a). This scheme is for a nonperiodic problem and here we show some typical results for an interior node and another node close to the inflow boundary, when RK_4 time integration strategy is used for solving Equation (8.1).

We provide here the SOUCS3 stencil (Dipankar and Sengupta 2006), which is a symmetrized version of OUCS3 compact scheme. The stencil for the basic OUCS3 scheme along with boundary stencils are given in the following for a uniform grid with spacing h. The interior stencil is given by

$$\alpha_{j-1}\, u'_{j-1} + \alpha_j\, u'_j + \alpha_{j+1}\, u'_{j+1} = \frac{1}{h}\sum_{k=-2}^{k=2} q_k\, u_{j+k}. \tag{8.31}$$

Primed quantities represent numerical derivatives obtained by simultaneous solution of linear algebraic equations, which are obtained by application of Equation (8.31) at all interior nodes. Furthermore, one requires one-sided boundary stencils at nodes $j = 1, 2$ and $N-1, N$, for solving nonperiodic problems. Such one-sided boundary closure schemes can violate the physical principle of information propagation. While obtaining solution using high accuracy compact schemes, inherent bias of these stencils can also accentuate the above mentioned problem of violating physical principle (Sengupta 2013, Sengupta *et al.* 2003a). Explicit boundary closure stencils are suggested for $j = 1$ and $j = 2$ as

$$u_1' = \frac{1}{2h}(-3u_1 + 4u_2 - u_3) \tag{8.32}$$

$$u_2' = \frac{1}{h}\left[\left(\frac{2\gamma_2}{3} - \frac{1}{3}\right)u_1 - \left(\frac{8\gamma_2}{3} + \frac{1}{2}\right)u_2 + (4\gamma_2 + 1)\,u_3 - \left(\frac{8\gamma_2}{3} + \frac{1}{6}\right)u_4 + \frac{2\gamma_2}{3}u_5\right], \tag{8.33}$$

where $\gamma_2 = -0.025$ is a parameter used to achieve better numerical properties of neutral stability, lesser phase and dispersion errors for the solution of convection dominated problems. Similarly, one can write boundary closure schemes for $j = N$ and $j = N-1$ using another parameter, $\gamma_{N-1} = 0.09$ by

$$u_N' = \frac{1}{2h}(3u_N - 4u_{N-1} + u_{N-2}) \tag{8.34}$$

$$u_{N-1}' = -\frac{1}{h}\left[\left(\frac{2\gamma_{N-1}}{3} - \frac{1}{3}\right)u_N - \left(\frac{8\gamma_{N-1}}{3} + \frac{1}{2}\right)u_{N-1} + (4\gamma_{N-1} + 1)\,u_{N-2} - \left(\frac{8\gamma_{N-1}}{3} + \frac{1}{6}\right)u_{N-3} + \frac{2\gamma_{N-1}}{3}u_{N-4}\right] \tag{8.35}$$

To achieve desired numerical properties, coefficients in Equation (8.31) are upwinded, as given by Sengupta (2013) and Sengupta *et al.* (2003a): $\alpha_{j\pm1} = 0.3793894912 \pm \eta/60$, $\alpha_j = 1$, $q_{\pm2} = \pm 0.183205192/4 + \eta/300$, $q_{\pm1} = \pm 1.57557379/2 + \eta/30$ and $q_0 = -11\eta/150$, where $\eta = -2.0$ is the upwind parameter for the interior stencil. Undesirable bias of the above OUCS3 scheme was removed using a symmetrization procedure (Dipankar and Sengupta 2006), which results in identical properties for all nodes located symmetrically about the central node.

To account for error contributing terms, we have compared $|G|$- and V_{gN}/c-contours between CD_2-CN and SOUCS3-RK4 methods, in Figures 8.3(b) to 8.3(e). In Figure 8.3(b), one observes neutral stability over the complete (N_c, kh)-plane, showing superiority of the CD_2-CN method, with respect to the third term on the right-hand side of Equation (8.30).

However, V_{gN}/c-contours for CD_2-CN method show less dispersion only very close to the origin in Figure 8.3(c), i.e. only for very small time steps and for very large length scales. Higher wavenumber components will not only propagate erroneously, but any wavenumber above

$kh = \pi/2$, will move in the opposite to physical direction. Such disturbances have been called q-waves (Sengupta *et al.* 2012a) and will be a major source of error altering the physical flow. Corresponding properties for the SOUCS3-RK4 scheme are shown in Figures 8.3(d) and 8.3(e). Figure 8.3(d) shows $|G_j|$-contours and indicates a very small range of N_c ensuring neutral stability. However, significant improvement is noted for V_{gN}/c-contours in Figure 8.3(e), which along with improved scale resolution in Figure 8.3(a) makes SOUCS3-RK4 method preferable over CD_2-CN method. Also for SOUCS3-RK4 scheme, we have a smaller q-wave region at significantly higher wavenumbers, as compared to CD_2-CN method.

Note the characterization of resolution is with respect to Fourier spectral method, which has the best resolution among all discrete method. Does this imply that one should use Fourier spectral method with any time discretization method and expect high accuracy results, irrespective of the problem solved? One such method is reported for direct numerical simulation (DNS) of homogeneous turbulence flows (Donzis *et al.* 2008, Donzis and Yeung 2010, Yeung *et al.* 2012), where Fourier spectral discretization in space has been used with second-order Runge-Kutta time integration, referred here as the RK_2-FS method. This general class of method is called pseudo-spectral method and is often cited as an example of high accuracy computing method.

We intend to highlight the importance of the analysis of schemes in Sengupta *et al.* (2007) by studying RK_2-FS method and show its numerical instability and poor dispersion properties. We also propose an alternative method that surmounts these shortcomings greatly. The RK_2-FS discretization scheme is severely restrictive in (N_c, kh)-space, as established here by spectral analysis using 1D model convection equation. It is obvious that any method which cannot solve this model convection equation will not be able to solve convection dominated flows given by Euler and Navier-Stokes equations correctly. The dynamical equilibrium in viscous fluid flows is determined by convection and diffusion processes which have to be captured with negligible error. Any shortcoming of representing these basic physical operators will take the flow to a wrong equilibrium state.

In Figure 8.4, contours of $|G|$, c_N/c and V_{gN}/c are plotted for the RK_2-FS scheme. One expects maximum resolution using Fourier spectral method for spatial derivatives. However for space-time dependent governing equations, performance of combined space-time discretization scheme is more relevant. The $|G|$ contours for RK_2-FS method graphically show this effect for the combined scheme's instabilities in the top frame of Figure 8.4, i.e. $|G| > 1$ in the entire (N_c, kh)-plane, except the very narrow shaded region near the origin. This implies that the amplitude of the solution of Equation (8.1) will increase unbounded with time for grid spacing and time steps having values outside the neutrally stable region. It is noted that the dispersion indicated by V_{gN}/c contours and the phase error indicated by $(1 - c_N/c)$ represent, for the chosen value of $N_c = 0.5$ in the reference (Yeung *et al.* 2012), are unacceptably high for the RK_2-FS scheme. Energy propagation speed given by V_{gN}/c will be erroneous by more than 10% for higher wavenumber components. Similar error would be noted for the phase error at higher wavenumbers. As we will see shortly that such a choice of N_c will not allow us to even solve Equation (8.1).

There has been significant progress made in recent times, in developing high accuracy compact schemes which are dispersion relation preserving (DRP) (Rajpoot *et al.* 2010, Sengupta *et al.* 2011). Any scheme for which the numerical and physical dispersion relations remain identical over a significant region in (N_c, kh)-plane and which is neutrally stable in the same region is called a DRP scheme. The DRP compact scheme developed in Sengupta *et al.*

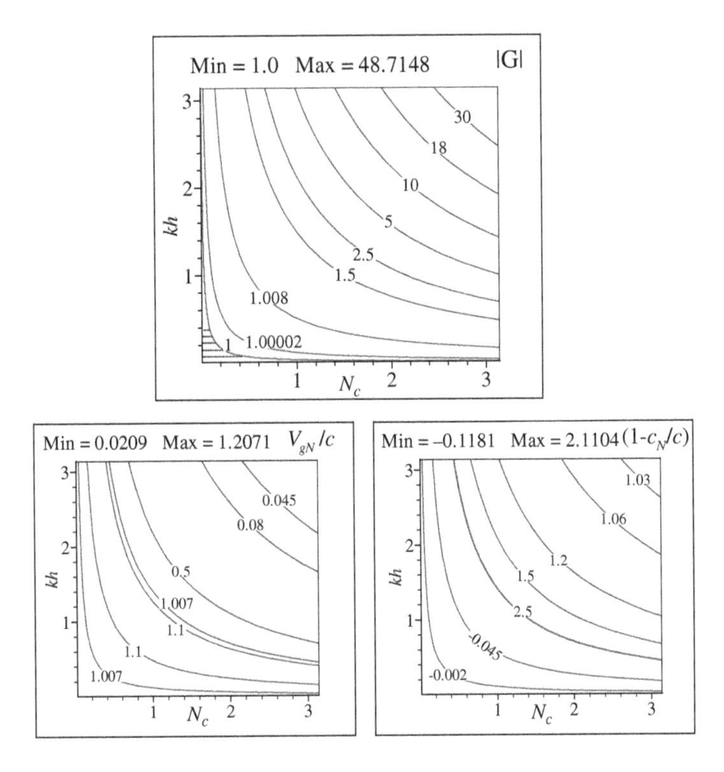

Figure 8.4 Contour plots showing numerical amplification factor $|G|$, scaled numerical group velocity (V_{gN}/c) and phase error $(1 - c_n/c)$ in (N_c, kh)-plane for 1D convection equation using second order two-stage Runge-Kutta-Fourier spectral $(RK_2 - FS)$ scheme

(2011) has been used to solve many practical inhomogeneous flows and is termed the $OCRK_3$ scheme. This method has been used in recent times for the DNS of inhomogeneous flows from the receptivity to fully developed turbulent stage, which has been reported for the first time (Sengupta and Bhaumik 2011, Sengupta *et al.* 2012). The computed fully developed two-dimensional turbulent flow stage, displayed k^{-3} spectrum for the energy, as has been shown theoretically (Batchelor 1969, Kraichnan and Montgomery 1980).

In Figure 8.5, corresponding numerical properties of $OCRK_3$ scheme are shown and one notices the presence of a range of N_c (up to $N_c = 0.21$) for which the scheme is neutrally stable, for all resolved values of kh (the Nyquist limit). The contours also reveal that this scheme remains stable up to $N_c = 0.99$, as opposed to nonexistent stable range of N_c for RK_2-FS scheme. In the lower right frame of Figure 8.5, contours of $(1 - c_N/c)$ or normalized numerical phase speed have been plotted and one notes significant improvement in the shaded region admitting error only up to 0.2%. In the bottom left frame, normalized numerical group velocity contours are plotted and the shaded region indicates the values of kh and N_c for which the error is less than 0.5% for the solution of Equation (8.1). Once again the comparison between $OCRK_3$ scheme with RK_2-FS scheme establishes that the latter method is inherently flawed, on both counts of numerical instability and dispersion/phase error properties.

The reason that the pseudo-spectral method was noted to work for homogeneous turbulence simulation (Yeung *et al.* 2012) is due to multiple reasons. First, the numerical instability

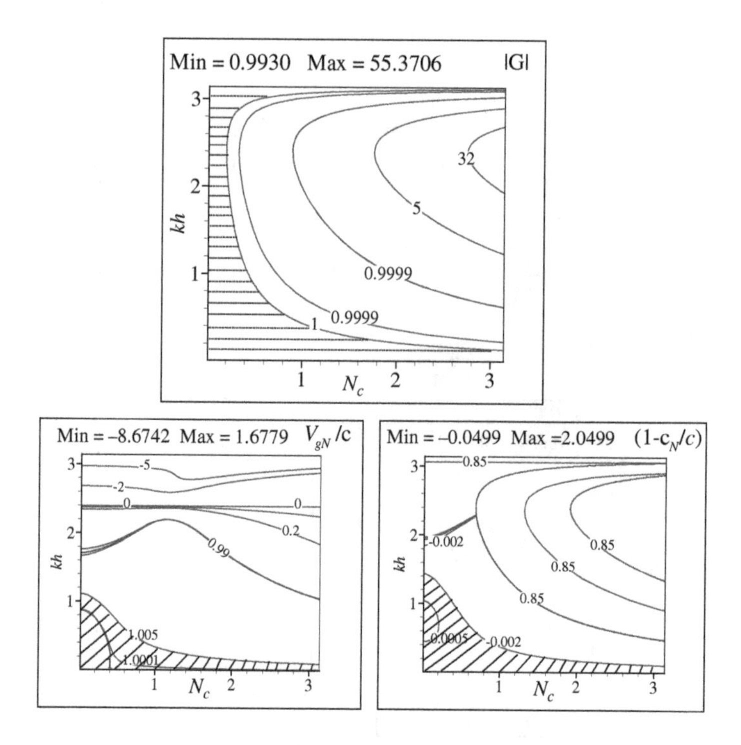

Figure 8.5 Contour plots showing numerical amplification factor $|G|$, scaled numerical group velocity (V_{gN}/c) and phase error $(1 - c_N/c)$ in (N_c, kh)-plane for 1D convection equation using optimized compact four-stage Runge-Kutta $(OCRK_3)$ scheme

indicated in Figure 8.4 is mild for smaller N_c values, which does not lead to catastrophic breakdown. However the authors also used higher values of N_c with the pseudo-spectral method for solving homogeneous turbulence problem using hyperviscosity that stabilizes higher wavenumbers numerically. It is interesting to note that due to numerical instability of RK_2-FS method, no explicit excitation is also needed to trigger turbulence, with the growth of numerical error stimulating the flow. However, creation of turbulence by numerical noise can have different transitional flow extent and the integrated load over aerodynamic bodies will be method-dependent in such cases. Ideally in computing such flows, one must compute the equilibrium flow first and then subject such equilibrium to realistic ambient disturbances. This approach will produce the correct loads and moments acting over aerodynamic bodies.

The above analysis for the two schemes considered highlights the importance of choosing a scheme for DNS. One can easily conclude that $OCRK_3$ scheme is superior to RK_2-FS scheme for solving flows where convection is the dominant physical phenomenon, which is evident from the solution of an inhomogeneous flows by $OCRK_3$ method, as compared to the ability of RK_2-FS method in solving isotropic homogeneous turbulence flow problem.

To further justify the above observations, we note the solution of the convection equation for the propagation of a wave-packet given by

$$u(x, t = 0) = e^{-\alpha(x-x_0)^2} \cos(k_o(x - x_o)). \tag{8.36}$$

We have chosen $\alpha = 0.5$ with centre of packet at $x_0 = 5$ in the computational domain $0 \leq x \leq 100$ with 32 768 equi-spaced points with grid-spacing of $\Delta x = 0.00305$. For this case, convection speed in Equation (8.1) is chosen as $c = 0.1$ and the nondimensional central wavenumber of $k_o h = 0.8$ with a time step of $\Delta t = 0.003$. For the chosen parameters, corresponding value of CFL number is $N_c = 0.09830$. Numerical results obtained with RK_2-FS method at different time instants are shown in Figure 8.6. One can notice the formation of Gibbs' phenomenon at upstream side after $t = 4.5$, which gives rise to spurious oscillations on the upstream side and this phenomenon becomes more dominating at later times, as can be seen at $t = 6.0$. Thereafter the numerical result are so laced with errors that these are not shown. The error is due to poor phase and dispersion properties of RK_2-FS method as shown in Figure 8.4. Moreover, for the chosen value of CFL number N_c, RK_2-FS methods is unstable, as can be seen from the $|G|$-contours shown in Figure 8.4. Due to this, one can note from Figure 8.6 that wave-packet's amplitude exceeds the initial unit amplitude indicating numerical instability. The exact solution of wave-packet propagation at the same time instants in the respective subframes is also shown with dotted lines. Furthermore, numerical results corresponding to $OCRK_3$ scheme are shown in Figure 8.7 at various time instants with the exact solution shown again by dotted lines. For the $OCRK_3$ scheme chosen time step of 0.006 is twice the value used for the RK_2-FS method, while the rest of the numerical parameters are same. This case is computed up to $t = 400$ and beyond. In contrast to RK_2-FS scheme, due to good phase and dispersion properties of the $OCRK_3$ scheme (shown in Figure 8.5), a good match with the exact solution is possible without any dissipation, phase and dispersion errors even up to $t = 400$. This is due to the fact that for the chosen values of numerical parameters, the $OCRK_3$ method is neutrally stable with negligible phase and dispersion error – a feature of the DRP method.

8.4 An Improved Orthogonal Grid Generation Method for Aerofoil

Space-time accurate solution of full time-dependent Navier-Stokes equations is essential while solving viscous flow past an aerofoil. While obtaining a numerical solution of such flows, one has to ensure reduction of all possible error sources. In such simulations, use of a good-quality structured grid with smooth grid metrics (Thompson *et al.*, 1982) is necessary as grid shocks or discontinuities in grid metrics give rise to mesh related flow distortions and additional sources of high wavenumber errors contributing to aliasing error (Sengupta, 2013). To resolve all length scales in the flow, the grid generation method used should provide a complete control over grid spacing.

Orthogonal formulation of Navier-Stokes equation has simpler form in the transformed plane as compared to nonorthogonal formulation for incompressible flows. This results in fewer and faster computations along with lower numerical errors. Additionally, with orthogonal body-fitted grid, boundary conditions can be prescribed easily and accurately. This is a major motivation for using orthogonal grid.

While solving Navier-Stokes equation using finite difference formulation, one needs to map nonuniformly spaced grid points in physical (x, y)-plane to equi-spaced points in the computational (ξ, η)-plane by relating the two via, $x = x(\xi, \eta)$ and $y = y(\xi, \eta)$. Such mapping must be unique for which any point (x, y) in the physical space must be a linear scalar functional of the coordinate (ξ, η) in the computational plane (Sengupta 2013, Thompson *et al.* 1982).

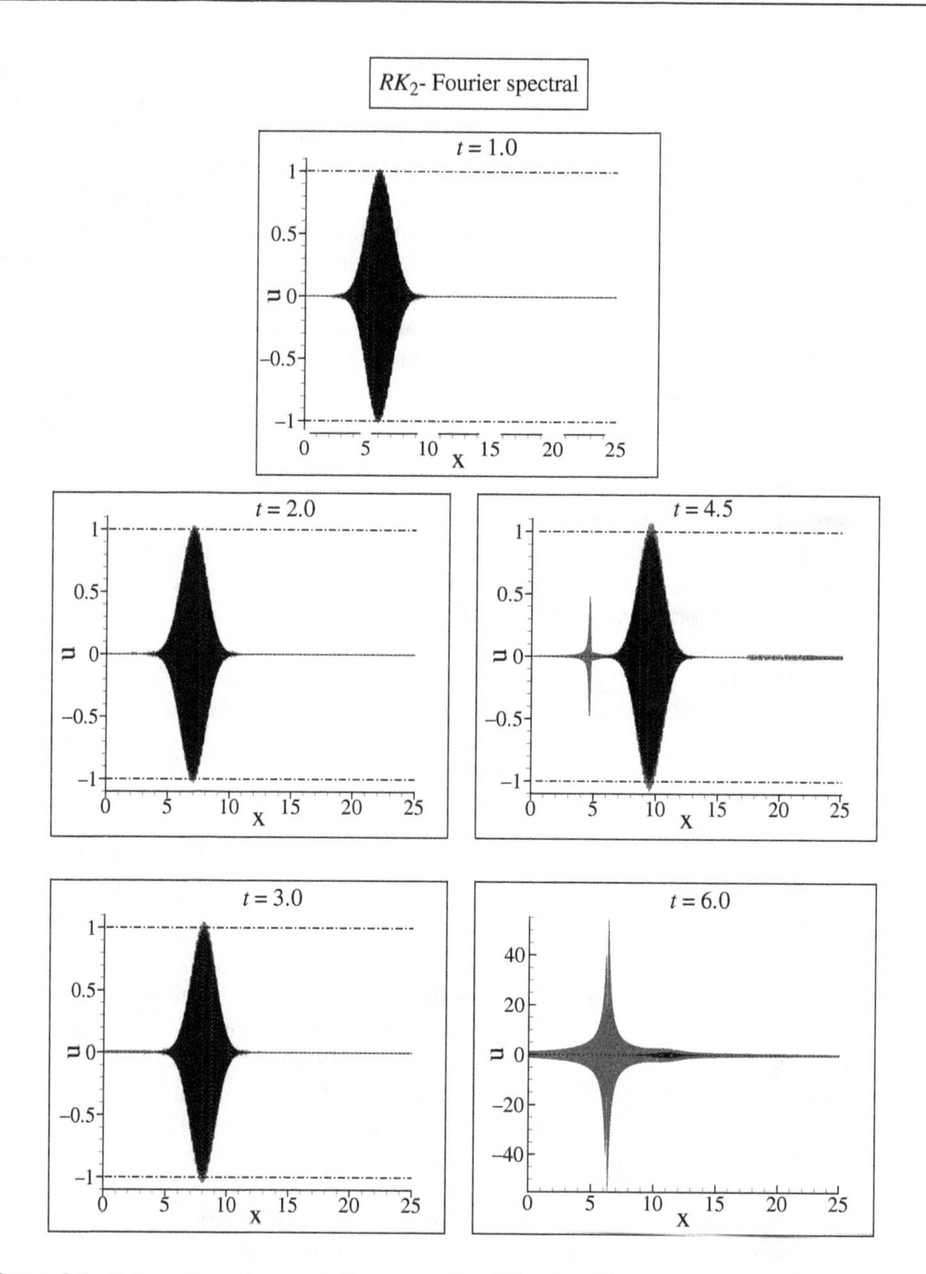

Figure 8.6 Propagation of computed wave packet following 1D convection equation with phase speed $c = 0.1$, $k_o h = 0.8$ and $N_c = 0.09830$ using second order two-stage Runge-Kutta-Fourier spectral $(RK_2 - FS)$ scheme. In all the subframes, the exact solution is shown by the dotted lines

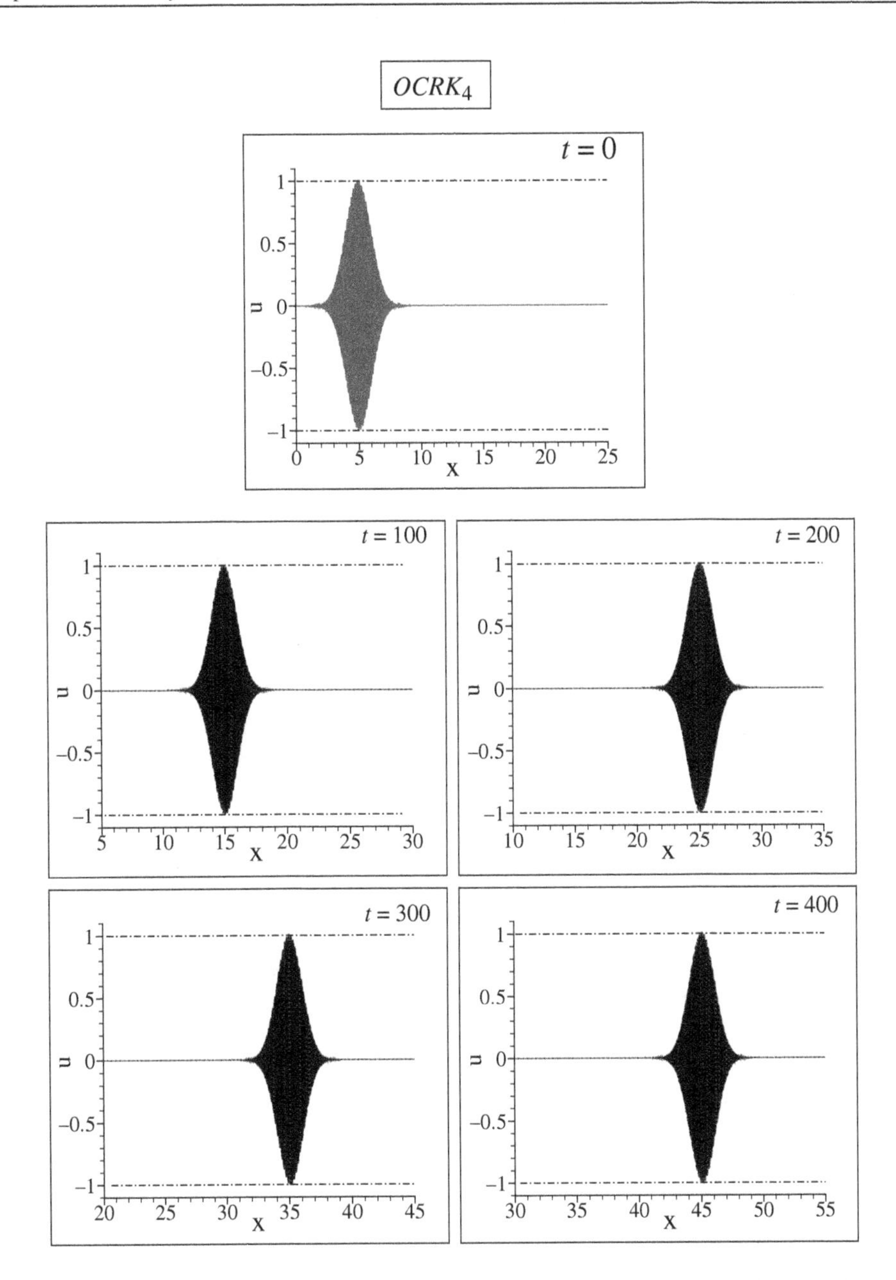

Figure 8.7 Propagation of computed wave packet following 1D convection equation with phase speed $c = 0.1$, $k_oh = 0.8$ and $N_c = 0.19660$ using optimized compact four-stage Runge-Kutta $OCRK_4$ scheme. In all the subframes, the exact solution is shown by the dotted lines

Thus, ∇x and ∇y are constant-valued vector fields which implies

$$\nabla^2 x = 0, \ \nabla^2 y = 0. \tag{8.37}$$

Included angle between the ξ = constant and η = constant lines in the physical plane is constrained to be orthogonal, if the numerical transformations satisfies

$$x_\xi x_\eta + y_\xi y_\eta = 0, \tag{8.38}$$

where subscripts indicate partial derivative with respect to it. Mehta and Lavan (1975) constructed conformal grids by satisfying the Cauchy–Riemann relation given as

$$x_\xi = -y_\eta, \quad x_\eta = y_\xi.$$

However while constructing orthogonal grid (Mehta and Lavan 1975), the trailing edge of the aerofoil was altered significantly, which may not be desirable. While maintaining orthogonality, strict condition of equality of scales imposed in conformal mapping can be relaxed, in nonconformal orthogonal grid. One can introduce a distortion function f as a ratio of scale factors $h_2 \left(= \sqrt{x_\eta^2 + y_\eta^2}\right)$ and $h_1 \left(= \sqrt{x_\xi^2 + y_\xi^2}\right)$ in η- and ξ-directions (Nair and Sengupta 1998, Ryskin and Leal 1983) as

$$f = \frac{h_2}{h_1}. \tag{8.39}$$

For general nonconformal orthogonal mapping (Ryskin and Leal 1983), following Beltrami equations are obtained using Equation (8.38) in Equation (8.39) as

$$x_\xi = -f y_\eta \tag{8.40}$$

$$x_\eta = f y_\xi. \tag{8.41}$$

Elimination of y and x from Equations (8.39) to (8.41) respectively, yields the following covariant Laplace equations (Ryskin and Leal 1983)

$$\frac{\partial}{\partial \xi}\left(f \frac{\partial x}{\partial \xi}\right) + \frac{\partial}{\partial \eta}\left(\frac{1}{f}\frac{\partial x}{\partial \eta}\right) = 0 \tag{8.42}$$

$$\frac{\partial}{\partial \xi}\left(f \frac{\partial y}{\partial \xi}\right) + \frac{\partial}{\partial \eta}\left(\frac{1}{f}\frac{\partial y}{\partial \eta}\right) = 0. \tag{8.43}$$

However, the solution of Equations (8.42) and (8.43) does not necessarily satisfy the orthogonality condition, Equation (8.38), as has been noted (Nair and Sengupta 1998) with grid generated by solving these elliptic PDEs. Generated grids have considerable deviations from orthogonality at many points in the domain.

In a marked departure, orthogonality has been maintained strictly in the generated grid by using the hyperbolic grid generation method (Steger and Chaussee 1980, Nair and Sengupta 1998). This has been achieved by using Equation (8.38) as one of the governing equations. In Steger and Chaussee (1980), prescription of cell volume given by $h_1 h_2$, has been used as the second governing equation for outwardly marching the equations from the boundary. In Nair and Sengupta (1998), one of the Beltrami Equations (8.40) and (8.41), has been used along with the orthogonality condition. It has been reported that for symmetric aerofoils this procedure (Nair and Sengupta 1998) produced superior grids. Despite satisfying Equation (8.38) numerically, orthogonality condition is not ensured strictly everywhere due to numerical error. In Steger and Chaussee (1980), the grid was generated for a highly cambered turbine blade, with results not at all satisfactory, as the worst nonorthogonality for a cambered turbine blade was as high as 20° at several points. Solution of Navier-Stokes equation has been presented using orthogonal formulation in Nair and Sengupta (1998), for flow past NACA 0015 aerofoil computed for 30° angle of attack, for $Re = 35\,000$. Using orthogonal grid formulation, computed results agreed well with experimental results. In Sengupta *et al.* (2007a) and Sengupta *et al.* (2005), orthogonal grid formulation was used to compute flow respectively past symmetric aerofoils for accelerated flow and when the aerofoil was flapping and hovering.

8.5 Orthogonal Grid Generation

The results presented here are from Bagade *et al.* (2014). First, we consider NACA 8504 aerofoil which has high camber (8%) and small thickness to chord ratio (4%), as shown in Figure 8.8(a). For this aerofoil, orthogonal grid has been generated with 456 points in the azimuthal direction and 250 points in the wall-normal direction. The grid has been generated following the orthogonal grid generation procedure in Nair and Sengupta (1998), up to a small distance from the aerofoil surface. The grid is generated by prescribing the gap between successive η = constant lines in the wall-normal direction following the tangent hyperbolic distribution given by

$$h_2 = H\left[1 - \frac{\tanh[\beta(1-\eta)]}{\tanh[\beta]}\right],$$

where β represents the clustering parameter. The grid-line increment in η-direction is represented by grid scale parameter h_2 and H is the nondimensional distance (in terms of chord of the aerofoil) of the outer boundary. The grid in Figure 8.8(a), has been constructed using first 50 out of total 250 points.

A region ABCD near the trailing edge of the aerofoil has been marked and is shown zoomed in Figure 8.8(b), with the outermost azimuthal terminal line is marked by a solid line. It is clear that if we generate the grid farther away from the aerofoil surface, then the wall-normal grid lines emerging from the bottom surface leading and trailing edge portions will eventually intersect. This is due to concavity of the bottom surface and one needs to modify the algorithm in Nair and Sengupta (1998), to generate grid around highly concave shaped bodies. The distinction between the concave and convex nature of the surface is visible from the variation of curvature ($\frac{d^2y}{dx^2}$) of the terminal line on the top and bottom surfaces, as shown in Figures 8.8(c) and 8.8(d), respectively. In these figures, negative value of curvature denotes convexity of the surface, and thus, the bottom surface has a positive curvature due to associated concavity. In order to remove this concavity progressively from successive grid lines generated in the bottom part, the grid spacing must be prescribed as a function of the azimuthal coordinate by progressively increasing the grid spacing in the wall-normal direction, in this part to fill the gap due to concavity of the aerofoil surface. Thus, one needs to obtain h_2 distribution as a function of both ξ- and η-direction to avoid grid intersection. Estimation of curvature on the

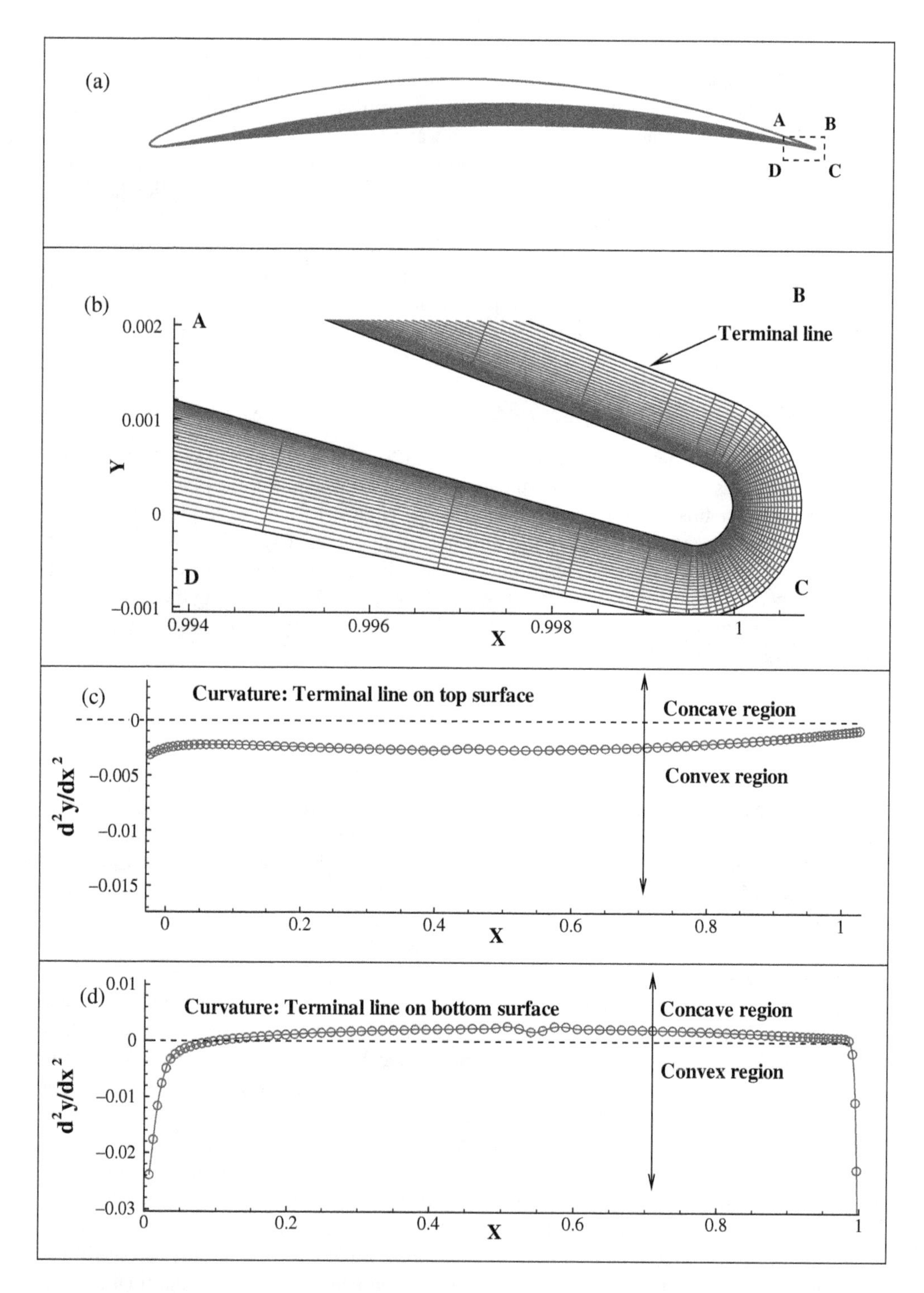

Figure 8.8 (a) Orthogonal grid near the surface of the NACA 8504 aerofoil has been generated following Nair and Sengupta (1998) along with marked ABCD region; (b) zoomed view of the ABCD region with a highlighted terminal line; curvature variations of the terminal line on top and bottom sides are shown in (c) and (d) respectively

terminal line helps one to locate the concave region and is an important parameter in the grid generation algorithm proposed next.

8.5.1 Grid Generation Algorithm

1. Generate an orthogonal grid around a given aerofoil following the solution of Equations (8.38) and (8.40) or (8.41), with Δs_η constant along azimuthal direction up to a small distance from aerofoil surface, so that the grid lines should emerge smoothly from the surface without the possibility of intersection, as shown in Figure 8.9(a).
2. Compute curvature of the terminal line and identify the concave region(s), as shown in Figures 8.8(c) and 8.8(d).
3. Increase the value of h_2 in the concave region by small amount Δh_2 as a function of ξ and once again generate a new grid for the modified h_2 distribution.
4. Compute curvature of the terminal line and repeat steps 3 and 4 unless concavity has been removed completely.

   At the end of this step one obtains a grid shown in Figures 8.9(a) and 8.9(b), with zero curvature on the bottom surface of the terminal line shown in Figure 8.9(d).
5. With complete removal of concavity from the terminal line, generate orthogonal grid farther away from the surface following the previous algorithm in Nair and Sengupta (1998), with results as shown in Figure 8.10.

While performing steps 3 and 4, a minimum limiting curvature on the terminal line is prescribed as $(\frac{d^2y}{dx^2}) = 0$. So the spacing in η-direction for a concave region is modified till one obtains a grid with a flat bottom terminal line, as shown in Figure 8.9(a). Increase of spacing in η direction causes $\xi = \text{constant}$ lines to bend, as shown in Figure 8.9(b). Thus the problem of possible intersection of $\xi = \text{constant}$ lines from leading and trailing edge of the bottom surface has been removed. Once concavity has been removed from the terminal line, one can generate orthogonal grid farther away from the surface by just following the algorithm described in Nair and Sengupta (1998). Figure 8.10(a) shows a zoomed view of the constructed grid, while Figure 8.10(b) shows a complete orthogonal grid around NACA 8504 aerofoil with 456×250 points. Deviation of included angle by the intersecting grid lines at any node, from orthogonality is shown in Figure 8.10(c) by contour plots. Maximum deviation $(\Delta\theta_{max})$ is 0.1039°, while minimum deviation $(\Delta\theta_{min})$ is −0.0730°. This maximum deviation is a significantly lesser deviation as compared to 0.524° reported for the case of NACA 0015 symmetric aerofoil earlier (Nair and Sengupta 1998). This establishes that the present grid generating algorithm has a capability to generate a grid around highly concave shaped bodies with very small deviation from orthogonality.

Although least deviation from orthogonality is a prime requirement for using orthogonal formulation of Navier–Stokes equations, one also has to ensure smooth variation of grid metrics. Here, this is discussed with respect to orthogonal formulation of Navier-Stokes equations in stream function-vorticity formulation. These equations are given by

$$\frac{\partial}{\partial\xi}\left[\frac{h_2}{h_1}\frac{\partial\psi}{\partial\xi}\right] + \frac{\partial}{\partial\eta}\left[\frac{h_1}{h_2}\frac{\partial\psi}{\partial\eta}\right] = -h_1 h_2 \omega \tag{8.44}$$

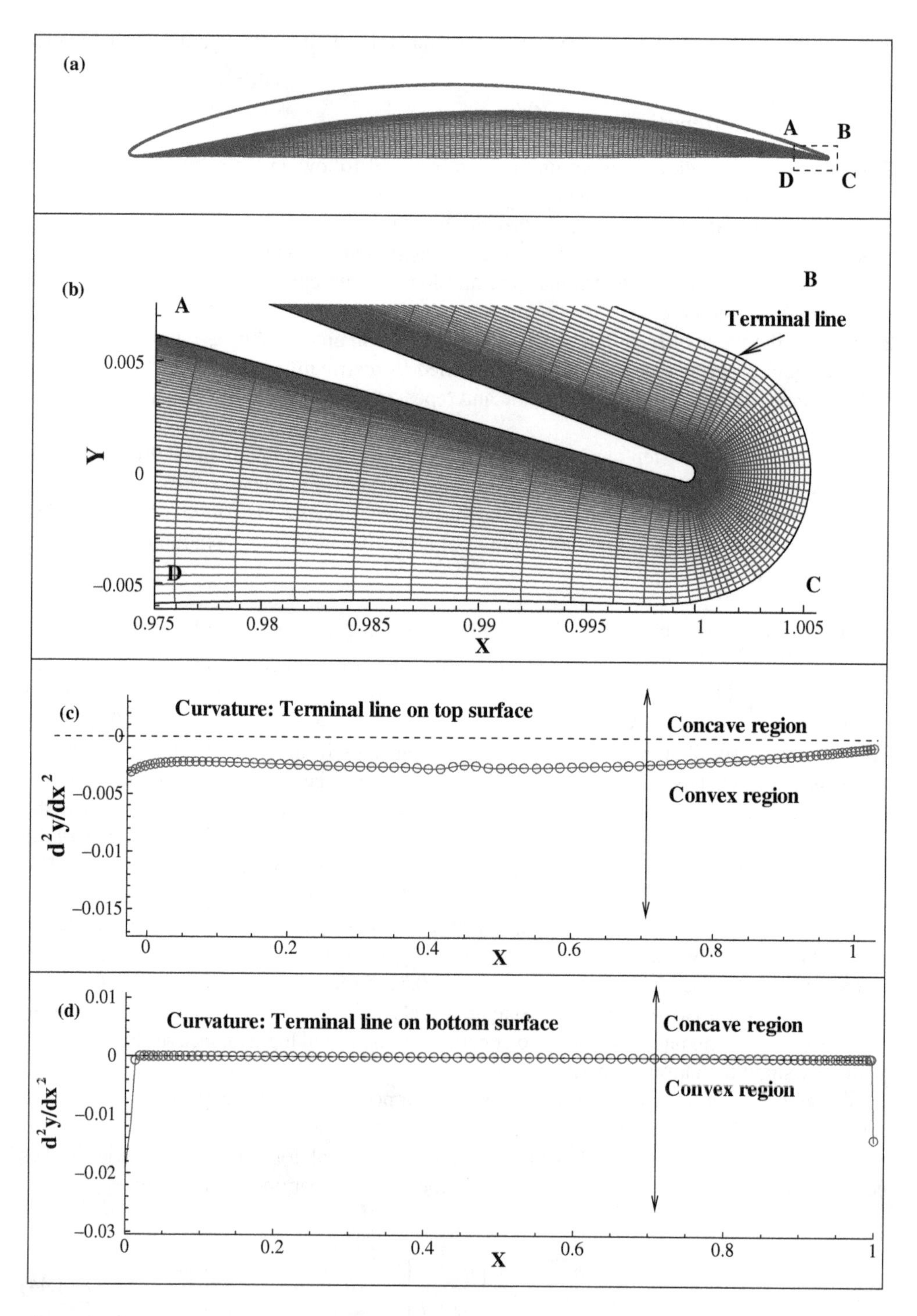

Figure 8.9 (a) Generated orthogonal grid at the end of step 4 of the proposed algorithm in Section 8.5.1; (b) zoomed view of the new ABCD region with a highlighted terminal line; curvature variations of the new terminal line on top and bottom sides are shown in (c) and (d), respectively. Note, concavity from the bottom side of the terminal line has been removed completely

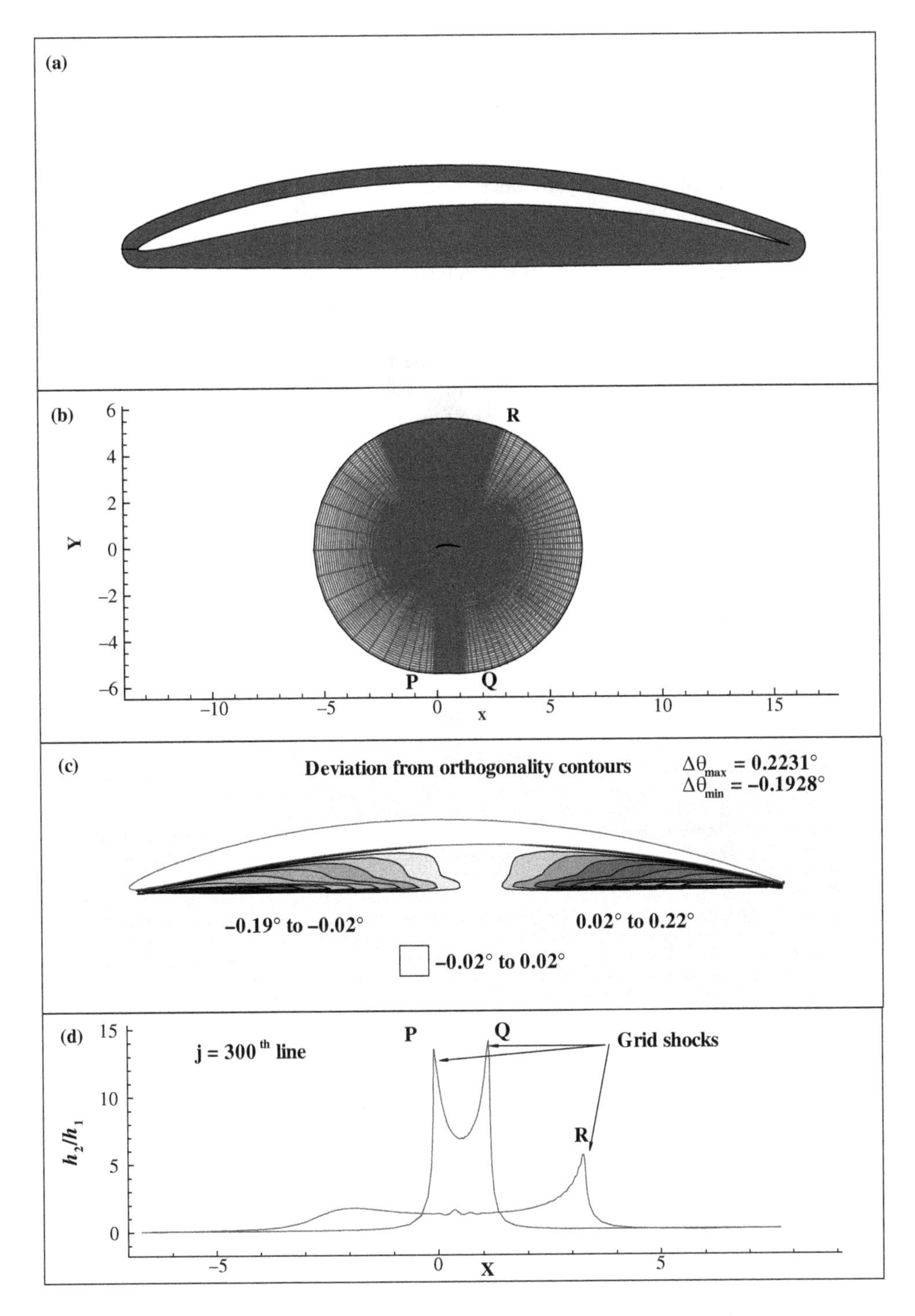

Figure 8.10 (a) Zoomed view of the orthogonal grid generated around NACA 8504 aerofoil; (b) complete grid; (c) deviation from orthogonality contours for the grid shown in Figure 8.10(b); (d) typical variation of grid scale parameter combination h_2/h_1 on the last azimuthal line. Note grid shocks at P, Q and R marked in frame (b). (See colour version of this figure in colour plate section)

$$h_1 h_2 \frac{\partial \omega}{\partial t} + h_2 u \frac{\partial \omega}{\partial \xi} + h_1 v \frac{\partial \omega}{\partial \eta} = \frac{1}{Re}\left[\frac{\partial}{\partial \xi}\left(\frac{h_2}{h_1}\frac{\partial \omega}{\partial \xi}\right) + \frac{\partial}{\partial \eta}\left(\frac{h_1}{h_2}\frac{\partial \omega}{\partial \eta}\right)\right], \quad (8.45)$$

where the scale factors have been defined before. Three combinations of scale factors $h_1 h_2$, h_1/h_2 and h_2/h_1 appearing in the above equations play important part in solving Navier–Stokes equation. In solving the stream function equation, one uses iterative procedure and the total number of iterations required to achieve convergence directly depend on these parameters. In the presence of grid-shocks (bunching of adjacent grid lines), diagonal dominance of the associated linear algebraic equation from discretized self-adjoint operator can be adversely affected, causing poor convergence. The trailing edge of the aerofoil is approximated with a smooth convex curve joining the top and bottom surface (Nair and Sengupta 1998), to make the slope variation continuous. Modification of the trailing edge is always restricted to the last 0.02% of the chord, which does not alter the basic geometry. Points on the ξ = constant grid lines open up more rapidly in the latter half, as compared to similar lines in the upstream part of the aerofoil. This creates a region of closely packed points in both the upper and lower surfaces of the aerofoil in the azimuthal direction. One can notice very high values of h_2/h_1 ratio on the last azimuthal line in the concave part of the aerofoil, as shown in Figure 8.10(d). One can clearly observe formation of grid-shocks near P, Q and R, as shown in Figure 8.10(b).

In order to reduce problems originating from grid-shocks, one can prescribe a minimum limiting curvature on the terminal line as a negative quantity, implying a convex terminal line on both the top and the bottom surfaces, while performing steps 3 and 4 . We have generated a completely new grid by prescribing a criteria on minimum limiting curvature on the terminal line as $(\frac{d^2y}{dx^2}) = -0.003$ and resulting grids shown in Figures 8.11(a) and 8.11(b). Zoomed view of the grid in Figure 8.11(a) shows a convex shape of a terminal line and the complete grid in Figure 8.11(b) shows reduction in grid-shock intensity at points P, Q and R, as compared to Figure 8.10(b). This is also evident from variation of h_2/h_1 ratio on the last azimuthal line as shown in Figure 8.11(d). However for such smooth variation of grid metrics, we have higher deviation from orthogonality, as seen in Figure 8.11(c), as compared to that shown in Figure 8.10(c). So after this exercise, the question naturally arises that out of the two requirements of grid smoothness and minimum deviation from orthogonality, which grid parameter should be considered as more important. This question can be answered by solving Navier-Stokes equation using these grids.

8.6 Orthogonal Grid Generation for an Aerofoil with Roughness Elements

Next, we have shown orthogonal grid generation around the natural laminar flow (NLF) SHM-1 Honda aerofoil (Fujino *et al.* 2003), with multiple roughness elements located on the top surface of the aerofoil near the leading edge. In Figure 8.12(a), we have shown the basic profile of SHM-1 aerofoil with surface roughness elements. These kind of surface irregularities may be present on the aircraft wing due to manufacturing problems, ice accretion or the presence of wing parting line. Zoomed view of the region A-B-C-D of Figure 8.12(a) with three successive roughness elements of the aerofoil is shown in Figure 8.12(b). It is important to study the change in the flow-field due to the presence of such roughness elements. Some numerical and experimental studies have been performed to study the effects of roughness elements for external flows. A few representative ones can be found in the literature (Huebsch and Rothmayer 2002,

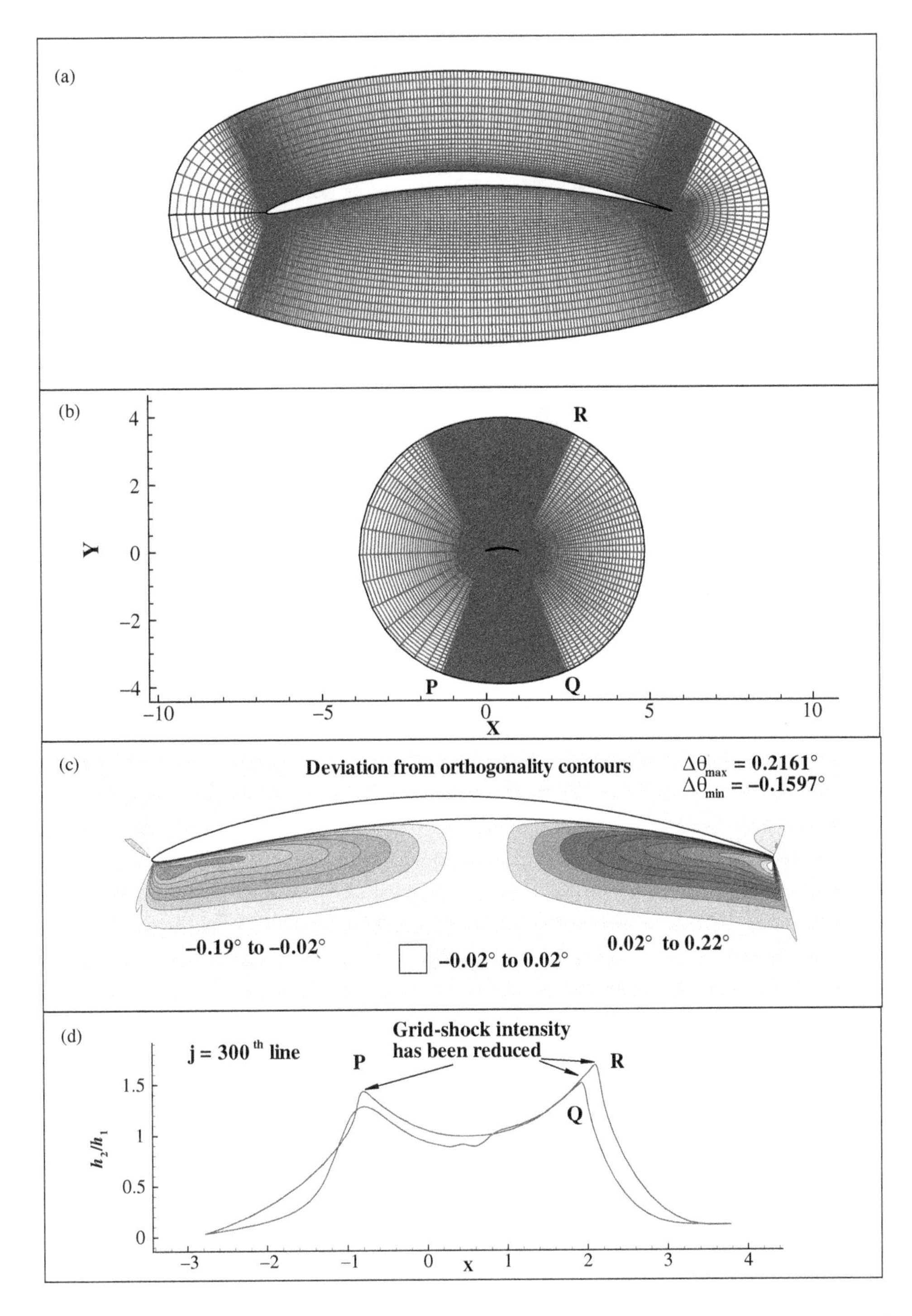

Figure 8.11 (a) Zoomed view of the orthogonal grid generated around NACA 8504 aerofoil; (b) complete grid; (c) deviation from orthogonality contours for the grid shown in Figure 8.11(b); (d) typical variation of grid scale parameter combination h_2/h_1 on the last azimuthal line. Note, intensity of grid-shocks at P, Q and R has been reduced. (See colour version of this figure in colour plate section)

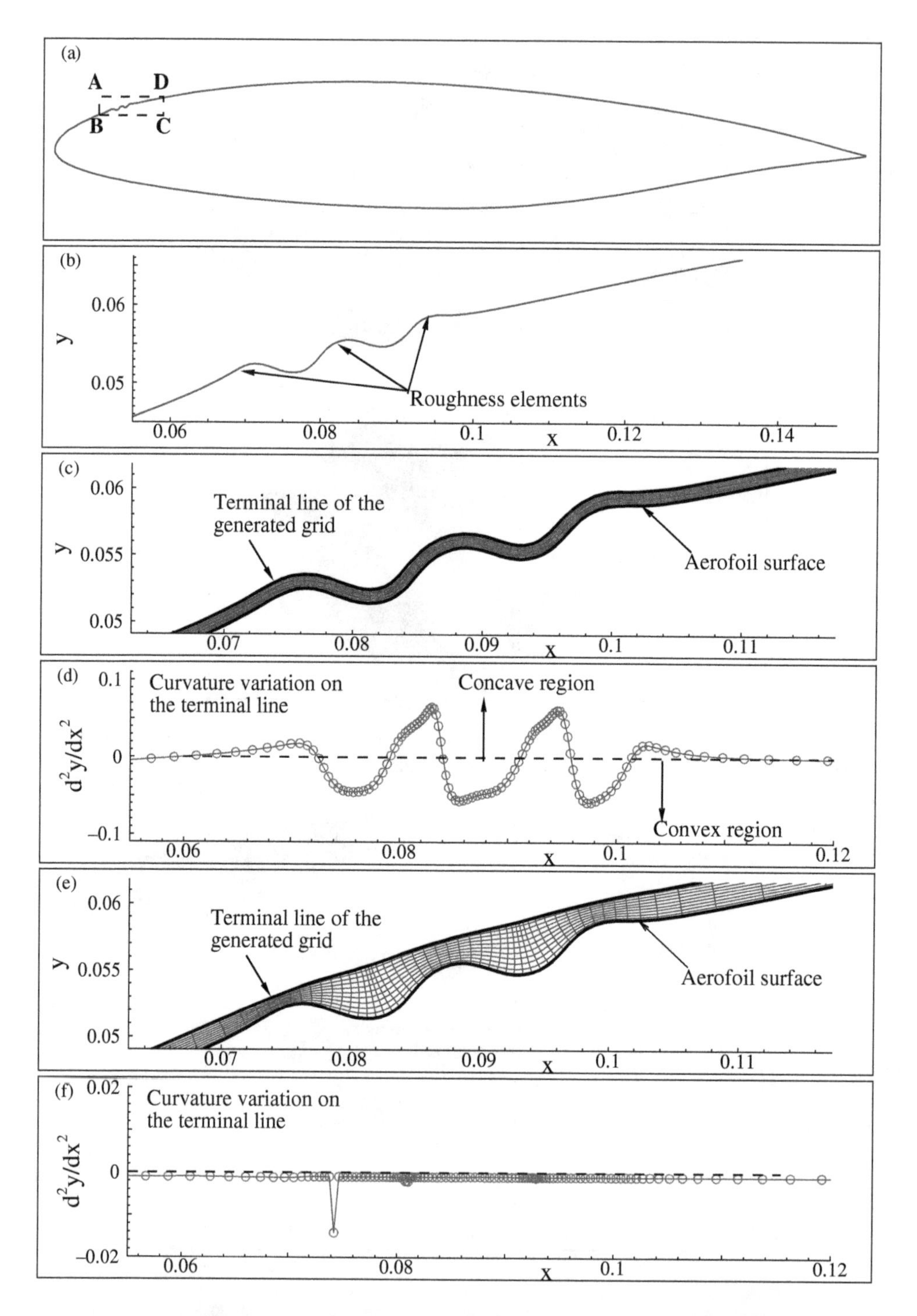

Figure 8.12 (a) Schematic of SHM-1 aerofoil (Fujino *et al.* 2003) with multiple surface roughness elements near leading edge; (b) zoomed view of surface roughness elements; (c) orthogonal grid generated by the method in (Nair and Sengupta 1998); (d) curvature variations of the terminal line shown in (c); (e) orthogonal grid generated using proposed algorithm and (f) curvature variations of the terminal line shown in (e)

2004, Kerho and Bragg 1997, Volino *et al.* 2009) and other references therein. In Huebsch and Rothmayer (2002, 2004), the two-dimensional Navier-Stokes equation has been used to investigate unsteady, incompressible viscous flow past an aerofoil with a surface roughness element at the leading edge. The roughness was simulated on the surface through the use of Prandtl's transposition (Huebsch and Rothmayer 2002, 2004) for altered wall boundary conditions. In Kerho and Bragg (1997), the effects of large distributed roughness near the leading edge of the aerofoil on the boundary layer development and transition were studied experimentally. An experimental study has also been performed in Volino *et al.* (2009), for two-dimensional roughness element on a zero pressure gradient boundary layer. It is noted that there is a difference between the scales of motion induced by two- and three-dimensional roughness elements.

For the present case, the generated grid has 601 points in the azimuthal direction and 700 points in the wall-normal direction. Figure 8.12(c) shows a zoomed view of the grid near the roughness elements. These surface roughness elements create two additional concave regions at the junction of the roughness element with the aerofoil, as shown in Figure 8.12(d). These concave regions are filled up using the same modified algorithm, during steps 3 and 4 described above. Concavity near roughness elements is much more localized as compared to the concavity of the bottom surface of the aerofoil. Concave portions on the top and bottom surfaces are removed using the modified algorithm discussed above. After execution of step 4, one obtains a grid as shown in Figure 8.12(e) with a criteria on minimum limiting curvature on the terminal line as $(\frac{d^2y}{dx^2}) = -0.001$ which is also observed from Figure 8.12(f). Note that the present procedure causes the terminal line to have no concave portion.

A close-up view of the grid near the roughness elements is shown in Figure 8.13(a), while the close-up view of complete grid is shown in Figure 8.13(b). As discussed earlier, the maximum deviation from orthogonality occurs in the concave portion and in this case it is mainly restricted to the region near the roughness elements. Figure 8.13(c) shows deviation from orthogonality contours with $\Delta\theta_{max} = 0.1315°$ and $\Delta\theta_{min} = -0.1159°$. As shown in Figure 8.11, one can control grid-shock intensity by prescribing a higher convex curvature to the terminal line at the cost of increased deviation from orthogonality.

8.7 Solution of Navier–Stokes Equation for Flow Past AG24 Aerofoil

Construction of orthogonal grid allows one to perform accurate calculations involving fewer and faster computations. Use of numerically generated orthogonal grid is demonstrated here by studying flow past a AG24 aerofoil for a Reynolds number of $Re = 4 \times 10^5$, a value commonly observed in glider flights. We have constructed a grid with 528 points in the azimuthal direction and 497 points in the wall-normal direction.

Figures 8.14(a)–8.14(c) show streamline contours of the flow past AG24 aerofoil at the indicated time instants and angles of attack (AOA). For zero AOA, flow near the leading edge of the aerofoil experiences acceleration due to the favourable pressure gradient and beyond the maximum thickness location, it experiences adverse pressure gradient. Near the trailing edge on the top and bottom surfaces, unsteady separation bubbles are formed due to adverse pressure gradient. However, for the case of negative and positive AOA, there is higher adverse pressure gradient experienced by the flow on the bottom and top surfaces, respectively. This results in larger separation bubbles on the respective surfaces, as shown in Figures 8.14(a) and 8.14(c). This is also observed from the variation of mean pressure coefficient on the top and the bottom surfaces at different AOA as shown in Figure 8.14(d).

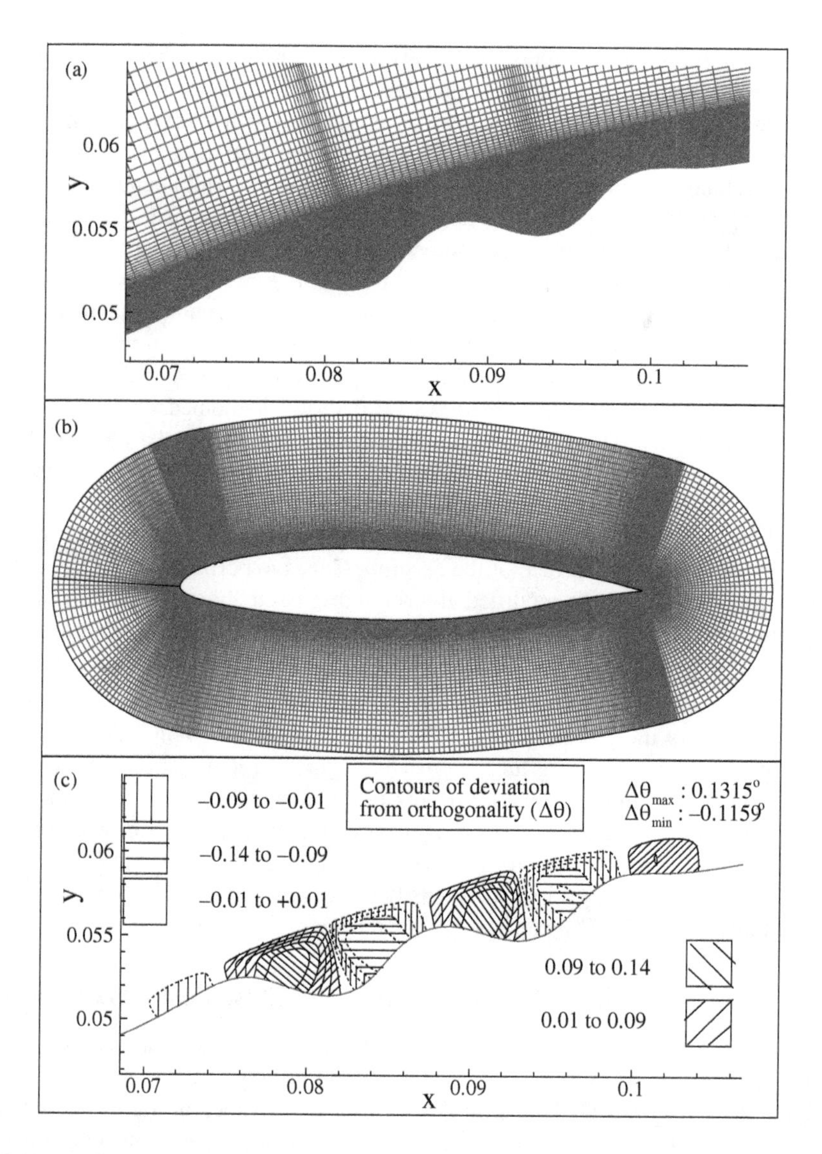

Figure 8.13 (a) Zoomed view of the grid near roughness elements; (b) zoomed view of the grid around complete aerofoil and (c) deviation from orthogonality contours for the generated grid in (b). (See colour version of this figure in colour plate section)

Time variations of lift and drag coefficients for zero AOA are shown in Figures 8.15(a) and 8.15(b), respectively. We have obtained the time averaged lift and drag coefficients for different AOA cases for $Re = 4 \times 10^5$, and plotted the drag polar along with experimental data of Williamson *et al.* (2012) in this figure. Numerical results are in very good agreement with the experimental results of Williamson *et al.* (2012). Thus, the present orthogonal grid generation technique is found to be effective for high accuracy space-time accurate computations of flows around cambered aerofoils, with and without roughness elements.

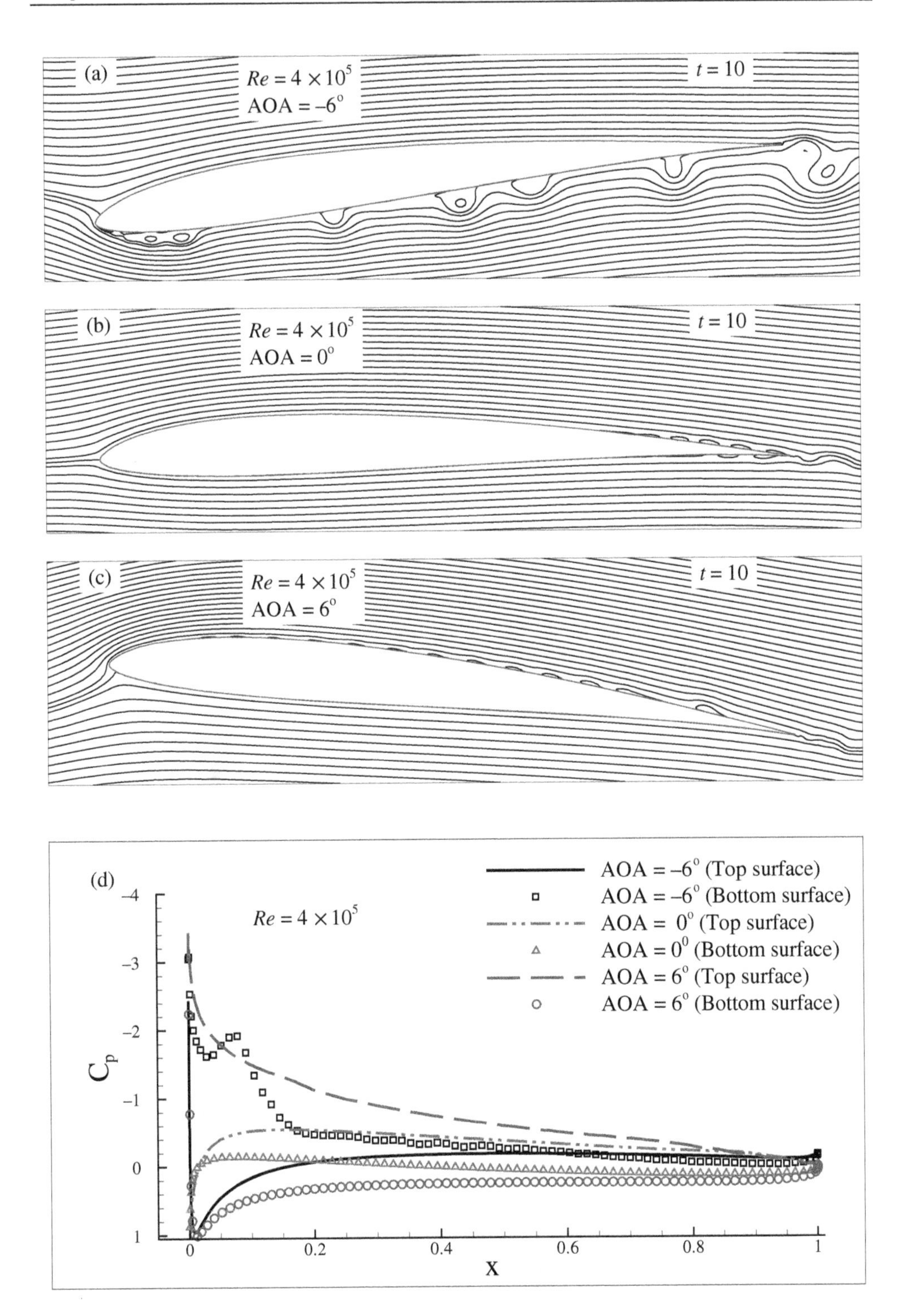

Figure 8.14 Streamline contours for flow past a AG24 aerofoil (Williamson *et al.* (2011), at the indicated AOA and instants for Reynolds number of $Re = 4 \times 10^5$ in (a), (b) and (c); variations of mean pressure coefficient on top and bottom surfaces for the AOA cases in (a), (b) and (c) are shown in (d)

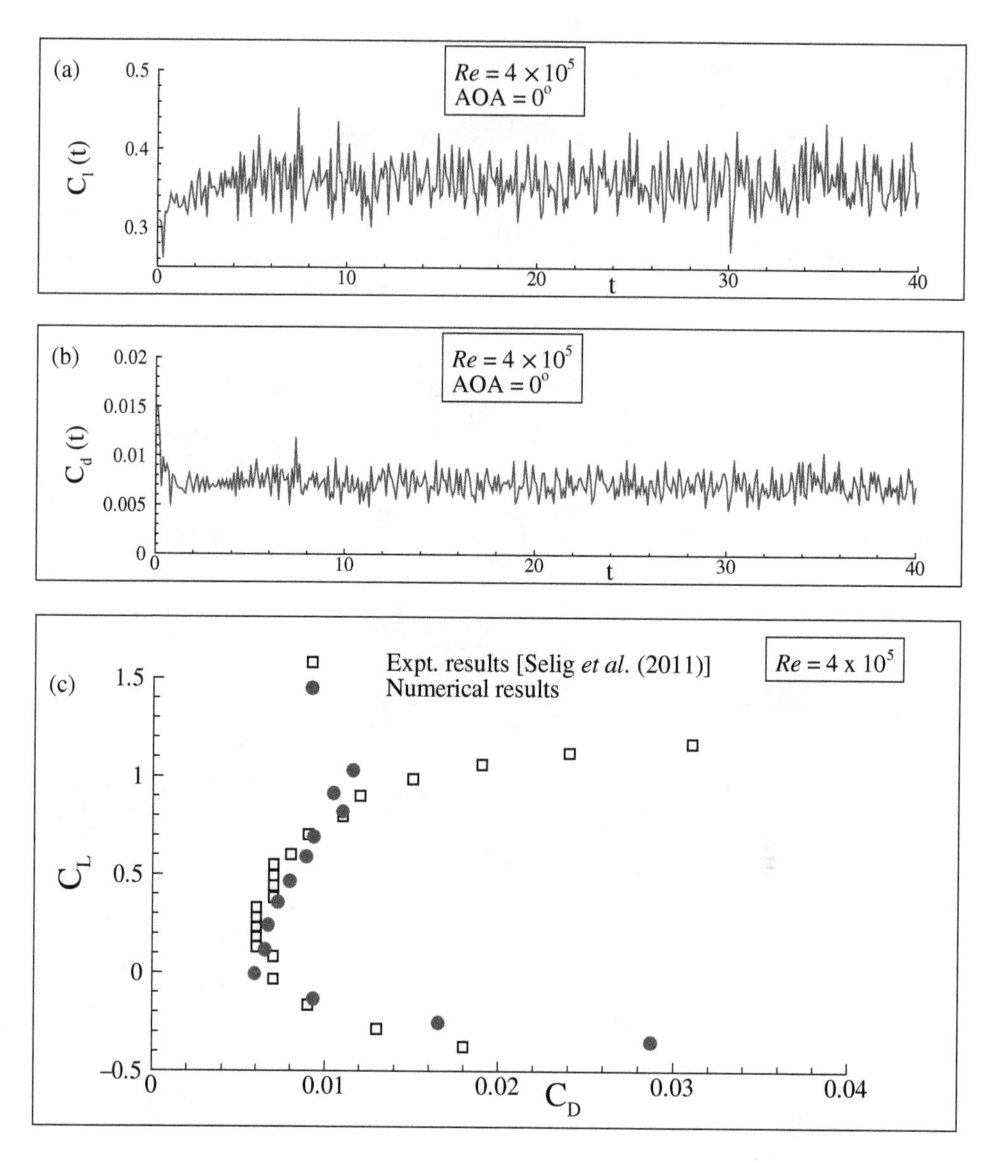

Figure 8.15 Variation of instantaneous lift and drag coefficients for zero AOA for $Re = 4 \times 10^5$ are shown (a) and (b); comparison of experimental (Williamson *et al.* 2012) and numerically obtained drag polar is shown in (c)

8.7.1 *Grid Smoothness vs Deviation from Orthogonality*

In Section 8.5, we are left with two conflicting choices for grid generation. Either one can produce a grid with strong grid-shocks with a lesser deviation from orthogonality, or one can generate a grid with smooth variation and more deviation from orthogonality. Here, we have constructed two different grids for AG24 aerofoil with stronger and weaker grid-shocks,

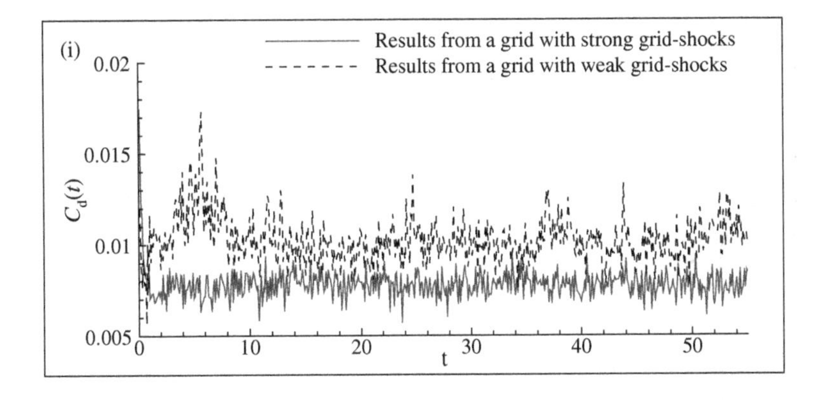

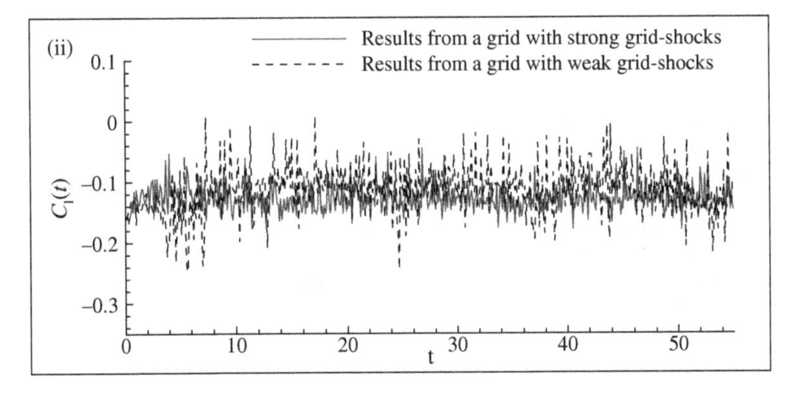

Figure 8.16 Variation of instantaneous drag and lift coefficients for −3.85° AOA for $Re = 4 \times 10^5$ are compared for the grids with strong and weak grid-shocks

similar to the grids shown in Figures 8.10 and 8.11. The grid with stronger grid-shock has deviation from orthogonality in the range $[-0.0730°, 0.1039°]$, while for the other grid with weaker grid-shock, deviation from orthogonality lies in the range $[-0.1597°, 0.2161°]$. These two grids are used to solve the flow at −3.85° AOA and $Re = 4 \times 10^5$. In Figure 8.16, we have compared instantaneous variation of drag and lift coefficients on the two grids. Results show that the smooth grid with more deviation from orthogonality has higher amplitude fluctuations than the grid with strong shocks and less deviation from orthogonality. Thus, the deviation from orthogonality is a considerable source of error in computations and can give rise to higher fluctuating flow parameters. Time averaged values of lift and drag coefficients for the strong shock grid are −0.1277 and 0.0078, respectively, while for the smooth grid these values are −0.1138 and −0.0102, respectively. Thus one does not observe much of a change in the mean values but considerable changes can be present in the fluctuating parameters.

Bibliography

Bagade, P.M., Bhumkar, Y.G. and Sengupta, T.K. (2014) An improved orthogonal grid generation method for solving flows past highly cambered aerofoils with and without roughness elements. Submitted to *Comput. Fluids*.

Batchelor, G.K. (1969) Computation of the energy spectrum in homogeneous two-dimensional decaying turbulence. *Phys. Fluids*, **12**, (suppl. II), 233–239.

Dipankar, A. and Sengupta, T.K. (2006) Symmetrized compact scheme for receptivity study of 2D transitional channel flow, *J. Comput. Phys.*, **215**, 245–273.

Davidson, P.A. (2004) *Turbulence: An Introduction for Scientists and Engineers*. Oxford University Press, UK.

Doering, C.R. and Gibbon, J.D. (1995) *Applied Analysis of the Navier-Stokes Equations*. Cambridge University Press, UK.

Donzis, D.A. and Yeung, P.K. (2010) Resolution effects and scaling in numerical simulation of passive scalar mixing in turbulence. *Physica D: Nonlinear Phenomena* **239**(14), 1278–1287.

Donzis, D.A., Yeung, P.K. and Sreenivasan, K.R. (2008) Dissipation and enstrophy in isotropic turbulence: resolution effects and scaling in direct numerical simulations. *Phys. Fluids* **20**, 045108.

Fujino, M., Yoshizaki, Y. and Kawamura, Y. (2003) Natural-laminar-flow airfoil development for a lightweight business jet. *J. Aircraft*, **40**(4), 609–615.

Huebsch, W.W. and Rothmayer, A.P. (2002) The effects of surface ice roughness on dynamic stall. *J. Aircraft*, **39**(6), 945–953.

Huebsch, W.W. and Rothmayer, A.P. (2004) Numerical prediction of unsteady vortex shedding for large leading-edge roughness. *Comput. and Fluids*, **33**, 405–434.

Kerho, M.F. and Bragg, M.B. (1997) Airfoil boundary-layer development and transition with large leading-edge roughness. *AIAA J.*, **35**(1), 75–84.

Kraichnan, R. and Montgomery, D. (1980) Two-dimensional turbulence. *Reports in Progress in Physics*, **43**, 547–619.

Mehta, U.B. and Lavan, Z. (1975) Starting vortex, separation bubbles and stall: A numerical unsteady flow around an airfoil. *J. Fluid Mech.*, **67**(2), 227–256.

Nair, M.T. and Sengupta, T.K. (1998) Orthogonal grid generation for Navier-Stokes computations. *Int. J. Num. Meth. Fluids*, **28**, 215–224.

Rajpoot, M.K., Sengupta, T.K. and Dutt, P.K. (2010) Optimal time advancing dispersion relation preserving schemes. *J. Comput. Phys.*, **229**(10), 3623–3651.

Ryskin, G. and Leal, L.G. (1983) Orthogonal mapping. *J. Comput. Phys.*, **50**, 71–100.

Sengupta, T.K. (2013) *High Accuracy Computing Methods: Fluid Flows and Wave Phenomena*. Cambridge University Press, USA.

Sengupta, T.K. and Bhaumik, S. (2011) Onset of turbulence from the receptivity stage of fluid flows. *Phys. Rev. Lett.*, 154501, 1-5.

Sengupta, T.K., Bhumkar, Y.G. and Sengupta, S. (2012) Dynamics and instability of a shielded vortex in close proximity of a wall. *Comput. Fluids*, **70**, 166–175.

Sengupta, T.K., Bhumkar, Y.G., Rajpoot, M., Suman, V.K. and Saurabh, S. (2012a) Spurious waves in discrete computation of wave phenomena and flow problems. *App. Math. Comput.*, **218**(18), 9035–9065.

Sengupta, T.K., De, S. and Sarkar, S. (2003) Vortex-induced instability of an incompressible wall-bounded shear layer. *J. Fluid Mech.*, **493**, 277–286.

Sengupta, T.K. and Dipankar, A. (2004) Comparative study of time advancement methods for solving Navier-Stokes equations. *J. Sci. Comp.*, **21**(2), 225–250.

Sengupta, T.K., Dipankar, A. and Sagaut, P. (2007) Error dynamics: Beyond von Neumann analysis. *J. Comput. Phys.*, **226**, 1211–1218.

Sengupta, T.K., Ganeriwal, G. and De, S. (2003a) Analysis of central and upwind compact schemes. *J. Comput. Phys.*, **192**, 677–694.

Sengupta, T.K., Lim, T.T., Sajjan, V.S., Ganesh, S. and Soria, J. (2007a) Accelerated flow past a symmetric airfoil: Experiments and computations. *J. Fluid Mech.*, **591**, 255–288.

Sengupta, T.K., Rajpoot, M.K. and Bhumkar, Y.G. (2011) Space-time discretizing optimal DRP schemes for flow and wave propagation problems. *Comput. Fluids*, **47**(1), 144–154.

Sengupta, T.K., Singh, H., Bhaumik, S. and Roy Chowdhury, R. (2013) Diffusion in inhomogeneous flows: Unique equilibrium state in an internal flow. *Comput. Fluids*, **88**, 440–451.

Sengupta, T.K., Vikas, V. and Johri, A. (2005) An improved method for calculating flow past flapping and hovering airfoils. *Theor. Comput. Fluid Dyn.*, **19**(6), 417–440.

Steger, J.L. and Chaussee, D.S. (1980) Generation of body-fitted coordinates using hyperbolic partial differential equations. *SIAM J. Sci. Stat. Comput.*, **1**, 431–437.

Tennekes, H. and Lumley, J.L. (1971) *A First Course in Turbulence*. MIT Press, Cambridge, MA, USA.

Thompson, J.F., Warsi, Z.U.A. and Mastin, C.W.J. (1982) Boundary-fitted coordinate systems for numerical solution of partial differential equations- a review. *J. Comput. Phys.*, **47**, 1–108.

Volino, R.J., Schultz, M.P. and Flack, K.A. (2009) Turbulence structure in a boundary layer with two-dimensional roughness. *J. Fluid Mech.*, **635**, 75–101.
Williamson, G.A., McGranahan, B.D., Broughton, B.A., Deters, R.W., Brandt, J.B. and Selig, M.S. (2012) Summary of Low Speed Airfoil Data, **5**, (http://www.ae.illinois.edu/m-selig/uiuc_lsat/Low-Speed-Airfoil-Data-V5.pdf).
Yeung, P.K., Donzis, D.A. and Sreenivasan, K.R. (2012) Dissipation, enstrophy and pressure statics in turbulence simulation at high Reynolds. *J. Fluid Mech.*, **700**, 5–15.

9

Instability and Transition in Aerodynamics

9.1 Introduction

Fluid dynamics equations of motion were first written down for inviscid flow by Euler in 1752. By 1840, the equation of fluid motion in the presence of friction, the Navier–Stokes equation, was published and few exact solutions for special cases were obtained. In one such case, Stokes compared theoretical prediction with available experimental data for pipe flow and found no agreement whatsoever. With hindsight we know this theoretical solution corresponded to undisturbed laminar flow, while the experimental data given to him corresponded to a turbulent flow. This prompted Osborne Reynolds to explain the mismatch by his famous pipe flow experiments (Reynolds 1883). It was reasoned that the basic flow, obtained as a solution to the Navier–Stokes equation, is unable to maintain its stability with respect to omnipresent ambient small disturbances. Mathematical existence of a solution does not guarantee its physical realization and observation. Existence of a mathematical solution shows plausibility of an equilibrium solution, which satisfies conservation laws. However, one needs to study the stability of every solution to establish their observability. Reynolds demonstrated experimentally that the equilibrium parabolic profile disintegrates into sinuous motion of water in the pipe, which eventually leads to turbulent or chaotic flow.

This prompted Landau and Lifshitz (Landau and Lifshitz 1959) to pronounce that

> *the flow that occurs in nature must not only follow the equations of fluid dynamics, but also be stable.*

If a solution cannot be *observed*, then it must be due to instability of the corresponding equilibrium flow. Here, *instability* implies deviation of equilibrium solution by growth of infinitesimally omnipresent small perturbations. The smallness of background disturbances allows one to study the problem of growth of disturbances by small perturbation approach. This helps immensely, if the governing nonlinear equations can be solved for the equilibrium solution and its instability studied by linearizing the governing equation(s) for the perturbation field.

Theoretical and Computational Aerodynamics, First Edition. Tapan K. Sengupta.

Companion Website: www.wiley.com/go/sengupta

The pipe flow experiments reported in Reynolds (1883) are the first recorded evidence of flow instability. He took pipes of different diameters fitted with a trumpet-shaped mouth-piece or bell-mouth, which accelerated the flow at entry. Such acceleration attenuates present disturbances due to the favourable pressure gradient. Reynolds visualized this instability in pipe flow by rapid diffusion of injected dye with surrounding fluid for a parameter combination, which is now known as the Reynolds number given by $Re = Va/\nu$, with V as the centre-line velocity in the pipe of diameter a, and ν is the kinematic viscosity. Reynolds designed his experiment such that the flow was orderly or laminar up to $Re = 12\,830$. He also noted that the critical Re is very sensitive to disturbances in the oncoming flow, which prompted him to note that

> *this at once suggested the idea that the condition might be one of instability for disturbance of a certain magnitude and stable for smaller disturbances.*

The relationship of instability with disturbance amplitude is a typical attribute of nonlinear instability. It is now well established that pipe and plane Couette flows are linearly stable for all Reynolds numbers, when the viscous linear stability analysis is performed. This suggests that either nonlinear and/or different unknown linear mechanism(s) of instabilities are at play for these flows.

Critical Re was further raised later for pipe flows, establishing that there is perhaps no upper limit above which transition to turbulence cannot be prevented. This example also suggests the importance of receptivity of flow to different types of input disturbances to the system. In the absence of input on a fluid dynamical system of a particular kind that triggers instability, response will be orderly, even if the system is unstable to that kind of input. Thus, if the basic flow is receptive to a particular disturbance, then the equilibrium flow will not be observed in the presence of such disturbances.

However, a major aspect which is not completely understood to date is the clear distinction between theoretical approaches and flow instability noted in controlled experiments in laboratory or in actual devices/settings. It is rather unfortunate that past efforts in theoretical analyses did not clearly distinguish between disturbance growth in space and time. Recent researches clearly indicate the disturbances to grow spatio-temporally in Sengupta and Bhaumik (2011), Bhaumik and Sengupta (2014) and Sengupta *et al.* (2012) obtained from DNS results by solving Navier–Stokes equation. However, the original linear stability theories are either in spatial or in temporal framework. Results obtained from either of the routes are interchangeably used to 'explain' the observed transition in actual flows. Such migration from one framework to another implicitly assumes the convective nature of the disturbance field. However, instability can occur where disturbances display temporal or spatio-temporal instabilities.

This topic has a rich history, with a recent work by Sengupta *et al.* (2013) on mixed convection flows have indicated that added extra strain rates due to body forces causes spatial and temporal instabilities, simultaneously. While the spatial instability is via a viscous mechanism, the temporal instability is by a new inviscid mechanism described in that reference. In this reference, presented DNS results show that the temporal instability dominates over spatial instability. Thus the inviscid mechanism dominates over viscous mechanism of instability, when both of these are present.

There were significant contributions initially by Helmholtz, Kelvin and Rayleigh (Rayleigh 1880, Rayleigh 1887) using inviscid analysis, which are still very relevant. Original inviscid analysis was justified by noting that viscous diffusion is dissipative and can only be stabilizing and need not be included in instability studies, even when the equilibrium flow is obtained by including viscous actions. However, this approach can not explain flow instability over a flat plate. Despite this, when Heisenberg (Heisenberg 1924) reported solution of perturbation equations including viscous terms (by solving the famous Orr-Sommerfeld equation), it received scant attention, as it was difficult to accept viscous action causing instability. It has been explained in Sengupta (2012) that viscous action can provide a correct phase shift for positive feedback in destabilizing the flow. Researchers from the Göttingen school led by Prandtl (Prandtl 1935) developing viscous linear stability theory also experienced similar opposition, till the onset of instability was verified by the famous vibrating ribbon experiments of (Schubauer and Skramstad 1947). The theory predicted creation of viscous tuned oscillation, known now as the Tollmien and Schlichting waves. The governing equation for viscous linear instability was developed by Orr and Sommerfeld (Orr 1907, Sommerfeld 1908). Like the experiments of Reynolds, vibrating ribbon experiment also did not explain the full chain of events which leads to eventual transition.

Primary flow instability of attached boundary layers has been predicted with some success and the corresponding empirical transition prediction methodologies are routinely used in aircraft industry. For a flat plate placed in a stream with moderately low ambient disturbance level (turbulence intensity below 0.5%), flow transition is noted to take place at a distance x from the leading edge given by

$$Re_{tr} = \frac{U_\infty x}{\nu} = 3.5 \times 10^5 \text{ to } 10^6.$$

Onset of viscous instability for zero pressure gradient flow over a flat plate is predicted theoretically at $Re_{cr} = 519$ (based on displacement thickness). It is important to realize that instability and transition are not synonymous. The actual process of transition begins with onset of instability, which is not a unique process, while completion of transition depends upon many factors; deciding secondary, tertiary and nonlinear instabilities, which makes prediction of Reynolds number at transition (Re_{tr}) more difficult compared to predicting the Reynolds number at the onset of instability (Re_{cr}) predicting critical Reynolds number.

A search for complete understanding on the origin and nature of turbulence continues with the hope that the numerical solution of the full Navier–Stokes equation by DNS will provide insight. Published DNS results however suffers from a major drawback, with most of them not requiring any explicit forcing of the flow via a definitive and realistic input disturbance field. Most of these depend upon computational noises and/or 'random noise.' Thus, the implicit assumption in these so-called DNS that these *numerical noise* sources will produce a *turbulent flow* which is yet to be established rigorously, although there are some recent curious results obtained by erroneous simulations in Sayadi *et al.* (2013), which obtain the mean flow closely, see the critique in Bhaumik and Sengupta (2014).

It is well known that for a zero pressure gradient flat plate boundary layer, the skin friction for laminar flow is given by $C_f = \frac{1.328}{\sqrt{Re}}$, which at a Reynolds number of 10^7 works out as 0.00043 (Schlichting 1979) and the same profile drag increases to 0.0035 for the equivalent turbulent flow (Van Driest 1951). This is the rationale for trying to keep a flow laminar so that one can obtain an order of magnitude drag reduction. Such drag reductions are also realizable for an airfoil. According to Viken (1983), flow past an airfoil at a moderate Reynolds number that is fully turbulent without any separation displays a profile drag coefficient of nearly 0.0085, which can be reduced to 0.0010 if the flow over the airfoil is maintained fully laminar.

Thus, transition delay for flow over aircraft wings takes on added importance when it is realized that by resorting to transition delay techniques on wings alone about 10 to 12% drag reduction is feasible on a modern transport aircraft. The flow in laminar or in turbulent state is governed by the same Navier–Stokes equation (Goldstein 1938). Late stages of transition and fully turbulent flows are not amenable to analytical solution due to the intractable nonlinearity of the Navier–Stokes equation (Morkovin 1991). However, the onset of the transition process (Morkovin 1958, Morkovin 1990) is bedevilled by our inability to catalogue and quantify the background omnipresent disturbances. Significant understanding of linear instability of flow has been used in aircraft design to reduce drag by transition delay, which relate to suppressing disturbance growth during the receptivity and primary instability stages. Late stages of transition and turbulent states of flows can be controlled by active means, an area of research not used in any operational aircrafts.

Traditionally, laminar boundary layer is maintained by tailoring favourable pressure gradients by contouring airfoil surface by passive means, which is practical and attractive. The resultant section is known as the NLF airfoil and is under renewed development over the last few decades. Early efforts in designing airfoils are given in (Abbot and Doenhoff 1959), which resulted in the six digit NACA series airfoils. However, these are found to be useful only for low Reynolds numbers. These airfoils exhibit low drag only for a narrow range of C_l's designed corresponding to cruise condition. However, such sections are not optimal in other flight segments. A practical modern NLF airfoil has low cruise drag, along with optimal performance for high lift necessary during landing, take off and climb.

Instabilities of three-dimensional flows over aircraft wing is complicated and seemingly contradictory to instabilities noted for two-dimensional flows. Instabilities for three-dimensional flows are more complex, as these occur for streamwise flow, along with instabilities of the cross flow component of the velocity profile. Additionally, one also must consider the problem of leading edge contamination along the attachment line of a sweptback wing (Sengupta and Dipankar 2005).

In Sengupta (2012) and Sengupta *et al.* (2013), it has been shown that a favourable pressure gradient due to buoyancy effects creates inflectional velocity profile, which trigger temporal instability over and above the spatial instability. Three-dimensional flows over aircraft wings show similar mechanisms, in the form of cross flow instabilities in a favourable pressure gradient region for a swept wing. Detailed understanding of these spatial and temporal mechanisms in isolation and in conjunction is necessary. For flight at high subsonic and transonic speeds, conflicting requirement for aircraft design forces one to keep the sweep-back angle as small as possible to delay cross flow instability, while reducing wave drag by sweep-back.

9.2 Temporal and Spatial Instability

Fluid flow instabilities are treated traditionally, as if the disturbance growth is either in space or in time. This approach is for expediency of analysis only and not rooted to actual physical events. Ideally disturbance growth even in linear regime is spatio-temporal, as shown in Sengupta *et al.* (1994, 2006, 2006a), by using Bromwich contour integral method for solving the Orr-Sommerfeld equation and by solving the Navier-Stokes equation using dispersion relation preservation (DRP) methods in Sengupta and Bhaumik (2011). If the disturbance originating from a fixed location in space grows as it moves downstream, then this is called the convective instability. An example of this type of instability was shown for wall-bounded shear layer by the classic vibrating ribbon experiment of Schubauer and Skramstad (1947). This has shown the existence of viscous instability, predicted theoretically earlier in Heisenberg (1924), Tollmien (1931) and Schlichting (1933). This experiment displayed receptivity of wall-bounded shear layers to vortical disturbances inside a shear layer, while showing inadequacy of acoustic excitation from outside the shear layer in creating TS waves. It has been noted that the classical route of convective instability is not correct and transition is created by spatio-temporal wave-fronts in Sengupta and Bhaumik (2011), Bhaumik and Sengupta (2014) and Sengupta *et al.* (2012). Also temporal instability is often more dominant, when the flow displays both temporal and spatial disturbances to be present, as has been shown in Sengupta *et al.* (2013) for mixed convection flows.

9.3 Parallel Flow Approximation and Inviscid Instability Theorems

The essentials of instability studies indicate the need to obtain the equilibrium flow first, which is then studied for its stability. Obtaining equilibrium flow is not always easy and it was compounded by the lack of a general procedure to study its instability, without making simplifying assumptions. Currently, there are instability theories based on Navier-Stokes equation without making any assumption (Sengupta *et al.* 2003, Sengupta 2012). Also with the availability of DNS results, in Sengupta *et al.* (2010, 2011a, 2012) nonlinear universal instability modes have been shown to exist, which are constructed in terms of proper orthogonal decomposition (POD) modes. This approach also requires no assumptions to be made. Despite all these recent advances, in this chapter we will mostly focus upon results of linear theories.

Instability studies have been facilitated by considering only those shear layers which grow very slowly, so that the streamlines within the shear layer can be approximated to be parallel to each other, the so-called parallel flow approximation. Boundary layer flows under mild pressure gradient remain unseparated and approximated as quasi-parallel flow to study instability, even when the equilibrium flow is obtained without such restrictions. A major part of linear (and weakly nonlinear) instability theories has been developed for quasi-parallel flows, starting with the pioneering work of Helmholtz (1868), Kelvin (1871) and Rayleigh (1880) for inviscid instability analysis, as demonstrated next. To study the stability of 2D parallel flow supporting 2D disturbance field, one considers the total flow field to be given by

$$u(x, y, t) = U(y) + \epsilon u'(x, y, t) \tag{9.1}$$

$$v(x, y, t) = \epsilon v'(x, y, t) \tag{9.2}$$

$$p(x, y, t) = P(x, y) + \epsilon p'(x, y, t). \tag{9.3}$$

The space-time dependence of the perturbation field is without any restrictions. In the absence of body force for constant density flows, governing equations for the disturbance field in small perturbation inviscid analysis is obtained from the linearized perturbation equations

$$\frac{\partial u'}{\partial x} + \frac{\partial v'}{\partial y} = 0 \tag{9.4}$$

$$\frac{\partial u'}{\partial t} + U\frac{\partial u'}{\partial x} + v'\frac{dU}{dy} = -\frac{1}{\rho}\left(\frac{\partial p'}{\partial x}\right) \tag{9.5}$$

$$\frac{\partial v'}{\partial t} + U\frac{\partial v'}{\partial x} = -\frac{1}{\rho}\left(\frac{\partial p'}{\partial y}\right). \tag{9.6}$$

One represents the perturbation quantities by their Laplace–Fourier transform via

$$u'(x,y,t) = \int \bar{u}(y;\alpha,\omega_0)\, e^{i(\alpha x - \omega_0 t)} d\alpha\, d\omega_0 \tag{9.7}$$

$$v'(x,y,t) = \int \bar{v}(y;\alpha,\omega_0)\, e^{i(\alpha x - \omega_0 t)} d\alpha\, d\omega_0 \tag{9.8}$$

$$\frac{p'(x,y,t)}{\rho} = \int \bar{p}(y;\alpha,\omega_0)\, e^{i(\alpha x - \omega_0 t)} d\alpha\, d\omega_0. \tag{9.9}$$

In these, α and ω_0 are the wavenumber and circular frequency, respectively. One can use Equations (9.7)–(9.9) in Equations (9.4)–(9.6) and eliminate $\bar{u}$ and $\bar{p}$ from these equations to get a single differential equation for $\bar{v}$ as

$$\left(U - \frac{\omega_0}{\alpha}\right)\left(\frac{d^2\bar{v}}{dy^2} - \alpha^2\bar{v}\right) - \frac{d^2U}{dy^2}\bar{v} = 0. \tag{9.10}$$

This is the **Rayleigh's stability equation** used for inviscid stability analysis. For this homogeneous equation, one can solve Equation (9.10) subject to the homogeneous boundary conditions for $\bar{v}$, for the eigenvalue problem. As noted before, for a fluid dynamical system admitting spatio-temporal growth of disturbances, both α and ω_0 are complex. However for ease of analysis, we study temporal growth by considering α as real and ω_0 as complex. If we write $c = \omega_0/\alpha$, then the complex phase speed ($= c_r + ic_i$) will determine the stability obtained as an eigenvalue of the equation given by

$$(U - c)\left(\frac{d^2\bar{v}}{dy^2} - \alpha^2\bar{v}\right) - \frac{d^2U}{dy^2}\bar{v} = 0. \tag{9.11}$$

The criterion for temporal instability then becomes: *There exists a solution with $c_i > 0$ for some positive α.*

9.3.1 Inviscid Instability Mechanism

Let $\bar{v}^*$ be a complex conjugate of $\bar{v}$, so that $\bar{v}\,\bar{v}^* = |\bar{v}|^2$. Multiplying Equation (9.11) by $\bar{v}^*$ and integrating over the possible limit (say, $-\infty$ to $+\infty$), we get

$$\int_{-\infty}^{+\infty} \bar{v}^*\left[\frac{d^2\bar{v}}{dy^2} - \alpha^2\bar{v} - \frac{U''}{U-c}\bar{v}\right] dy = 0. \tag{9.12}$$

This equation is nonsingular, as we are looking for solutions with $c_i \neq 0$, such that the denominator of the third term does not vanish. Following the analysis in Sengupta (2012), inviscid instability is indicated, if U'' changes sign in the interior of the domain of investigation. Let $U'' = 0$ for some point at $y = y_s$, within the limits of integration, then this is the inflection point, which leads to the following:

Rayleigh's Inflection Point Theorem: A *necessary condition* for instability is that the basic velocity profile should have an inflection point.

By this theorem, at the inflection point within a shear layer, the local velocity is given by U_s and the second derivative of the mean flow vanishes, i.e. $U'' = 0$. A stronger version of Rayleigh's theorem was given later in Fjϕrtoft (1950).

Fjϕrtoft's Theorem: A *necessary condition* for instability is that $U''(U - U_s)$ is less than zero somewhere in the flow field.

Both these theorems are necessary and do not provide any sufficient condition for instability. A fundamental difference between flows having an inflection point (such as in free shear layer, jets and wakes; cross flow component of some three-dimensional boundary layers) and flows without inflection points (as in wall-bounded flows in channel or in boundary layers) exists. Flows with inflection points are susceptible to temporal instabilities at very low Reynolds numbers (Betchov and Criminale 1967, Drazin and Reid 1981). In Sengupta *et al.* (2013), it is noted that temporal instability is also present in many flows with heat transfer, with the above theorems revised for the mixed convection flow. If a flow shows propensity for both viscous spatial and inviscid temporal instabilities, it is the latter which dominates the flow evolution. This has been established for both adiabatic and isothermal wall condition on a horizontal plate placed in a uniform flow.

9.4 Viscous Instability of Parallel Flows

Initially it was incorrectly thought that viscous action is strictly dissipative and inviscid instability mechanism was studied with Rayleigh's equation. However, this does not explain instability in many flows including the flow over a flat plate. This prompted many researchers to look for viscous instability, once again using parallel flow assumption. Flow past flat plate without heat transfer indicated viscous diffusion to be destabilizing. This is discussed next for the stability of 3D mean profile given by

$$U = U(y)\,\hat{i} \ \text{ and } \ W = W(y)\,\hat{k}. \tag{9.13}$$

The 3D mean flow is considered to be parallel and depends on wall-normal distance only. In this theory, equilibrium flow is usually obtained using the thin shear layer approximation or directly from the Navier–Stokes equation. However, stability of the fluid dynamical system is studied from time dependent Navier–Stokes equation. We obtain the disturbance equations in the Cartesian coordinate system by

$$\frac{\partial u}{\partial t} + u\frac{\partial u}{\partial x} + v\frac{\partial u}{\partial y} + w\frac{\partial u}{\partial z} = -\frac{\partial p}{\partial x} + \frac{1}{Re}\nabla^2 u \tag{9.14}$$

$$\frac{\partial v}{\partial t} + u\frac{\partial v}{\partial x} + v\frac{\partial v}{\partial y} + w\frac{\partial v}{\partial z} = -\frac{\partial p}{\partial y} + \frac{1}{Re}\nabla^2 v \tag{9.15}$$

$$\frac{\partial w}{\partial t} + u\frac{\partial w}{\partial x} + v\frac{\partial w}{\partial y} + w\frac{\partial w}{\partial z} = -\frac{\partial p}{\partial z} + \frac{1}{Re}\nabla^2 w \tag{9.16}$$

$$\frac{\partial u}{\partial x} + \frac{\partial v}{\partial y} + \frac{\partial w}{\partial z} = 0. \tag{9.17}$$

Above equations are written in nondimensional form with shear layer edge velocity, U_e as the velocity scale and L as the length scale, so that $Re = U_e L/\nu$ is the Reynolds number. For linearized spatial stability analysis of steady mean flow, we split all flow quantities q, into a mean (Q) and an unsteady disturbance term ($\epsilon q'$), which is an order of magnitude smaller than the mean quantities, i.e.

$$q(x, y, z, t) = Q(x, y, z) + \epsilon\, q'(x, y, z, t). \tag{9.18}$$

The mean velocity field is further assumed as parallel/ quasi-parallel, i.e. $V = 0$ and other components as given in Equation (9.13). The splitting of variables as in Equation (9.18) is used in Equations (9.14)–(9.17) and the $O(\epsilon)$ terms are collated to get the disturbance equations. For normal mode stability analysis, as the flow is inhomogeneous in the wall-normal direction, one expands disturbance quantities as

$$\lfloor u', v', w', p' \rfloor^T = \lfloor f(y), \varphi(y), h(y), \pi(y) \rfloor^T e^{i(\alpha x + \beta z - \omega_0 t)}. \tag{9.19}$$

The disturbance amplitudes f, ϕ, h and π are complex functions and ω_0 is the dimensionless circular frequency, $(= \omega_0^* L/U_e)$. When Equation (9.19) is substituted in $O(\epsilon)$ equations, following ODEs result

$$i[\alpha U + \beta W - \omega_0]f + U'\phi = -i\alpha\pi + \frac{1}{Re}[f'' - (\alpha^2 + \beta^2)f] \tag{9.20}$$

$$i[\alpha U + \beta W - \omega_0]\phi = -\pi' + \frac{1}{Re}[\phi'' - (\alpha^2 + \beta^2)\phi] \tag{9.21}$$

$$i[\alpha U + \beta W - \omega_0]h + W'\phi = -i\beta\pi + \frac{1}{Re}[h'' - (\alpha^2 + \beta^2)h] \tag{9.22}$$

$$i(\alpha f + \beta h) + \phi' = 0, \tag{9.23}$$

with primes denoting differentiation with respect to y. Above homogeneous equations are solved subject to homogeneous boundary conditions in eigenvalue analysis. At the wall, no-slip boundary conditions apply

$$f(0) = \phi(0) = h(0) = 0 \tag{9.24}$$

At the free stream, disturbance velocity decays to zero, i.e.

$$f(y),\ \phi(y),\ h(y) \to 0 \quad \text{as } y \to \infty. \tag{9.25}$$

Nontrivial solutions of Equations (9.20)–(9.23), subject to boundary conditions in (9.24) and (9.25), exist for particular combinations of the parameters α, β, ω_0 and Re, which produces the dispersion relation

$$g(\alpha, \beta, \omega, Re) = 0. \tag{9.26}$$

Equations (9.20)–(9.23) can be replaced by the single equation for ϕ as

$$\phi^{iv} - 2(\alpha^2 + \beta^2)\phi'' + (\alpha^2 + \beta^2)^2\phi$$
$$= i\, Re[(\alpha U + \beta W - \omega_0)[\phi'' - (\alpha^2 + \beta^2)\phi] - (\alpha U'' + \beta W'')\phi]. \tag{9.27}$$

This is the well-known Orr–Sommerfeld equation, which for 2D disturbance field of a 2D mean flow simplifies to

$$\phi^{iv} - 2\alpha^2\phi'' + \alpha^4\phi = i\, Re[(\alpha U - \omega_0)[\phi'' - \alpha^2\phi] - \alpha U''\phi]. \tag{9.28}$$

The Orr–Sommerfeld equation is a fourth order ordinary differential equation and has the same form whether the mean flow is 3D or 2D.

9.4.1 Temporal and Spatial Amplification of Disturbances

If α, β and ω_0 are all real, then from Equation (9.23) one notes the disturbance to propagate with constant amplitude at all times. The case where the disturbance changes with time is dealt in *temporal amplification theory* and in *spatial amplification theory* disturbance grow in space. If α, β and ω_0 are complex, then the disturbances grow in both space and time and are dealt with as *spatio-temporal growth of disturbances*, which should be ideally studied – as explained in Sengupta (2012).

In temporal amplification theory, $\omega_0 = \omega_r + i\omega_i$ and (α, β) being real, the disturbance field is given by

$$q'(x, y, z, t) = [\hat{q}(y)\, e^{\omega_i t}]\, e^{i(\alpha x + \beta z - \omega_r t)}. \tag{9.29}$$

The magnitude of the wavenumber vector is $\bar{\alpha} = (\alpha^2 + \beta^2)^{1/2}$ and its inclination to x-axis is given by $\psi = \tan^{-1}(\beta/\alpha)$. The phase speed of the disturbance field is given by $c_{ph} = \omega_r/\bar{\alpha}$. If A represents the magnitude of the disturbance at a particular height y given by the quantity in square bracket in Equation (9.29), then from above

$$\frac{1}{A}\frac{dA}{dt} = \omega_i. \tag{9.30}$$

Solution of Equation (9.30) is $A = A_0\, e^{\omega_i t}$, with ω_i as the *amplification rate*, when $\omega_i > 0$, otherwise the disturbance attenuates with time for $\omega_i < 0$.

In spatial amplification theory, ω_0 is real and α and β are complex, i.e. $\alpha = \alpha_r + i\alpha_i$ and $\beta = \beta_r + i\beta_i$, so that the disturbance is given by

$$q'(x, y, z, t) = [\hat{q}(y)e^{-(\alpha_i x + \beta_i z)}]\; e^{i(\alpha_r x + \beta_r z - \omega_0 t)}. \tag{9.31}$$

If one defines

$$\bar{\alpha}_r = [\alpha_r^2 + \beta_r^2]^{1/2} \quad \text{and} \quad \psi = \tan^{-1}(\beta_r/\alpha_r), \tag{9.32}$$

then the phase speed of the disturbance field is given by $c_{ph} = \omega_0/\bar{\alpha}_r$. Additionally, if one defines $\bar{\alpha}_i = [\alpha_i^2 + \beta_i^2]^{1/2}$ and $\bar{\psi} = \tan^{-1}(\beta_i/\alpha_i)$, then one can define two new directions, $\tilde{x}$

along $\bar{\alpha}_r$ and $\bar{x}$ along $\bar{\alpha}_i$ and rewrite Equation (9.31) as

$$q'(x, y, z, t) = \hat{q}(y)\, e^{-\bar{\alpha}_i \bar{x}}\, e^{i(\bar{\alpha}_r \bar{x} - \omega_0 t)} \tag{9.33}$$

Thus, one can similarly write a spatial amplification rate in the particular direction given by $\bar{\psi}$ as

$$\frac{1}{A}\frac{dA}{d\bar{x}} = -\bar{\alpha}_i. \tag{9.34}$$

The amplification rates are different in different directions and different $\bar{\alpha}_i$'s are functions of ψ. For 3D waves, spatial theory comes with the added complication of wave orientation angle ψ to be defined, which is different from the amplification direction $\bar{\psi}$. If $\bar{\alpha}_i < 0$, then we have *instability*.

The relationship between temporal and spatial theories can be obtained by introducing the concept of group velocity from the dispersion relation as

$$\omega_0 = \omega_0(\alpha, \beta, Re, \ldots). \tag{9.35}$$

For 3D disturbance field, the group velocity components in x- and z-directions are

$$\vec{V}_g = \left(\frac{\partial \omega_0}{\partial \alpha}, \frac{\partial \omega_0}{\partial \beta}\right). \tag{9.36}$$

For parallel flows, spatial and temporal amplification rates are related by

$$\frac{d}{dt} = \vec{V}_g \frac{d}{d\bar{x}}, \tag{9.37}$$

where $\bar{x}$ is chosen in the direction of $\vec{V}_g$. Consequently

$$\bar{\alpha}_i = -\frac{\omega_i}{|\vec{V}_g|^2}\, \vec{V}_g \tag{9.38}$$

and the direction of $\bar{\alpha}_i$ is obtained from

$$\bar{\psi} = \tan^{-1}\left[\frac{\partial \omega_r / \partial \beta}{\partial \omega_r / \partial \alpha}\right] \tag{9.39}$$

For 2D flows, $\partial \omega_i / \partial \alpha_i$ can be approximated by noting that ω_i decreases from its temporal value to zero in the spatial theory, as α_i goes from zero to its value obtained in the spatial theory. If the amplification rates are small (as in linear theory), then the above variations are linear and thus $\partial \omega_i / \partial \alpha_i \simeq -\omega_i / \alpha_i$. Therefore, $[V_g]_x = -\omega_i / \alpha_i$.

9.5 Instability Analysis from the Solution of the Orr–Sommerfeld Equation

For 2D spatial instability of 2D mean flow, disturbance quantities are given by Equation (9.29) with $\beta = 0$. For a fixed *Re*, a complex α is sought with the shear layer excited for prescribed ω_0. If *Re* is defined in terms of the displacement thickness δ^* by $Re = U_e \delta^* / \nu$, then the results

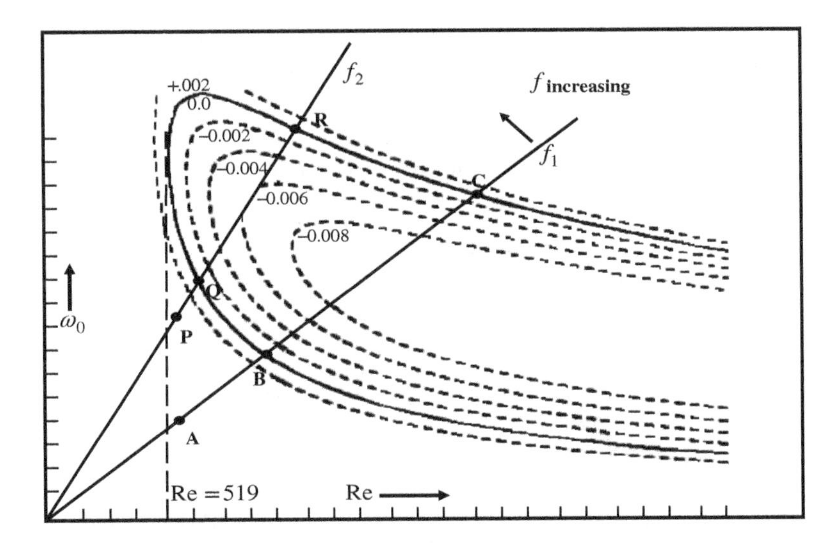

Figure 9.1 Contour plots of asymptotic growth rate in the Reynolds number-circular frequency plane

are plotted as contours of constant amplification rates α_i in the (Re, ω_0)-plane, as shown in Figure 9.1 for zero pressure gradient flow.

The solid line in the figure corresponds to $\alpha_i = 0$ and represents the condition of neutral stability and called the *neutral curve*. The area inside the neutral curve corresponds to $\alpha_i < 0$ and represents flow instability for these parameter combinations. The area outside the *neutral curve* corresponds to stable disturbances. The leftmost point of the *neutral curve* occurs for $Re = 519.23$ and implies the flow to be spatially stable below this value of Re for any frequency. Hence, this value represents the *critical Reynolds number*. The lower part of the *neutral curve* is the *branch I*, while the upper part of the neutral curve is called the *branch II*. We also note that the neutral curve is the same for both the spatial and temporal linear stability studies.

In Figure 9.1, ω_0 is nondimensional, based on the length (δ^*) and velocity (U_e) scales. Thus

$$\omega_0 = 2\pi f_i \frac{\delta^*}{U_e} = \left(\frac{2\pi\nu\, f_i}{U_e^2}\right)\frac{U_e\delta^*}{\nu} = F\, Re. \tag{9.40}$$

Here F is the nondimensional frequency of excitation. Thus, a constant physical frequency (f_i) line in the (Re, ω_0)-plane is a straight line through the origin and different constant-frequency lines have slopes $F = \omega_0/Re$. Figure 9.1 is interpreted in the following manner: Consider a physical frequency of excitation f_1 with the exciter at point A outside the neutral curve, created disturbance will decay according to spatial theory following the constant-frequency line, ABC. Once the disturbance crosses B on *branch I*, there onwards disturbance will amplify as it propagates downstream till the disturbance reaches *branch II* at C. Thenceforth, the disturbance is damped according to linear spatial theory, giving rise to a wave-packet which will pulsate at f_1. Thus, in this framework with parallel flow approximation, disturbances adjust locally to prevalent wave properties.

The validation of spatial instability theory came from experiments in Schubauer and Skramstad (1947), where background disturbances were minimized and 2D controlled disturbances were introduced in the flow. Measurements were done at fixed x by hot wire, whose signal was fed to oscillograph. From oscillogram trace, transition was indicated at any station when the signal represented wide-band fluctuations. Without explicit excitation, location of transition would progressively shift downstream by reducing the turbulence intensity (Tu) all the way down to 0.08%, where Tu is defined in terms of fluctuating components of velocity by

$$Tu = \frac{1}{U_\infty}\left[\frac{(u'^2 + v'^2 + w'^2)}{3}\right]^{1/2}. \tag{9.41}$$

Further reduction in Tu had no effects on the position of the *natural transition*. Figure 9.2 taken from Schubauer and Skramstad (1947) reveals *natural transition*, in the absence of explicit excitation, for a flow speed of 53 ft/sec.

First few frames of Figure 9.2 have been amplified in the figure for ease of viewing. Oscillations noted up to 9 ft from the leading edge of the flat plate is not purely *monochromatic*. Monochromatic TS waves were created by perfectly controlled 2D disturbances by electromagnetically vibrating a ribbon inside the shear layer. In this experiment, the neutral curve was charted out and compared with Schlichting's theoretical results. This was considered very good and thought to provide complete justification of the viscous stability theory.

9.5.1 *Local and Total Amplification of Disturbances*

The Reynolds number at an observed transition location (defined as a location where the intermittency factor is about 0.1, i.e. the flow is 10% of the time turbulent and the rest of

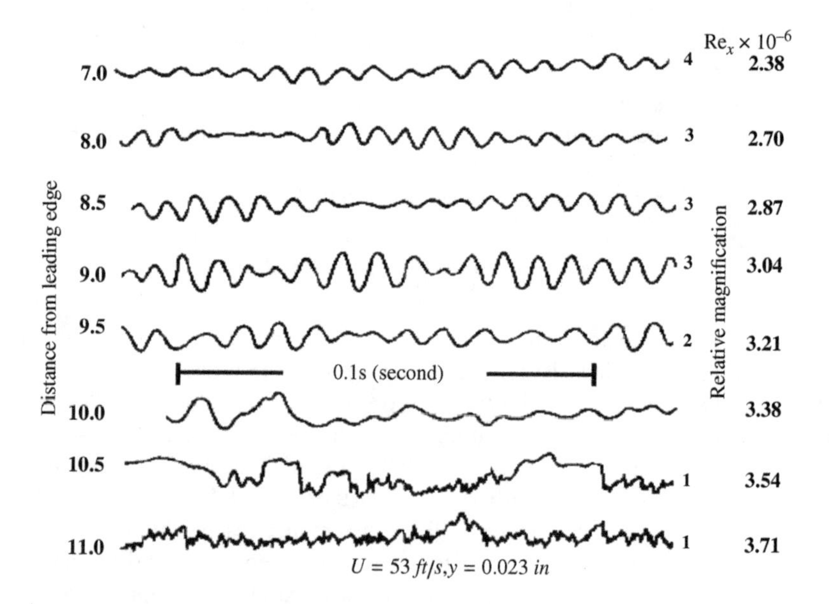

Figure 9.2 Hot-wire oscillogram traces showing natural transition from laminar to turbulent flow on a flat plate. *Source:* Schubauer and Skramstad (1947). Reproduced with perrmission from the the American Institute of Aeronautics and Astronautics, Inc.

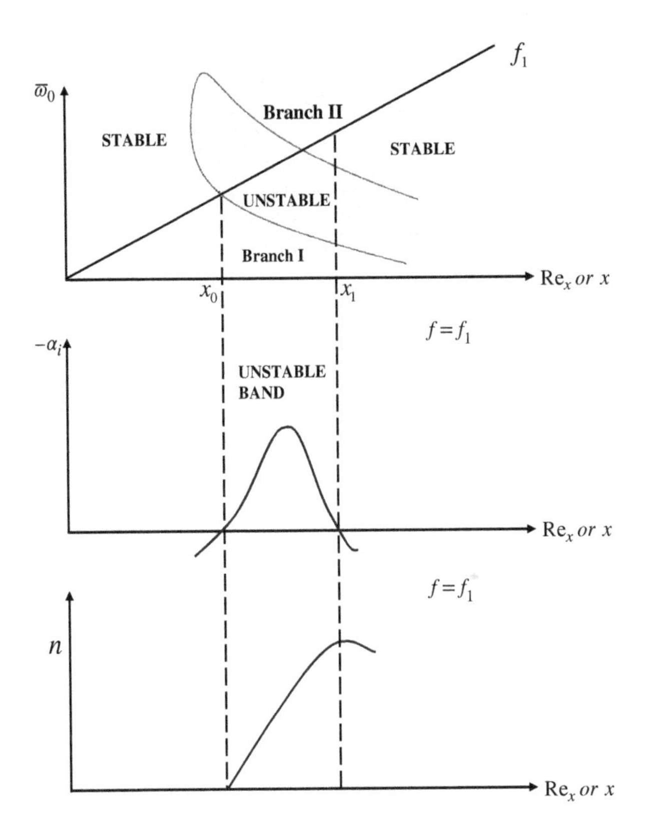

Figure 9.3 Sketch of local and total amplification of the disturbance field

the time it is laminar) for Blasius boundary layer is about 3.5×10^5, which corresponds to $Re_{\delta^*} = 950$. The distance between the points of instability and transition depends on the degree of amplification and the kind of disturbance present with the oncoming flow. This has been used to empirically estimate local and total amplification of disturbances, described in Arnal (1984) for 2D incompressible flows. On the top panel of Figure 9.3, the neutral curve in the (Re, ω_0)-plane is shown with attention focused on a frequency f_1, by following the ray emanating from the origin. According to this figure, this disturbance will decay up to $x = x_0$ from the exciter and thereafter it will grow till $x = x_1$ located on the branch II.

The amplification rate $(-\alpha_i)$ suffered by disturbances inside neutral curve is shown in the middle panel of Figure 9.3, with negative values plotted along the positive ordinate. For 2D disturbances in 2D mean flow the amplification rate is expressed as

$$\frac{1}{A}\frac{dA}{dx} = -\alpha_i. \tag{9.42}$$

If disturbance amplitude at $x = x_0$ is A_0, then the amplitude at x is given by

$$A(x) = A_0 e^{-\int_{x_0}^{x} \alpha_i(x)\,dx}. \tag{9.43}$$

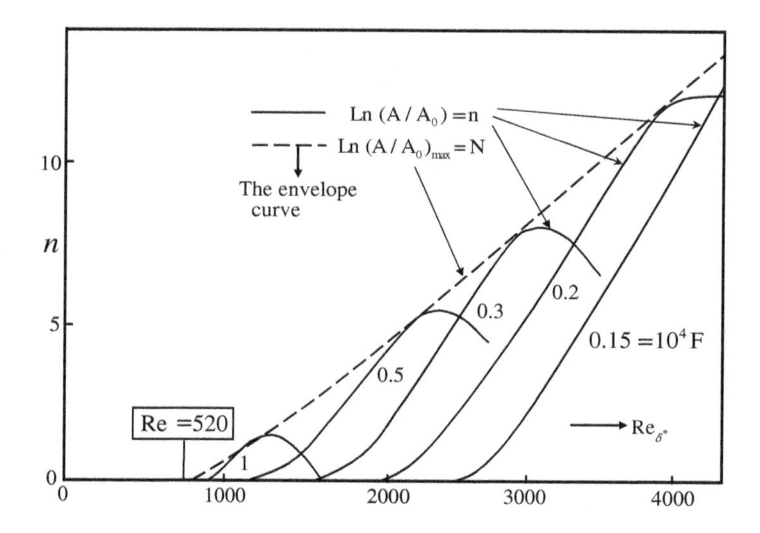

Figure 9.4 Total amplification rates for different frequencies for the Blasius profile. *Source:* Arnal (1984). Copyright NATO-CSO

In the lower panel of Figure 9.3, the exponent n is shown as a function of x where

$$n = -\int_{x_0}^{x} \alpha_i(x)\, dx. \tag{9.44}$$

This factor $n(x)$ is calculated for a particular frequency from x_0 onwards, which can be repeated for range of frequencies and a composite plot is shown in Figure 9.4.

The total amplification suffered by each frequency follows the solid lines in Figure 9.4, while the envelope of these curves shown by the dotted line represents maximum amplification suffered by different frequencies after entering the unstable zone and is designated by

$$N = \left[\text{Ln}\frac{A}{A_0}\right]_{max} = \max_{f}\left[\int_{x_0}^{x} -\alpha_i(x)\, dx\right]. \tag{9.45}$$

Different frequencies become unstable entering the *branch I* at different x_0. Also, lower frequencies suffer instability for longer distances and amplify the most. In this interpretation, it is assumed that disturbance switch frequency as it propagates downstream. As we will see soon that this is merely a model, which chooses different values of N to predict transition. Actual transition in many cases do not even depend upon TS wave-packet amplification, instead Orr-Sommerfeld operators support spatio-temporal wave-fronts which are responsible for transition (Sengupta and Bhaumik 2011, Sengupta 2012, Bhaumik and Sengupta 2014).

9.5.2 *Effects of the Mean Flow Pressure Gradient*

The total amplification tracking method of previous subsection has been used to study transition for flows with pressure gradient using quasi-parallel flow assumption. Arnal (Arnal 1984) used equilibrium solution of Falkner-Skan family for this purpose. In Figure 9.5, the neutral curves

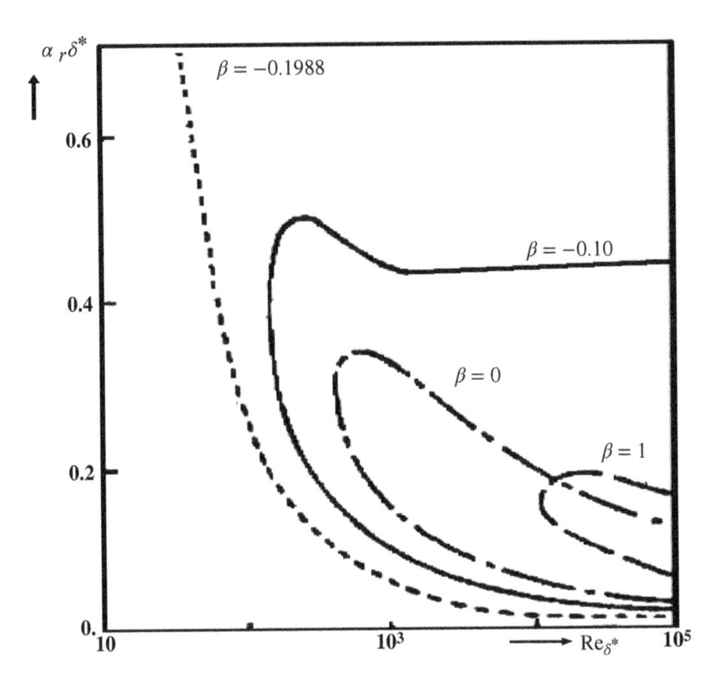

Figure 9.5 Neutral curves of some shear layer velocity profiles shown in Figure 7.7

for the velocity profiles of Figure 7.7 are compared. The main trend is that as H increases or β decreases, Re_{cr} decreases and for flows with $\frac{dp}{dx} > 0$, Re_{cr} is significantly lower.

The fact that such profiles are prone to instability earlier is the reason for calling these adverse pressure gradient flows. Along with this, the growth rate of unstable waves becomes larger as the shape factor increases. Also, for decelerated flows, the neutral curve does not close, as seen for the $H = 2.80$ ($\beta = -0.10$) case. For separated profiles, there is no *branch II* of the neutral curve at all. The stagnation region on a body is a site of high stability, as can be seen for the Hiemenz flow (stagnation point flow profile with $\beta = 1$) which is more stable than the Blasius flow ($\beta = 0$).

The envelope curve method of Figure 9.4 for a β can be repeated for other pressure gradients and in Figure 9.6, the N-curves are shown for a few representative adverse pressure gradients and compared with the Blasius profile ($H = 2.591$) showing also the incipient separated flow ($H = 4.032$). As H increases, Re_{cr} decreases. The slope of these envelope curves (dN/dRe_{cr}) also increases with H. The shape factor H is a very important parameter which indicates flow separation, as well as indicate change in stability property of the flow.

Thus, by plotting H variation in the streamwise direction, one obtains indication of flow transition, as at transition onset H variation exhibits a sudden negative slope. The results in Figure 9.7 from Arnal (1984) show that flow transition in an adverse pressure gradient shear layer always precedes theoretical location of laminar separation, indicated by vertical bars in the frames.

Frame (A) of Figure 9.7 represents a zero pressure gradient flow which does not separate theoretically. Sudden decrease of H is an indicator of flow transition. In frames (B) through (F), the pressure gradient is progressively more adverse and consequently the theoretical

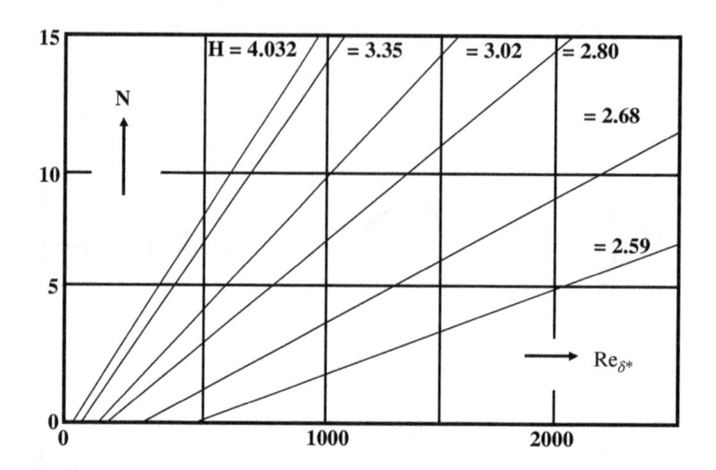

Figure 9.6 Envelope curves for total amplification for Falkner–Skan similarity profiles. *Source:* Arnal (1984). Copyright NATO-CSO

location of the laminar separation point moves forward and the H variation becomes sharper with larger fall in the value, at the location of transition. For severe adverse pressure gradient flow, point of separation is almost coincident with point of transition and this input is often used in severe adverse pressure gradient flow calculations. However, this is not always the case for predicting transition location. For example, at low Re, flow separates first and then transition occurs subsequently. Separated flow being very unstable, the resulting transition is very quick – one has practically a free shear layer with a point of inflection. Turbulent flows withstand much higher adverse pressure gradients, and such separated turbulent flows reattach to the wall, forming a separation bubble. Beyond the reattachment point, the flow is turbulent

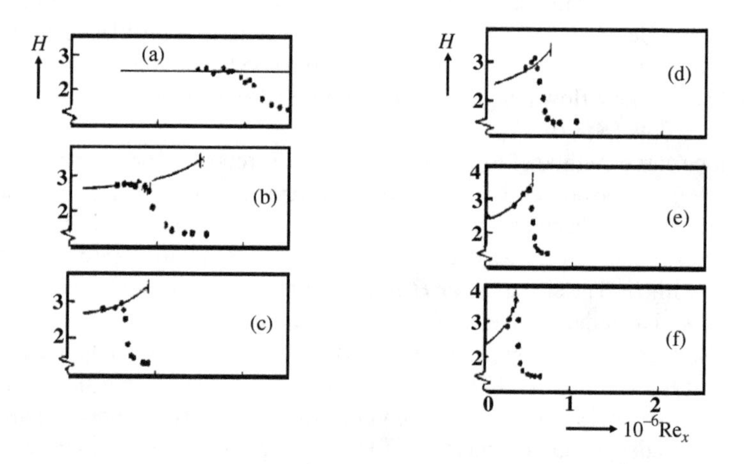

Figure 9.7 Effects of a positive pressure gradient on separation and transition. Discrete symbols are experimental data points and continuous lines are laminar calculations. The vertical bar in the laminar calculation indicates the location of the point of laminar separation. *Source:* Arnal (1984). Copyright NATO-CSO

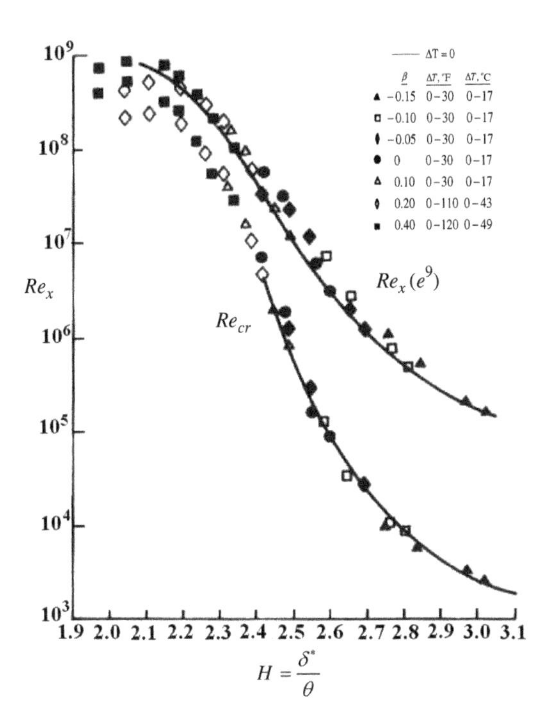

Figure 9.8(a) Correlation of critical and transitional Reynolds number versus boundary layer shape factor. *Source:* Adapted from Wazzan *et al.* (1981)

and such flows are observed for uniform flow past circular cylinders in Re ranges of 3×10^5 to 3×10^6, on the leeward side. The flow after turbulent reattachment separates again and remains fully separated. Separation bubbles are also seen in flow past airfoils downstream of the suction peak at moderate Reynolds numbers and at nonzero angles of attack. There is an intense interaction of the viscous flow in the shear layer with the outer inviscid flow in such cases, and boundary layer approximation fails.

Importance of H has been further utilized in predicting transition in Wazzan *et al.* (1981), by the H-Rex method where by noting similar correlations to exist for parameters such as pressure gradient, suction/blowing, heating/cooling, etc., if $(Re_{\delta^*})_{crit}$ is plotted in terms of H at transition. In Figure 9.8(a), this correlation is shown by the upper curve obtained by e^9 method (described in next subsection). Lower curve gives Re_{crit} as computed by converting Re_{δ^*} to Re_x. To use Figure 9.8(a), computed $H(x)$ by any laminar boundary layer code is used to predict transition, when the local $H(x)$ intersects the upper curve. This curve is also given by the following empirical fit

$$\log_{10}(Re_{x,tr}) = -40.4557 + 64.8066H - 26.7538H^2 + 3.3819H^3, \tag{9.46}$$

which is valid in the range $2.1 < H < 2.8$.

Reader's attention is drawn to the fact that in Figure 9.8(a), the ordinate is plotted in logarithmic scale and the presented data in the figure may suggest a wrong inference that with increase in value of H (i.e. the flow suffering more adverse pressure gradient) the different between the critical and transitional Reynolds number also increases. This is indeed not

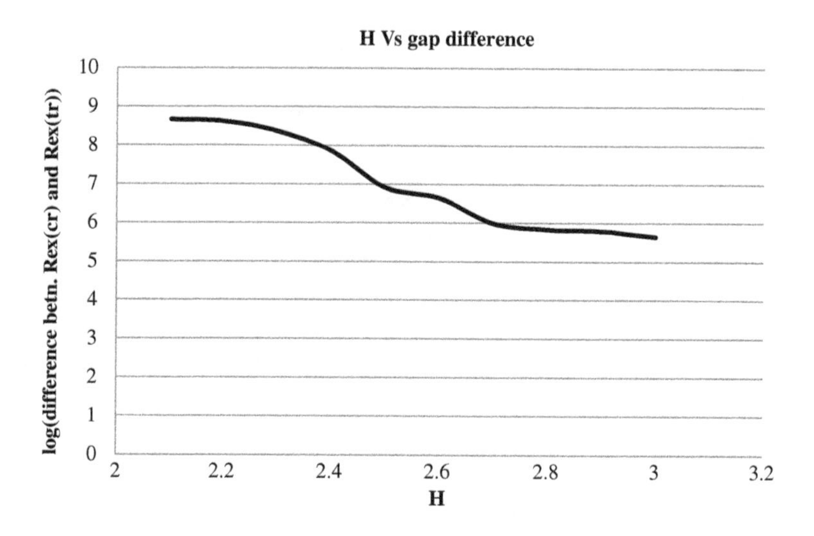

Figure 9.8(b) Difference in Reynolds numbers (critical and transitional values) plotted as function of boundary layer shape factor

the case. With adverse pressure gradient increasing, the flow disturbance amplifies faster and both the critical and transitional Reynolds numbers are lower. This is amply justified in Figure 9.8(b), where the difference between these Reynolds numbers is plotted as a function of H. The exact values are tabulated in Table 9.1, as quantified data extracted from Figure 9.8(a).

9.5.3 Transition Prediction Based on Stability Calculation: e^N Method

Following the experiments of Schubauer and Skramstad (1947), efforts have been made to link the stability theory with transition. Michel (Michel 1952) reported first that his compiled data showed the transition to be indicated when the total amplification of TS waves corresponded to $A/A_0 \approx 10^4$, where A_0 is the disturbance amplitude at the onset of instability. This motivated Smith and Gamberoni (1956) and Van Ingen (1956) to use viscous temporal theory to show that at transition the total amplification is given by

$$\frac{A}{A_0} = exp\left[\int_{x_{cr}}^{x_{tr}} \alpha\ c_i\ dt\right]. \tag{9.47}$$

Note that the integration ranges from the critical point (x_{cr}) to the point of transition (x_{tr}). As the right-hand side roughly equals e^9 at transition (Smith and Gamberoni 1956), this method is also known as the e^9 method. In Van Ingen (1956), value of the exponent was reported to be between 7 and 8. Disparity of the exponent arises due to different amplitude and spectrum of background disturbances. Thus, it is now known as the e^N method, where the exponent value depends on other factors affecting transition. Later Jaffe *et al.* (1970) performed spatial stability calculations and reported the exponent as 10. Performing 2D Navier-Stokes solution, Sengupta *et al.* (2012) have shown that for transition the TS wave-packet remains stationary; instead

Table 9.1 The variation of the difference between $(\text{Re}_x)_{critical}$ and $(\text{Re}_x)_{transition}$ with the shape factor H is shown below

H	Actual difference	Log (difference)
2.1	460000000	8.662758
2.2	415000000	8.618048
2.3	233000000	8.367356
2.4	72000000	7.857332
2.5	8220000	6.914872
2.6	4400000	6.643453
2.7	973000	5.988113
2.8	684000	5.835056
2.9	624000	5.795185
3	445000	5.64836

the spatio-temporal wave-front (STWF) nonlinearly grows to create transition, while the TS wave-packet remains unaffected. The receptivity calculations clearly shown that nonlinearity and spot formation can occur when the above amplitude ratio takes an exponent value of N less than 3 even for high amplitude excitation cases, where TS wave-packet is not formed (so-called bypass transition)! Thus, e^N method should be strictly viewed as an empirical tool, although some researchers still prefers to call it a transition prediction method based on physical foundation.

Experiments on transition for 2D boundary layers have revealed that the onset process is dominated by low frequency disturbance, when the free stream turbulence (FST) level is low and the later stages of transition process are dominated by nonlinearity. This phase spans a very small streamwise stretch and this observation has been used as a justification for using linear stability analysis to determine the extent of transitional flow. This is cited as the success of linear stability based transition prediction methods. However, it must be emphasized that nonlinear, nonparallel and multi-modal interaction processes are more important in all cases.

Despite the so-called success of linear stability theory in predicting transition, it is important to underscore its limitations. This should help one to look for other mechanisms which play bigger role in transition than might have been suspected for a long time. The envelope method does not require any information about the frequency spectrum of oncoming flow disturbances, independent of A_0. Such a procedure always predicts transition based on lower frequency events, the reason can be found in the nonlinear receptivity calculations in Sengupta *et al.* (2012). If oncoming disturbance does not contain low frequency events, then will the nonlinearity create lower frequencies first, which will subsequently amplify? In such a reverse scenario, transitional region will be prolonged as compared to *natural* cases and transition prediction based on normal mode analysis will be inadequate. Also, it is noted that e^N method works for low FST levels. One should use full Navier-Stokes equation based high fidelity solvers to predict transition for high FST cases. Some low accuracy method results have been published in recent times by Wu and Moin (2010) and such results need critical scrutiny before the results can be accepted. As the transition process does not show distinct presence of TS wave-packet, this is considered a case of bypass transition, as described in Sengupta (2012). Next, we discuss an empirical method based on e^N method developed for low-level FST.

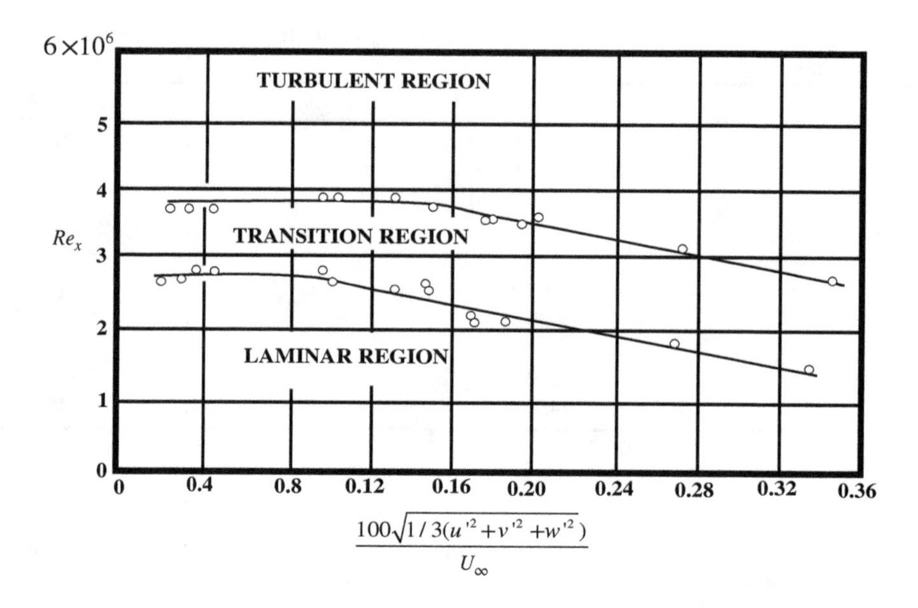

Figure 9.9 Effect of free stream turbulence on transition for flat plate boundary layers. *Source:* Schubauer and Skramstad (1947). Reproduced with perrmission from the American Institute of Aeronautics and Astronautics, Inc.

9.5.4 *Effects of FST*

The *success* of the e^N method depends on experimental results obtained in low Tu tunnels. The effects of Tu is strong on transition. For example, in the experiments of Schubauer and Skramstad (1947), Re_{tr} dropped by 50% when Tu was increased to 0.35% from its highest value for $Tu = 0.04\%$, as shown in Figure 9.9, reproduced from Schubauer and Skramstad (1947).

As Re_{tr} decreases rapidly with Tu greater than 0.10%, this variation could not be explained by the e^N method. Mack (Mack 1977) has suggested the following empirical correlation, linking Tu with N of the e^N method in the range $0.0007 \leq Tu \leq 0.00298$

$$N = -8.43 - 2.4\ \mathrm{Ln}(Tu). \tag{9.48}$$

This correlation is based on experimental results of Dryden (1959). In Arnal (1986), its application for flows with adverse pressure gradients is reported. At the upper end of the range of application for $Tu = 0.00298$, one notes $N \approx 0$, implying that the transition would occur right at the location of Re_{cr}. This is without any physical justification and in applying this correlation, the following needs to be noted:

- For $Tu < 0.10\%$, the transition location is insensitive to Tu. Such low levels of disturbances are thought to be typical of acoustic noise controlling transition, rather than vortical disturbances. This distinction is not made in the literature clearly between the two and everything is included in the catch-all terminology of FST.
- For higher values of Tu (greater than 2 to 3%), transition often occurs without the appearance of TS waves at $Re < Re_{cr}$ for wall-bounded flows. It is said that the linear process of instability is *bypassed* and such transition processes are called *bypass transition.*

9.5.5 Distinction between Controlled and Uncontrolled Excitations

There seems to be confusion between instability and receptivity in the research community arising from the way the instability problem is formulated as a homogeneous governing equation with homogeneous boundary condition. This makes practitioners completely unprepared in comprehending how disturbances are received and amplified to cause flow transition. For example, Gaster and Grant (1975) claimed that they could simulate natural transition due to FST by designing an experiment where vortical disturbances are created at the wall by impulse excitation. In the traditional instability studies, the formulations imply wall excitation with disturbance decaying with height from the wall, which has been appropriately termed as wall mode in Sengupta (2012). However, free stream excitation creates free stream mode whose asymptotic properties are distinctly different and hence the route of transition is also different. This aspect was verified by the experiment of Schubauer and Skramstad (1947) when they failed to create TS wave-packets by acoustic excitation created in the free stream.

To create monochromatic TS wave-packet, one needs to remove background noise, as much as possible, and then provide a deterministic time-harmonic wall excitation, as shown experimentally by Schubauer and Skramstad (1947). However, instability theory cannot provide all the details of localized time-harmonic excitation noted in the experiment. Various elements of the response field arising due to localized excitation (known as the impulse response of the boundary layer) were demonstrated for the first time by receptivity analysis from the solution of Orr-Sommerfeld equation in Sengupta (1991). The same elements were also shown from the solution of Navier-Stokes equation in Sengupta *et al.* (2006). A typical response field obtained from the solution of Navier-Stokes equation is shown in Figure 9.10, which displays the presence of a TS wave-packet (not TS waves!), similar to that predicted by spatial stability theory. One of the advantages of studying the flow excited deterministically, also displays a local solution near the exciter. More importantly, one notes the presence of a spatio-temporal wave-front (STWF) in Figure 9.10.

Figure 9.11 shows the evolution of u_d plotted against x for the indicated time instants, obtained from the solution of Navier-Stokes equation in 2D, with the nondimensional excitation

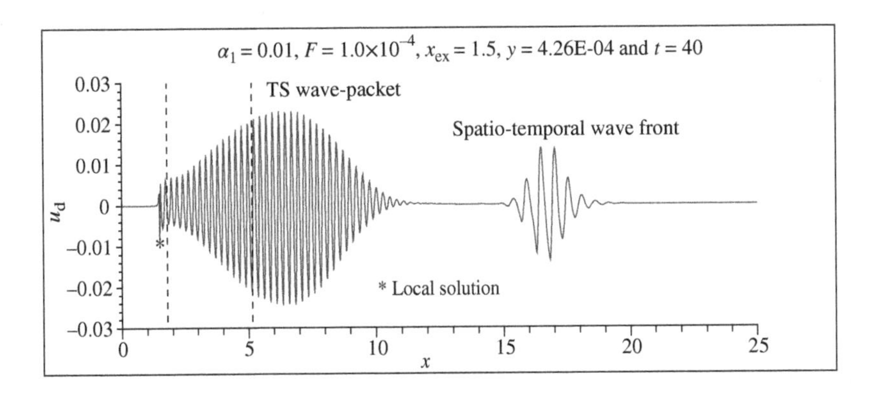

Figure 9.10 Streamwise disturbance velocity plotted as function streamwise distance, at $t = 40$ and $y = 4.26 \times 10^{-4}$ for an amplitude of excitation which is 1% of the free stream speed. Also, $F = 10^{-4}$ for an exciter located at $x_{ex} = 1.5$. Dashed vertical lines indicate entry and exit location the constant F ray into and from the neutral curve

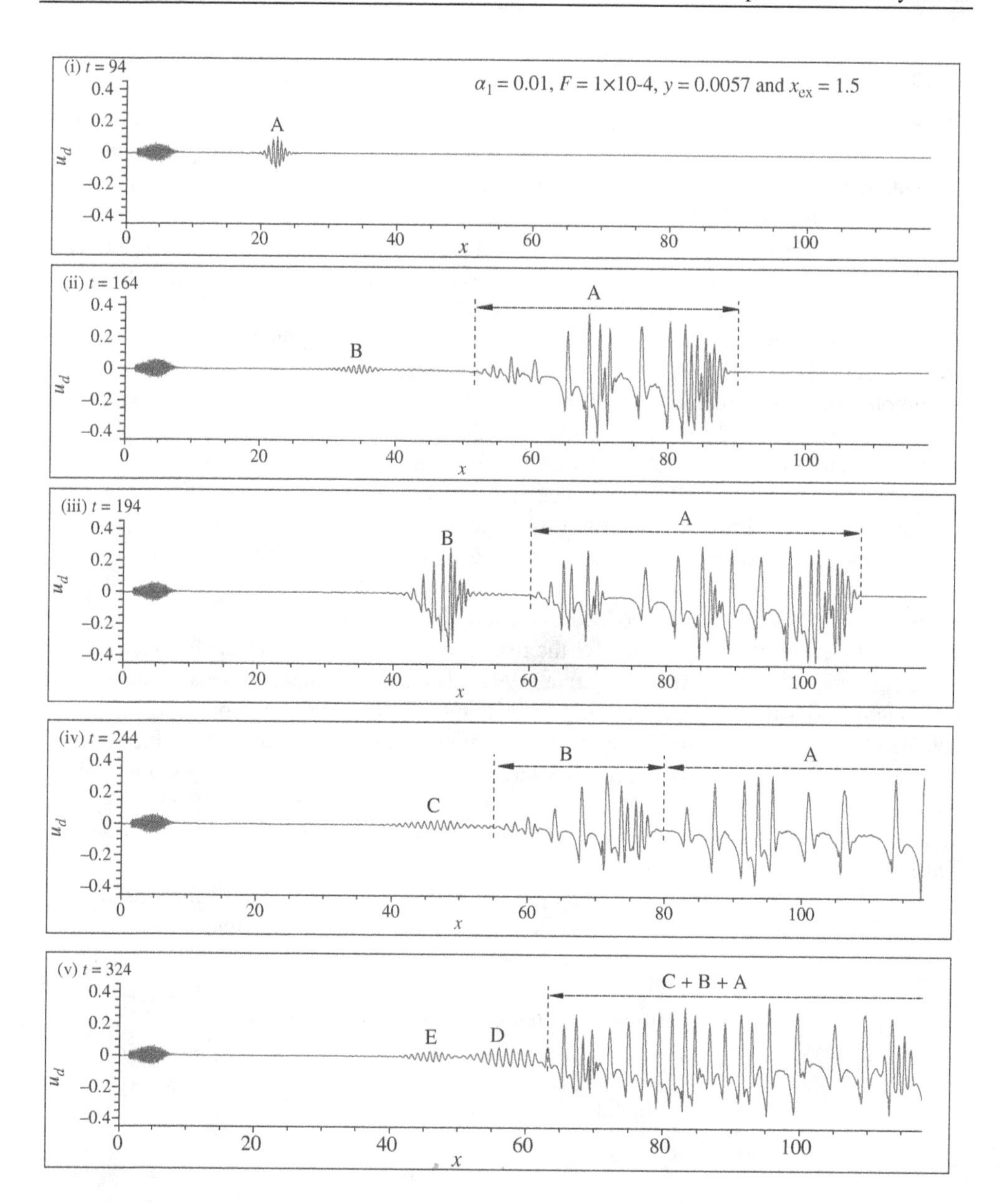

Figure 9.11 Streamwise disturbance velocity plotted as function streamwise distance at $y = 0.0057$ and indicated times for amplitude of excitation which is 1% of the free stream speed. Also, $F = 10^{-4}$ for an exciter located at $x_{ex} = 1.5$

frequency given by $F = 10^{-4}$ and a wall exciter located at $x_{ex} = 1.5$ creating a peak wall-normal velocity of 1% of the free stream speed. Plotted data are along a line at the height of $y = 0.0057$. The top frame shown for $t = 94$ indicates the TS wave-packet and a STWF marked as *A*. In the next frame at $t = 164$, one notices that the STWF at *A* exhibiting nonlinear growth and significant dispersion. This STWF also induces another STWF at *B*. There are two features worth noting here: First, the transition process is seen to be associated with upstream

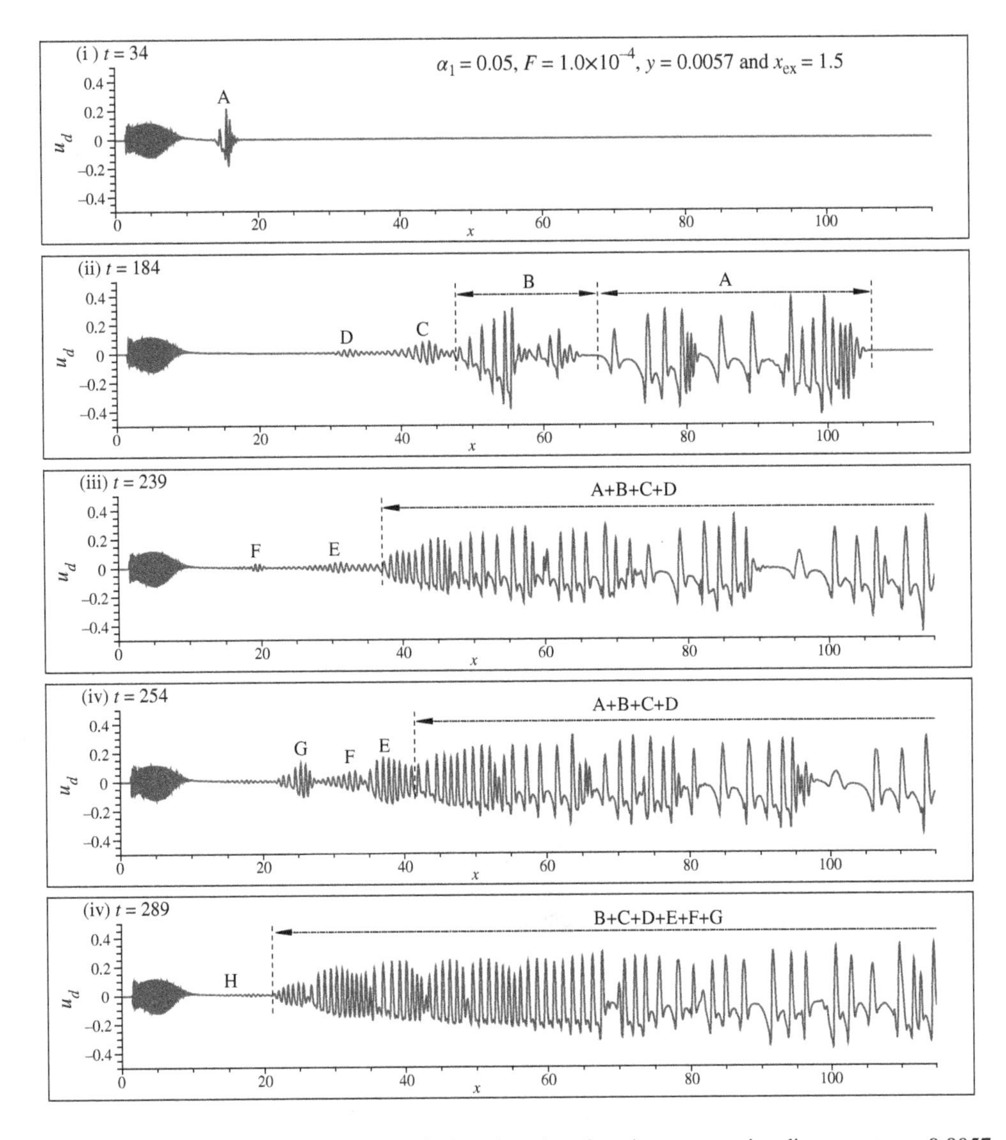

Figure 9.12 Streamwise disturbance velocity plotted as function streamwise distance at $y = 0.0057$ and indicated times for amplitude of excitation which is 5% of the free stream speed. Also, $F = 10^{-4}$ for an exciter located at $x_{ex} = 1.5$

induction of STWF. Secondly, the growth of the second STWF is very rapid, as noted by the amplitude of B at $t = 164$ and $t = 194$. Thereafter, the flow becomes very intermittent near the front of the disturbance leading edge, with continuous upstream spawning of fresh STWF marked as C, D and E in the last two frames of this figure.

The fact that the transitional flow was shown to be created by STWF in Figure 9.11 is generic, can be shown by considering another case with even higher excitation level of wall-normal velocity given by 5% of the free stream speed, while the other physical and excitation parameters remaining the same. The results are shown in Figure 9.12 for u_d plotted against x for different times. In comparison to the case of Figure 9.11, here the STWF is noted at a much earlier time and even at $t = 34$, one notices nonlinear distortion of the STWF, marked as A. There is also another factor noted in this high amplitude excitation case, where the TS wave-packet amplitude is also higher, but proportionately lesser than the factor by which the excitation is increased five-fold. Subsequent frames in Figure 9.12 clearly show the nonlinear growth, dispersion and induction of upstream STWF marked as B, C, D, E etc. The last frame at $t = 289$ shows the intermittent transitional flow with upstream edge of the composite disturbance front is noted to start earlier from $x = 20$ at this advanced time. Details of the excitation field and the nonlinear nature of the transition process caused by the growth of STWF can be found in Sengupta (2012) and the reader is referred to the source material.

9.6 Transition in Three-Dimensional Flows

Two different classes of 3D boundary layers are found in many external or internal flows of engineering interest: *Boundary sheets* and *boundary regions*, with the former exemplified by a boundary layer forming over a wing (excluding the root and tip regions), as shown in Figure 9.13. The boundary layer over the wing is thin and parallel to the surface. The free shear layer in the wake is a continuation of the boundary layer over the wing and is also classified as the *boundary sheet*. The *boundary region* noted at the wing-fuselage junction and near the wing tips is characterized by confluence of *boundary sheets* coming from the constituent geometries as a typical example.

In *boundary sheets*, length scales in the x- (streamwise) and z-directions (spanwise) are both of order one and so the Navier-Stokes equation in the x- and z-directions can be simplified. The governing equations for different types of 3D boundary layers are given in Cebeci and Bradshaw (1977), Cebeci and Cousteix (2005), Crabtree *et al.* (1963) and Prandtl (1946). In the following, we will mostly keep our attention focused on the *boundary sheet* flow and refer to it as boundary layer. When such a boundary layer forms over the wing, it is instructive to find out how the cross flow originates.

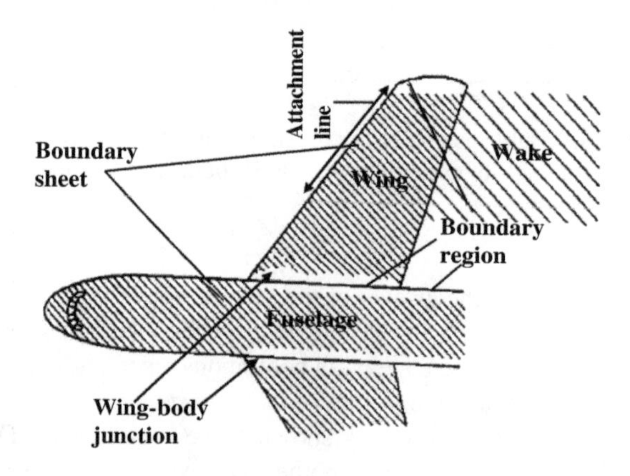

Figure 9.13 Typical three-dimensional boundary layers over a transport aircraft

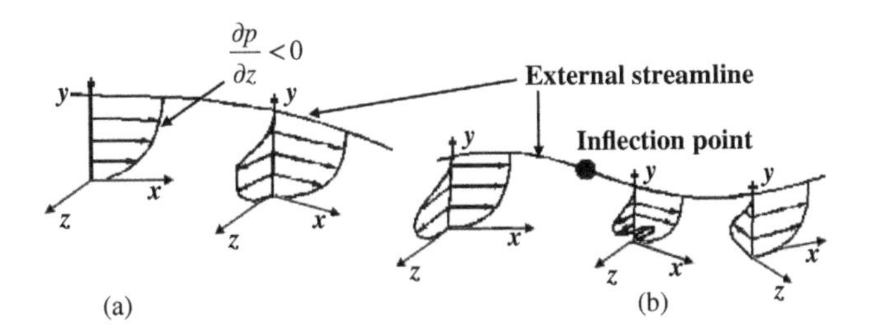

Figure 9.14 Secondary flow generated by different spanwise pressure gradients. (a) Flow with $\frac{\partial p}{\partial z} < 0$ (b) Flow where $\frac{\partial p}{\partial z}$ changes sign

For 3D flows, as the external velocity vector is not in the direction of the pressure gradient, the external streamlines near the surface are curved, which gives rise to the cross flow. Consider the flow shown in Figure 9.14, where a negative pressure gradient ($\partial p/\partial z < 0$) is applied in the spanwise direction. This pressure gradient is related to the curvature of the external streamline and is dynamically balanced by a centrifugal force, i.e.

$$\partial p/\partial z = -\rho u_e^2/r_e,$$

where r_e is the radius of curvature of the external streamline. Closer to the wall, but still outside the shear layer, the velocity is u, we must have $\partial p/\partial z = -\rho u^2/r$, with r as the radius of curvature of the local streamline.

The pressure and hence $\partial p/\partial z$ do not vary in the wall-normal direction, and hence r is less than r_e, as u is less than u_e. Thus, the crosswise pressure gradient $\partial p/\partial z$ skews the streamlines differently at different heights, causing a component of the flow shown in Figure 9.14a, with the no-slip condition ensuring zero cross flow at the wall. Prandtl (Prandtl 1946) termed this the secondary flow of the first kind or skew-induced secondary flow. When $\partial p/\partial z$ changes sign, the external flow displays an inflection point, as shown in Figure 9.14b. When Equation (9.52) given below is viewed at the wall, one has $u = v = w = 0$ and thus

$$\frac{\partial p}{\partial z} = \frac{\partial \tau_{zy}}{\partial y}.$$

As a consequence, the cross flow displays reverse flow near the wall in its direction with S-shaped profile, because near the wall flow reacts more rapidly, with inertia playing a subdominant role. This is the mechanism of a pressure driven 3D boundary layer, as one would find in flow over a swept wing.

In the following, 3D boundary layer equations are stated in the local Cartesian coordinate system. For more details on 3D boundary layer equations in a streamline, body-oriented coordinate system one can refer to Cebeci and Cousteix (2005). Here the equations are written for laminar flow, as would be required for the equilibrium flow whose stability or receptivity is studied:

$$\frac{\partial u}{\partial x} + \frac{\partial v}{\partial y} + \frac{\partial w}{\partial z} = 0. \tag{9.49}$$

$$u\frac{\partial u}{\partial x}+v\frac{\partial u}{\partial y}+w\frac{\partial u}{\partial z}=-\frac{1}{\rho}\frac{\partial p}{\partial x}+\nu\frac{\partial^2 u}{\partial y^2} \tag{9.50}$$

$$\frac{\partial p}{\partial y}=0 \tag{9.51}$$

$$u\frac{\partial w}{\partial x}+v\frac{\partial w}{\partial y}+w\frac{\partial w}{\partial z}=-\frac{1}{\rho}\frac{\partial p}{\partial z}+\nu\frac{\partial^2 w}{\partial y^2}. \tag{9.52}$$

These equations are solved subject to the wall ($y = 0$) boundary conditions

$$u = w = 0 \quad \text{and} \quad v = v_w(x, z). \tag{9.53a}$$

And at $y = \delta$:

$$u = u_e \quad \text{and} \quad w = w_e. \tag{9.53b}$$

Boundary conditions at the wall show the possibility of mass transfer, which is often needed to obtain equilibrium flow in the presence of suction. Receptivity studies of such an equilibrium flow can be pursued either in a linear or nonlinear framework, similar to 2D studies in Sengupta (2012). This is required for laminar flow control applications by suction. At the edge of the boundary layer, the edge velocity components are related to the pressure gradients by

$$u_e\frac{\partial u_e}{\partial x}+w_e\frac{\partial u_e}{\partial z}=-\frac{1}{\rho}\frac{\partial p}{\partial x} \tag{9.54a}$$

$$u_e\frac{\partial w_e}{\partial x}+w_e\frac{\partial w_e}{\partial z}=-\frac{1}{\rho}\frac{\partial p}{\partial z}. \tag{9.54b}$$

Above equations for a 3D boundary layer can be further simplified under certain conditions. Problems of practical engineering interests arise where the flow becomes independent of one variable. Such degenerate cases of 3D flows having three nonzero velocity components dependent on two independent variables are found, for example, in flow near the leading edge of a swept body with constant cross section and *infinite* span in the attachment line plane of Figure 9.13. Similar simplification is possible for flow over a radial diffuser (Wheeler and Johnston 1972) or flow over a straight tapered swept wing (Bradshaw *et al.* 1976). Of specific interest is flow past an infinite swept wing (Arnal 1986) or yawed cylinder (Poll 1985) which is discussed next.

9.7 Infinite Swept Wing Flow

Here we define two coordinate systems with respect to the sketch of a wing shown in Figure 9.15. The wing fixed coordinate system (WFCS) (X, Y, Z), is fixed to the wing with the Z-axis along the swept leading edge; X is in the plane of the wing orthogonal to the Z-direction and Y is perpendicular to the plane of the wing in the wall-normal direction. Additionally, we introduce a coordinate system which is the external streamline fixed coordinate system (ESFCS) indicated by the (x, y, z)-system. One notices that Y and y are essentially the same coordinate.

For flows over a yawed cylinder or infinite swept wings of constant cross section, the flow is independent of the spanwise coordinate, Z. For such a flow, there will be a constant spanwise

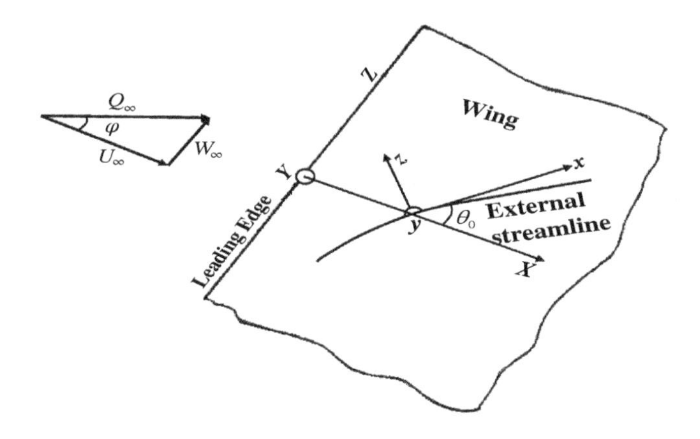

Figure 9.15 Coordinate systems over an infinite swept wing

component of velocity, but the derivatives of all physical variables with respect to Z are set equal to zero. The coordinate system is as shown in Figure 9.15, where φ is the sweep angle.

The governing boundary layer equations given in Equations (9.49) to (9.52) simplify for an infinite swept wing to

$$\frac{\partial u}{\partial X} + \frac{\partial v}{\partial Y} = 0 \tag{9.55}$$

$$u\frac{\partial u}{\partial X} + v\frac{\partial u}{\partial Y} = -\frac{1}{\rho}\frac{\partial p}{\partial X} + \nu\frac{\partial^2 u}{\partial Y^2} \tag{9.56}$$

$$u\frac{\partial w}{\partial X} + v\frac{\partial w}{\partial Y} = \nu\frac{\partial^2 w}{\partial Y^2}. \tag{9.57}$$

These are supplemented by the reduced equations

$$\frac{\partial P}{\partial Y} = \frac{\partial P}{\partial Z} = 0.$$

These equations show that u and v are independent of w and the resultant equations are just like that for a 2D boundary layer. At the edge of the boundary layer, u and w attain values u_e and w_e. From Equation (9.57), it is noted that at the free stream $\frac{\partial w}{\partial x} = 0$ and thus w_e is a pure constant, fixed as w_∞. There is, however, a difference between the solution procedures of general 3D and 2D boundary layers. For the solution of 3D boundary layers, one needs initial conditions on two intersecting planes. For some external flows, one can take advantage of available symmetry conditions from physical considerations. For example, one can use the conditions near the attachment line of bodies, as explained next.

9.8 Attachment Line Flow

The line joining stagnation points on each section of the body constitutes the symmetry plane. On this plane of symmetry, the flow is 2D except for the cross flow derivative. This is called the attachment line flow, with the attachment line as a streamline on the body, on which both the cross flow velocity and the cross flow pressure gradient are identically zero. If we represent the flow in the vicinity of the stagnation region, then w and $\frac{\partial p}{\partial z}$ are identically zero, making the z-momentum equation singular along the attachment line.

However, this degeneracy can be removed by writing another equation obtained by differentiating the z-momentum equation with respect to z and treating $w_z = \partial w/\partial z$ as the new dependent variable with additional symmetry conditions on the attachment line, i.e.

$$\frac{\partial u}{\partial z} = \frac{\partial v}{\partial z} = \frac{\partial^2 w}{\partial z^2} = 0.$$

The resultant equations are given by

$$\frac{\partial u}{\partial x} + \frac{\partial v}{\partial y} + w_z = 0 \tag{9.58}$$

$$u\frac{\partial u}{\partial x} + v\frac{\partial u}{\partial y} = u_e\frac{\partial u_e}{\partial x} + \frac{1}{\rho}\frac{\partial}{\partial y}\left(\mu\frac{\partial u}{\partial y}\right) \tag{9.59}$$

$$u\frac{\partial w_z}{\partial x} + v\frac{\partial w_z}{\partial y} + w_z^2 = u_e\frac{\partial (w_z)_e}{\partial x} + (w_z)_e^2 + \frac{1}{\rho}\frac{\partial}{\partial y}\left(\mu\frac{\partial w_z}{\partial y}\right). \tag{9.60}$$

The boundary conditions for these equations at $y = 0$:

$$u = v = 0 = w_z. \tag{9.61a}$$

And at $y = \delta$:

$$u = u_e(x, z) \quad \text{and} \quad w_z = w_{ze}. \tag{9.61b}$$

The boundary layer equations are not solved in the form given above. Instead these are transformed first and solved in the transformed plane, which allows taking larger steps in the x- and z-directions. The scaled variables in the transformed plane vary slower as compared to their variations in the physical plane. Next, we obtain the boundary layer equations in the transformed plane.

9.9 Boundary Layer Equations in the Transformed Plane

The boundary layer equations, Equations (9.49) to (9.52), are expressed using a Falkner-Skan type transformation for the independent variables used for 2D flows

$$x \equiv x, \; z \equiv z \; \text{and} \; \eta = y\sqrt{\frac{u_e}{\nu x}}. \tag{9.62}$$

This is a nonorthogonal transformation that stretches the $0(\delta)$ shear layer in the y-variable to a thickness that is $0(1)$ in the η-variable. Apart from stretching the wall-normal coordinate, this transformation also removes the singularity of the governing parabolic PDE at $x = 0$ and $z = 0$.

For dependent variables, two-component vector potentials $(\phi,\ \hat{\psi})$ are introduced such that

$$u = \frac{\partial \hat{\psi}}{\partial y} \tag{9.63a}$$

$$w = \frac{\partial \phi}{\partial y}. \tag{9.63b}$$

The continuity equation provides

$$v = -\left(\frac{\partial \hat{\psi}}{\partial x} + \frac{\partial \phi}{\partial z}\right). \tag{9.63c}$$

Furthermore, one introduces nondimensional vector potential components f and g by

$$\hat{\psi} = (u_e \nu x)^{1/2}\ f(x, z, \eta) \tag{9.64a}$$

$$\phi = w_e \left(\frac{\nu x}{u_e}\right)^{1/2} g(x, z, \eta), \tag{9.64b}$$

so that $u = u_e f'$ and $w = w_e g'$, with a prime indicating a derivative with respect to η. The governing equations for f and g are obtained as (Cebeci and Cousteix 2005):

$$f''' + m_1 f f'' + m_2(1 - f'^2) + m_5(1 - f'g') + m_6 g f''$$
$$= x\left[f'\frac{\partial f'}{\partial x} - f''\frac{\partial f}{\partial x} + m_7\left(g'\frac{\partial f'}{\partial z} - f''\frac{\partial g}{\partial z}\right)\right] \tag{9.65}$$

$$g''' + m_1 f g'' + m_4(1 - f'g') + m_3(1 - g'^2) + m_6 g g''$$
$$= x\left[f'\frac{\partial g'}{\partial x} - g''\frac{\partial f}{\partial x} + m_7\left(g'\frac{\partial g'}{\partial z} - g''\frac{\partial g}{\partial z}\right)\right], \tag{9.66}$$

where

$$m_2 = \frac{x}{u_e}\frac{\partial u_e}{\partial x};\ m_1 = \frac{m_2 + 1}{2};\ m_3 = \frac{x}{u_e}\frac{\partial w_e}{\partial z};\ m_4 = \frac{x}{w_e}\frac{\partial w_e}{\partial x};$$

$$m_5 = \frac{w_e}{u_e}\frac{x}{u_e}\frac{\partial u_e}{\partial z};\ m_6 = m_3 - \frac{m_5}{2}\ \text{ and }\ m_7 = \frac{w_e}{u_e}.$$

Boundary conditions for the solutions of Equations (9.65) and (9.66) at $\eta = 0$:

$$f, f', g, g' = 0. \tag{9.67a}$$

And at $\eta = \eta_\infty$:

$$f', g' \rightarrow 1. \tag{9.67b}$$

Once again, to solve these equations we need initial conditions on two intersecting planes. Thus, we start from the symmetry plane: the flow at the attachment line that has been described in the last section. The attachment line equations can also be transformed by a set of transformations via a different two-component vector potential as

$$u = \frac{\partial \hat{\psi}}{\partial y} \quad \text{and} \quad \frac{\partial w}{\partial z} = \frac{\partial \phi}{\partial y}. \tag{9.68}$$

So that $v = -\left(\frac{\partial \hat{\psi}}{\partial x} + \phi\right)$. Once again, one can introduce nondimensional vector potential f with Equation (9.64a) and the second nondimensional vector potential is introduced via

$$\phi = \left(\frac{dw}{dz}\right)_e \left(\frac{\nu x}{u_e}\right)^{1/2} g_1(x, z, \eta). \tag{9.69}$$

Substitution and simplification yield the following equations for the attachment line boundary layer

$$f''' + m_1 f f'' + m_2(1 - f'^2) + m_5(1 - f' g_1') + m_3 g_1 f'' = x\left[f' \frac{\partial f'}{\partial x} - f'' \frac{\partial f}{\partial x}\right] \tag{9.70}$$

$$g_1''' + m_1 f g_1'' + m_4(1 - f' g_1') + m_3(1 - g_1'^2) + m_3 g_1 g_1'' = x\left[f' \frac{\partial g_1'}{\partial x} - g_1'' \frac{\partial f}{\partial x}\right]. \tag{9.71}$$

Here the definition of all the terms remains the same except

$$g_1' = \frac{(\partial w/\partial z)}{(\partial w/\partial z)_e} \quad \text{and} \quad m_4 = \frac{x}{(\partial w/\partial z)_e} \frac{\partial}{\partial x}\left(\frac{\partial w}{\partial z}\right)_e.$$

The corresponding boundary conditions at $\eta = 0$ are:

$$f, f', g_1, g_1' = 0. \tag{9.72a}$$

And at $\eta = \eta_\infty$:

$$f', g_1' \to 1. \tag{9.72b}$$

9.10 Simplification of Boundary Layer Equations in the Transformed Plane

All the above flow equations in the transformed plane (Cebeci and Bradshaw 1977) can be further simplified if external flow takes special forms. Introducing nondimensional vector potential f, g of Equations (9.64) and if external flow is of the form (Cebeci and Cousteix 2005)

$$u_e = Ax^m z^n \text{ and } w_e = Bx^r z^s, \tag{9.73}$$

with $m_2 = m$; $r = m_2 - 1$; and $s = n + 1$, then the governing equations (9.65) and (9.66), simplify to the following equations, assuming the corresponding laminar flow to be similar

$$f''' + \frac{m+1}{2} f f'' + m(1 - f'^2) + \frac{B}{A}\left\{\frac{n+2}{2} g f'' + n(1 - f'g')\right\} = 0 \tag{9.74}$$

$$g''' + \frac{m+1}{2} f g'' + (m-1)(1 - f'g') + \frac{B}{A}\left\{\frac{n+2}{2} g g'' + (n+1)(1 - g'^2)\right\} = 0. \tag{9.75}$$

These equations were solved in Hansen and Yohner (1958) for different values of B, A, m and n. Cebeci and Cousteix (Cebeci and Cousteix 2005) point out that such external flows are rotational and exhibit *overshoots* in u profile due to nonuniformities in total pressure. Further simplifications arise when $r = s = n = 0$ and arbitrary m makes the external flow irrotational, which results in the well-known Falkner–Skan–Cooke equations given by

$$f''' + \frac{m+1}{2} f f'' + m(1 - f'^2) = 0 \tag{9.76}$$

$$g''' + \frac{m+1}{2} f g'' = 0. \tag{9.77}$$

These equations further simplify for $m = 1$, which represent the equations from which the similarity solution of the attachment line flow can be obtained. These equations were used to obtain the equilibrium flow for LEC studied in Sengupta and Dipankar (2005) to study bypass transition caused by a convecting vortex outside the shear layer, but which remains strictly in the attachment line plane.

Similarly one can obtain a nonsimilar solution for flow past an infinite swept wing by noting the independence of flow properties with z from Equations (9.65) and (9.66) as

$$f''' + m_1 f f'' + m_2(1 - f'^2) = x\left[f'\frac{\partial f'}{\partial x} - f''\frac{\partial f}{\partial x}\right] \tag{9.78}$$

$$g''' + m_1 f g'' = x\left[f'\frac{\partial g'}{\partial x} - g''\frac{\partial f}{\partial x}\right]. \tag{9.79}$$

Note that Equation (9.78) is independent of g. This makes Equation (9.79) a linear equation. In the above, nonsimilarity arises due to right-hand side terms in Equations (9.78) and (9.79). Having obtained various 3D equilibrium solutions, it is possible to study their stability.

9.11 Instability of Three-Dimensional Flows

One can expect to see some similarities between instabilities of 3D flows with 2D flows due to similarities of the mean flow obtained via the Falkner–Skan–Cooke and Falkner–Skan similarity transforms, respectively. However, during flight tests on swept wing aircraft in RAE during 1951 and 1952, it was found that the flow instability and transition zone moved towards the wing leading edge abruptly, which is not compatible with what one would expect from 2D flow analysis. Details of these experiments and a plausible explanation can be found in Gray (1952), Owen and Randall (1952, 1953). Beyond a certain speed the transition point moved near the leading edge and it was observed that the leading edge radius and sweep angle are the determining factors. Two aspects stood out, with the first that in some cases, the instability was noted in the accelerated part of the flow, which according to 2D instability theory should be stable. Secondly, in some cases the point of transition moved right at the attachment line itself, which according to 2D spatial instability has a Re_{cr} that is an order of magnitude higher than the zero pressure gradient boundary layer. These flight experiments performed on experimental aircraft AW52 at RAE led to the conclusion in Gray (1952) that

> *no laminar flow is present on normal wings of any appreciable size and speed if the sweep angle exceeds roughly* 20°.

Even today in selecting sweepback, this is used as a rule of thumb.

Furthermore, sublimation patterns on the surface indicated external streamline aligned streaks which can be interpreted as stationary waves. These streaks are also regularly spaced in the spanwise direction – vividly demonstrated experimentally in the laboratory in Arnal *et al.* (1984a). All these evidences establish the destabilizing effects of sweep-back and bring to the fore the observation that transition occurs in a region of strong negative pressure gradients occurring at large free stream speeds. Also, note the transition occurs downstream of the streaks.

Overall, instability of flow over a swept wing has similarities with mixed convection flow over a heated horizontal plate, in terms of exhibiting both spatial and temporal instabilities. This is due to the presence of an inflection point for the streamwise velocity component for mixed convection flow, with an inflection point in the cross flow profile of flow over swept wing. Such inflection points give rise to temporal instability, following similar theorems laid out in Sengupta *et al.* (2013) for mixed convection flows for inviscid instability mechanism at the onset.

9.11.1 Effects of Sweep-back and Cross Flow Instability

Once the destabilizing effect of a sweptback wing was recognized, further experiments followed using wing tunnels (Anscombe and Illingworth 1952, Gregory *et al.* 1955), confirming the flight test results at RAE. These were followed by further flight experiments reported in Allen and Burrows (1956) and Burrows (1956) on a straight swept wing. Subsequently, Boltz *et al.* (1960) carried out extensive wind tunnel experiments focusing on effects of Reynolds number, angle of attack and sweep angle on transition location. Further experiments on this topic were reported in Arnal *et al.* (1984a), Arnal and Jullien (1987) and Michel *et al.* (1984), where the authors studied flow over an ONERA D aerofoil profile. Although the model had the usual ONERA D profile over the top surface between $x/c = 0.2$ and $x/c = 1$, it had a special cambered leading edge and lower surface, as shown in Figure 9.16. Note that the original ONERA D profile is a symmetric section and this modified section was used to highlight the cross flow instability.

As the cross flow instability was noted in that part of the wing where the flow was accelerating in this ONERA D model, the instability was mimicked by appropriately created accelerated flow by choosing negative angles of attack with a specific sweep angle. This was the motivation for the design of the leading edge of the ONERA D wing shown in Figure 9.16. This is also seen in the velocity profiles shown for the suction side at the indicated angles of attack in Figure 9.16. Following this interesting idea of simulating sweptback wing flow on a straight wing, Saric and Yeates (1985) created swept wing flow on a flat plate with a swept leading edge and the accelerated flow was created by a bump on the roof of the tunnel. While this last experiment avoided curvature effects, this also precluded the attachment line problem.

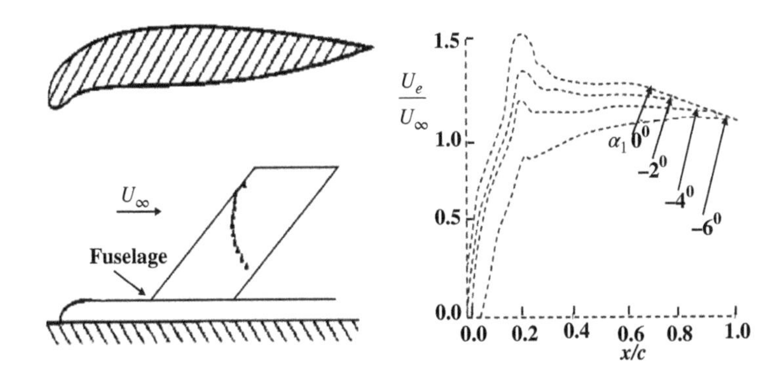

Figure 9.16 ONERA D airfoil fitted with a cambered leading edge for the wing. The edge velocity distribution is shown on the right-hand side. *Source:* Arnal and Jullien (1987). Reproduced with permission from the American Institute of Aeronautics and Astronautics, Inc.

In all these cases for sufficiently large *Re*, transition was seen to occur in regions of a strong negative pressure gradient. Also, wall visualization techniques (sublimation, china-clay, oil flow, etc.) indicated the presence of streaks upstream of the transition location. The fact that the flow becomes unstable in favourable pressure gradient regions indicates that the instability is not related to the usual instability that arises for the flow in the streamwise direction. Thus, researchers have focused their attention on linking the ensuing instability of the flow component which is created by skewing the profile by the imposed pressure gradient, as discussed in Section 9.2. We have already noted the presence of a cross flow (in the perpendicular direction to the external streamlines) created by such skewing by a spanwise pressure gradient and it is natural to investigate its instability. Near the leading edge of a swept wing, both the surface and the streamlines are highly curved due to the combined effects of the pressure gradient and sweep angle, turning the flow inboard. A schematic of the flow is shown in Figure 9.17, taken from Reed and Saric (1989).

Such sharp changes of curvature of flow are also noted near the wing trailing edge, as well. Lower inertia of flow close to the body causes these changes to be more pronounced in creating the cross flow, as shown in Figure 9.15. Also as shown in Figure 9.14, the pressure gradient

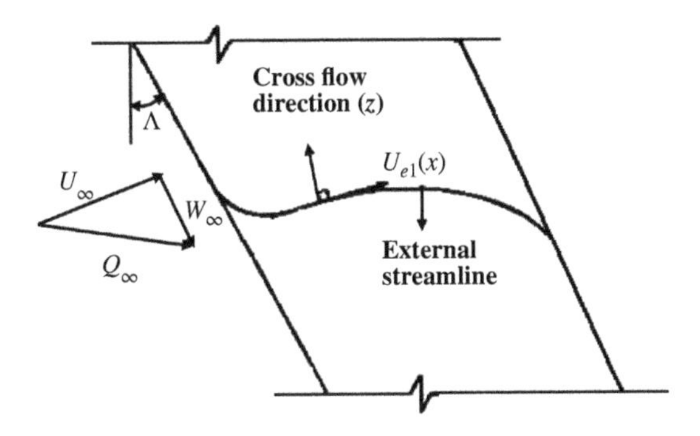

Figure 9.17 Schematic of streamline at the edge of a three-dimensional shear layer over a swept wing (Reed and Saric 1989)

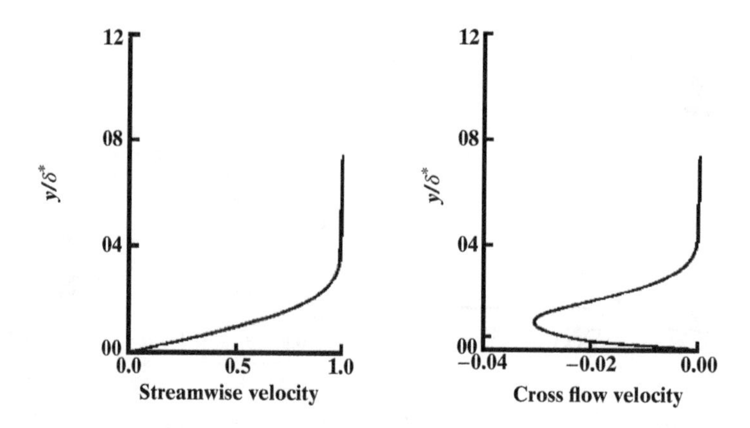

Figure 9.18 Computed streamwise and cross flow components of velocity inside a three-dimensional shear layer. *Source:* Reed (1987). Reproduced with permission from the American Institute of Physics

dictates whether this cross flow is S-shaped or not. Irrespective of the spanwise pressure gradient sign, this component of the flow displays a maximum in the interior of the boundary layer and goes to zero at the wall and at the edge of the shear layer. In Figure 9.18, computed velocity profiles for the experiments in Saric and Yeates (1985) are taken from Reed (1988) to give one an idea about the relative magnitudes of the streamwise and cross flow components in 3D boundary layer.

The velocity components normalized by the edge velocity show that the maximum cross flow velocity is only 3% and this is typical of all reported measured velocities in tunnels and in flight. Despite the fact that this component is significantly small as compared to the streamwise flow component, its effect on the flow field is more profound due to temporal inviscid instability caused by the presence of an inflection point in the velocity profile. This is qualitatively explained by Rayleigh's and FjØrtoft's theorems. It has also been explained in Reed and Saric (1989) that such instability causes cross flow vortex structures similar to Kelvin's cat's eye, with the axes aligned along the streamwise direction. While these vortices are of the dimension of boundary layer thickness, they all rotate in the same direction, and thus these are unlike Goertler vortices. These cross flow vortex structures were computed in Reed (1988) for the experimental results reported in Saric and Yeates (1985). Thus, according to Reed and Saric, cross flow instability was brought into focus in Gray (1952) for the first time and was interpreted in Owen and Randall (1952, 1953), by Squire (as an addendum to Owen and Randall 1952) and in Stuart (1953), independently. Much of the work on instability of 3D flows has been reviewed in Arnal (1986), Poll (1984) and Reed and Saric (1989).

9.12 Linear Viscous Stability Theory for Three-Dimensional Flows

In Equation (9.27), we developed a general linear stability equation for a 3D disturbance field in a 3D parallel mean flow as

$$\phi^{iv} - 2(\alpha^2 + \beta^2)\phi'' + (\alpha^2 + \beta^2)^2\phi$$

$$= iRe\{(\alpha U_1 + \beta W_1 - \omega_0)[\phi'' - (\alpha^2 + \beta^2)\phi] - (\alpha U_1'' + \beta W_1'')\phi\}, \tag{9.80}$$

where U_1 and W_1 are the streamwise and cross flow profiles; α and β are the wavenumber components in the streamwise and cross flow directions and ω_0 is the circular frequency of the disturbance field.

This equation is written in a special Cartesian coordinate system which is defined in Section 9.11. As mentioned, the cross flow instability is related to severe turning of streamlines near the leading and trailing edges of the wing; thus it has often been noted that one should write the equilibrium and/or the stability equations retaining the curvature terms. However, neglect of curvature effects in solving the first order boundary layer equation seems justifiable, as the curvature terms are of the same order of magnitude as the neglected higher order terms in the boundary layer equation. Neglecting higher order terms and retaining curvature terms do not appear to be consistent. This is not so for the stability equations, which have been written for orthogonal curvilinear coordinate systems in Arnal (1986), Cebeci (2004), Cebeci *et al.* (1992), Collier *et al.* (1985) and Malik and Poll (1985). According to Cebeci (2004), despite earlier confusion about curvature effects originating from the work reported in Malik and Poll (1985), the work by Cebeci *et al.* (1992) has established the correctness of stability solutions without curvature effects. In discussing the following, we therefore refer to stability equations developed in the Cartesian coordinate system without curvature terms.

9.12.1 Temporal Instability of Three-dimensional Flows

Stuart, in writing a special section in Gregory *et al.* (1955), has shown that the Orr–Sommerfeld equation, Equation (9.80), can be reduced to a 2D problem in studying the temporal stability problem by introducing the following

$$\alpha_\psi^2 = \alpha^2 + \beta^2 \text{ and } U_\psi = U_1 \cos\psi + W_1 \sin\psi, \tag{9.81}$$

where $\tan\psi = \beta/\alpha$ relates the real wavenumbers in the streamwise and cross flow directions, with U_1 and W_1, as defined later in Equations (9.93). The Orr–Sommerfeld equation simplifies to

$$\phi^{iv} - 2\alpha_\psi^2\phi'' + \alpha_\psi^4\,\phi$$

$$= iRe\left\{(\alpha_\psi U_\psi - \omega_0)[\phi'' - \alpha_\psi^2\,\phi] - \alpha_\psi \frac{d^2U_\psi}{dy^2}\phi\right\}. \tag{9.82}$$

Here U_ψ represents the projection of velocity in the direction of the wavenumber vector, $\vec{\alpha}_\psi$ and Equation (9.82) represents a 2D Orr–Sommerfeld equation in the direction of the wavenumber vector. All benefits of the above transformation disappear in spatial theory, where complex α, β do not allow such manipulations. The above temporal approach is used in the COSAL code developed in Malik (1982), where one has five real parameters $(Re, \omega_r, \omega_i, \alpha, \beta)$. For a given Reynolds number Re and real frequency ω_r, if the wavenumber angle ψ is fixed, then the other two remaining parameters can be obtained from the dispersion relation as an eigenvalue. After obtaining the growth rate ω_i for different values of ψ, one can calculate the amplification rate $n = \text{Ln}\,(\frac{A}{A_0})$ by using the generalized Cauchy–Riemann transformation from

$$n = \int_{x_0}^{x} \frac{\omega_i}{|V_g|} dx', \tag{9.83}$$

where the group velocity is given by

$$V_g = \left[\left(\frac{\partial \omega_0}{\partial \alpha} \right)_{\psi, Re}, \left(\frac{\partial \omega_0}{\partial \beta} \right)_{\psi, Re} \right]$$

and x' is the arc length along the group velocity direction, which most of the time is close to the external streamline direction. In the above envelope method, wave propagation direction ψ is sought along which ω_i is maximum and this maximum value is then integrated according to Equation (9.83). In order to reduce computer time, Srokowski and Orszag (1977) in their SALLY code looked at only stationary waves, $\omega_r = 0$.

9.12.2 *Spatial Instability of Three-dimensional Flows*

Note that spatial stability problems are more complicated as we have six parameters now $(Re, \omega_r, \alpha_r, \alpha_i, \beta_r, \beta_i)$, as compared to five parameters in the temporal formulation. In Arnal *et al.* (1989) and Mack (1988), the authors tried to circumvent this by a special form of spatial theory for 3D problems. In these formulations, β is assumed real and along with the real frequency ω_r for a fixed Re, one is left with the task of finding the complex α as the eigenvalue. A similar approach was also used in Sengupta *et al.* (1997) in explaining the Klebanoff mode, which originates as a 3D disturbance field due to very low frequency of excitation. Arnal *et al.* (Arnal *et al.* 1989) improved the procedure by maximizing the amplification rate α_i, with respect to all possible β, because in the presence of all possible directions of excitations (as given by various values of β), only the maximum growth will prevail in an asymptotic sense. However, this approach is not strictly a spatial instability investigation.

True spatial calculations were later formulated and reported in Cebeci and Stewartson (1980) and Nayfeh (1980) by deriving a propagation condition. The eigenvalue formulation in these studies is based on complete spatial theory, where the relationship between the two wavenumbers α and β is not assumed as in Arnal *et al.* (1989) and Mack (1988), but is computed making use of group velocity and the saddle point of the dispersion relation, as given in these references. In the normal mode form, the disturbance is assumed to be of the form:

$$q'(x, y, z, t) = q(y)\, e^{i(\alpha x + \beta z - \omega_0 t)}. \tag{9.84}$$

In the spatial theory developed in Cebeci and Stewartson (1980) and Nayfeh (1980), propagation of a wave packet is considered instead of normal mode analysis of instability theory, such that the disturbance field is expressed by

$$Q(x, y, z, t) = \int q(y)\, e^{i(\alpha x + \beta z - \omega_0 t)} d\beta. \tag{9.85}$$

Note that in the above equation, one is considering a fixed frequency disturbance and the integration range is defined in terms of β only in the complex plane along the Bromwich contour, the other wavenumber being an analytical function of β and related to it via

$$\alpha = \alpha(\beta). \tag{9.86}$$

Using this relation between the two wavenumbers, Equation (9.85) can be alternately written as

$$Q(x, y, z, t) = \int q(y)\, e^{ix(\alpha(\beta)+\beta\frac{z}{x}-\frac{\omega_0}{x}t)}\, d\beta. \tag{9.87}$$

For large values of x (away from the excitation source) following a ray given by $z/x =$ constant and $t/x =$ constant, one can use the saddle point method or the method of stationary phase (Bender and Orszag 1987) to find the leading contribution to Q. This occurs when the phase of the exponential term is stationary and would occur when the derivative of the phase with respect to β is zero for a fixed frequency disturbance field, i.e.

$$\left(\frac{\partial\alpha}{\partial\beta}\right)_{\omega_0,Re} = \left(\frac{\partial\alpha_r}{\partial\beta_r}\right)_{\omega_0,Re} + i\left(\frac{\partial\alpha_i}{\partial\beta_r}\right)_{\omega_0,Re} = -\frac{z}{x}. \tag{9.88}$$

As α is an analytic function of β, one can use the Cauchy–Riemann relation for the above to get

$$\left(\frac{\partial\alpha_r}{\partial\beta_r}\right)_{\omega_0,Re} = \frac{\partial\alpha_i}{\partial\beta_i} = -\frac{z}{x}. \tag{9.89a}$$

$$\left(\frac{\partial\alpha_i}{\partial\beta_r}\right)_{\omega_0,Re} = \frac{\partial\alpha_r}{\partial\beta_i} = 0. \tag{9.89b}$$

This immediately implies, according to the saddle point method, that the real part of $(\partial\alpha/\partial\beta)_{\omega_0,Re}$ is related to wave propagation angle ϕ_1 by

$$\left(\frac{\partial\alpha}{\partial\beta}\right)_{\omega_0,Re} = -\tan\phi_1. \tag{9.90}$$

Equations (9.89) define the complex quantity $\beta^* = \beta_r^* + i\beta_i^*$ in terms of z/x. Moreover, along a given ray in the (x, z)-plane, β^* is constant.

Angle ϕ_1 also provides the group velocity direction, since

$$\left(\frac{\partial\alpha}{\partial\beta}\right)_{\omega_0,Re} = -\frac{\partial\omega_0/\partial\beta|_{\alpha,Re}}{\partial\omega_0/\partial\alpha|_{\beta,Re}},$$

with α and β related by Equations (9.89), the amplitude of the disturbance is dominated by the exponential factor $\exp[-(\alpha_i x + \beta_i z)]$ or $\exp[-x(\alpha_i + \beta_i z/x)]$. Using Equation (9.88), one can see the equivalent streamwise growth as given by

$$\Gamma = \alpha_i - \beta_i\left(\frac{\partial\alpha}{\partial\beta}\right)_{\omega,Re}.$$

The sign of Γ and the direction of propagation of the disturbances would indicate whether the waves are growing, decaying or staying neutral with streamwise distance. Operationally, following steps are followed:

1. For a given frequency and at a streamwise station x, one chooses a given ray, z/x, i.e. the disturbance propagation angle is fixed; then we have $(\partial\alpha/\partial\beta)_{\omega_0,Re}$ along that direction.
2. Divide the β_r^* axes into intervals and identify these by their mid-points. Using the mid-point value, one iterates on the value of β_i^* which satisfies Equation (9.89b).
3. Once α and β are found for a particular disturbance propagation angle, the same procedure can be repeated for other angles to obtain the maximum amplification rate $\Gamma(x)$ for that x. This amplification rate can be integrated with respect to x to obtain the total amplification factor.
4. As in 2D calculations, here also the same exercise is repeated for different frequencies to find the critical frequency that leads to the highest integrated amplification rate.

The above method is similar to that described for 2D flows in Figures 9.17 and 9.18. However, we emphasize that the above approaches are all based on the signal problem and do not include the spatio-temporal approach as advocated in Sengupta *et al.* (2006) and Sengupta and Bhaumik (2011). This is an area which has not been undertaken for linear studies. In comparison, nonlinear studies based on the Navier–Stokes equation require use of DRP methods, as described in Chapter 8. This would require also using massive computational resources.

There is an alternative method called the envelope or the zarf method which was suggested in Cebeci and Stewartson (1980), where a neutral curve is defined and called the zarf. This is to mimic the procedure for 2D flows to track constant frequency disturbances (signal problem), as shown in Figure 9.4. While *Re* and ω_0 are real, α and β are in general complex and only on the neutral curve will these be real. Despite this, there are a plethora of possible neutral curves in 3D space. To circumvent this complication of nonuniqueness of such curves, in Cebeci and Stewartson (1980) it was suggested to take a special curve that the authors have termed the *absolute neutral curve*, which has the properties of α, β and $\frac{\partial\alpha}{\partial\beta}$ to be all real along the zarf. Before we close this section, we note that these complications can be easily avoided by resorting to time-dependent nonlinear receptivity analysis instead, as noted for 2D flows using the Bromwich contour integral method developed in Sengupta (1991), Sengupta *et al.* (1994) and Sengupta *et al.* (2006, 2006a).

9.13 Experimental Evidence of Instability on Swept Wings

Collected experimental evidence so far has established that there are many mechanisms in action for 3D flows as compared to 2D flows. For a swept wing, the flow can suffer primarily the following instabilities: (a) streamwise instability (SI), (b) cross flow instability (CFI), (c) leading edge contamination (LEC). It is noted that this classification scheme is an idealized one, as the instability mechanisms would not appear in isolation and more importantly the spatio-temporal route has not been addressed even today. Despite this, experiments have been carefully designed to highlight individual mechanisms one at a time, as described next.

The elaborate experiments in Boltz *et al.* (1960) using the NASA Ames 12 ft tunnel facility with very low turbulent intensity (0.02%), an *infinite* swept back wing of 1.22 metre chord

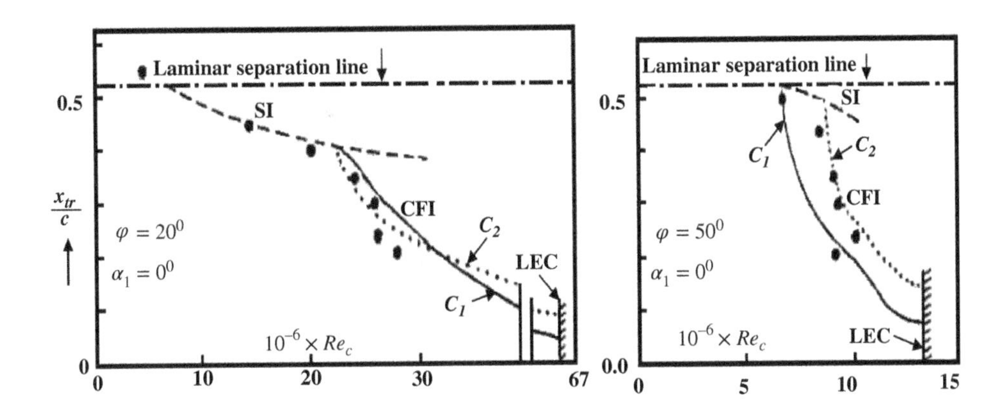

Figure 9.19 Experimental and calculated transition locations for flow over a swept wing. Experiments are as reported in Boltz *et al.* (1960) and corresponding calculations are as in Arnal (1984). SI indicates streamwise instability, CFI refers to cross flow instability and LEC indicates leading edge contamination. Note that there are two CFI empirical criteria following C_1 and C_2

was tested which had a sweep angle of 20^0 with a basic profile of NACA 64_2A015 aerofoil section. In Figure 9.19 from Arnal (1986), the experimenta locations of transition points are marked as discrete points along with theoretical predictions of the same by various criteria for the above three mechanisms discussed in a later section.

The x-axis in the figure is the Reynolds number based on chord (Re_c) and the non-dimensional transition point location (x_{tr}) is shown along the ordinate. At low Re_c, transition occurs beyond mid-chord in the region of a positive (adverse) pressure gradient corresponding to streamwise instability (SI). However, beginning at $Re_c = 20 \times 10^6$, transition point shifts forward, settling rather closer towards the leading edge in the negative (favourable) pressure gradient zone. As explained before, this corresponds to cross flow instability (CFI). In Boltz *et al.* (1960), another set of experiment was performed using a wing with the higher sweep-back angle of 50°. These results are also shown in Figure 9.19, which indicates no transition due to streamwise instability (SI). Instead, it was noticed that transition point moved more rapidly upstream towards the leading edge, as compared to the 20° sweep angle case. At $Re_c > 14 \times 10^6$, the transition point moved at the leading edge, which is attributed to leading edge contamination (LEC).

The implications of these mechanisms of instability were reported in Arnal (1986), with the help of flow past an infinite swept wing. The reason for undertaking studies with an infinite swept wing is not only for the ease of analysis it affords, but this is due to its canonical status, the same accorded to the Blasius boundary layer or to the Falkner–Skan boundary layer for 2D flows. In Figure 9.20, it is shown how one can use infinite swept wing results for any real finite swept wing. For a given section, an equivalent infinite swept wing is constructed with the local chord, while the sweep angle of the original wing is maintained by aligning the quarter chord.

In the figure at any spanwise station, an equivalent infinite swept wing is constructed with the sweep angle determined by the quarter chord line. This is akin to using a quasi-parallel approximation for predicting instability and transition for 2D flow over an airfoil. For this reason, in the following we discuss infinite swept wing flow and its stability.

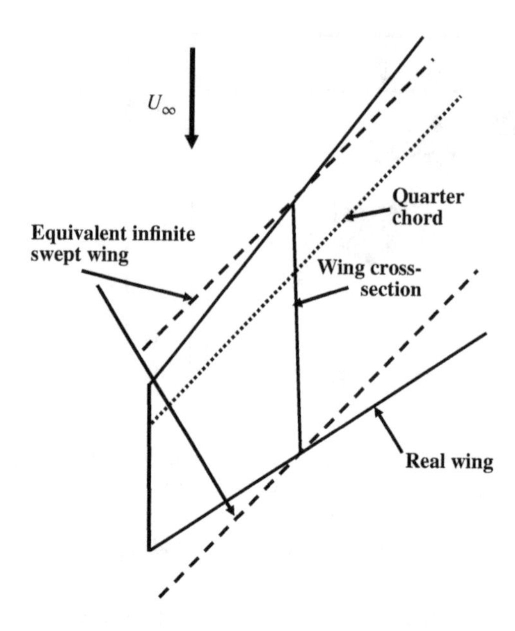

Figure 9.20 Equivalent infinite swept wing approximation for an arbitrary finite swept wing

9.14 Infinite Swept Wing Boundary Layer

Referring to Figure 9.15, for the flow past a swept wing we noted the sweep angle φ and the two axes systems for the flow near the leading edge. The free stream velocity Q_∞ is split into a component normal to the leading edge, U_∞, and a component parallel to the leading edge, W_∞. One associates the WFCS with the (X, Y, Z)-axes system and the coordinate system (x, y, z) is attached to the ESFCS. As noted earlier, $Y \equiv y$ and hence no distinction is made between these two. In these two coordinate systems, velocity $Q(y)$ inside the boundary layer is split into $U(y)$ and $W(y)$ in WFCS and $U_1(y)$ and $W_1(y)$ in ESFCS. It is these last components (streamwise and cross flow profiles) which are used in the stability equations, Equations (9.27) and (9.82). At the edge of the boundary layer (y_e), these components are given as $U(y_e) = U_e$, $W(y_e) = W_e$ and $U_1(y_e) = U_{1e} = (U_e^2 + W_e^2)^{1/2}$, $W_1(y_e) = 0$. The assumption of an infinite swept wing is expressed by an invariance of flow variables along the span. The mean flow is obtained from the solution of the following equations.

$$\frac{\partial U}{\partial X} + \frac{\partial V}{\partial Y} = 0 \tag{9.91}$$

$$U\frac{\partial U}{\partial X} + V\frac{\partial U}{\partial Y} = U_e\frac{dU_e}{dX} + \nu\frac{\partial^2 U}{\partial Y^2} \tag{9.92}$$

$$U\frac{\partial W}{\partial X} + V\frac{\partial W}{\partial Y} = \nu\frac{\partial^2 W}{\partial Y^2}. \tag{9.93}$$

One notes that the first two equations are decoupled from the third and can be solved in that sequence. Also, these equations are the same as given in Section 9.6, where we proved following Equation (9.93) that at the free stream $W_e = W_\infty$.

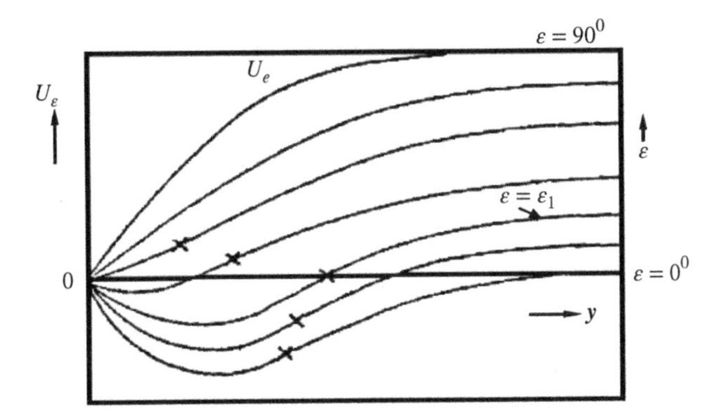

Figure 9.21 Projected velocity profiles according to Equation (9.95). *Source:* Arnal (1986). Copyright NATO-CSO

To solve Equations (9.91) to (9.93), one needs the velocity profile at $x = 0$, which generally is obtained from the attachment line equations corresponding to an infinite swept wing. At any station, having obtained $U(y)$ and $W(y)$ from the solution of the above equations, one obtains the streamwise and cross flow profiles from

$$U_1(y) = U\cos\theta_0 + W\sin\theta_0 \tag{9.94a}$$

$$W_1(y) = -U\sin\theta_0 + W\cos\theta_0, \tag{9.94b}$$

where the angle between the X and x axes is obtained from $\tan\theta_0 = \frac{W_e}{U_e}$. If ψ represents the angle formed by the wavenumber vector and the external streamline, then one can define U_ψ as in Equation (9.81) and solve the Orr–Sommerfeld equation given by Equation (9.82). In Arnal (1986), a complementary angle ϵ has been defined with respect to the cross flow direction, i.e. $\epsilon = \pi/2 - \psi$, and used to study the stability of the velocity vector projected in all directions from $\epsilon = 0^0$ to 90^0, and these profiles are obtained from

$$U_\epsilon = U_1\sin\epsilon + W_1\cos\epsilon. \tag{9.95}$$

By varying ϵ from 0° to 90°, one moves continuously from the cross flow to the streamwise profile. At a given streamwise location, a complete temporal calculation using Equation (9.82) reduces the 3D instability calculation to a sequence of 2D calculations on the family of profiles U_ϵ. In Figure 9.21, we show a set of such profiles for different values of ϵ from 0° (cross flow profile) to 90° (streamwise profile).

In this figure, a cross indicates the location of an inflection point for the profile, with the streamwise profile representing the typical 2D boundary layer profile with the negative pressure gradient and hence its first derivative is monotonically changing without any inflection point. There is a range of low values of ϵ starting from the cross flow profile for which an inflection point is always present and represents vulnerability to inviscid instability. Of all the angles in this range, one would be interested in one particular angle (ϵ_I) for which the inflection point occurs at a height where the local mean velocity is zero. This profile is termed the critical profile in Arnal (1986).

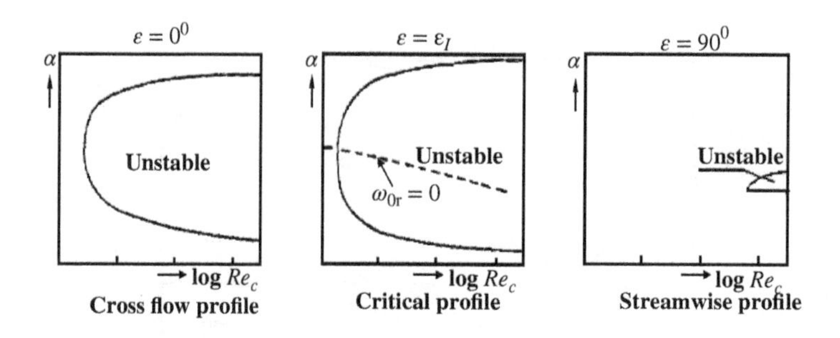

Figure 9.22 Calculated neutral curves for some of the representative velocity profiles shown in Figure 9.21. *Source:* Arnal (1986). Copyright NATO-CSO

Figure 9.22 shows representative stability diagrams for $\epsilon = 0°$, ϵ_I and $90°$. One can draw the following conclusions from the frames in Figure 9.22.

(i) The streamwise profile is very stable, as one notes a very high value of Re_{cr}. Also, as Re tends to large values, the neutral curve collapses, indicating no unstable frequencies, and thus the observed instability is caused by viscous action only at finite Reynolds numbers.
(ii) However, due to the presence of inflection points in the velocity profiles, the other two cases are highly unstable, with very low Re_{cr}. Also, the amplification rates are significantly higher and there is always a wide range of unstable wavenumbers at very large Re_c, implying the instability is dominated by an inviscid mechanism which does not go away, even in the limit of $Re \rightarrow \infty$.
(iii) The critical profile turns out to be the most unstable among the three profiles shown and more importantly, one can see that zero frequency waves are amplified, which corresponds to standing waves. Such unstable standing waves are seen for a range of ϵ around ϵ_I, say $\epsilon = 2$ to 6 degrees. Note that the experiments in Arnal *et al.* (1984a) showed the presence of streaks in wall visualization, which are the standing wave patterns amplified by the profiles for values of ϵ in the neighbourhood of ϵ_I.

Thus, the results from this infinite swept wing study clearly explain the observed experimental phenomenon at large enough Re_c for swept wing flow, near the leading edge in the negative pressure gradient region. While the above description tells us about the onset of instability near the leading edge of an infinite swept wing, one may also like to inquire about the downstream development of such initial unstable disturbances.

A description of it can be followed by noting that the equation of the external streamline is given by

$$\frac{dz}{W_e} = \frac{dx}{U_e}.$$

Since for the infinite swept wing flow, $W_e = W_\infty$, hence the equation of the streamline can be alternately written as $\frac{dz}{dx} = \frac{W_\infty}{U_e}$, which in turn implies that the streamline curvature $(\frac{d^2z}{dx^2})$ changes sign at an abscissa location (x_M), where U_e is maximum (i.e. where $\frac{dU_e}{dx} = 0$). Thus, x_M is also the location where the pressure gradient changes sign. That in turn implies

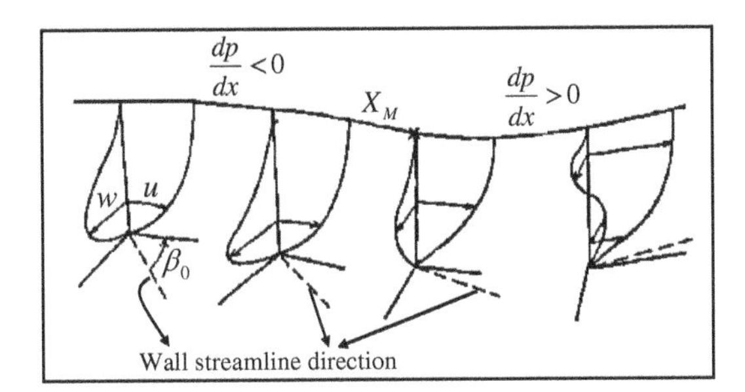

Figure 9.23 Sketch of a laminar boundary layer developing over a swept wing. The angle β_0 represents the orientation of external streamline with respect to wall streamline

that across x_M the pressure gradient changes from a negative pressure gradient to a positive pressure gradient. This is demonstrated in Figure 9.23, showing the evolution of laminar flow developing over a swept wing.

In this figure, β_0 is the angle between the external streamline and the wall streamline directions. Near the leading edge, the negative pressure gradient displays a full cross flow that increases with height initially and then decreases back to zero at the edge of the shear layer, with nowhere becoming negative. This causes the external streamline to curve inwards in the first quadrant of the xz-plane. However, as x_M is approached, the pressure gradient reduces to zero and that makes the cross flow component also weaker. This causes the wall streamline to deviate towards the streamwise direction, as indicated by the reduction of β_0. When the longitudinal pressure gradient changes sign, somewhere downstream of that location the cross flow velocity component $W_i(y)$ itself can switch sign, close to the wall, giving rise to an S-shaped velocity profile. Thus, the S-shaped cross flow profile occurs towards the end of the negative pressure gradient zone and may not necessarily be the most critical profile. This aspect is highlighted again while discussing the Falkner–Skan–Cooke profile in the next section. When the positive pressure gradient is very strong, the cross flow velocity can become fully negative inside the shear layer. It is in this region, $U_1(y)$ can exhibit an inflection point which can imply strong streamwise instability in that part of the flow. Once again, this region is dominated by streamwise instability and not by cross flow instability.

In Section 9.6, we noted the equilibrium flows expressed in the transformed plane, with Equations (9.65) and (9.66) providing equations for an infinite swept wing. If one considers the external flow description given by Equation (9.73), then by the additional choices of $r = s = n = 0$, these equations further reduce to the Falkner–Skan–Cooke equations given by Equations (9.76) and (9.77).

9.15 Stability of the Falkner–Skan–Cooke Profile

This is a similarity profile describing an infinite swept wing and the attachment line flow and is helpful in explaining the passage from streamwise to cross flow instability. For the infinite swept wing, the outer inviscid velocity components in the normal and parallel to the leading edge are given by $U_e = kx^m$, $W_e = W_\infty$. In this description, we use the following similarity variables

$$\eta = \left(\frac{m+1}{2}\right)^{1/2}\left(\frac{U_e}{\nu x}\right)^{1/2} y$$

$$\frac{U}{U_e} = \bar{F}'(\eta) \text{ and } \frac{W}{W_e} = \bar{G}(\eta) \tag{9.96}$$

With the above transformations, boundary layer equations reduce to the following ordinary differential equations

$$\bar{F}''' + \bar{F}\bar{F}'' + \beta(1 - \bar{F}'^2) = 0 \tag{9.97}$$

$$\bar{G}'' + \bar{F}\bar{G}' = 0, \tag{9.98}$$

where $\beta = \frac{2m}{m+1}$ and the attachment line flow can be further obtained corresponding to $m = \beta = 1$. The angle between the external streamline and the normal to the leading edge is $\theta_0 = \tan^{-1}\frac{W_e}{U_e}$ and so the streamwise and cross flow profiles are given by

$$U_1 = U\cos\theta_0 + W\sin\theta_0 \text{ and } W_1 = -U\sin\theta_0 + W\cos\theta_0.$$

Note that

$$\cos\theta_0 = U_e/(U_e^2 + W_e^2)^{1/2};\ \sin\theta_0 = W_e/(U_e^2 + W_e^2)^{1/2}$$

and

$$U_{1e} = (U_e^2 + W_e^2)^{1/2};\ W_{1e} = 0.$$

Thus

$$U_1/U_{1e} = \bar{F}'\cos^2\theta_0 + \bar{G}\sin^2\theta_0 \tag{9.99}$$

$$W_1/U_{1e} = (\bar{G} - \bar{F}')\sin\theta_0\cos\theta_0. \tag{9.100}$$

Thus, it is possible to construct streamwise and cross flow profiles depending upon two quantities: (i) the pressure gradient parameter β and (ii) the cross flow parameter θ_0. Note that W_1 is maximum for $\theta_0 = \pi/4$. Also for $\beta = 0$, from Equation (9.97) one notes that $\bar{F}' = \bar{G}$, which leads one to the Blasius profile: $U_1/U_{1e} = \bar{F}_0'$ and $W_1/U_{1e} = 0$. Figure 9.24 shows the cross flow velocity profile for $\theta_0 = \pi/4$ and for different β.

For $\beta < 0$, the whole cross flow profile changes sign. The Falkner–Skan-Cooke (FSC) profiles cannot represent S-shaped profiles. As discussed in the previous section, the S-shaped profile and fully reversed profiles occur in the downstream location, where cross flow instability is not the most critical event. In Arnal (1986) and Mack (1984), stability properties of the FSC profiles have been reported. Of specific interest, are the results shown in Figure 9.25, where the critical Reynolds number is plotted as a function of β for the velocity profile with the cross flow parameter $\theta_0 = \pi/4$, only for zero frequency disturbances.

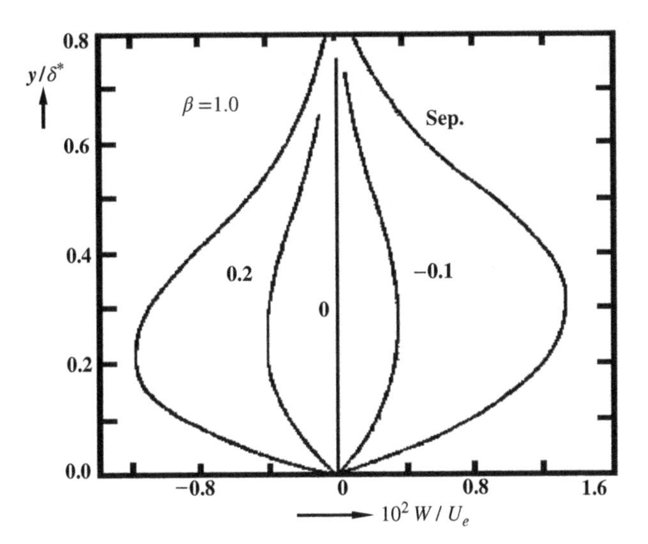

Figure 9.24 Cross flow velocity profile obtained from solution of the Falkner–Skan–Cooke equation. *Source:* Arnal (1986). Copyright NATO-CSO. The extreme right curve is for the case of β for which separation is incipient

The dotted line in Figure 9.25 is for the 2D Falkner–Skan profile ($\theta_0 = 0$) and it is noted that for all adverse pressure gradient flows ($\beta < 0$), the FSC profiles are more stable than the 2D Falkner–Skan profile. However, there is a critical $\beta = 0.07$, above which 3D flow is more unstable than the corresponding 2D flow. The Reynolds number is defined with respect to x

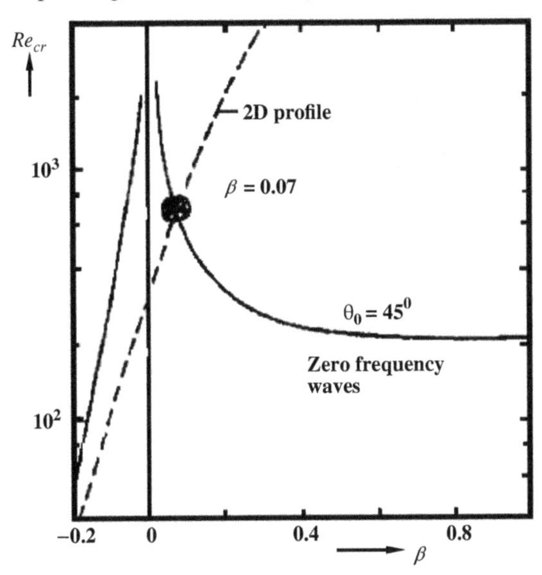

Figure 9.25 Critical Reynolds number variation with pressure gradient parameter, β, for zero frequency disturbances. The dotted line shows the corresponding two-dimensional critical Reynolds number for the Falkner–Skan profile. *Source:* Arnal (1986). Copyright NATO-CSO

in these calculations. This also shows that there is a possible existence of stationary waves for 3D flows, as compared to 2D flows, which do not support stationary waves. This aspect is discussed next for 3D flows, mainly from the point of view of unstable stationary waves.

9.16 Stationary Waves over Swept Geometries

There is sufficient experimental and theoretical evidence for the existence of unstable stationary waves for 3D flows. This has been shown for the Falkner–Skan–Cooke profile in Figure 9.25. In Arnal *et al.* (1984a) and Michel *et al.* (1984), experimental results have been reported for a swept wing with an ONERA D profile; special features of the model are also discussed and shown in Figure 9.16. The sweep angle for the model was 40°, while the angle of attack was −8° and the wing chord was 35 cm. These angles were decided so that the external velocity normal to the leading edge (U_e) is always accelerating over all possible x/c, as shown in Figure 9.26, for $Q_\infty = 81$ m / s.

The large acceleration over the top surface is obtained via the large sweep angle, as well as the negative angle of attack. The measured spanwise variation of the external velocity component is shown for different streamwise stations in Figure 9.27.

When the measured velocity is plotted at a height very close to the wall and in the laminar part of the flow, one notes spanwise waviness with a spanwise wavelength corresponding to the streak spacing noted in the wall visualization shown in Figure 9.23 of Arnal (1986). In the transitional part of the flow with an onset at $x/c = 0.30$, the flow evolution in time is noted to be completely chaotic. The external flows displayed in Figure 9.27 are time averaged values. Also, time averaged flow velocity in the transitional part of the flow is higher than the measured one in the turbulent part of the flow.

9.17 Empirical Transition Prediction Method for Three-Dimensional Flows

Here one begins with the assumption that transition will be caused by any one of the dominant mechanisms discussed in this chapter. Thus, transition criteria are applied for each of these mechanisms and it is assumed that the boundary layer will cease to be laminar when any one of these is satisfied.

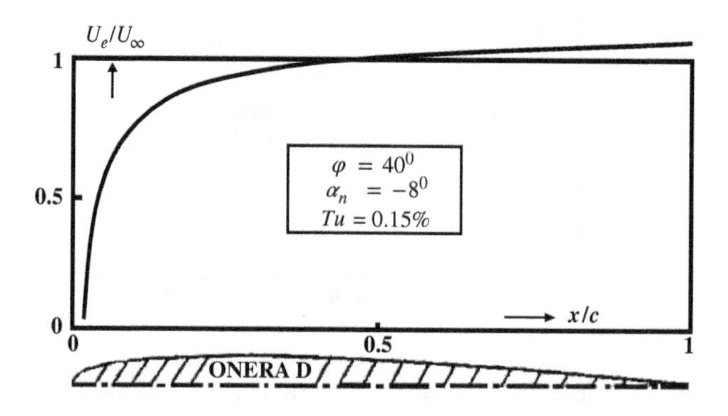

Figure 9.26 Edge velocity over a swept wing with an ONERA D profile. *Source:* Arnal (1986). Copyright NATO-CSO

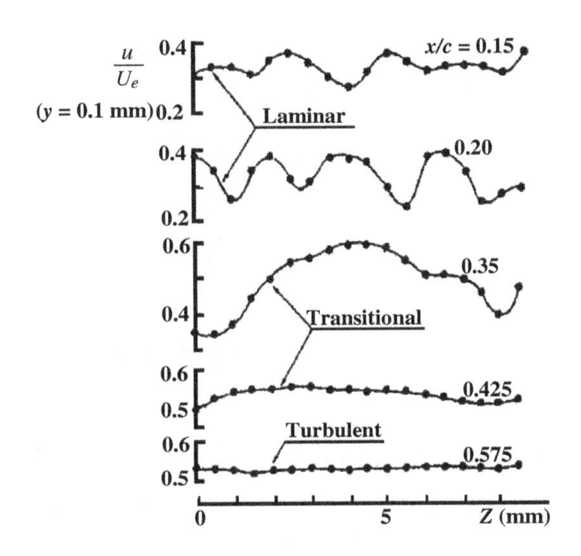

Figure 9.27 Streamwise velocity variation with spanwise coordinate for the experimental results reported in. *Source:* Arnal (1986). Copyright NATO-CSO

9.17.1 Streamwise Transition Criterion

Streamwise velocity profiles are closer to two-dimensional flow profiles and hence the criterion would be identical when applied along the external streamlines. One can apply any one of the methods described for two-dimensional flows or the modified method due to Granville, as given in Arnal (1986).

9.17.2 Cross Flow Transition Criteria

An early attempt in providing a transition criterion due to cross flow instability was in Owen (1952, 1953). However, it overestimates the instability and is not used any more. Beasley (Beasley 1973) developed a criterion based on a cross flow profile W_1 and a Reynolds number defined from $\delta_2 = \int_0^\infty \frac{W_1}{U_{Ie}} dy$ as $Re_{\delta 2} = \frac{U_{1e}\delta_2}{\nu}$. Beasley (Beasley 1973) proposed that transition would occur whenever $Re_{\delta 2} \geq 150$. While this is a simple criterion, the following separate criteria proposed in Arnal *et al.* (1984a) are found more satisfactory.

a) C1 Criterion:

The above criterion of Beasley ($Re_{\delta 2} \geq 150$) does not involve any parameter depending upon streamwise location, and hence its applicability is limited. It has been reported in Arnal *et al.* (1984a), by comparing experimental data with a computed laminar boundary layer, that transition does not depend uniquely upon $Re_{\delta 2}$. Better results were obtained by correlating a transition Reynolds number with a parameter linked to the streamwise velocity profile. When shape factor (H) is used to represent the latter, the following empirical relation represents the experimental data quite well.

$$[Re_{\delta 2}]_{tr} = \frac{300}{\pi} \tan^{-1}\left[\frac{0.106}{(H_{tr} - 2.3)^{2.05}}\right]. \tag{9.101}$$

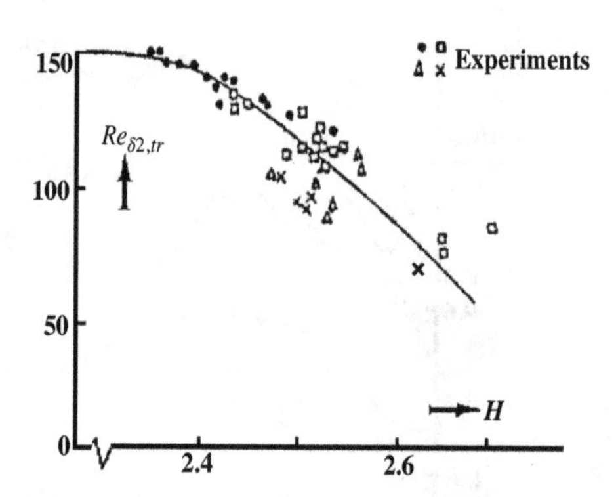

Figure 9.28 Correlation of experimental transition location point with the calculation using the $C1$ criterion. *Source:* Arnal (1986). Copyright NATO-CSO

This criterion remains valid for $H < 2.7$, beyond which the streamwise criterion is found to be important. Figure 9.28 shows general agreement with the ensemble of experimental data and the criterion given by Equation (9.101).

b) C2 Criterion:

This method uses linear stability theory along with Stuart's theorem. To study the stability along a perturbation direction, making an angle ϵ with the cross flow direction, we define a velocity parameter

$$\frac{U_\epsilon}{U_{\epsilon e}} = \frac{U_1}{U_{1e}} + \frac{W_1}{U_{1e}} \cot \epsilon, \tag{9.102}$$

where $U_{\epsilon e} = U_{1e} \sin \epsilon$ with $\epsilon = 0$ and π excluded. Similarly, define a parametric displacement thickness and a Reynolds number by

$$\delta_{1\epsilon} = \int_0^\infty \left[1 - \frac{U_\epsilon}{U_{\epsilon e}}\right] dy \tag{9.103}$$

$$Re_{\delta_{1\epsilon}} = \frac{U_{\epsilon e}\delta_{1\epsilon}}{\nu} = Re_{\delta_1} \sin \epsilon + Re_{\delta_2} \cos \epsilon. \tag{9.104}$$

Next, one describes the stability of the U_ϵ profiles and obtains a $(Re_{\delta_{1\epsilon}})_{cr}$ by stability analysis. For any experimental case, one can define another function

$$g(\epsilon) = \frac{(Re_{\delta 1\epsilon})_{cr}}{Re_{\delta 1\epsilon}}. \tag{9.105}$$

Obviously if $g(\epsilon) > 1$, then we have a stable condition. However, if $g(\epsilon) < 1$, then the flow in that direction is unstable. For a three-dimensional flow at any given abscissa, g depends on ϵ and on the Reynolds number based on chord (Re_c). Indeed, $Re_{\delta_{1\epsilon}}$ varies as $\sqrt{Re_c}$, whereas $(Re_{\delta_{1\epsilon}})_{cr}$ is invariant. Thus, if $g_1(\epsilon)$ and $g_2(\epsilon)$ correspond to Re_{c1} and Re_{c2}, then

$$\frac{g_1}{g_2} = \sqrt{\frac{Re_{c2}}{Re_{c1}}}. \tag{9.106}$$

That is, when the chord Reynolds number increases, g decreases and the range of unstable directions widens.

9.17.3 Leading Edge Contamination Criterion

In Section 9.15, we found the momentum thickness of the attachment line boundary layer to be given by Equation (9.102):

$$\theta = \int_0^\infty \frac{U_1}{U_{1e}}\left(1 - \frac{U_1}{U_{1e}}\right)dy = 0.404\left(\frac{\nu}{k}\right)^{1/2}, \tag{9.107}$$

where $k = \left(\frac{dU_{1e}}{dx}\right)_{x=0}$. Therefore, the Reynolds number based on the momentum thickness is given by

$$Re_\theta = \frac{Q_\infty \theta \sin\varphi}{\nu}. \tag{9.108}$$

Experimentally, it has been noted that transition by leading edge contamination occurs when $Re_\theta > 100$. This can be used as the criterion for leading edge contamination. In Figure 9.19, various estimates of transition given by the criteria discussed above are shown for the experimental data of Boltz *et al.* (1960) for the two sweep-back angle cases. The two cross flow criteria give results which are close to experimental values, as well as to themselves. The lower sweep angle case showed a smooth passage from streamwise to cross flow instability with the increase of the chord Reynolds number. Streamwise instability occurred where the pressure gradient was positive, while cross flow instability occurred in a negative pressure gradient region. There are no available experimental data by which the leading edge contamination criterion could be tested. For the higher sweep angle case, the instability is due to cross flow only for Re_c greater than 6 million – as seen for the $\varphi = 50°$ case shown in Figure 9.19. For the straight wing ($\varphi = 0°$) case in Figure 9.19, instability is seen to be due to streamwise instability for $Re_c \leq 20 \times 10^6$.

Bibliography

Abbot, I.H. and von Doenhoff, A.E. (1959) *Theory of Wing Section*. Dover Publications, New York, USA.

Allen, L.D. and Burrows, F.M. (1956) Flight experiments on the boundary layer characteristics of a sweptback wing. *Cranfield, College of Aero. Rept.* **104**.

Anscombe, A. and Illingworth, L.N. (1952) Wind tunnel observations of boundary layer transition on a wing at various angles of sweep back. *British ARC* R & M **2968**.

Arnal, D. (1984) Description and prediction of transition in two-dimensional incompressible flow. *AGARD Report* 709, **709**, 2.1–2.71.

Arnal, D. (1986) Three-dimensional boundary layer: Laminar-turbulent transition. *AGARD Report* **741**, 1–34.

Arnal, D., Casalis, G. and Jullien, J.C. (1989) Experimental and theoretical analysis of natural transition on infinite swept wing. In the IUTAM Symp. Proc. on *Laminar-Turbulent Transition*, (Eds.: D. Arnal and R. Michel), Springer-Verlag, Germany.

Arnal, D., Coustols, R. and Jullien, J.C. (1984a) Experimental and theoretical study of transition phenomena on infinite swept wing. *La Recherche Aerospatiale*, **1984-4**, 275–290.

Arnal, D., Habiballah, M. and Coustols, E. (1984) Laminar instability theory and transition criteria in two- and three-dimensional flow. *La Recherche Aerospatiale*, **1984-2**.

Arnal, D. and Jullien, J.C. (1987) Three-dimensional transition studies at ONERA/CERT. *AIAA Paper* 87-1335.

Bagade, P.M., Krishnan, S.B. and Sengupta, T.K. (2014) DNS of low Reynolds number aerodynamics for MAV applications. Unpublished results.

Beasley, J.A. (1973) Calculation of the laminar boundary layer and the prediction of transition on a sheared wing. *British ARC* R & M **3787**.

Bender, C.M. and Orszag, S.A. (1987) *Advanced Mathematical Methods for Scientists and Engineers.* McGraw-Hill Book Co., International edn., Singapore.

Betchov, R. and Criminale, W.O. Jr, (1967) *Stability of Parallel Flows.* Academic Press, New York, USA.

Bhaumik, S. and Sengupta, T.K. (2014) Precursor of transition to turbulence: Spatiotemporal wave front. *Phys. Rev. E,* **89**, 043018.

Boltz, F.W., Kenyon, G.C. and Allen, C.Q. (1960) Effect of sweep angle on the boundary layer stability characteristics of an untapered wing at low speeds. *NASA TN D*-**338**.

Bradshaw, P., Mizner, G. and Unsworth, K. (1976) Calculation of compressible turbulent boundary layers on straight-tapered swept wing. *AIAA J.,* **14**, 399–400.

Burrows, F.M. (1956) A theoretical and experimental study of the boundary layer flow on a 45° swept back wing. *Cranfield College of Aero. Rept.* **109**.

Cebeci, T. (2004) *Stability and Transition: Theory and Application.* Horizons Publishing Inc. Springer, Long Beach, CA, USA.

Cebeci, T. and Bradshaw, P. (1977) *Momentum Transfer in Boundary Layers.* Hemisphere Publishing Corp., Washington DC, USA.

Cebeci, T., Chen, H.H. and Kaups, K. (1992) Further consideration of the effect of curvature on the stability of three-dimensional flows. *Comput. Fluids,* **21**(4), 491–502.

Cebeci, T. and Cousteix, J. (2005) *Modelling and Computation of Boundary-Layer Flows.* Second Edn. Horizon Publishing Inc., Springer, Long Beach, CA, USA.

Cebeci, T. and Stewartson, K. (1980) Stability and transition in three-dimensional flows. *AIAA J.,* **18**, 398–405.

Collier, F.S., Bartlett, D.W., Wagner, R.D., Tat, V.V. and Anderson, B.T. (1985) Correlation of boundary-layer stability analysis with flight transition data. In IUTAM Symp. Proc. on *Laminar-Turbulent Transition.* Eds. Michel R and Arnal J, Springer Verlag, Germany.

Crabtree, L.F., Kuchemann, D. and Sowerby, L. (1963) Three-dimensional boundary layers. In *Laminar Boundary Layers* (Ed.: L. Rosenhead). Clarendon Press, UK.

Drazin, P.G. and Reid, W.H. (1981) *Hydrodynamic Stability.* Cambridge University Press, UK.

Dryden, H.L. (1959) Transition from laminar to turbulent flow. In *Turbulent Flows and Heat Transfer: High Speed Aerodynamics and Jet Propulsion.* Edited by C.C. Lin, Princeton, USA.

FjØrtoft, R. (1950) Application of integral theorems in deriving criteria of stability for laminar flows and for baroclinic circular vortex. *Geofys. Publ. Oslo,* **17**(6), 1–51.

Gaster, M. and Grant, I. (1975) An experimental investigation of the formation and development of a wave packet in a laminar boundary layer. *Proceedings Royal Society, London Series A.* **347**, 253–269.

Goldstein, S. (1938) *Modern Development in Fluid Mechanics.* 1 and 2, Clarendon Press, Oxford, UK.

Gray, W.E. (1952) The effect of wing sweep on laminar flow. *RAE TM Aero.* **255**.

Gregory, N., Stuart, J.T. and Walker, J.S. (1955) On the stability of three dimensional boundary layer with application to the flow due to a rotating disc. *Phil. Trans. Roy. Soc. London,* Series A, **248**, 155–199.

Hansen, A.G. and Yohner, P.L. (1958) Some numerical solutions of similarity equations for three-dimensional, laminar incompressible boundary-layer flows. *NACA TN* **4370**.

Heisenberg, W. (1924) Über stabilityät und turbulenz von flüssigkeitsströmen. *Ann. Phys. Lpz.,* **379**, 577–627. (Translated as 'On stability and turbulence of fluid flows'. *NACA TM* **1291** (1951).

von Helmholtz, H. (1868) On discontinuous movements of fluids. *Phil. Mag.,* **(4)36**, 337–346.

Jaffe, N.A., Okamura, T.T. and Smith, A.M.O. (1970) Determination of spatial amplification factors and their application to predicting transition. *AIAA,* **8**(2), 301–308.

Landau, L.D. and Lifshitz, E.M. (1959) *Fluid Mechanics,* vol. 6. Addision-Wesley. Pergamon Press, London, UK.

Lord Kelvin (1871) Hydrokinetic solutions and observations. *Phil Mag.(4),* **42**, 362–377.

Lord Rayleigh (1887) On the stability or instability of certain fluid motions. In *Scientific Papers,* **3**, 17–23, Cambridge University Press, Cambridge, UK.

Lord Rayleigh (1890) *Scientific Papers.* **2**, Cambridge University Press, Cambridge, UK.

Mack, L.M. (1984) Boundary layer stability theory. Special course on. *Stability and Transition of Laminar Flow. AGARD Report* **709**.

Mack, L.M. (1977) Transition and laminar instability. *JPL Publication,* 77–15.

Mack, L.M. (1988) Stability of three-dimensional boundary layers on swept wings at transonic speeds. In IUTAM Symp.Proc. of *Transonic III,* Goettingen, Germany.

Malik, M.R. (1982) COSAL – a black box compressible stability analysis code for transition prediction in three-dimensional boundary layer. *NASA CR* bf 165 925.

Malik, M.R. and Poll, D.I.A. (1985) Effect of curvature on three-dimensional boundary-layer stability. *AIAA J.,* **23**, 1362–1369.

Michel, R., Arnal, D., Coustols, E. and Jullien, J.C. (1984) Experimental and theoretical studies of boundary layer transition on a swept infinite wing. In Proc. of IUTAM Symp. *On Laminar-Turbulent Transition*, Springer Verlag, Novosibirsk, USSR.

Morkovin, M.V. (1958) Transition from laminar to turbulent flow – a review of the recent advances in its understanding. *ASME Trans.*, **80**, 1121–1128.

Morkovin, M.V. (1990) On receptivity to environmental disturbances. In *Instability and Transition* (Eds.: Hussaini MY and Voigt RG). 272–280, Springer Verlag, New York, USA.

Morkovin, M.V. (1991) Panoramic view of changes in vorticity distribution in transition instabilities and turbulence. In *Transition to Turbulence* (Eds. Reda DC, Reed HL and Kobayashi R). *ASME FED Publ.*, **114**, 1–12.

Nayfeh, A.H. (1980) Stability of three-dimensional boundary layers. *AIAA J.*, **18**, 406–416.

Orr W. McF, (1907) The stability or instability of the steady motions of a perfect liquid and of a viscous liquid. Part I: A perfect liquid. Part II: A viscous liquid. *Proc. Roy. Irish Acad.* **A27**, 9–138.

Owen, P.R. and Randall, D.G. (1952) Boundary layer transition on a sweptback wing. *RAE TM* **277**.

Owen, P.R. and Randall, D.G. (1953) Boundary layer transition on a sweptback wing: a further investigation. *RAE TM* **330**.

Poll, D.I.A. (1984) Transition description and prediction in three-dimensional flows. *AGARD Rept.* **709.** (Special Course on Stability and Transition of Laminar Flows, VKI, Belgium).

Poll, D.I.A. (1985) Some observations of the transition process on the windward face of a long yawed cylinder. *J. Fluid Mech.,* **150**, 329–356.

Prandtl, L. (1935) In *Aerodynamic Theory*. (Ed. W.F. Durand). **3**, 178–190. Springer, Berlin, Germany.

Prandtl, L. (1946) On boundary layers in three-dimensions. *British ARC Rept.* **9828**.

Reed, H.L. (1988) Wave interactions in swept-wing flow. *Phys. Fluids,* **30**(11), 3419–3426.

Reed, H.L. and Saric, W.S. (1989) Stability of three-dimensional boundary layers. *Ann. Rev. Fluid Mech.,* **21**, 235–285.

Reynolds, O. (1883) An experimental investigation of the circumstances which determine whether the motion of water shall be direct or sinuous and of the law of resistance in parallel channels. *Phil. Trans. Roy. Soc.,* **174**, 935–982.

Saric, W.S. and Yeates, L.G. (1985) Experiments on the stability of cross flow vortices in swept-wing flows. *AIAA Paper* 85-0493.

Sayadi, T., Hamman, C.W. and Moin, P. (2013) Direct numerical simulation of complete H-type and K-type transitions with implications for the dynamics of turbulent boundary layers. *J. Fluid Mech.*, **724**, 480–509.

Schlichting, H. (1933) Zur entstehung der turbulenz bei der plattenströmung. *Nach. Gesell. d. Wiss. z. Gött., MPK,* 181-208.

Schlichting, H. (1979) *Boundary Layer Theory.* Seventh Edn., McGraw-Hill, New York, USA.

Schubauer, G.B. and Skramstad, H.K. (1947) Laminar boundary layer oscillations and the stability of laminar flow. *J. Aero. Sci.*, **14**(2), 69–78.

Sengupta, T.K. (1991) Impulse response of laminar boundary layer and receptivity. In *Proc.* 7th *Int. Conf. Numerical Meth. for Laminar and Turbulent Layers,* (Ed. C. Taylor), Pineridge Publ., UK.

Sengupta, T.K. (2012) *Instabilities of Flows and Transition to Turbulence.* CRC Press, Florida, USA.

Sengupta, T.K. and Bhaumik, S. (2011) Onset of turbulence from the receptivity stage of fluid flows. *Phys. Rev. Lett.*, **154501**, 1-5.

Sengupta, T.K., Bhaumik, S. and Bhumkar, Y.G. (2011) Nonlinear receptivity and instability studies by proper orthogonal decomposition. In the *Proc. 6th AIAA Theoretical Fluid Mechanics Conf. AIAA*, **1**, 373–415 (Curran Associates Inc., USA).

Sengupta, T.K., Bhaumik, S. and Bhumkar, Y.G. (2012) Direct numerical simulation of two-dimensional wall bounded turbulent flow from receptivity stage. *Phys. Rev. E*, **85**, 026308.

Sengupta, T.K., Bhaumik, S. and Bose, R. (2013) Direct numerical simulation of two-dimensional transitional mixed convection flows: Viscous and inviscid instability mechanisms. *Phys. Fluids*, **25**, 094102.

Sengupta, T.K. and Dipankar, A. (2005) Subcritical instability on the attachment-line of an infinite swept wing. *J. Fluid Mech.,* **529**, 147–171.

Sengupta, T.K., Ballav, M. and Nijhawan, S. (1994) Generation of Tollmien-Schlichting waves by harmonic excitation. *Phys. Fluids,* **6**(3), 1213–1222.

Sengupta, T.K., Kasliwal, A., De, S. and Nair, M. (2003) Temporal flow instability for Robins-Magnus effect at high rotation rates. *J. Fluids Struct.*, **17**, 941–953.

Sengupta, T.K., Nair, M.T. and Rana, V. (1997) Boundary layer excited by low frequency disturbances – Klebanoff mode. *J. Fluids Struct.,* **11**, 845–853.

Sengupta, T.K., Rajpoot, M.K. and Bhumkar, Y.G. (2011) Space-time discretizing optimal DRP schemes for flow and wave propagation problems. Comput. Fluids, **47**(1), 144–154.

Sengupta, T.K., Rao, A.K. and Venkatasubbaiah, K. (2006) Spatio-temporal growing wave fronts in spatially stable boundary layers. *Phys. Rev. Lett*, **96**(22), 224504.

Sengupta, T.K., Rao, A.K. and Venkatasubbaiah, K. (2006a) Spatio-temporal growth of disturbances in a boundary layer and energy based receptivity analysis. *Physics of Fluids,* **18**, 094101.

Sengupta, T.K., Singh, N. and Suman, V.K. (2010) Dynamical system approach to instability of flow past a circular cylinder. *J. Fluid Mech.,* **656**, 82–115.

Sengupta, T.K., Vijay, V.V.S.N. and Singh, N. (2011a) Universal instability modes in internal and external flows. *Comput. Fluids,* **40**, 221–235.

Smith, A.M.O. and Gamberoni, N. (1956) Transition, pressure gradient and stability theory. *Douglas Aircraft Rept.* **ES-26338**.

Sommerfeld, A. (1908) Ein Beitrag zur hydrodynamiscen Erklarung der turbulenten Flussigkeitsbewegung. In *Proc. 4th Int. Cong. Mathematicians,* Rome, 116–124.

Srokowski, A.J. and Orszag, S.A. (1977) Mass flow requirements for LFC wing design. *AIAA Paper* 77-1222.

Stuart, J.T. (1953) The basic theory of the stability of three-dimensional boundary layer. *British ARC Report* **15904**.

Tollmien, W. (1931) Über die enstehung der turbulenz. I, English translation. *NACA TM* **609**.

Van Driest, E.R. (1951) Turbulent boundary layer in compressible fluids. *J. Aero. Sci.,* **18**, 145–160.

Van Ingen, J.L. (1956) A suggested semi-empirical method for the calculation of the boundary layer transition region. *Inst. Of Technology, Dept. of Aeronautics and Eng. Rept.* **VTH-74**.

Viken, J.K. (1983) Aerodynamic design considerations and theoretical results for a high Reynolds number natural laminar flow airfoil. M.S. thesis. The School of Engineering and Applied Science, George Washington Univ., USA.

Wazzan, A.R., Gazley, Jr C. and Smith, A.M.O. (1981) $H - R_x$ method for predicting transition. *AIAA J.,* **19**(6), 109–114.

Wheeler, A.J. and Johnston, J.P. (1972) Three-dimensional turbulent boundary layers – data sets for two-space-coordinate flows. *Stanford Univ. Rept.,* **MD 32**.

White, F.M. (1991) *Viscous Fluid Flow.* McGraw-Hill Int. Edn., New York, USA.

Wu, X. and Moin, P. (2010) Transitional and turbulent boundary layer with heat transfer. *Phys. Fluids,* **22**, 085105.

10

Drag Reduction: Analysis and Design of Airfoils

10.1 Introduction

The design of an efficient aircraft depends on drag experienced by the vehicle, among the major aerodynamic properties. For a given engine power, drag reduction translates into better range and endurance. In Table 10.1, profile drag breakdown for a complete transport jet is provided (Bushnell and Heffner 1990) under two different hypothetical scenarios. In the first set of drag data, assumption is made that the flow over all the surfaces are fully turbulent, with contributions coming mainly from wing, fuselage, empannages, and the rest contributed by the nacelle and other miscellaneous causes. It is noted that the fuselage contributes 48.7% of the drag, while the wing and empannages contribute almost the same amount. It is readily evident that the lifting surfaces (wing, horizontal and vertical stabilizers) are streamlined bodies and holds the potential of delaying transition from laminar to turbulent flow over these. A fictitious case is considered in the second set of results posted in Table 10.1, where flow is considered to be fully laminar over the lifting surfaces. It is presumed that the fuselage and the nacelle are not candidates for viscous drag reduction. Under these hypothetical state of affairs, the wing drag reduces by a factor of three and accounts for only 15.3% of the total drag. We will discuss about the possible drag reduction for fuselage and nacelle in the following.

Drag breakdown of another subsonic transport airplane is shown in Figure 10.1 (Vijgen *et al.* 1987), highlighting the importance of fuselage drag after the flow over the wing is rendered laminar by some passive means. This brings down wing contribution to overall drag from roughly 27% of total viscous drag to 15%. This seems feasible (Viken 1983) as it has been noted that the profile drag of 0.0085 for a fully turbulent airfoil at a Reynolds number of 10^7 can come down to 0.0010, if the flow over the full airfoil is completely laminar. Such a lifting element has been termed aptly as natural laminar flow (NLF) airfoil. For regional aircrafts, desired aerodynamic drag with an NLF airfoil is obtained for a chord Reynolds number range of 4 to 17 million, when the cruise Mach number is subcritical ($0.4 \leq M_\infty \leq 0.6$).

Theoretical and Computational Aerodynamics, First Edition. Tapan K. Sengupta.

Companion Website: www.wiley.com/go/sengupta

Table 10.1 Profile drag break-up of a transport jet aircraft (Bushnell and Hefner (1990))

	Fully turbulent flow condition for all surfaces		Fully laminar flow on lifting surfaces only	
	Drag count	Percentage	Drag count	Percentage
Wing	0.0060	31.8%	0.0020	15.3%
Fuselage	0.0092	48.7%	0.0092	70.2%
Empennage	0.0027	14.3%	0.0009	6.9%
Nacelle/misc.	0.0010	5.2%	0.0010	7.6%
Total profile drag	0.0189		0.0131	

Source: Courtesy of NASA

For larger transport aircrafts, it is inconceivable that the flow over the lifting surfaces can be maintained fully laminar. In Figure 10.2, a realistic estimate is shown over an existing aircraft with areas over wing and tailplanes shaded to indicate where flow can be kept laminar, which can result in overall 12% drag reduction over the wing and 3% reduction over the tail planes. An additional 1% overall drag reduction is considered feasible by redesigning the engine nacelle. In this projection, it is noted that the potential for natural laminar flow by passive means is feasible for smaller aircrafts, when wing leading edge sweep angle can be restricted below 20°. However, it is also appreciated that such a theoretical projection may not hold promise in actual aircraft development due to poor prediction of transition currently used in research and development, which assumes linear spatial instability as the main precursor to transition. Any aircraft performance calculations based on such prediction is not sufficiently reliable. One of the major bottleneck in transition research is that experimental validation is even more unreliable. We have noted in the previous chapter that transition does not occur through the mechanism of growth predicted by linear spatial theory, which is often used for transition prediction. Even the usage of parabolized stability equation (PSE) is debatable, as in the previous chapter we noted that disturbances are induced upstream of wave-packets

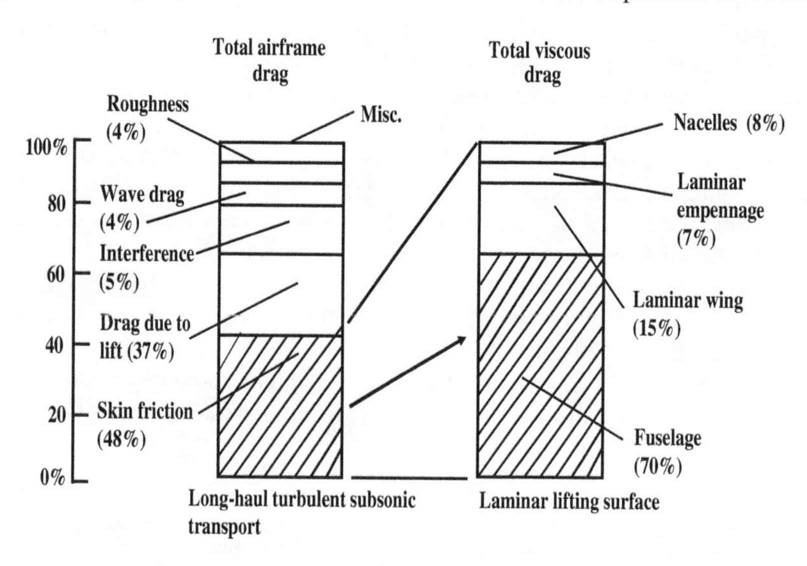

Figure 10.1 Drag breakdown of a subsonic transport airplane

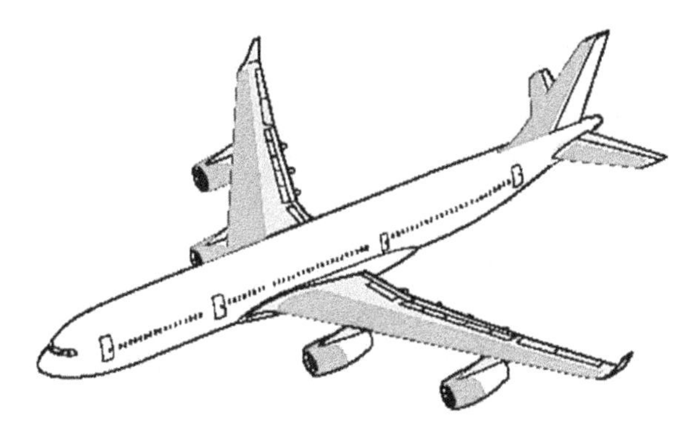

Figure 10.2 A commercial swept wing aircraft isometric view, indicating the target of flow laminarization over lifting surfaces and nacelle

whose spatio-temporal growth causes actual transition. As this is due to disturbance growth via spatio-temporal route and not strictly upon spatial growth of Tollmien-Schlichting waves. For this reason, the route of transition defined in the previous chapter is termed the bypass route. We have noted that only for low frequency excitations the spatio-temporal wave front remains connected to the Tollmien-Schlichting wave-packet at all times, which apparently is compatible with the traditional viewpoint of Tollmien-Schlichting wave-packet leading to eventual transition. However, we emphasize that this TS wave-packet also grows temporally, as shown from the solution of full Navier-Stokes equation. This route of growth for TS wave is also associated with the streamwise instability experienced for sweptback wings. Additionally, flow over sweptback wings suffer instability experienced by the cross flow components. In traditional analysis for transition over sweptback wings, these two routes have been mostly studied independently and transition criteria for each is considered to act independently, which is really not the case.

Thus, it is essential to understand how transition occurs over an aircraft wing. However, if the transition over the airfoil section constituting the wing is more critical, then one must understand the airfoil drag contribution first and ways to keep sectional drag at a lower level is a legitimate goal. In this regard, an NLF airfoil maintains extensive laminar flow solely by means of favourable pressure gradient obtained by profiling the shape by passive means. Active control for keeping flow laminar in aerodynamics has still not matured and considered reliable to use on operational aircrafts.

In designing NLF airfoil, the aerodynamic performance must consider different requirements for the major segments of a typical flight history. For example, for regional transport aircraft, one must consider landing, take off and climb performance equally important, often more than the cruise performance. For long haul aircrafts, cruise condition dominates and defines the objective function for the purpose of optimization. Such design considerations for short haul aircrafts, in the vocabulary of design and optimization, would be termed multi-objective optimization. Additionally, constraints arising from considerations of structural design, manufacturing practices, other geometric restrictions and aero-elasticity must be factored in. Such multi-objective, multi-constrained optimization lead to a conflict for the objective function with respect to the constraint(s). From the perspective of drag reduction by transition delay over aerofoil section is still developing when viewed from the scientific approaches by computing the flow field by solving Navier-Stokes equation, as described in the previous chapter. As these DNS are expensive, it is not immediately conceivable that such solver can be used in design cycle. Present approaches discussed in this chapter is based on increasing our understanding of the transition flow field and coming out with some guidelines by which NLF aerofoil can be designed.

10.2 Laminar Flow Airfoils

The most widely known sections is the family of NACA six-series laminar flow airfoils designed by keeping the minimum pressure point as aft as possible without flow separation, during the pressure recovery stage in the rear of the airfoil. The first digit in the nomenclature identifies the series; the second digit gives the location of minimum pressure in tenth of chord from the leading edge; the third digit is the design lift coefficient in tenths and the last two digits give the maximum thickness in hundredths of chord. The NACA 67-314 airfoil (which served as the baseline section for the design of the NLF section NLF(1)-0414F) has the first digit 6 as the series identifier; the minimum pressure occurs at 0.7c for the basic symmetric thickness distribution at zero lift; the design lift coefficient is 0.3 and the airfoil is 14% thick. Abbott and von Doenhoff (Abbott and Doenhoff 1959) is a repository of airfoil data of NACA series airfoils up to 1949, containing coordinates and properties of sections based on extensive experimental data. There is uncertainty about the effectiveness of all six digit series airfoil in this reference. For NACA 67,1-215 airfoil there are sectional data, but for this section

> 'there are no data above $Re = 6 \times 10^6$, presumably because the higher Reynolds number data produced little laminar flow'

(Viken *et al.* 1987) – implying the shortcoming of these laminar airfoils at higher Reynolds numbers, even in the controlled confines of a wind tunnel. This is due to extreme sensitivity of the designed sections to ambient disturbance conditions, even the trace amount inside the wind tunnels enough to trigger transition. Thus, the purpose of designing these sections with cruise condition defining the objective function remained unfulfilled due to limitation of the section at higher flight speeds. These are still good candidates for gliders and sailplanes applications, which are used for lower Reynolds numbers only.

The aerodynamic characteristics of an aerofoil section is represented by a plot of lift coefficient plotted against the drag coefficient and is called the drag polar. In Figure 10.3, the drag polar and pitching moment curves are shown for NACA $66_2 - 015$, which has a symmetric section (Abbot and Doenhoff 1959), which shows the well known *drag bucket*, implying a very narrow range of lift coefficient for which drag shows optimum lower value. Note that the *drag bucket* disappeared, when the airfoil was tested with standard roughness, as practised in NACA (Abbot and Doenhoff 1959), for a Reynolds number of six million. The standard roughness relates to applying a strip with grit near the leading edge of the airfoil. The adverse effect of the roughness also leads to severe degradation of $c_{l_{max}}$ with the standard roughness (Abbot and Doenhoff 1959), as noted in Figure 10.3.

The NACA series airfoils designed before World War II were based exclusively on experiments and analysis by conformal mapping (Theodorson's method). The 6-digit series NACA airfoils are based on wind tunnel experiments, which are costly, time consuming and restrictive, with details of development available in literature (Abbot and Doenhoff 1959, Jacobs

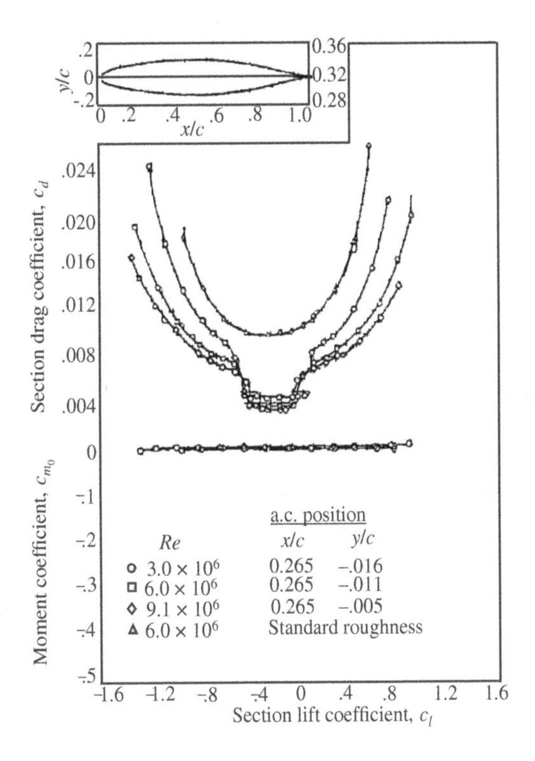

Figure 10.3 Drag polar and pitching moment coefficient of the NACA 66_2-015 airfoil

1939, Pfenninger 1947, Tani 1952). Apart from the above reason of difficulties in maintaining the flow to be laminar at higher Reynolds number, the other reasons are due to difficulties of manufacturing and operations. The manufacturing techniques of the 1930s and 1940s used rivets and surface waviness of the finished wing surfaces gave rise to a much larger drag value than expected.

Presence of rivets and surface roughness causes nonoptimum drag behaviour for higher Reynolds numbers, as noted in flight tests at RAE (Gray and Fullam 1950, Gray 1950, Smith and Higton 1950), using a King Cobra aircraft with a smoothed and nonwavy wing surface. According to Viken (Viken 1983).

> 'roughness induces streamwise disturbance vortices, which add a cross flow component to the mean boundary layer flow and tends to distort the two-dimensional Tollmien–Schlichting (T-S) disturbances three dimensionally. Once these oscillations are distorted three-dimensionally they amplify much quicker to the critical level which causes transition. This effect necessitates the minimization of all avoidable surface roughness, whether from poor construction techniques or from contamination during use (insects, dirt etc.)'.

Without providing details, we note that the viscous instability as predicted by Orr-Sommerfeld equation is much weaker in comparison to the inviscid mechanism according to the Rayleigh-FjØrtoft theorem via the existence of inflection point in the equilibrium flow velocity profile for incompressible flows. In the presence of discrete roughness elements and/

or wavy walls, the mean flow displays inflectional velocity profile. In the presence of both viscous and inviscid mechanisms, the latter dictates transition with the corresponding disturbance growth rate more violent and all pervasive. This is the reason for the predominant instability in the presence of cross flow over a swept wing, which displays inflectional profile and leads to early transition.

Also, wing surfaces made of metal by conventional techniques buckle easily, setting small scale waviness on wing surface zones subjected to compressive loads. This was also confirmed in the above flight experiments at RAE, which recorded a large increase of the width of the *drag bucket* when the amplitude to wavelength ratio of the airfoil surface was reduced from 0.00125 to 0.0005. Composite material wing can reduce surface roughness and waviness. Also, some metal wings have been manufactured from a single block by using numerically controlled milling procedure, which leads to avoidance of surface protuberances and waviness. Absence of rivets and associated structures inside the wing also leads to lower wing weight.

10.2.1 The Drag Bucket of Six-Digit Series Aerofoils

Reason for the *drag bucket* in six-digit NACA aerofoils can be explained by looking at the calculated squared normalized velocity distribution shown in Figure 10.4 for the NACA 67,1-015 aerofoil. For this symmetric section, the middle curve corresponds to both the upper and lower surfaces at zero angle of attack. The case of $\alpha = 2^0$ corresponds to the lift coefficient of 0.2, for which a relatively flat favourable pressure gradient exists up to $x = 0.62c$, for the zero angle of attack case. For such accelerated flows, the disturbance in boundary layer is damped and this delays flow transition significantly and explains the reduced drag for this and very small angles of attack around this. However for the lifting case, on the upper surface the pressure gradient flattens implying the flow not to accelerate, as compared to zero angle of

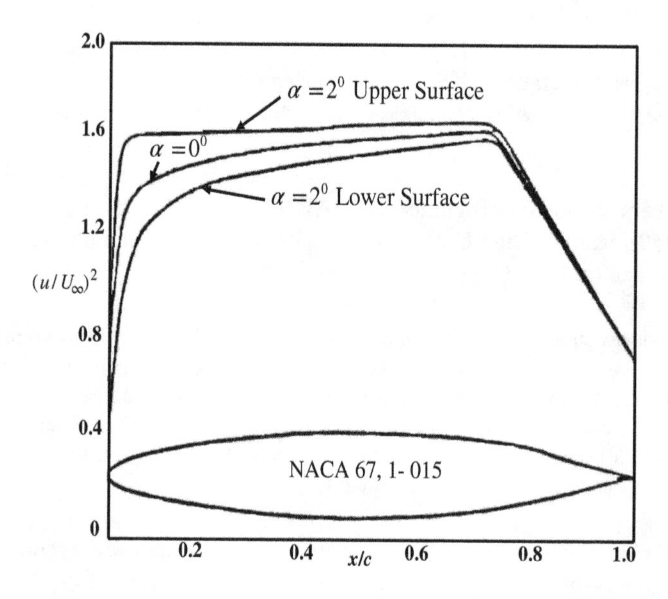

Figure 10.4 Normalized streamwise velocity squared distribution, indicating pressure variation over the NACA 67,1-015 airfoil

attack case. Absence of favourable pressure gradient makes the flow susceptible to undergo transition earlier as compared to $\alpha = 0°$ case, explaining the narrow range of lift coefficients for which the *drag bucket* exists.

The second reason for the abrupt increase in drag coefficient at higher angles of attack is due to the design pressure recovery beyond $x/c = 0.62$ being linear. It is to be understood that flow transition would be inevitable for such pressure distribution. Also, after the transition such linear pressure recovery will lead to flow separation of the corresponding turbulent flow. This observation is phenomenological and somewhat simplistic, as all the analysis and design tools are not sophisticated enough to sense the unsteadiness of the flow field beyond maximum thickness, as shown in the last chapter from the solution of full time-dependent Navier-Stokes equation for a contemporary NLF aerofoil.

Another aspect of these NACA six-digit airfoils designed by Theodorsen's method (Abbot and Doenhoff 1959) is that the loading is fixed first as the design criterion and the profile is subsequently obtained by an inverse method. The design is such that the load near the trailing edge is low for nonzero angles of attack. This apparent low aft-loading would yield near-zero pitching moment, provided the flow is not separated. Thus, the attractive proposition of inviscid analysis results may be lost due to bad design pressure recovery, which leads to both separation and transition simultaneously. This is the reason for the shallow drag bucket and has been attempted to be rectified in newer laminar flow aerofoils.

10.2.2 Profiling Modern Laminar Flow Aerofoils

During the 1970s, NASA designed a series of low speed airfoils with the idea of improving the performance over NACA laminar flow airfoils. These were designed computationally by the panel method and boundary layer calculations, with subsequent wind tunnel tests conducted verifying the design. The first such section designed for general aviation purposes was called GA (W)-1 section, which has been re-designated as LS(1)-0417 airfoil, with LS specifically representing low speed applications, 04 stands for the design cruise lift coefficient (i.e. $c_l = 0.4$) for the 17% thick aerofoil. One notes that the inviscid pressure distribution for GA (W)-1 section is similar to supercritical airfoils designed by R. Whitcomb in 1965 for high speed applications. The earliest general aviation airfoils are derivatives of this and the letter W in GA (W) is a recognition of Whitcomb's original contribution in this field. These aerofoils are characterized by large leading edge radius (0.08c), introduced to flatten the peak in C_p distribution near the nose. A large reflex camber near the trailing edge forms a cusp, providing significant aft loading for the section.

Some of the other airfoils in this series are the LS(1)-0409 and LS(1)-0413 sections (McGhee and Beasley 1973, McGhee *et al.* 1980). As compared to earlier NACA flow aerofoils with the same thickness XX, these new LS(1)-04XX airfoils have (i) approximately 30% higher $c_{l,max}$, (ii) approximately 50% increase in L/D at a lift coefficient of 1.0 (Anderson 1991). This latter design point has been introduced with a view to having lower drag for a value of lift coefficient typically used during climb section of the flight envelope.

Thus, the six-digit aerofoils have been designed for optimal performance for single flight condition at lower Reynolds numbers. The change of philosophy in redesigning NLF aerofoils by NASA in 1970s and 1980s has been well explained in Somers (1981, 1981a) and is explained below with the help of Figure 10.5(a) showing a typical drag polar of the NACA six-series airfoil (shown by the dotted line) and its desired improvement (shown by the solid line). For the drag polar of the improved section, one can sketch the corresponding desired pressure distribution over the airfoil for different lift coefficients. Point A corresponds to cruise condition ($c_l = 0.4$) for which one can expect the aerofoil to experience the lowest profile drag. In NACA designs, the upper surface favourable pressure gradient was constrained to be sustained up to $(x/c)_{us} \leq 0.3$, while the lower surface pressure gradient was kept unconstrained. The lowest drag point fixes the camber of the airfoil and if c_l for cruise condition is pushed downwards, then $c_{l,max}$ point would also be lowered proportionately from point C in Figure 10.5(a). The change in design philosophy to multi-point optimization for the aerofoil is indicated by the choice of point B (at $c_l = 1.0$), as compared to a single design point of older NACA six-series designs. This operating point corresponds to the requirement of maximum climb rate. This necessitates the transition point on the upper surface to move slowly and steadily towards the leading edge with increasing c_l. This feature of smooth variation of transition point with increasing α can be brought about by changing the leading edge shape of the airfoil.

In Figure 10.5(b), the drag polar of a conceptual high-altitude, long-endurance remotely piloted vehicle is shown for its typical design features (Maughmer and Somers 1989). Such a vehicle must operate at fairly high lift coefficients and relatively low Reynolds numbers. In this reference, a typical vehicle is designed for an operational altitude of 20 km and endurance of 90 hrs with a range of 32 000 km. The design goal here is to achieve c_l's for the key operational points noted in the figure. To achieve these goals with least possible profile drag coefficient, the polar should be as much to the left as possible, while increasing the width of the bucket as wide as possible. Note the high maximum lift coefficient needed for take-off and landing performances. Instead of what is being taught in flight dynamics optimization of maximizing the three-dimensional endurance parameter, $C_L^{3/2}/C_D$, Maughmer and Somers (Maughmer and Somers 1989) maximized the objective function or the figure of merit,

$$\frac{c_{l,max}}{c_d \ at \ c_l = 1.5}.$$

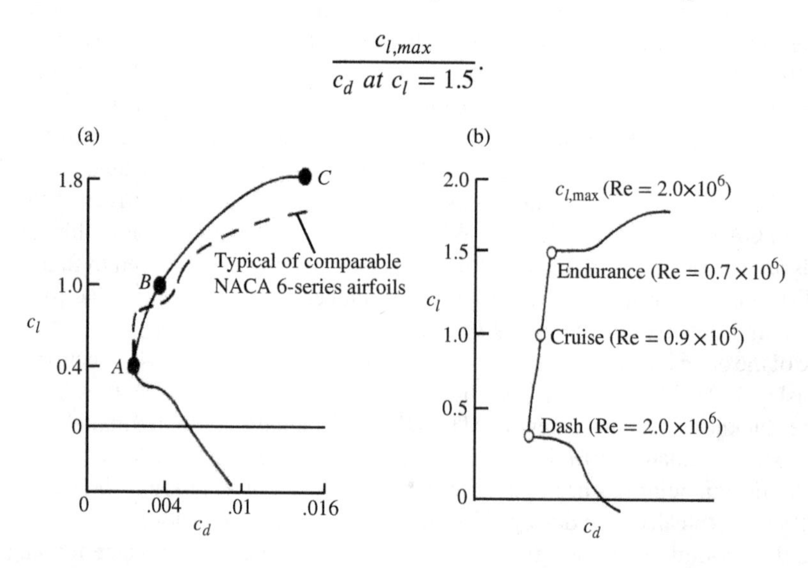

Figure 10.5 (a) Sketch of conceptual drag polar for newer NLF airfoils. *Source:* Somers (1981). Courtesy of NASA; (b) conceptual drag polar for a high-altitude, long-endurance remotely piloted vehicle. *Source:* Maughmer and Somers (1989). Reproduced with permission from the American Institute of Aeronautics and Astronautics, Inc.

Here, the reduced wing profile drag is obtained by decreasing the wing area with the help of higher $c_{l,max}$ and reducing section profile drag at the operational c_l for maximum endurance, indicated in Figure 10.5(b). It is perfectly possible to reduce the profile drag further, at the cost of reducing the bucket width, i.e. by reducing c_l for endurance point. Such an airfoil for high-altitude, long-endurance RPV, a single element airfoil is considered, as the added weight of flap system will degrade the performance.

Typical pressure distributions for operational points A and B in Figure 10.5 are sketched in Figure 10.6. For point A, favourable pressure gradient on the upper surface is determined by the applied constraint. Beyond 0.30c, a short region of adverse pressure gradient is purposely built in to promote efficient transition and further aft a steeper concave pressure recovery is introduced which produces lesser drag, as compared to six-digit series airfoils shown in Figure 10.4. Pressure recovery in the aft portion of the airfoil upper surface, where the flow is turbulent, has been a subject of detailed research (Stratford 1959). The lower surface pressure distribution determines many things, although it is treated as unconstrained. While the upper and lower surface pressure distributions are described separately, these are not shielded from one another. It is for this reason that one cannot constrain these pressure distribution simultaneously. The quest for maximum lift coefficient requires an optimum pressure distribution on the lower surface which should have lower profile drag by having a shallow favourable pressure gradient over the forward portion, followed by an abrupt and sharper concave pressure recovery (with respect to the top surface) obtained by significant aft camber. Absolute magnitude of this camber is constrained by the pitching moment requirement $c_m \geq -0.10$, an absolute requirement for operation at higher Mach numbers.

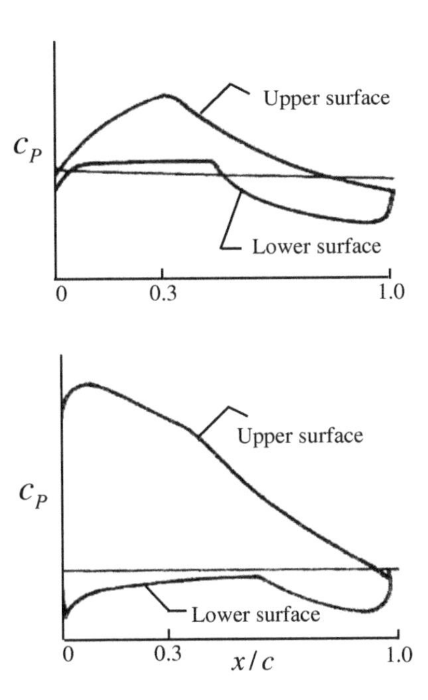

Figure 10.6 Typical pressure distribution for the design points of Figure 10.5 (top figure for point A and bottom figure for point B)

It is important to realize the central role of computed pressure distribution by inviscid analysis, even when viscous effects are included with flow undergoing transition to turbulence. This is excluding the case of flow separation, whether the flow is laminar or turbulent. This aspect of flow behaviour prediction by inviscid analysis needs further explanation. For unseparated flows with boundary layer assumptions holding, the pressure is transmitted through the shear layer for both laminar and turbulent flows. For turbulent flows, variation of pressure across the shear layer is additionally due to wall-normal fluctuating component of the velocity field; yet the pressure distribution computed by inviscid analysis is very qualitatively similar to the actual values. Thus, when one visualizes the pressure distribution over an aerofoil, it is understood that the C_p distribution for design condition is composed of a steady part in the front part of the aerofoil and an unsteady part corresponding to the transitional and turbulent flow part. This will be shown later for a modern NLF SHM-1-aerofoil.

Also, notice that at higher lift coefficients the pressure distribution does not display a suction spike (that is typical of thin and sharp leading edges of older airfoils). This is incorporated by designing airfoils with thicker leading edges and detailed tailoring of leading edge thickness distribution. These features of pressure distribution are explained with the help of two aerofoil sections, DESB154 and DESB165 designed by Viken (Viken 1983) and shown in Figure 10.7. The design proceeded with hand drawing an airfoil (based on experience!) and then calculating corresponding (i) pressure distribution, (ii) boundary layer and (iii) its stability properties, in

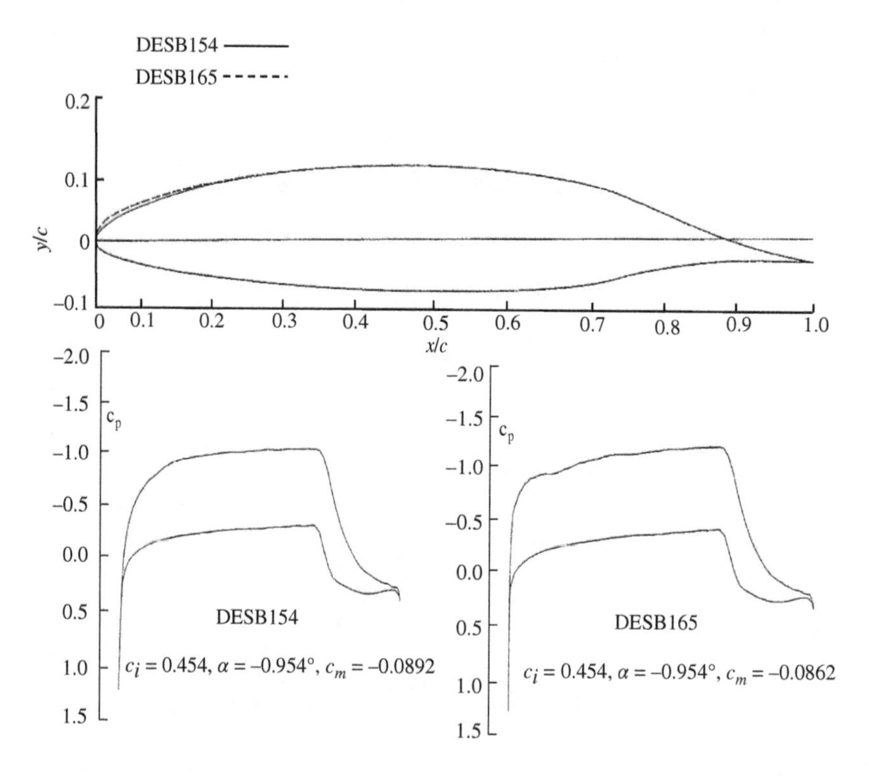

Figure 10.7 Comparison of airfoil profiles DESB154 and DESB165 aerofoils and pressure distribution for the same angle of attack. *Source:* Viken (1983)

that sequence. This procedure was carried through iteratively until the design goals were met for a Mach number $M = 0.4$. Both the airfoils are 14.3% thick and designed to have accelerated flow all the way up to $x/c = 0.70$ on both the surfaces for cruise configuration for a Reynolds number of 10 million. Maximum thickness occurs at approximately 45% of the chord and the section pitching moment coefficient about the quarter chord point is −0.0882 at the design point $c_l = 0.45$. This pitching moment coefficient varied from −0.833 to −0.0934 when c_l was varied from 0.15 to 0.95. The DESB154 airfoil was designed with a much sharper leading edge than is normally seen on airfoils of this type to reduce high negative C_p values at low angles of attack, giving a wider low drag c_l range. Also to retain lower drag at cruise, the trailing edge has a sharp edge to give a thinner boundary layer at the trailing edge. However, the thin leading edge compromised $c_{l,max}$ performance. This led to the development of the DESB165 airfoil, which is a modification of the DESB154 airfoil obtained by thickening the leading edge portion (only up to $0.2c$). Thickness was superposed directly onto the leading edge, while retaining the aft shape as much as possible. Both the airfoils are shown in Figure 10.7 to indicate the changes made at the leading edge in the DESB154 airfoil to obtain the DESB165 airfoil.

In Figure 10.7, pressure distributions for the same angle of attack for these two airfoils are compared, obtained by an inviscid calculations for $\alpha = -0.954°$. Calculated results show a difference in the third decimal place for c_l and c_m. Thus, this indicates that the lift coefficient does not change much for lower angles of attack, while the relative change in c_m is higher and the changes at higher angles of attack will be more profound for this marginal change in leading edge shape.

Further evidence for the effects of changing the airfoil leading edge shape is provided in (Viken *et al.* 1987), which explains the design of the NLF(1)-0414F airfoil starting from the baseline design of the DESB159 airfoil. In Figure 10.8, the leading edge of the NLF(1)-0414F airfoil is compared with the DESB159 and NACA 67-314 airfoils. The leading edge of the DESB159 airfoil was modified to achieve an acceptable low drag c_l range for the zero flap

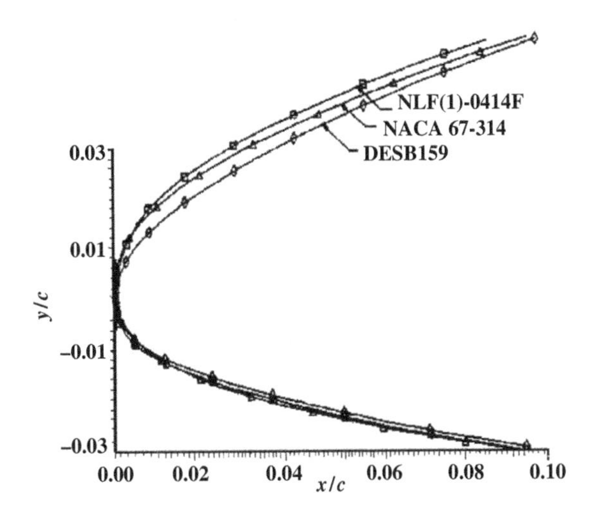

Figure 10.8 Comparison of indicated airfoil leading edges shown in. *Source:* Viken, Viken, Pfenninger, Morgan and Campbell (1987). Courtesy of NASA

deflection case. For lower angles of attack which provide low drag operation for the DESB159 airfoil, the sharp leading edge helps suppress large negative pressure peaks. However when the angle of attack is increased, the same sharp leading edge causes large negative pressure peaks as a result of the centripetal forces to turn the flow around the bend. Here, an additional thickness is superposed on the baseline aerofoil profile, merging the two at x/c =0.15–0.20, to reduce negative pressure peak. The idea is to turn the flow at lower speeds by a smaller radius of curvature. For DESB159 and NACA 67-314 aerofoils, the smallest radius of curvature is at the leading edge, while for the NLF(1)-0414F aerofoil the smallest radius of curvature is placed on the lower surface.

10.3 Pressure Recovery of Some Low Drag Airfoils

Having discussed the differences in shape near the leading edge which exist between NACA six-series and NLF aerofoils, we focus upon the differences in aerodynamic properties brought by the shape and thickness distribution near the trailing edge for these aerofoils. In Figure 10.9, we see the comparison between NLF(1)-0414F and NACA 67-314 aerofoils in terms of geometry and pressure distribution, mainly in the aft portion of the aerofoils (Viken *et al.* 1987).

Results are shown for the same Mach number (0.4) and lift coefficient ($c_l = 0.46$) for slightly different angles of attack in Figure 10.9, with both the aerofoils having same maximum thickness. The favourable pressure gradient regions are similar, with the NLF(1)-0414F aerofoil having slightly higher acceleration on both surfaces. The upper surface acceleration of NLF(1)-0414F aerofoil was optimized by the use of a flat pressure distribution to have

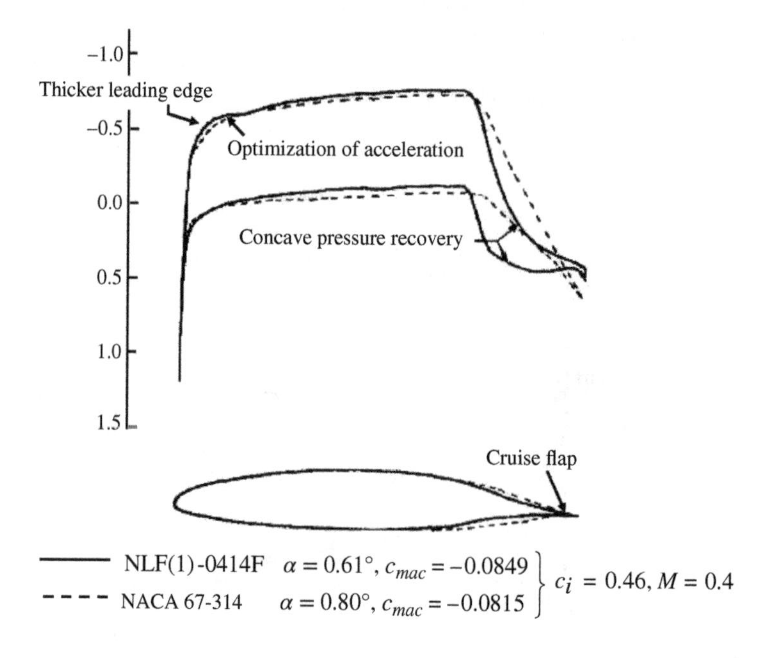

Figure 10.9 Comparison of geometry and pressure distribution between NLF(1)-0414F and NACA 67-314 aerofoils. *Source:* Viken, Viken, Pfenninger, Morgan and Campbell (1987). Courtesy of NASA

beneficial effects on TS waves (Viken *et al.* 1987) around $x/c = 0.10$. This aerofoil deploys a small chord trailing edge (cruise) flap for enhanced low drag performance. The major difference between the two aerofoils relates to the reduced thickness for the NLF section in the rear part, affecting pressure recovery. The concave pressure recovery of NLF aerofoil should be contrasted with the NACA 67-314 aerofoil's linear pressure recovery. Such different aft geometry of aerofoils results in significantly different behaviour related to the problem of turbulent separation in the pressure recovery region for the six series aerofoil.

For low drag, the pressure gradient of NLF aerofoil is kept favourable up to the point of transition. After the minimum pressure point, boundary layer suffers transition due to adverse pressure gradient, which increases the growth rate of the boundary layer and increases drag. This is understood by inspecting the von Kármán momentum integral equation (Cebeci and Bradshaw 1977) to reveal differences among various pressure recoveries for aerofoil designs. This equation is given by

$$\frac{d\theta}{dx} + \frac{dU_e}{dx}\frac{\theta}{U_e}(H + 2 - M^2) = \frac{\tau_0}{\rho_e U_e^2}. \tag{10.1}$$

Here the momentum thickness (θ) growth rate is given by the first term on the left-hand side. The second term incorporates the effects of pressure gradient expressed in terms of boundary layer edge velocity (U_e) gradient in the streamwise direction. The skin friction term on the right-hand side has a secondary role in the steep pressure rise region, as is the case in the pressure recovery region near the trailing edge of the aerofoil. To retain lower drag, Stratford (Stratford 1959) has shown that the boundary layer should be kept on the verge of separation throughout the pressure recovery (with the shape factor $H = 2.0$). Perfect pressure recovery of the Stratford type is not recommended, as the resultant turbulent boundary layer will separate at off-design conditions. Thus, a milder concave pressure recovery resembling the Stratford type is preferred, providing lower drag without separation at off-design conditions.

In the top frame of Figure 10.10, a concave pressure recovery typical of NLF designs is marked as 2 and compared with a linear recovery typical of six-series aerofoil (marked as 1) via boundary layer calculations for $Re = 10^6$. These results are as in Hoerner (1950). Up to a certain distance the linear pressure recovery shows lesser drag as compared to the concave pressure recovery (closer to $x/c = 0.88$), as indicated by the θ plot. The shape factor (H) shows spectacular growth for case 1 above $x/c = 0.8$. In the bottom frame of Figure 10.10, separation is indicated for $H = 1.8$ and the case 2 is in no danger of flow separation, while case 1 indicates separation around $x/c = 0.92$. Schubauer and Spangenberg (Schubauer and Spangenberg 1958) have noted that incompressible turbulent boundary layers separate when H grows to a value of 2.

In Figure 10.11, the pressure distribution and shape factor are compared between NLF(1)-0414F and NACA 67-314 aerofoils for $M = 0.4$, $Re = 10^6$ and $c_l = 0.46$, when the flows are assumed to be turbulent right from the leading edge. In Figure 10.8, it is shown that the six-series aerofoil has a linear pressure recovery, while the NLF aerofoil displays concave pressure recoveries near the trailing edge. In Figure 10.11, flows are accelerated up to $x/c = 0.7$ for both the aerofoils. However, H distribution for NLF(1)-0414F aerofoil shows growth to a maximum value of 1.9 at $x/c = 0.875$ and thereafter it reduces to 1.825 at the trailing edge. For the same conditions of the NLF aerofoil, NACA 67–314 aerofoil suffers separation at $x/c = 0.90$ for the linear pressure recovery. The last correct calculated H is 1.83

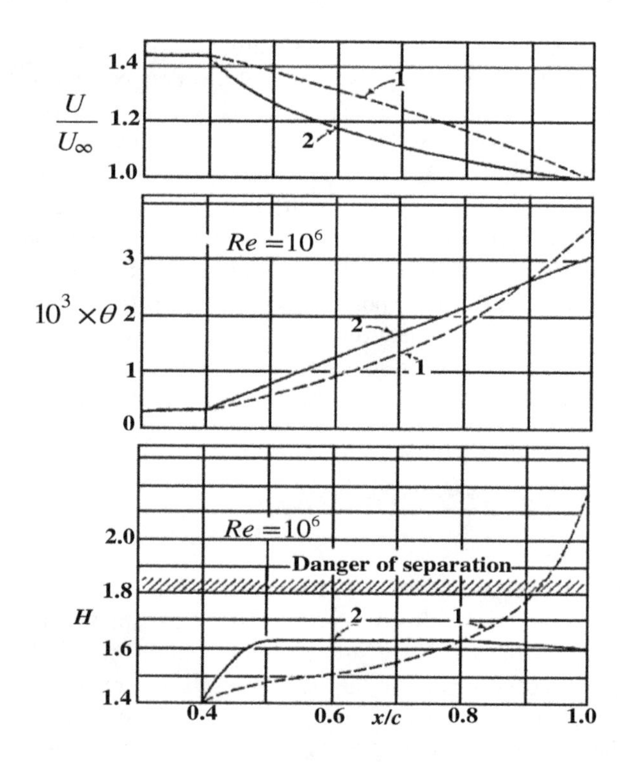

Figure 10.10 Effects of different pressure recoveries near the trailing edge on a turbulent boundary layer. *Source:* Hoerner (1950). Reproduced with permission from the American Institute of Aeronautics and Astronautics, Inc.

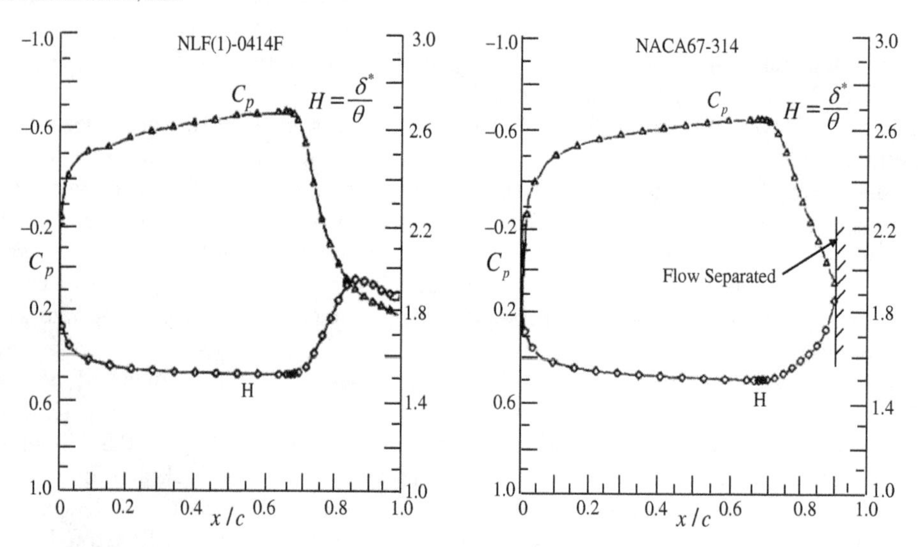

Figure 10.11 Comparison of pressure distribution and shape factor for NLF(1)-0414F and NACA 67-314 aerofoils for identical flow condition of $M = 0.4$, $Re = 10^7$ and $c_l = 0.46$. *Source:* Viken, Viken, Pfenninger, Morgan and Campbell (1987). Courtesy of NASA. The same lift is obtained by $\alpha = 0.61°$ for NLF(1)-0414F and $\alpha = 0.793°$ for NACA 67-314 aerofoils

for NACA 67-314 aerofoil (Viken *et al.* 1987). Such flows shown in Figure 10.11 correspond to the case as might happen due to leading edge contamination responsible for instantaneous transition.

10.4 Flap Operation of Airfoils for NLF

For NLF aerofoils, a new type of flaps are used as shown in top of Figure 10.12, where a small chord cruise flap is used to alleviate the problem of turbulent separation in the pressure recovery region at off-design c_l conditions. This cruise flap allows lift by geometric angle of attack replaced by lift due to camber (Viken 1983). Downward deflection of this flap adds aft camber, which achieves same lift at a lower angle of attack, resulting in a milder pressure recovery region and thereby preventing turbulent separation. In addition, one can employ the traditional flaps to generate higher and higher $c_{l,max}$ with higher flap deflection (δ_f), as shown in the lower part of Figure 10.12. Effects of flap deflection on NLF(1)-0414F aerofoil aerodynamics performance is shown on the drag polar, which indicates a lateral shift of optimum drag value for higher c_l, with flap deflection increasing.

For the NLF(1)-0414F airfoil, the drag polar is shown in Figure 10.12, for very low speed operation (Mach number less than 0.12) for $Re = 10^7$, for flap deflection in the range $-10^0 \leq \delta_f < 12.5^0$. This figure shows that the use of a cruise flap helps to achieve a wider

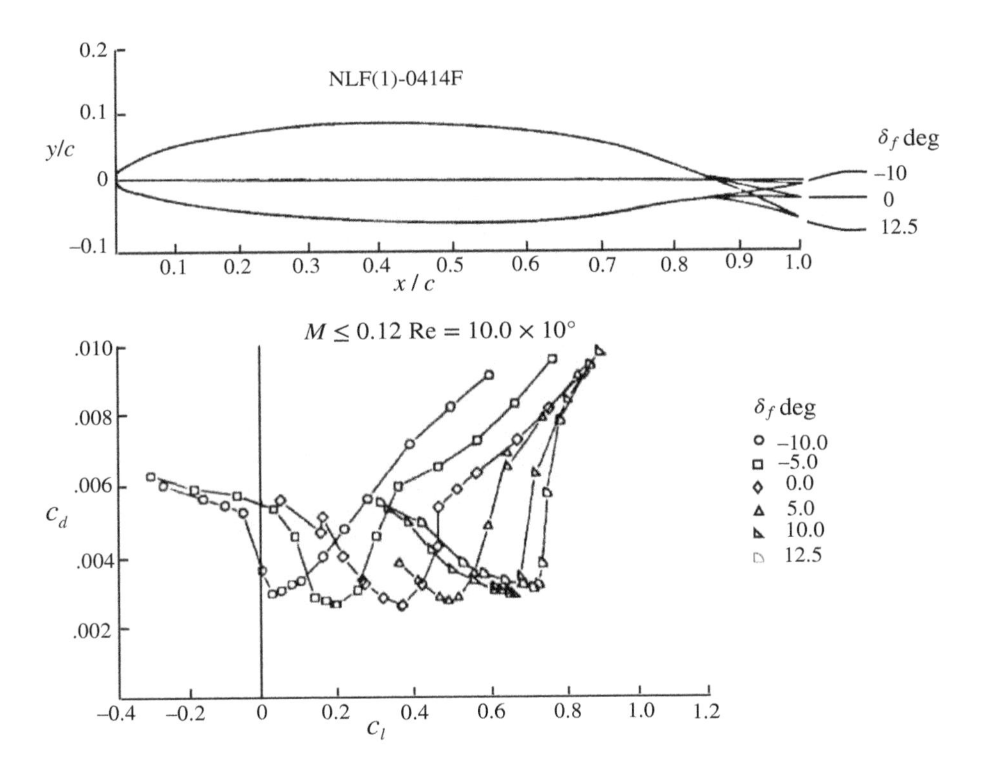

Figure 10.12 Effects of flap deflection on the NLF(1)-0414F airfoil as shown for the drag polar. *Source:* Viken, Viken, Pfenninger, Morgan and Campbell (1987). Courtesy of NASA

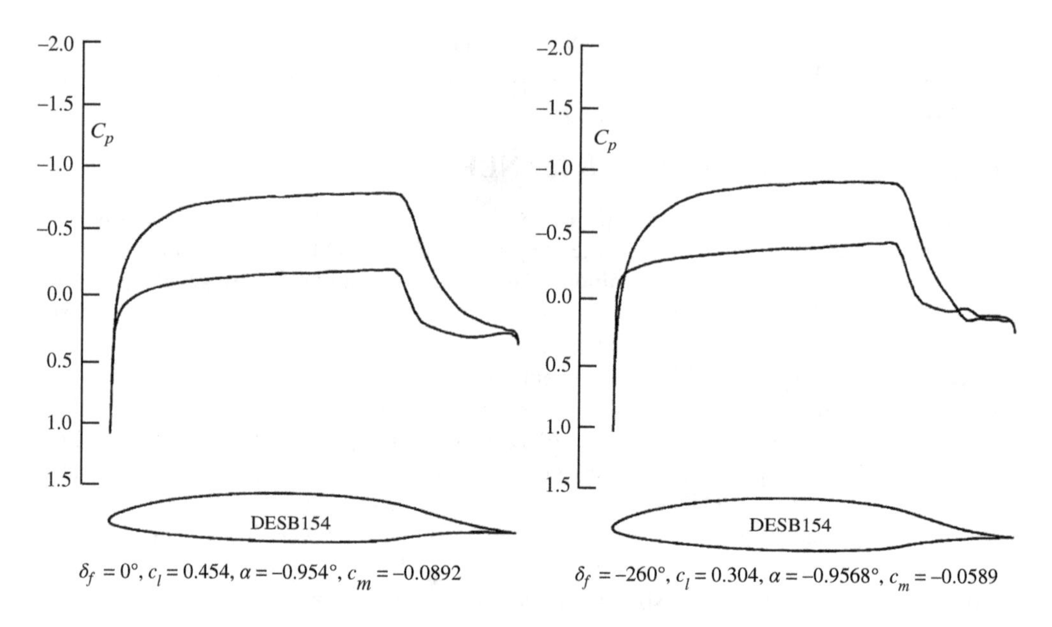

Figure 10.13 Comparison of inviscid pressure distribution, with (right) and without (left) the flap deflection for DESB154 airfoil. *Source:* Viken (1983)

range of c_l for low drag operation, especially when viewed with respect to undeflected flap drag polar, for which one notes $c_{d,min} = 0.0027$ occurring at $c_l = 0.41$. For $\delta_f = -10°$, minimum drag increases to 0.0030, occurring at $c_l = 0.01$. When the flap is deflected to 12.5°, then the minimum drag coefficient increases to only 0.0033 at $c_l = 0.81$ yielding $L/D = 245$.

The traditional flaps are used primarily as high-lift devices via improving performance by (i) augmenting the section lift coefficient and mainly to achieve $c_{l,max}$, (ii) altering the section pitching moment coefficient, to meet any constraint on c_m, e.g., one might impose a constraint ($c_{m,cruise} \geq -0.05$), (iii) shifting the low drag range for NLF aerofoils to higher lift values. For high speed applications, this can also be used to change M_{cr}. Here, first and second requirements conflict, in general, and the flap is used to alleviate this by deflecting the flap upwards. This allows an aerofoil to be designed with significantly higher camber to achieve $c_{l,max}$ while retaining the ability to keep c_m smaller at cruise flight conditions.

This effect is noted from the inviscid pressure distribution shown in Figure 10.13 for the DESB154 airfoil (from Viken 1983) for almost the same angle of attack, with ($\delta_{f'} = -2.6^0$) and without ($\delta_{f'} = 0$) the cruise flap deflection. Notice the difference in lift coefficient values, but more importantly the sectional pitching moment is changed significantly by the flap deflection. This suggests that the flap is used to alter the pitching moment and thereby reduce trim drag.

10.5 Effects of Roughness and Fixing Transition

The effects of fixing transition on section properties are shown in Figure 10.14 for NLF(1)-0414 aerofoil with the flap undeflected for $M = 0.10$ for two Reynolds numbers of $Re = 10^7$ and $Re = 6 \times 10^6$ (Viken *et al.* 1987). For both the Reynolds numbers, c_l and c_m do not change

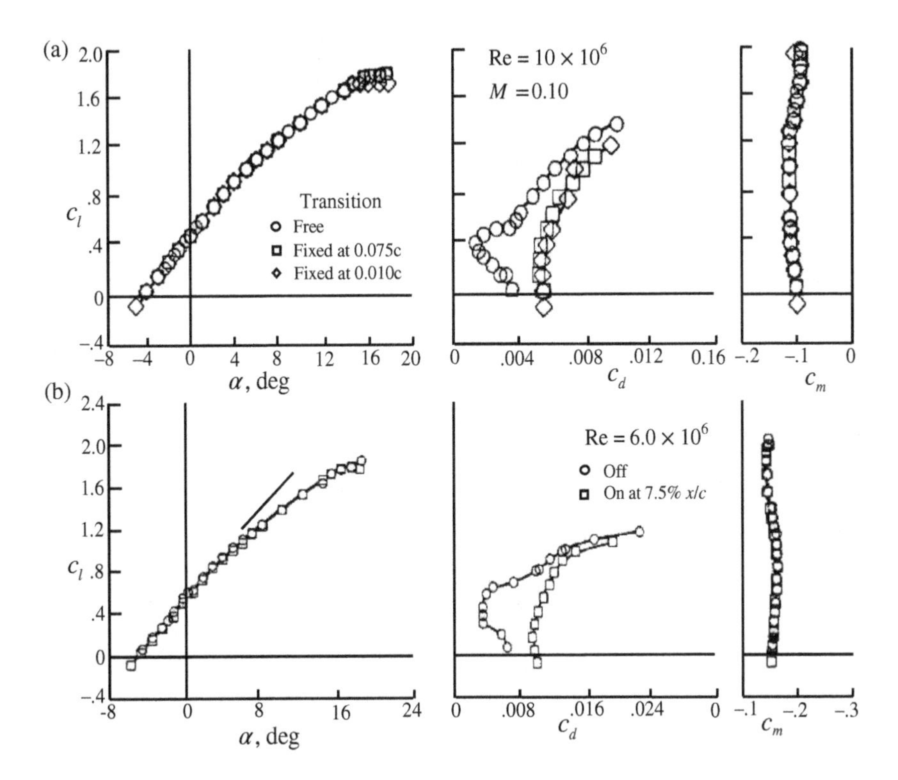

Figure 10.14 Effects of fixing transition for the NLF(1)-0414F aerofoil for (a) $Re = 10^7$ and (b) $Re = 6 \times 10^6$. Perceptible effects are seen in the drag polar. *Source:* Viken, Viken, Pfenninger, Morgan and Campbell (1987) & Murri, McGhee, Jordan, Davis and Viken (1987). Courtesy of NASA

very much between the free transition and fixed transition cases, with the latter fixed near the leading edge. However, the drag polar changes when one compares the free transition case with fixed transition cases. For the free transition case at $Re = 10^7$, the minimum profile drag coefficient is 0.0027 at $c_l = 0.41$. This profile drag is only 38% of an unseparated fully turbulent airfoil drag ($c_d = 0.0083$). The maximum lift coefficient is marginally better at 1.83 for the free transition case at $\alpha = 18°$. In contrast to lift and drag coefficients, the pitching moment coefficient is unaffected by fixing transition location on the airfoil. Transition is usually fixed in experiments by using a standard NACA roughness strip (Abbot and Doenhoff 1959) or by fixing cylindrical trip wires fixed on identified locations.

In Figure 10.14(b), effects of roughness applied near the leading edge on section properties are shown for the NLF(1)-0414F aerofoil (Murri *et al.* 1987) for $Re = 6 \times 10^6$ obtained experimentally in the NASA Langley Low Turbulence Pressure Tunnel. For this case also c_l and c_m do not change appreciably with roughness element located at $0.075c$. However, c_d changes significantly for this lower Reynolds number, with the drag bucket width reducing with increase in Re. There is also an interesting characteristic for $Re = 6 \times 10^6$, which is independent of roughness effects. This is related to reduction of lift curve slope near 4^0 for NLF(1)-0414F aerofoil. This reduction in $c_{l\alpha}$ is due to trailing edge separation in the pressure recovery region. This has been purposely used to control c_m in (Fujino and Kawamura 2003) for the design of the SHM-1-airfoil. However if this is undesirable, then it can be eliminated by using boundary layer re-energizer or vortex generators placed on the upper surface of the aerofoil (Murri *et al.* 1987).

10.6 Effects of Vortex Generator or Boundary Layer Re-Energizer

In Figure 10.14(b), we have noted the presence of an angle of attack ($\alpha = 4°$) above which $c_{l\alpha}$ decreased for the NLF(1)-0414F aerofoil due to flow separation in the turbulent pressure recovery region. This can be avoided by forced turbulent mixing, as demonstrated in (Schubauer and Spangeberg 1958) by employing a simple plow and vortex generator whose function is to scour the boundary layer by creating an alternate strip of higher and lower velocity patterns. Murri *et al.* (Murri *et al.* 1987) report results of using a vortex generator in an experimental investigation for NLF(1)-0414F aerofoil shown in Figure 10.15, comparing the cases of with and without the vortex generator. Vortex generators are small low aspect ratio wings positioned at 0.6*c*, which are 0.2 inches high, spaced 1.6 inches apart, positioned on the upper surface of the aerofoil at high α with respect to the local flow and presented results are for $Re = 3 \times 10^6$. These delta wings have attached leading edge vortex energizing the flow locally via turbulent mixing. Such forced mixing causes the pressure recovery region to spread over twice the streamwise distance, causing the concave pressure recovery milder, precluding turbulent separation.

Trailing edge separation is indicated by reduced $c_{l\alpha}$ for the case without the vortex generator. For the case with the vortex generator this reduction in the value of $c_{l\alpha}$ disappears, while $c_{l,max}$ remains more or less same. The profile drag for α greater than 4° also improves. But at a lower lift coefficient range, a drag penalty is shown for the case with vortex generator.

10.7 Section Characteristics of Various Profiles

In the following, we compare properties of various other NLF airfoils. In Figure 10.16, we compare the maximum lift coefficient of the NLF(1)-0416 section with two NACA six-series airfoils of similar design lift coefficient and thickness, at various Reynolds numbers. There is a

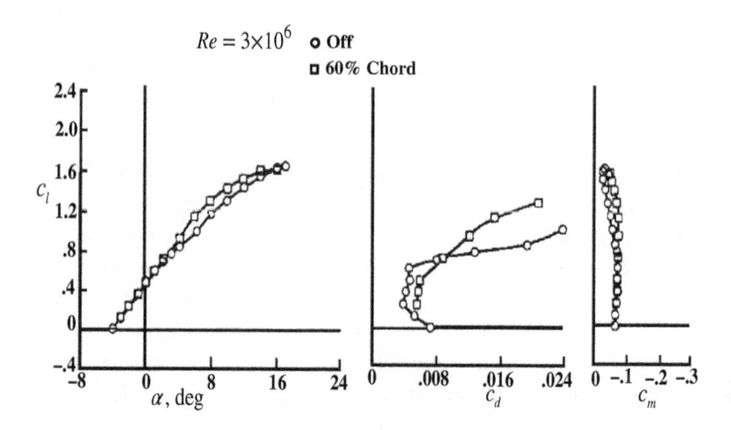

Figure 10.15 Effects of vortex generator used in the turbulence recovery zone for NLF(1)-0414F aerofoil for $Re = 3 \times 10^6$. *Source:* Murri, McGhee, Jordan, Davis and Viken (1987). Courtesy of NASA

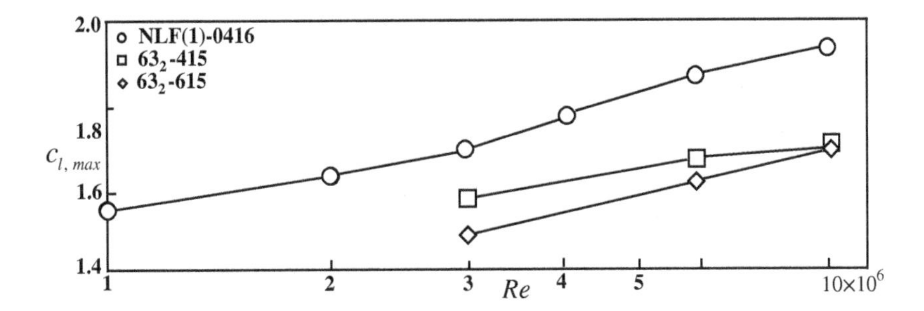

Figure 10.16 Comparison of maximum lift coefficients of the NLF(1)-0416 airfoil with two NACA six-series airfoils. *Source:* Somers (1981). Courtesy of NASA

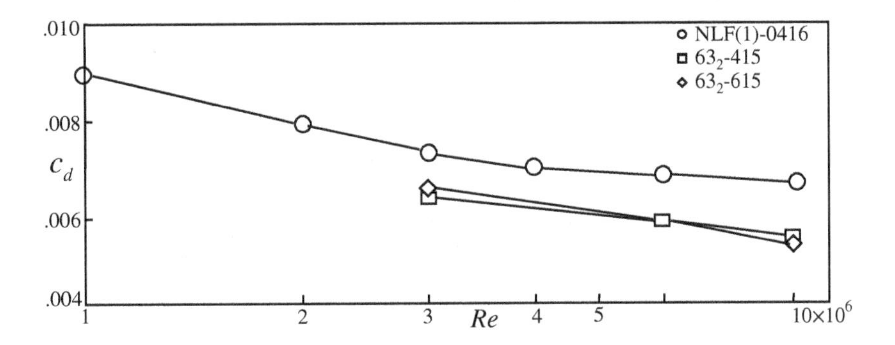

Figure 10.17 Comparison of the drag coefficient of the NLF(1)-0416 airfoil with two NACA six-series airfoils at $c_l = 0.4$. *Source:* Somers (1981a). Courtesy of NASA

significant improvement for this NLF section with respect to the six-series sections. The drag coefficient of the NLF(1)-0416 airfoil is compared with the NACA $63_2 - 415$ and $63_2 - 615$ airfoils in Figures 10.17 for $c_l = 0.4$. For $c_l = 0.4$, six-series airfoils have a smaller drag coefficient. Also, the benefits of NLF airfoils do not persist if the airfoils are used for turbulent flows. However, if the flow suffers premature transition due to leading edge contamination etc., there are no severe degradations of aerodynamic properties, as compared to traditional aerofoil sections.

10.8 A High Speed NLF Aerofoil

It is known that laminar compressible boundary layers are more stable than incompressible boundary layers, so high acceleration is not needed to keep the flow laminar. Also, as lift increases, overall acceleration increases instead of negative pressure peaks forming at the leading edge (Viken *et al.* 1987). However with added acceleration, comes the problem at the recovery region, with the transition point moving upstream. Also, with an increase in Mach number, the acceleration can lead to formation of shocks. Up to a Mach number of 0.4, section properties do not change appreciably, as has been shown experimentally for the NLF(1)-0416F airfoil (Somers 1981a) for moderate Reynolds numbers, without the flap deployed. However, when the Mach number increases beyond 0.4, significant compressible effects come into play in determining aerodynamic section properties.

Apart from NASA series NLF aerofoils, there is another NLF aerofoil that has been actually used in the design of a business jet. This airfoil is designated SHM-1 and reported in Fujino *et al.* (2003), with its additional features which makes it suitable for application in a general aviation jet class aircraft. The SHM-1 aerofoil has a thickness of 15% and at the same time it has a high value of drag divergence Mach number, so that its cruise Mach number is fixed at 0.69 for $Re = 11.7 \times 10^6$, with a cruise lift coefficient of 0.26. The larger thickness of the SHM-1 airfoil allows a wing design with minimum planform area, thereby achieving low profile drag obtained through the NLF feature of the section. The low profile drag is desired over a range of $c_l = 0.18$ for $Re = 11.7 \times 10^6$, $M = 0.69$ to $c_l = 0.35$ for $Re = 13.6 \times 10^6$, $M = 0.31$. Note that to provide operational margin, the lower limit of the low drag range is reduced from the actual cruise lift coefficient. Higher values of lift coefficient and Reynolds number correspond to sea-level climb conditions. Additionally, the section has a low nose-down pitching moment tendency, docile stall characteristics and no significant maximum lift degradation due to insect contamination problem. The sectional maximum lift coefficient without flap is set to a minimum of 1.6 for $Re = 4.8 \times 10^6$, $M = 0.134$ and this should not degrade by more than 7% due to leading edge contamination (LEC). The pitching moment constraint is given by $c_m \geq -0.04$ at $c_l = 0.38$ for $Re = 7.93 \times 10^6$ and $M = 0.70$ to minimize the trim-drag penalty at high altitude and high Mach number cruise condition. For this condition, the drag divergence Mach number should be higher than 0.7.

In Figure 10.18, the profile and a typical potential flow pressure distribution is shown for this aerofoil. According to results in Fujino *et al.* (2003), the flow accelerates on the upper surface up to 0.42*c*, followed by a very mild concave pressure recovery and that represents a compromise among the requirements of maximum lift, pitching moment and drag divergence. The pressure gradient on the lower surface is typical of all NLF aerofoils and is favourable up to $x/c = 0.63$ for cruise conditions.

Also, the leading edge of the airfoil is so designed that there is flow transition near it, at high angles of attack. This alleviates the problem of loss of lift due to LEC. Additionally, the trailing

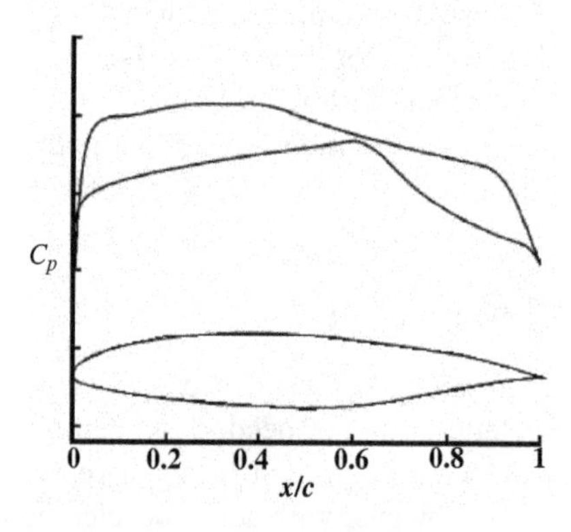

Figure 10.18 The SHM-1 airfoil profile and typical pressure distribution at design conditions. The SHM-1 airfoil profile and typical pressure distribution at design conditions. *Source:* Fujino, Yoshizaki and Kawamura (2003). Reproduced with permission from the American Institute of Aeronautics and Astronautics, Inc.

edge portion of the airfoil is also redesigned, which confines the movement of the separation point at high angles of attack. It is furthermore claimed that such a trailing edge design actually induces a small separation near the trailing edge, resulting in reduction of the pitching moment. An associated drag penalty caused by the separation is claimed to be negligible, as it induces a short and shallow bubble, which is verified later from Navier-Stokes solution.

In Figures 10.19 and 10.20, the transition location and lift curve are shown for the indicated Reynolds and Mach numbers. The transition point was located in the wind-tunnel experiment as well as by flight tests via an infrared flow-visualization technique, on the upper surface. The lift curves also indicate that barring $Re = 2.8 \times 10^6$, at all higher Reynolds numbers the design constraint of a minimum value of 1.6 for $c_{l,max}$ was met from the wind tunnel test results.

When a roughness element of a 0.3 mm-high strip was installed on the upper surface at $0.05c$, the loss of maximum lift was about 5.6% compared to the clean case for $Re = 4.8 \times 10^6$. In Figure 10.21, the drag polar of the SHM-1 airfoil is shown for $Re = 10.3 \times 10^6$, $M = 0.27$, which is for a cruise flight condition.

A very interesting set of results given in Fujino *et al.* (2003) relates to the effects of steps and surface roughness on drag obtained from a flight test. A 0.2 mm high step located at $0.2c$ on the upper surface causes a drag increase of about 3 counts for $Re = 13 \times 10^6$ and $M = 0.62$. The drag increase is ten times greater if the same step is located at $0.1c$, demonstrating that the chordwise location of the step is critical with respect to drag. This can be taken as a guideline to position the upper skin parting line for wing structural design. It is also reported that the drag increases by 17 counts when the surface is roughened by sandpaper S.P. 600, as compared to matte paint surface for $Re \leq 14 \times 10^6$. When the roughened surface was polished with wax, the drag decreased to the same prior level before roughening. Drag divergence characteristics of the airfoil are shown in Figure 10.22 for the indicated lift coefficient values (Fujino *et al.* 2003).

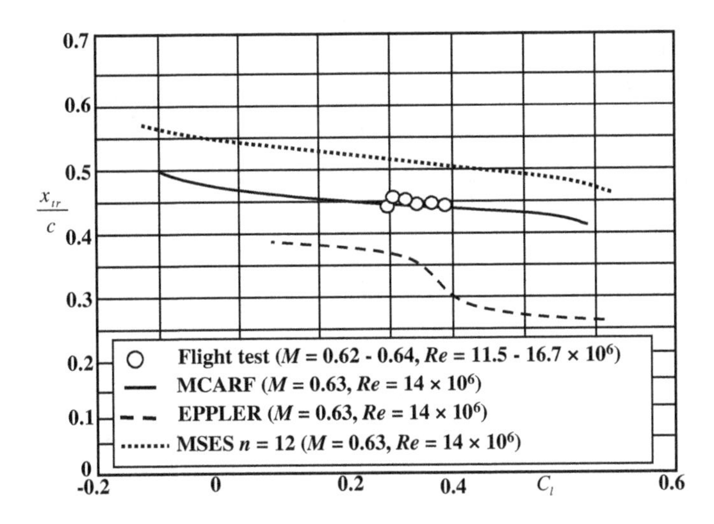

Figure 10.19 Experimental (flight test data) and theoretical prediction of transition location of the SHM-1 airfoil. *Source:* Fujino, Yoshizaki and Kawamura (2003). Reproduced with permission from the American Institute of Aeronautics and Astronautics, Inc.

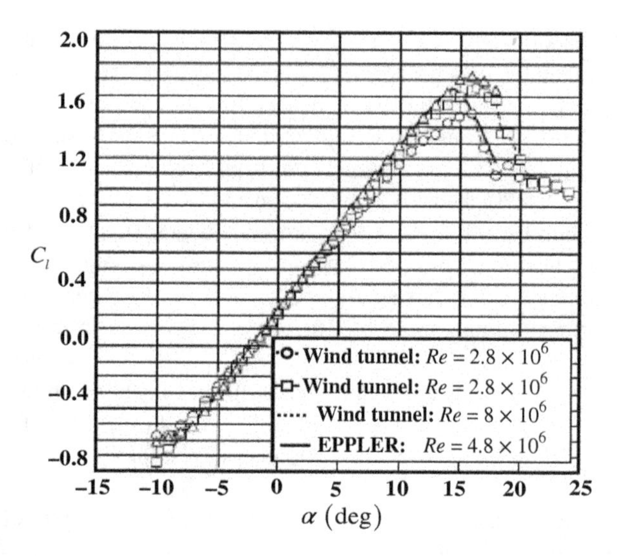

Figure 10.20 Wind tunnel data for the lift curve for the SHM-1 airfoil for the indicated Reynolds number. *Source:* Fujino, Yoshizaki and Kawamura (2003). Reproduced with permission from the American Institute of Aeronautics and Astronautics, Inc.

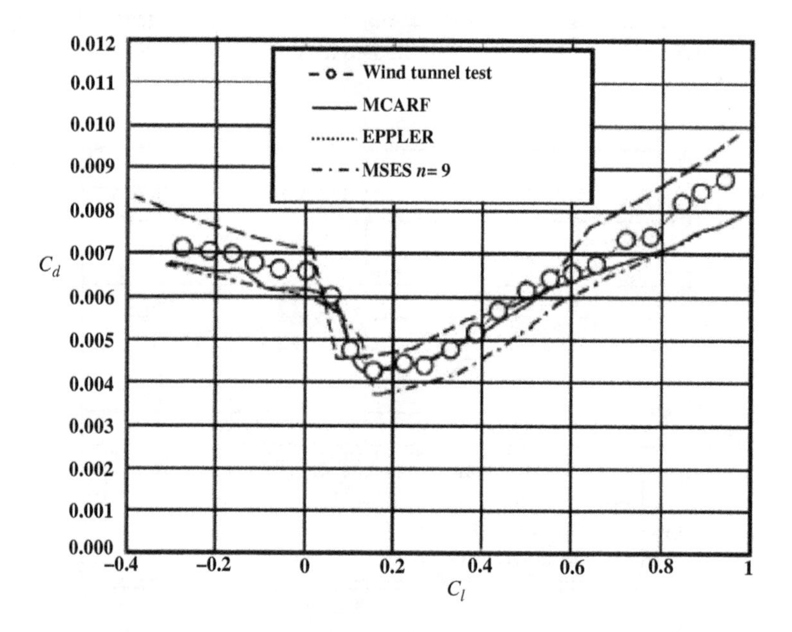

Figure 10.21 Experimental and theoretical drag polar for the SHM-1 airfoil for near-climb flight conditions–see text. *Source:* Fujino, Yoshizaki and Kawamura (2003). Reproduced with permission from the American Institute of Aeronautics and Astronautics, Inc.

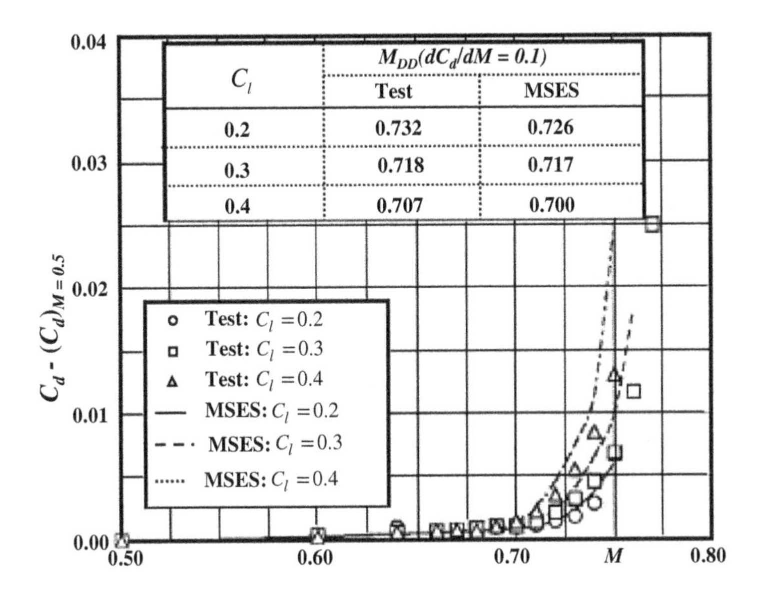

Figure 10.22 Experimental estimation of drag divergence Mach number for the SHM-1 airfoil. *Source:* Fujino, Yoshizaki and Kawamura (2003). Reproduced with permission from the American Institute of Aeronautics and Astronautics, Inc.

In discussing the above, we have not talked about the various design and analysis tools that have been used for most of the results presented here. Inviscid flows have been calculated either by solving the Laplace equation using panel methods for incompressible flows and by solving the Euler equation for compressible flows. In solving the boundary layer equations, both the differential and integral equations have been solved. Many of the successful design codes actually depend upon solving the integral equation form of the boundary layer for quick calculations. However, the differential formulations would be preferred for higher accuracy. Transition prediction is solely based upon the linear stability theory for the evolution of TS waves and their exponential growth. A growth exponent is chosen for different codes and although they are backed by some experimental data, it is still very empirical in nature. In contrast, there are only a very few design tools based on the solution of the full Navier–Stokes equations currently in use.

10.9 Direct Simulation of Bypass Transitional Flow Past an Airfoil

Flow past an SHM-1 aerofoil is studied by solving the full Navier–Stokes equation for $Re = 10.3 \times 10^6$, for this NLF aerofoil design. In Fujino *et al.* (2003), this aerofoil was designed using commercial codes for boundary layer analysis and transition prediction, as indicated in the legend of Figure 10.19. However, transition prediction is based on the classical view that it is caused by the growth of Tollmien–Schlichting waves and the design of the aerofoil is by delaying transition on the top and bottom surfaces of the airfoil. This also assumes that disturbance growth should be tracked independently on the top and bottom surface, starting from the front stagnation point.

Thus, it is necessary to compute the flow past the aerofoil by solving the Navier-Stokes equation without any models for transition and turbulence. As the corresponding experimental results are obtained from wind tunnel tests designed specifically for two-dimensional flows, it would be all right to compute the two-dimensional Navier-Stokes equation, as described next.

10.9.1 Governing Equations and Formulation

We have obtained numerical solution of Navier-Stokes equation using stream function–vorticity (ψ, ω)-formulation. This formulation ensures solenoidality of velocity and vorticity field simultaneously and is found to be accurate, as compared to other formulations. Using an orthogonal grid for the flow allows explicit use of orthogonal formulation resulting in fewer numerical computations per time step and reduce numerical errors by discretizing fewer terms with efficient discretization. This aspect is best demonstrated by the isotropic treatment of the diffusion operators in orthogonal formulation, as compared to nonorthogonal formulations. Also, physically the governing Navier-Stokes equation is decoupled into kinematics and kinetics of the flow in (ψ, ω)-formulation and expressed in the transformed plane by the following equations

$$\frac{\partial}{\partial \xi}\left[\frac{h_{22}}{h_{11}}\frac{\partial \psi}{\partial \xi}\right] + \frac{\partial}{\partial \eta}\left[\frac{h_{11}}{h_{22}}\frac{\partial \psi}{\partial \eta}\right] = -h_{11}h_{22}\,\omega \tag{10.2}$$

$$h_{11}h_{22}\frac{\partial \omega}{\partial t} + h_{22}u\frac{\partial \omega}{\partial \xi} + h_{11}v\frac{\partial \omega}{\partial \eta} = \frac{1}{Re}\left[\frac{\partial}{\partial \xi}\left(\frac{h_{22}}{h_{11}}\frac{\partial \omega}{\partial \xi}\right) + \frac{\partial}{\partial \eta}\left(\frac{h_{11}}{h_{22}}\frac{\partial \omega}{\partial \eta}\right)\right], \tag{10.3}$$

where h_{11} and h_{22} are the scale factors used in mapping physical (x, y)-plane to a computational (ξ, η)-plane (Sengupta 2013), where ξ coordinate is in azimuthal direction and η is normal to airfoil surface. For the orthogonal mapping, the scale factors are given by $h_{11} = \sqrt{x_\xi^2 + y_\xi^2}$ and $h_{22} = \sqrt{x_\eta^2 + y_\eta^2}$.

The leading and trailing edge portions of the aerofoil have a sharp slope variation and to simulate the flow correctly, distribution of points on the aerofoil is performed by judiciously clustering more points there. Also to resolve the boundary layer over the aerofoil, the points in the wall-normal η-direction uses a tangent hyperbolic distribution given by

$$h_{22} = H\left[1 - \frac{\tanh[\beta(1-\eta)]}{\tanh[\beta]}\right], \tag{10.4}$$

where β represents the clustering parameter. The grid spacing in η-direction is represented by grid scale parameter h_{22} and H is the nondimensional (in terms of chord of the aerofoil) distance of the outer boundary.

Following initial and boundary conditions are used in solving the governing equations. On the aerofoil surface, following no-slip conditions are used

$$\psi = \text{constant}; \quad \frac{\partial \psi}{\partial \eta} = 0. \tag{10.5}$$

These conditions also help fix the wall vorticity, which is required as the boundary condition for the vorticity transport equation, Equation (10.3). In O-grid topology, one introduces a cut starting from the leading edge of the aerofoil to the outer boundary. Periodic boundary conditions apply at the cut, which are introduced to make the computational domain simply-connected. For the stream function equation, Equation (10.2), at the outer boundary, Sommerfeld boundary condition is used on the η-component of the velocity field. Resultant value of stream function is used to calculate the vorticity value at the outer boundary. From the stream function equation the wall vorticity is calculated as

$$\omega|_{body} = -\frac{1}{h_2^2}\frac{\partial^2\psi}{\partial\eta^2}\bigg|_{body}. \tag{10.6}$$

10.9.2 Results and Discussion

Present numerical results are validated with experimental results provided in Fujino *et al.* (2003) for two Reynolds number cases, $Re = 2.8\times10^6$ and $Re = 10.3\times10^6$ using a (597×397) grid as given in Sengupta and Bhumkar (2012). In Figure 10.23(a), numerically obtained mean lift coefficient (c_l) with angle of attack for $Re = 2.8\times10^6$ are compared with experimental results provided in Fujino *et al.* (2003) and a good match between the two is noted. Calculations are for strictly two-dimensional flow without any explicit forcing, while experimental results involve three-dimensionality and background tunnel noise. Time variation of lift and drag coefficient for $Re = 10.3\times10^6$ and zero angle of attack case is shown in Figure 10.23(b) and (c), respectively. Time averaged values of c_l and c_d are shown with horizontal dashed lines in these frames. Once again a good match between experimental and numerically obtained values are noted. These have been presented in Sengupta and Bhumkar (2012) as validation of the numerical procedure.

Next, numerical results obtained using a (5169×577) grid is presented for a careful study of flow transition over the NLF aerofoil. Variation of RMS value of azimuthal component of disturbance velocity over top and bottom surfaces are shown in Figure 10.23(d) at a distance of 0.000056 from the aerofoil surface. This RMS value shows rapid rise in fluctuations on the top surface indicating flow transition after 60% of chord, similar to the experimental results shown in Fujino *et al.* (2003). On the bottom surface, one notes even a sharper variation of the RMS component of azimuthal velocity at $x = 0.70c$. Correct prediction of transition location highlights importance of highly space-time accurate numerical solutions reported in Sengupta and Bhumkar (2012).

Computed flow past SHM-1 aerofoil for $Re = 10.3\times10^6$ at zero angle of attack are shown in Figure 10.24(a) with stream function contours at $t = 4.50$. Top frame shows the flow field around the complete aerofoil, while in frame (ii) of Figure 10.24(a), the flow field near the trailing edge of the aerofoil is shown at the same instant. Flow near the leading edge of the aerofoil experiences acceleration due to the favourable pressure gradient and near the trailing edge, it experiences progressive adverse pressure gradient. Due to this varying adverse pressure gradient, small unsteady separation bubbles are formed on the aerofoil surface near trailing edge, which move downstream along with the flow. These separation bubbles are observed in the zoomed view of a trailing edge as shown in frame (ii) of Figure 10.24(a). As these bubbles move towards the trailing edge, these further excite the flow field. As shown earlier in Figure 10.23(d), flow transition on top surface starts after $0.60c$, these separation bubbles also first appear around the same location. In Figure 10.24(b), variations of displacement thickness (δ^*) on top and bottom surfaces of SHM-1 aerofoil are shown in frame up to 60% of chord of SHM-1 aerofoil at $t = 4.50$. Additionally, variation of a steady flow separation

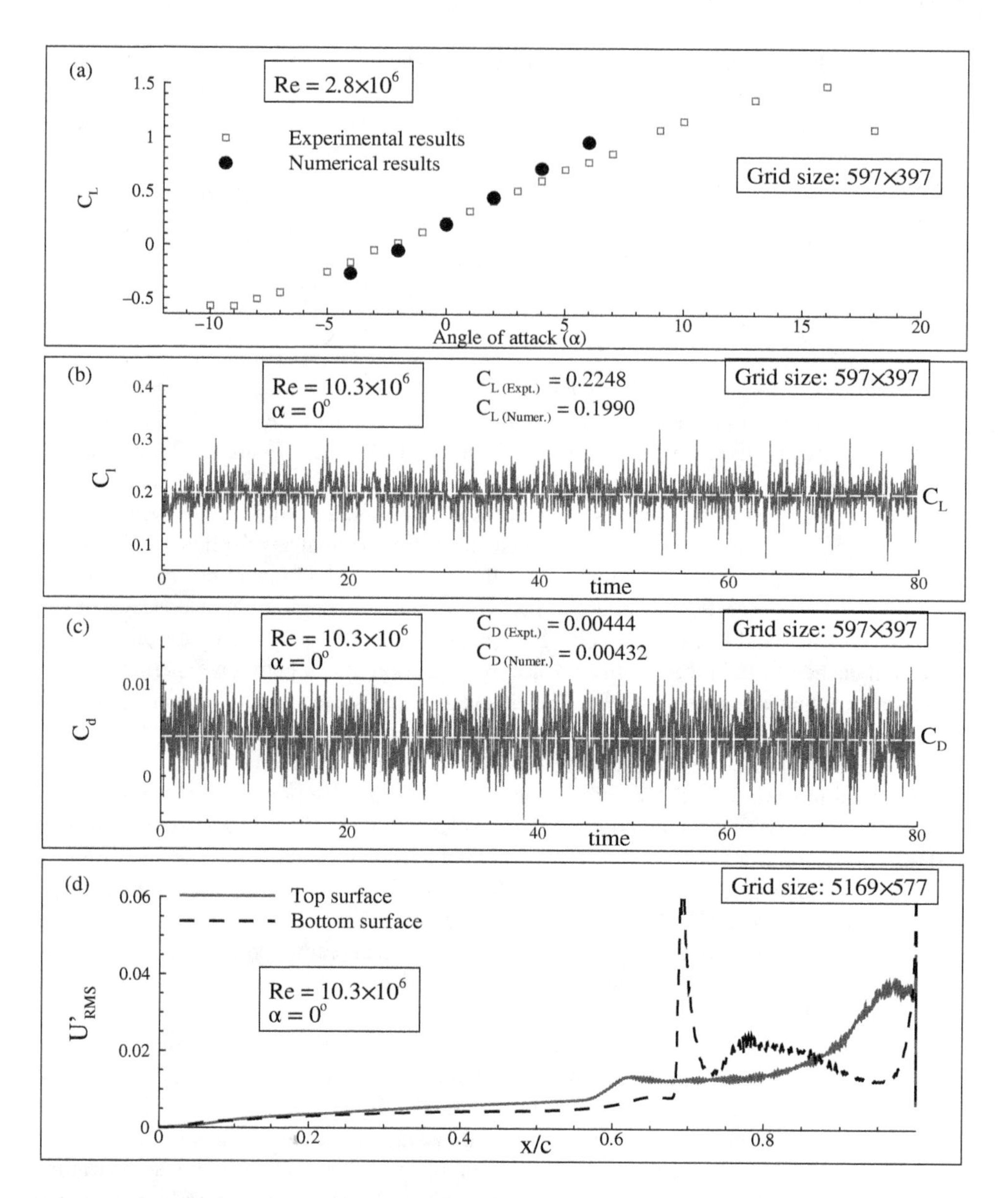

Figure 10.23 Numerically obtained c_l with angles of attack is shown along with experimental results in (Fujino *et al.* 2003); variation of instantaneous C_l and C_d are shown in frames (b) and (c), respectively; (d) variation of RMS value of the azimuthal component of the velocity over aerofoil for $Re = 10.3 \times 10^6$ on a line close to the aerofoil

parameter used in Falkner-Skan analysis, $m = \frac{x}{U_e}\frac{\partial U_e}{\partial x}$, is shown for the top and bottom surfaces of SHM-1 aerofoil in frame (ii) of Figure 10.24(b), at $t = 4.50$. A horizontal dashed line corresponding to steady separation flow criterion ($m = -0.09$) is marked for comparison purpose. This figure shows that on the top surface near the leading edge, flow experiences

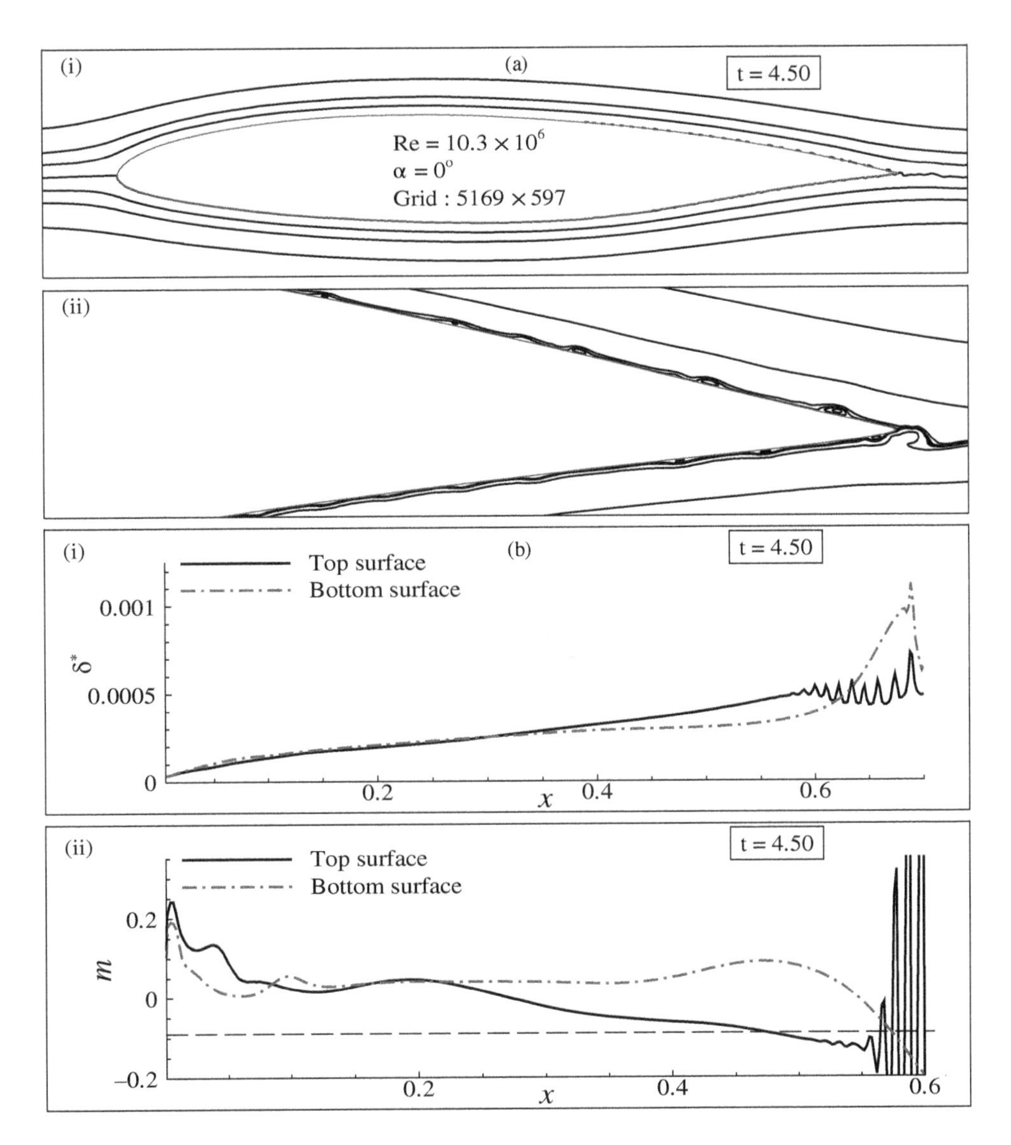

Figure 10.24 (a) Stream function contours over the complete aerofoil as well as near the trailing edge are shown for the indicated parameters; (b) variation of the displacement thickness (δ^*) and steady flow separation parameter m, on top and bottom surfaces of SHM-1 aerofoil are shown in frames (i) and (ii), respectively

higher favourable pressure gradient, as compared to the bottom surface. Pressure gradient parameter (m) smoothly decreases on the top surface after the initial peak near the leading edge. However, on the bottom surface, pressure gradient parameter remains favourable to a longer distance from the leading edge of the aerofoil and then it has a rapid variation aft of the mid-chord location.

Formation and propagation of separation bubbles on the top surface is shown in Figure 10.25(a) by showing variation of the azimuthal component of the velocity (u) at a distance

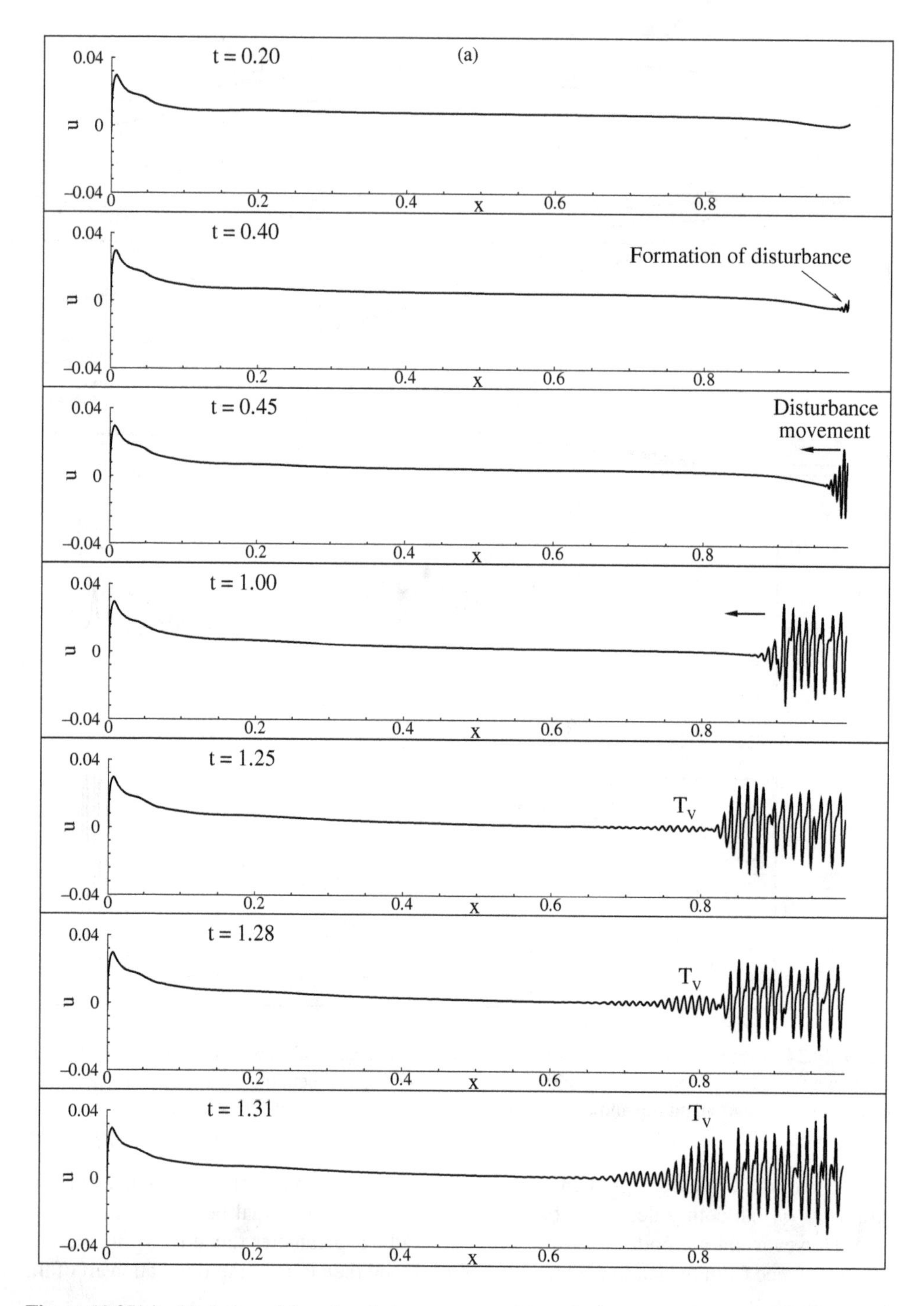

Figure 10.25(a) Variation of the azimuthal component of the velocity (u) at a distance of $1.242 \times 10^{-6}c$ from the top surface of SHM-1 aerofoil has been shown for the case of $Re = 10.3 \times 10^6$ and $\alpha = 0°$, at the indicated times

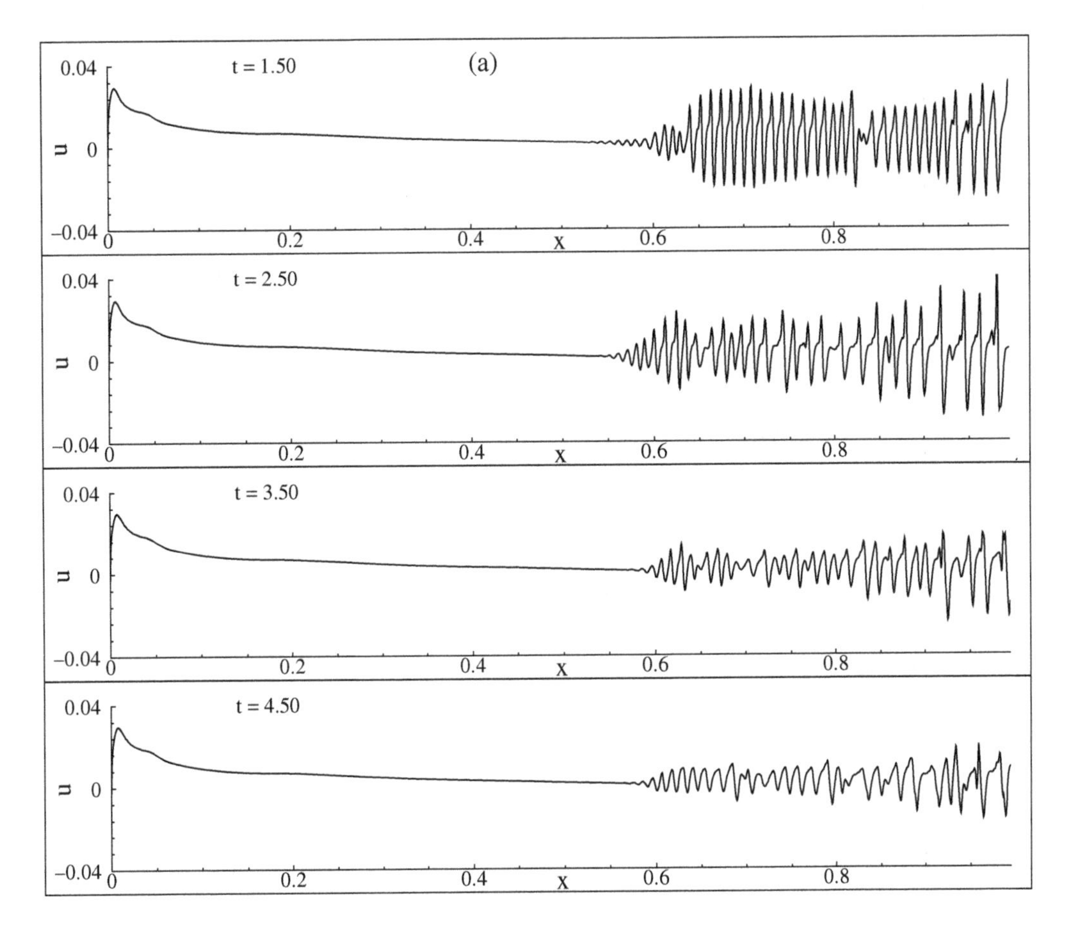

Figure 10.25(a) (*Continued*)

of $1.242 \times 10^{-6}c$ from the top surface of the SHM-1 aerofoil with time. This is at the second azimuthal line (η = constant line) from the aerofoil surface. At $t = 0.40$, small wavy disturbances are noted which originate from the trailing edge. These disturbances move upstream with time, as noted in subsequent frames of the figure. Adverse pressure gradient near the trailing edge magnifies the disturbances, which propagate further upstream with time. These calculations performed on the finer grid, resolve length scales correctly. The upstream propagation of the disturbances suggest physical bypass transition phenomenon (Sengupta 2012). Figure 10.25(a) shows how additional wave-packets are introduced upstream of this propagating main disturbance, marked as T_{υ} in the frames at $t = 1.25$ and later. This induced wave packet (T_{υ}) grows quite rapidly, as noted from the frames from $t = 1.25$ to $t = 1.31$, while convecting downstream. With time, upstream migrating disturbances fill up the top surface up to about $x = 0.55c$ by $t = 1.50$. Thereafter, disturbances formed on the top surface stop upstream propagation and continue to convect downstream, as shown till $t = 4.50$.

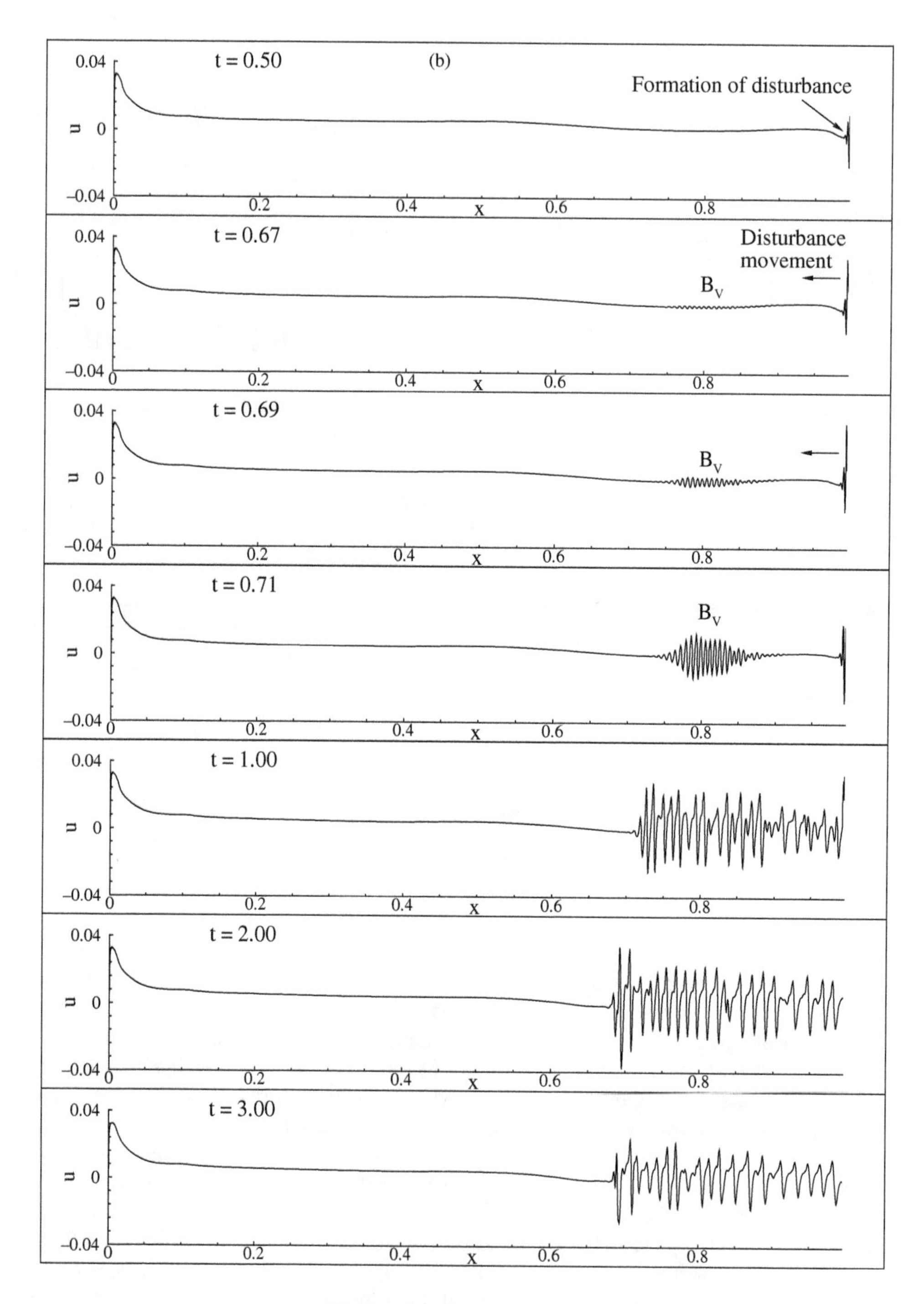

Figure 10.25(b) *(Continued)*

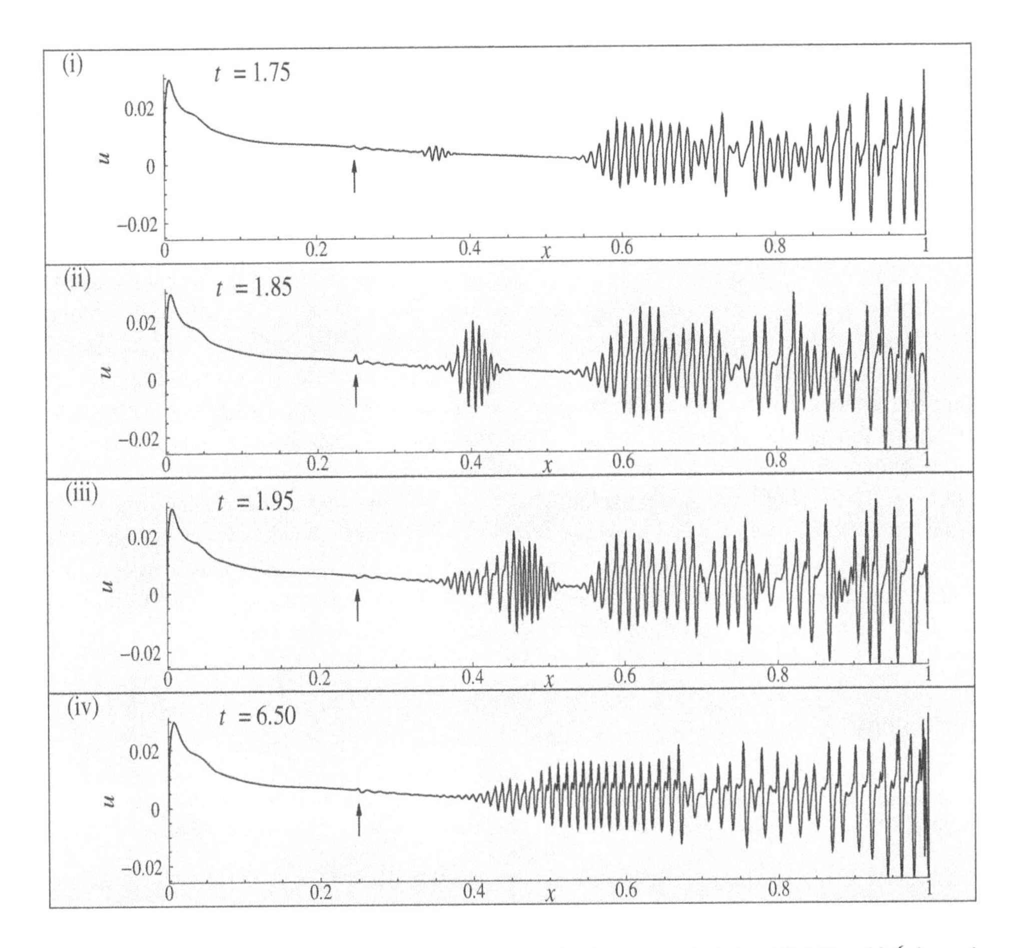

Figure 10.26 Variation of azimuthal component of velocity (u) at a height of 1.242×10^{-6} from the top surface of the SHM-1 airfoil at the indicated time instants. Wall excitation corresponds to an SBS frequency of $F = 1.11441 \times 10^{-5}$ with an amplitude of 0.001

Similar formation and propagation of disturbances on the bottom surface is shown in Figure 10.25(b). In this figure, once again the variation of azimuthal component of the velocity (u) is shown at a distance of $1.242 \times 10^{-6}c$ from the bottom surface. Similar to top surface, wavy disturbances originate from the trailing edge and can be observed as early as at $t = 0.50$. By $t = 0.67$, an induced small disturbance packet is noted close to $x = 0.8c$, which is marked as B_v. Its rapid growth is traced in the subsequent time frames and one can note the nonlinear distortion by $t = 0.71$. Trailing edge portion near $x = 0.8c$ has large concavity imposing strong adverse pressure gradient, which results in drastic amplification of disturbances. The resultant unsteady separation bubbles cause increased drag. As compared to the top surface, the upstream location of the disturbance is restricted up to $x = 0.70c$, as noted in the bottom frames of Figure 10.25(b).

NLF airfoils are seen to suffer bypass transition without any excitation. However, the computed bypass transition location is far aft of what has been noted in Figure 10.19. The difference is caused due to the fact that in computed flows, we consider a perfectly smooth geometry placed in a uniform flow and the transition is caused by the numerical disturbances acting as the seed for the adverse pressure gradient region over the airfoil surface. It is noted in Bhumkar (2012) and Sengupta and Bhumkar (2012) that the bypass transition location shifts upstream by deterministic disturbances. In Figure 10.26, results for a case is shown which has simultaneous blowing suction (SBS) harmonic excitation strip applied with a nondimensional frequency $F = 1.1144 \times 10^{-5}$, at the location indicated in the figure by an arrowhead. Here, the streamwise velocity is plotted as a function of x/c for the indicated time instants. A little before $t = 1.75$, a spatio-temporal packet is created downstream of the harmonic exciter. With time this grows and convects downstream and a little after $t = 1.95$, this packet merges with main packet which existed without the SBS excitation. This enlarges the region over which bypass transition is seen to be active. The frame at $t = 6.50$ shows that the transition point moves to $x = 0.40c$, which is noted experimentally in Figure 10.19. This, once again, establishes the fact that if one is using a high accuracy method, then bypass transition is seen to occur computationally downstream, as compared to that is observed in experiments. To match experimental value of drag, one must know the level of background disturbance environment in the experimental facility and simulate the flow accordingly. Such attempts at modelling free stream turbulence (FST) have been made in Sengupta *et al.* (2009) and Sengupta *et al.* (2001a) based on wind tunnel noise data in Frisch (1995). In Sengupta *et al.* (2009), additional data have been obtained from flight tests to statistically characterize FST.

Bibliography

Abbot, I.H. and von Doenhoff, A.E. (1959) *Theory of Wing Section.* Dover Publications, New York, USA.

Anderson, J.D. Jr, (1991) *Fundamentals of Aerodynamics.* McGraw-Hill, New York, USA.

Bhumkar, Y.G. (2012) High performance computing of bypass transition. Ph.D. thesis, Dept. of Aerospace Engineering, IIT Kanpur, India.

Bushnell, D.M. and Hefner, J.N. (1990) *Viscous Drag Reduction in Boundary Layers. AIAA Prog. in Aeronaut. and Astronaut.* Series, **123**.

Cebeci, T. and Bradshaw, P. (1977) *Momentum Transfer in Boundary Layers.* Hemisphere Publishing Corp., Washington DC.

Frisch, U. (1995) *Turbulence.* Cambridge Univ. Press, Cambridge, UK.

Fujino, M. and Kawamura, Y. (2003) Wave drag characteristics of an over-the-wing nacelle business-jet configuration. *J. Aircraft,* **40**(6), 1177–1184.

Fujino, M., Yoshizaki, Y. and Kawamura, Y. (2003) Natural-laminar-flow airfoil development for a lightweight business jet. *J. Aircraft,* **40**(4), 609–615.

Gray, W.E. (1950) Transition in flight on a laminar-flow wing of low waviness (King Cobra). *RAE Report* **2364**.

Gray, W.E. and Fullam, P.W.J. (1950) Comparison of flight and wind tunnel measurements of transition on a highly finished wing (King Cobra). *RAE Report* **2383**.

Hoerner, S.F. (1950) Base drag and thick trailing edges, *J. Aero. Sciences,* **17**(10), 622–628.

Jacobs, E.N. (1939) Preliminary report on laminar-flow airfoils and new methods adopted for airfoil and boundary layer investigations. *NACA WR L-345.*

McGhee, R.J. and Beasley, W.D. (1973) Low-speed aerodynamic characteristics of a 17-percent thick airfoil section designed for general aviation applications. *NASA TN D* **7428**.

McGhee, R.J., Beasley, W.D. and Whitcomb, R.T. (1980) NASA low- and medium-speed airfoil development. *Advanced Technology Airfoil Research,* II, *NASA CP* **2046**.

Maughmer, M. and Somers, D. (1989) The design of an airfoil for a high-altitude, long-endurance remotely piloted vehicle. In *NASA Conf. Pub., Part 3* on *Research in Natural Laminar Flow and Laminar-Flow Control,* NASA Langley, Virginia, USA.

Murri, D.G., McGhee, R.J., Jordan, F.L., Davis, P.J. and Viken, J.K. (1987) Wind tunnel results of the low speed NLF(1)-0414F airfoil. *NASA CP* **2487**(3), 673–695.

Pfenninger, W. (1947) Investigations on reductions of friction on wings, in particular by means of boundary layer suction. *NACA TM* **1181**.

Schubauer, G.B. and Spangenberg, W.G. (1958) Forced mixing in boundary layers. *National Bureau of Standards Rept.,* **6107**.

Sengupta, T.K. (2012) *Instabilities of Flows and Transition to Turbulence,* CRC Press, USA.

Sengupta, T.K. (2013) *High Accuracy Computing Methods: Fluid Flows and Wave Phenomena.* Cambridge University Press, USA.

Sengupta, T.K. and Bhumkar, (2013) Direct numerical simulation of transition over a NLF aerofoil methods and validation. *Frontiers in Aerospace Engineering,* **2**(1), 39–52.

Sengupta, T.K., Das, D., Mohanamuraly, P., Suman, V.K. and Biswas, A. (2009) Modelling free-stream turbulence based on wind tunnel and flight data for instability studies. *Int. J. Emerging Multidisc. Fluid Sci.,* **1**(3), 181–200.

Sengupta, T.K., De, S. and Gupta, K. (2001a) Effect of free-stream turbulence on flow over airfoil at high incidences. *J. Fluids Struct.,* **15**(5), 671–690.

Smith, F. and Higton, D.J. (1950) Flight test on a King Cobra FZ-440 to investigate the practical requirements for the achievement of low profile drag coefficients on a /low drag/ airfoil. *British ARC* R & M **2375.** Also *RAE Report* **2043**.

Somers, D. (1981) Design and experimental results for a natural-laminar-flow airfoil for general aviation applications. *NASA Tech. Mem.* **1861**.

Somers, D. (1981a) Design and experimental results for a natural-laminar-flow airfoil for general aviation applications. *NASA Tech. Mem.* **1865**.

Stratford, B.S. (1959) The prediction of separation of the turbulent boundary. *J. Fluid Mech.,* **5**(1), 1–16.

Tani, I. (1952) On the design of airfoils in which the transition of the boundary layer is delayed. *NACA TM* **1351**.

Vijgen, P.M.H.W. and Holmes, B.J. (1987) Experimental and numerical analysis of laminar boundary-layer flow stability over an aircraft fuselage forebody. In *Research in Natural Laminar Flow and Laminar-Flow Control. NASA CP* **2487**(3), 861–886.

Viken, J.K. (1983) Aerodynamic design considerations and theoretical results for a high Reynolds number natural laminar flow airfoil. M.S. thesis. The School of Engineering and Applied Science, George Washington Univ., USA.

Viken, J.K., Viken, S.A., Pfenninger, W., Morgan, H.L. and Campbell, R.L. (1987) Design of the low-speed NLF(1)-0414F and the high-speed HSNLF(1)-0213 airfoils with high lift systems. Symp. Proc. *Research in Natural Laminar Flow and Laminar-Flow Control, NASA CP* **2487**(3).

11

Direct Numerical Simulation of 2D Transonic Flows around Airfoils

11.1 Introduction

Transonic flows represent many additional complex phenomena, such as formation of shocks and contact discontinuities. The evanescent shock interacts with the boundary-layer and with strong interactions aerodynamic surfaces experience buffeting. These flows are time-dependent and can be well understood by invoking viscous nature of the flow, as emphasized by Moulden (1984) noting that the viscous terms are important interacting with convective process. In real flows, changes occur in nonmonotonic fashion producing surface fluctuations, even in the absence of unsteady boundary conditions (Maybey 1989). Therefore, transonic flows must be described by the time-dependent, compressible Navier-Stokes equations, as has been shown in Sengupta *et al.* (2013). Present day numerical methods have reached a level of sophistication where solving time accurate Navier-Stokes equations for 2D airfoil is not too difficult. In Jameson and Ou (2011), authors summarize developments in CFD ranging from the solution of full potential equation to Reynolds Averaged Navier-Stokes equation (RANS) by use of packages for transonic flows for engineering analysis. Use of numerical diffusion and turbulence models are common in RANS for transonic flows. In Visbal (1990) and Garbaruk *et al.* (2003) turbulence models have been used, whereas direct numerical simulation (DNS) of transonic flows can be found in Alshabu *et al.* (2006). This reports results indicating upstream moving pressure waves over a supercritical airfoil. Sengupta *et al.* (Sengupta *et al.* 2013) reports various schemes for calculation of compressible flows; Allaneu and Jameson (2009) proposed kinetic energy preserving scheme; Chiu *et al.* (2011) have used a conservative meshless scheme. Other notable older references are with the implicit factored scheme (Beam and Warming 1978), the essentially nonoscillatory (ENO) shock capturing scheme (Chakravarthy *et al.* 1986), the weighted essentially nonoscillatory (WENO) scheme (Liu and Osher 1994). High accuracy, dispersion relation preserving (DRP), optimized upwind

---

Theoretical and Computational Aerodynamics, First Edition. Tapan K. Sengupta.

Companion Website: www.wiley.com/go/sengupta

compact schemes (OUCS), originally developed for incompressible flows (Sengupta *et al.* 2003, 2004), Sengupta (2013), have been shown to be equally applicable for transonic flows in Sengupta *et al.* (2013) by solving and validating two-dimensional transonic flows around NACA 0012 and SHM-1 airfoils using two OUCS schemes. DNS of two-dimensional wall bounded turbulent flow is also reported in Sengupta *et al.* (2012), where OUCS3 scheme as described in Chapter 8, have been used. In Sengupta *et al.* (2007), a symmetrized version of the OUCS4 scheme has been used in parallel computing framework, to solve three-dimensional unsteady compressible Navier-Stokes equations for supersonic flow past a cone-cylinder for $M_\infty = 4$ and $Re_\infty = 1.12 \times 10^6$.

Since the validity of computation is based on available experimental results, knowledge of experimental conditions and possible errors of data is essential. In Garbaruk *et al.* (2003) and Binion (1979), problems of wall interference, 3D effects in 2D tests, aero-elastic effects, flow unsteadiness have been discussed. Such corrections of tunnel wall effects are uncertain and not directly usable for CFD. As discussed in Maybey (1976), flow unsteadiness is noted for velocity, pressure and temperature as fluctuations. Most transonic tunnels have high level of unsteadiness, in particular pressure fluctuations, causing significant adverse effects. In case of free-transition, such noise has influence on transition location (Pate 1968). Free stream turbulence can also have influence on attached boundary layers to cause separation (Otto 1976). Authors in Garbaruk *et al.* (2003) concluded that available experimental data on transonic airfoils are insufficient for turbulence model validation. In Mercer *et al.* (1981), authors have also concluded that comparison with wind tunnel have been limited to cases with insignificant wall effects due to lack of appropriate data.

The careful experimental results for 2D NACA 0012 airfoil in Harris (1981) are often used for validation. Although the wind tunnel used was not for 2D testing, the aspect ratio of the model with large span-to-chord ratio of 3.43, is useful for 2D testing. Also, the width of the tunnel helped to minimize sidewall boundary layer effects making conditions suitable for 2D test. The correction due to wall-induced lift interference is suggested as incremental change in angle of attack in Barnwell (1979).

11.2 Governing Equations and Boundary Conditions

The governing equations of motion are the conservation laws for mass, momentum and energy given here by 3D unsteady, compressible Navier-Stokes equation given by Equations (2.5), (2.21) and (2.91), respectively. In nondimensionalized strong conservation form, these are given in Equation (2.92). In the present chapter, we only focus on 2D viscous transonic flow past airfoils.

All terms in the following governing equations are nondimensionalized with scale as the chord (c) for length, the free stream density (ρ_∞) for density, the free stream velocity ($U_\infty = M_\infty \sqrt{\gamma R T_\infty}$) for velocity, twice the free stream dynamic pressure ($\rho_\infty U_\infty^2$) for the pressure, the free stream temperature (T_∞) for temperature and (c/U_∞) for the time, where γ denotes the ratio of specific heats, R denotes the universal gas constant and M_∞ as the free stream Mach number. The nondimensional parameters, Reynolds number (Re_∞), Prandtl number (Pr) and Mach number (M_∞) are defined as,

$$Re_\infty = \frac{\rho_\infty U_\infty c}{\mu_\infty}, \qquad Pr = \frac{\mu C_p}{\kappa}, \qquad M_\infty = \frac{U_\infty}{a_\infty},$$

where κ denotes the thermal conductivity and a_∞ denotes the speed of sound at the free stream. The two coefficients of viscosity are related by Stokes hypothesis: $3\lambda + 2\mu = 0$.

Additional information is given by Sutherland's viscosity law relating the coefficient of viscosity (μ) with T as, $\mu = \frac{1.45T^{3/2}}{T+110} \times 10^{-6}$. The closure of system of equations is provided by the ideal gas equation given by $p = \rho RT$ and it is used to define e_t as, $e_t = c_v T + \frac{(u^2+v^2)}{2}$ where c_v is the specific heat at constant volume and is given by $c_v = \frac{R}{\gamma-1}$. Governing equations are transformed to the generalized curvilinear coordinates, by introducing transformation $[\xi = \xi(x, y), \eta = \eta(x, y)]$. The transformed equation in the strong conservative form for 2D can be written as

$$\frac{\partial U}{\partial t} + \frac{\partial F}{\partial \xi} + \frac{\partial G}{\partial \eta} = \frac{\partial F_v}{\partial \xi} + \frac{\partial G_v}{\partial \eta}, \tag{11.1}$$

where corresponding state variables and flux vectors are given as

$$U = \hat{U}/J, \quad F = \left(\xi_x \hat{F} + \xi_y \hat{G}\right)/J, \quad G = \left(\eta_x \hat{F} + \eta_y \hat{G}\right)/J$$

$$F_v = \left(\xi_x \hat{F}_v + \xi_y \hat{G}_v\right)/J, \quad G_v = \left(\eta_x \hat{F}_v + \eta_y \hat{G}_v\right)/J.$$

The grid metrics (ξ_x, ξ_y, η_x and η_y) and the Jacobian of the transformation are defined as

$$\xi_x = J y_\eta, \quad \xi_y = -J x_\eta, \quad \eta_x = -J y_\xi, \quad \eta_y = J x_\xi$$

$$J = \frac{1}{(x_\xi y_\eta - x_\eta y_\xi)}.$$

The corresponding governing equations of motion in nondimensionalized strong conservation form is expressed in the physical (x, y)-plane as

$$\frac{\partial \hat{U}}{\partial t} + \frac{\partial \hat{F}}{\partial x} + \frac{\partial \hat{G}}{\partial y} = \frac{\partial \hat{F}_v}{\partial x} + \frac{\partial \hat{G}_v}{\partial y},$$

where the conservative variables are given by

$$\hat{U} = \lfloor \rho, \ \rho u, \ \rho v, \ \rho e_t \rfloor^T.$$

The convective flux vectors $\hat{F}$ and $\hat{G}$ are given as

$$\hat{F} = \lfloor \rho u, \ \rho u^2 + p, \ \rho u v, \ (\rho e_t + p) u \rfloor^T$$

$$\hat{G} = \lfloor \rho v, \ \rho u v, \ \rho v^2 + p, \ (\rho e_t + p) v \rfloor^T$$

and the viscous flux vectors $\hat{F}_v$ and $\hat{G}_v$ are given as

$$\hat{F}_v = \lfloor 0, \ \tau_{xx}, \ \tau_{xy}, \ (u\tau_{xx} + v\tau_{xy} - q_x) \rfloor^T$$

$$\hat{G}_v = \lfloor 0,\ \tau_{yx},\ \tau_{yy},\ (u\tau_{yx} + v\tau_{yy} - q_y) \rfloor^T.$$

The viscous shear stress components are given by

$$\tau_{xx} = \frac{\mu}{Re_\infty}\left[\frac{4}{3}\left(\xi_x u_\xi + \eta_x u_\eta\right) - \frac{2}{3}\left(\xi_y v_\xi + \eta_y v_\eta\right)\right]$$

$$\tau_{yy} = \frac{\mu}{Re_\infty}\left[\frac{4}{3}\left(\xi_y v_\xi + \eta_y v_\eta\right) - \frac{2}{3}\left(\xi_x v_\xi + \eta_x u_\eta\right)\right]$$

$$\tau_{xy} = \tau_{yx} = \frac{\mu}{Re_\infty}\left[\xi_y u_\xi + \eta_y u_\eta + \xi_x v_\xi + \eta_x v_\eta\right]$$

and heat conduction terms in the transformed plane are given by

$$q_x = -\frac{\mu}{PrRe_\infty\,(\gamma - 1)\,M_\infty^2}\left(\xi_x T_\xi + \eta_x T_\eta\right)$$

$$q_y = -\frac{\mu}{PrRe_\infty\,(\gamma - 1)\,M_\infty^2}\left(\xi_y T_\xi + \eta_y T_\eta\right).$$

On the airfoil surface, no-slip condition is imposed on the velocity components and adiabatic wall condition is imposed on heat conduction terms

$$u = 0, \quad v = 0 \quad \text{and} \quad q_x = 0, \quad q_y = 0.$$

A typical sketch for the flow past an airfoil is shown in Figure 11.1, and the problem is solved in an O-grid topology. This requires introduction of a cut, as marked in the figure, along which periodic boundary condition implemented for all flow variables. Free stream initial condition is implemented for all the flow properties with respective free stream values. At any time, the far-field boundary condition can be either of the following: a subsonic inflow or a supersonic inflow; a subsonic outflow or a supersonic outflow boundary condition. Inflow and outflow boundary conditions are applied at the boundary, depending upon the sign of normal component of velocity (V_n). Here we have used an orthogonal grid, and this velocity component is the same as the contravariant component of velocity ($\eta_x u + \eta_y v$) in the transformed plane. Additionally, the far-field boundary conditions are determined based on eigenvalues of the locally one-dimensional description of the flow, details of which can be found in Hoffmann and Chiang (2000).

11.3 Numerical Procedure

A computational domain with outer boundary segment located at approximately $10c$ from the airfoil is used, to ensure accurate implementation of far-field boundary condition and to reduce the acoustic wave reflections from the boundary. An orthogonal, body-conforming O-grid is generated using hyperbolic grid generation technique (Sengupta 2013, Nair and Sengupta 1998), as given in Section 8.4.

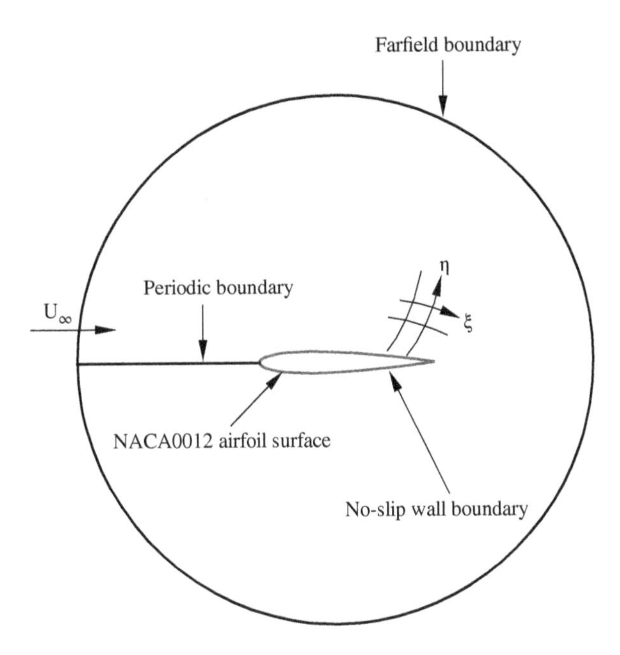

Figure 11.1 Computational domain and boundary segments

In the presented results, high accuracy optimized upwind DRP schemes and their symmetrized versions, OUCS3 and S-OUCS3 schemes (Dipankar and Sengupta 2006), are used for evaluating the convective flux terms in η- and ξ-directions (given in Equations (8.31) to (8.35)). These schemes have also been used in parallel computing framework by domain decomposition technique (Sengupta *et al.* 2007) to solve the problem in smaller sub-domains independently for parallel computing.

In the presented results, another fifth order upwind compact scheme, OUCS2, is also used. The interior stencil of the OUCS2 scheme (Sengupta *et al.* 2003, 2004, 2006) to evaluate first derivative (d') is given by

$$b_1 d'_{j-1} + b_2 d'_j + b_3 d'_{j+1} = \frac{1}{h} \sum_{k=-2}^{2} a_k d_{j+k}, \tag{11.2}$$

where, $b_1 = \frac{b_2}{3} - \frac{\alpha_1}{12}$; $b_3 = \frac{b_2}{3} + \frac{\alpha_1}{12}$; $a_{\pm 2} = \mp\frac{b_2}{36} + \frac{\alpha_1}{72}$; $a_{\pm 1} = \mp\frac{7b_2}{9} + \frac{\alpha_1}{9}$ and

$a_0 = -\frac{\alpha_1}{4}$. The free parameter α_1 is the coefficient of $(\frac{\partial^6 d}{\partial x^6})$, which is treated as the free parameter to fix the order of representation. The interior stencil of the OUCS2 is a fifth order accurate upwind scheme for $\alpha_1 < 0$ and a sixth-order accurate central scheme results for $\alpha_1 = 0$.

In applying Equations (8.33) and (8.35), the value of γ_j is chosen as 0.02 for $j = 2$ and 0.09 for $j = (N - 1)$ for OUCS2 scheme. Similarly, value of γ_j is chosen as −0.025 for $j = 2$ and 0.09 for $j = (N - 1)$, while using OUCS3 scheme.

Artificial diffusion terms due to Jameson, Schmidt and Turkel (JST) (Jameson *et al.* 1981) are added explicitly to the numerically evaluated convective flux terms, in order to damp higher wavenumber components, which arise due to nonlinearities and discontinuities in the solution and to accurately capture discontinuities. The artificial diffusion terms (D_ξ) and (D_η) added in ξ- and η-directions, respectively, are given by

$$D_\xi = \sigma_\xi J^{-1}[(d_{\xi_2}) - (d_{\xi_4})] \tag{11.3}$$

$$D_\eta = \sigma_\eta J^{-1}[(d_{\eta_2}) - (d_{\eta_4})], \tag{11.4}$$

where, σ_ξ and σ_η are the spectral radii of the flux Jacobian matrices (Hoffmann and Chiang 2000). These terms are blend of second and fourth diffusion terms, d_{ξ_2}, d_{ξ_4}, d_{η_2} and d_{η_4} added selectively at various points of the computational domain and are defined as

$$d_{\xi_2} = \epsilon_{2_\xi} \nabla_\xi \Delta_\xi (JU)$$

$$d_{\xi_4} = \epsilon_{4_\xi} \left(\nabla_\xi \Delta_\xi\right)^2 (JU)$$

$$d_{\eta_2} = \epsilon_{2_\eta} \nabla_\eta \Delta_\eta (JU)$$

$$d_{\eta_4} = \epsilon_{4_\eta} \left(\nabla_\eta \Delta_\eta\right)^2 (JU),$$

with ϵ_{2_ξ}, ϵ_{4_ξ}, ϵ_{2_η} and ϵ_{4_η} are defined at points i, j in the computational domain as

$$\epsilon_{2_\xi} = \kappa_2 \, max \left(\Upsilon_{i+1,j}, \Upsilon_{i,j}, \Upsilon_{i-1,j}\right)$$

$$\epsilon_{4_\xi} = \kappa_4 \, max \left(0, \kappa_4 - \epsilon_{2_\xi}\right)$$

$$\epsilon_{2_\eta} = \kappa_2 \, max \left(\Upsilon_{i,j+1}, \Upsilon_{i,j}, \Upsilon_{i,j-1}\right)$$

$$\epsilon_{4_\eta} = \kappa_4 \, max \left(0, \kappa_4 - \epsilon_{2_\eta}\right),$$

where κ_2 and κ_4 are the parameters which controls the amount of numerical diffusion to be added. Values of $\kappa_2 = 0.25$ and $\kappa_4 = 0.15$ have been used. The terms $\Upsilon_{i,j}$, are the pressure based switch defined at the point (i, j) as given in Jameson *et al.* (1981) by

$$\Upsilon_{i,j} = \frac{|p_{i+1,j} - 2p_{i,j} + p_{i-1,j}|}{|p_{i+1,j}| + |2p_{i,j}| + |p_{i-1,j}|}. \tag{11.5}$$

The term $\Upsilon_{i,j}$ has very high value near a shock. Hence, it acts as a pressure switch activating the second difference term, wherever there are discontinuities in pressure. The fourth difference term is switched to zero near shock. These artificial diffusion terms are added to the numerically evaluated convective flux derivative terms for numerical stability and better dispersion properties. Viscous flux terms represent diffusion phenomenon, which is isotropic.

Hence, the derivatives are calculated using second order central discretization (CD2) in self-adjoint form, preserving isotropic nature. Classical RK4 scheme is used for time integration.

The parallel computing is performed by domain decomposition technique, developed using the above-mentioned numerical methods, which require overlap of subdomains. Motivation behind the overlap is to overcome problems due to the nonphysical growth and attenuation of the solution near the interfaces of neighboring subdomains, due to the effects of boundary closures (Sengupta *et al.* 2003) of compact schemes. Selective addition of artificial second and fourth order diffusion terms (Jameson *et al.* 1981) help to use smaller number of overlapping points (Sengupta *et al.* 2007). Here eight overlapping points have been used.

11.4 Some Typical Results

Here, we report results for transonic flows past NACA 0012 and SHM-1 airfoils displaying mixed flow nature, i.e. there are regions where the flow speed exceeds sonic speed locally, while in other parts the flow Mach number remain subsonic. To compare with experimental results, the computed solution is time averaged taking significant time intervals.

11.4.1 Validation of Methodologies for Compressible Flow Calculations and Shock Capturing

In computing transonic flow past NACA 0012 airfoil, 636 points in the azimuthal and 350 points in the wall-normal directions have been taken in O-grid topology, for the solution of compressible Navier-Stokes equation for $Re_\infty = 3 \times 10^6$ and three Mach numbers, $M_\infty = 0.6$, 0.758 and 0.779, and compared with available experimental results for validation. For the SHM-1 airfoil, 663 points have been taken in ξ- and 450 points in η-directions, respectively. The close-up views of grids used for these airfoils are shown in Figures 11.2(a) and 11.2(b).

SHM-1 airfoil has higher (15%) thickness, yet it has better performance at cruise Mach number ($M_\infty = 0.69$), as compared to NACA 0012 airfoil, as seen by comparing the drag polar of SHM-1 airfoil in Fujino *et al.* (2003) with results from literature on NACA 0012 airfoil. For SHM-1 airfoil, computed results for $M_\infty = 0.62$ and $Re_\infty = 13.6 \times 10^6$ at an angle of attack, $\alpha = 0.27°$ are validated with experimental data. The near-wall resolution for NACA 0012 and SHM-1 airfoils used are given by $\Delta s_\eta = 1.89 \times 10^{-5}c$ and $\Delta s_\eta = 8.20 \times 10^{-6}c$, respectively. There are many reasons for choosing such values of wall-normal spacing with the used schemes. Employed compact schemes (Sengupta *et al.* 2003, Sengupta 2013) have near-spectral accuracy and these wall resolutions are found more than adequate. Secondly, the flow is not turbulent right from the leading edge of the airfoil for both the cases considered. Specifically, the flow over NLF airfoil remains laminar up to and beyond the mid-chord, as per the design requirement. For example, at the locations where the flow pressure gradient is zero, these spacing are obtained in wall unit as, $\Delta y^+|_w = 0.1669$ for the NACA 0012 airfoil and $\Delta y^+|_w = 0.1312$ for the SHM-1 airfoil. The wall units are defined traditionally for characterizing Reynolds-averaged turbulent flow as $y^+ = yu_\tau/\nu$, where the wall friction is used to define a velocity scale given by $u_\tau = \sqrt{\tau_w/\rho}$. We emphasize that for downstream locations past the shock location, wall friction value decreases due to flow separation and there $\Delta y^+|_w$ values reduce even further for both the airfoils. Such behaviour of wall friction with streamwise distance is also reported in Pirozzoli *et al.* (2010) and is due to shock-boundary layer interaction.

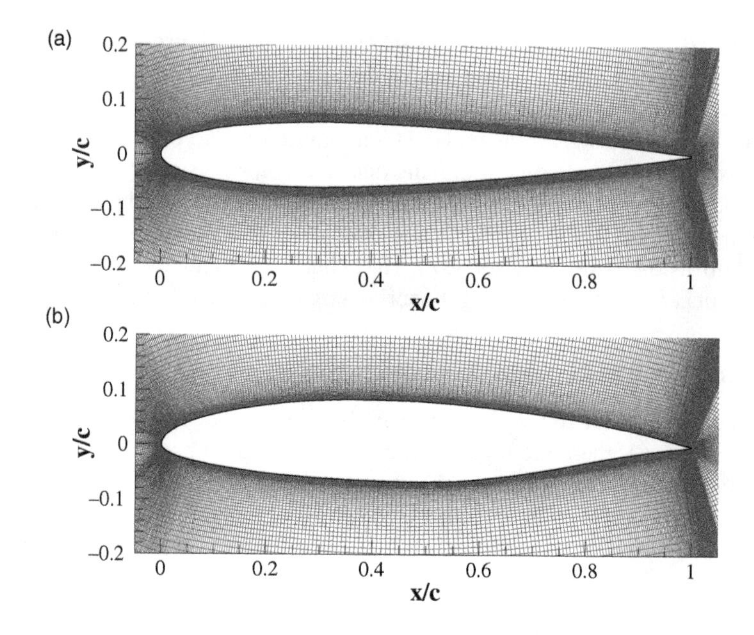

Figure 11.2 Close-up view of orthogonal grid generated around (a) NACA 0012 and (b) SHM-1 airfoils

All computations reported here using OUCS3-RK4 scheme (unless otherwise stated) are with a small time step of $\Delta t = 10^{-6}$, which is necessary for DRP properties. Such restriction of time step for explicit time integration schemes are global in nature. We emphasize that these numerical schemes have excellent DRP properties, measured by calibrating the methods in capturing discontinuities, without dissipation and dispersion errors. Thus, the developed methods produce accurate results for incompressible and compressible flows, without requiring any flow-specific fixes.

In Figure 11.3(a), comparison between computed and experimental (Harris 1981) pressure coefficient (C_p) distribution is shown for flow over NACA 0012 airfoil for $M_\infty = 0.6$, $Re_\infty = 3 \times 10^6$ and $\alpha = -0.14°$. Experimental correction for angle of attack ($\Delta\alpha = -1.55C_L$) is implemented in the code, which is due to wall-induced lift interference effects (Harris 1981), where C_L denotes sectional normal-force coefficient. Figure 11.3(b) shows comparison between computed and experimental (Harris 1981) C_p distribution for flow over NACA 0012 airfoil for $M_\infty = 0.758$, $Re_\infty = 3 \times 10^6$ at $\alpha = -0.14°$. One notices good match for higher M_∞ case, between numerical and experimental C_p distribution. The performance of the methods in capturing the shock is further validated by simulating flow for $M_\infty = 0.779$ with corresponding results shown in Figure 11.4, which exhibit better agreement between computed and experimental (Harris 1981) C_p-distribution for the same Reynolds number and angle of attack. Here one notices the shock being captured with the pre-shock dip in C_p noted upstream of the shock.

One of the most redeeming features of the present calculations is that no models have been used for transition and /or turbulence. The near-spectral accuracy of DRP compact schemes enable one to capture shock with good resolution. To understand the distinctive feature of the developed methodology to simultaneously capture the laminar, transitional and turbulent

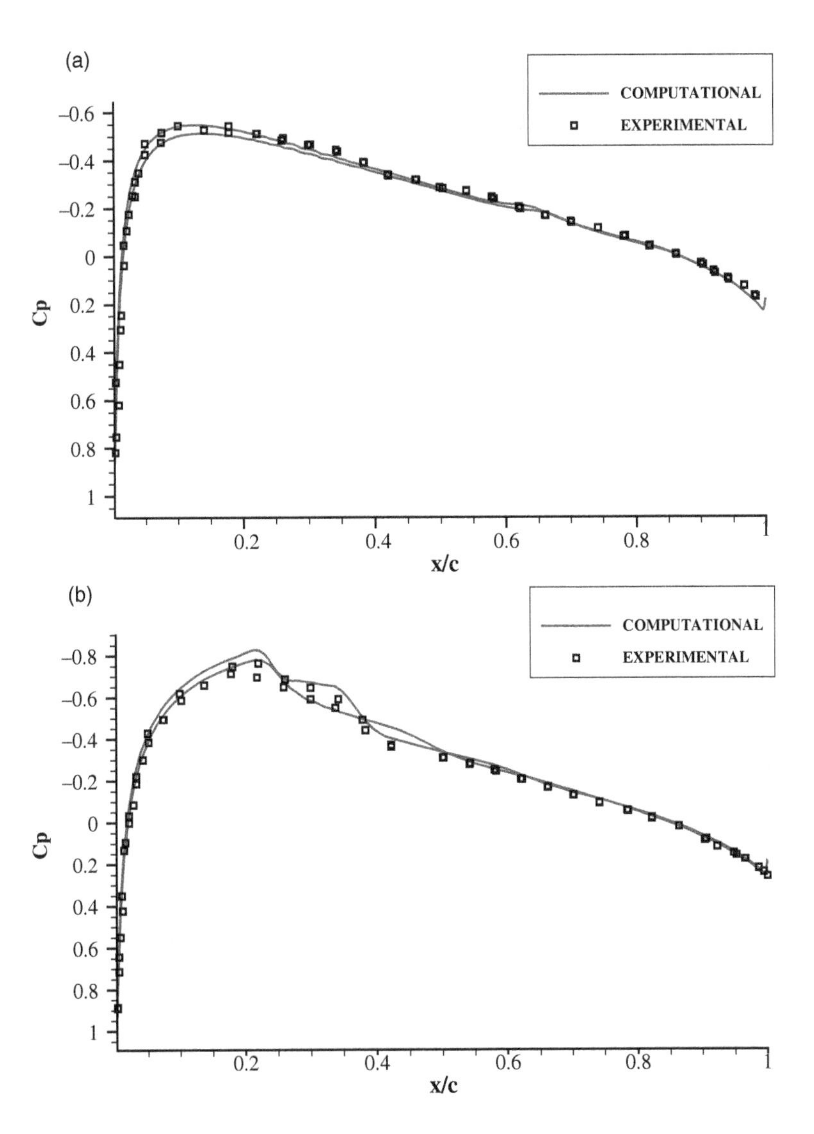

Figure 11.3 Comparison of computed and experimental C_p-distribution for flow around NACA 0012 airfoil with $Re_\infty = 3 \times 10^6$ at $\alpha = -0.14°$ and (a) $M_\infty = 0.6$ and (b) $M_\infty = 0.758$. Computed data in both cases is time-averaged over $t = 10$ to 40

flows, the velocity profiles of the computed flow past NACA 0012 airfoil are shown in Figure 11.5 in wall unit, for the case of Figure 11.4. Such a representation is preferred for turbulent flows, where one notices the presence of a viscous sublayer near the wall followed by the inertial log law region, with a comprehensive description of the profile given by the Spalding law (Panton 2005). In Figure 11.5, one can clearly notice profiles representing laminar, transitional and turbulent flows for this $Re = 3 \times 10^6$ flow, one notes the flow to become turbulent near the trailing edge with the distinct match shown by the plotted time-averaged profile with that given by the Spalding law.

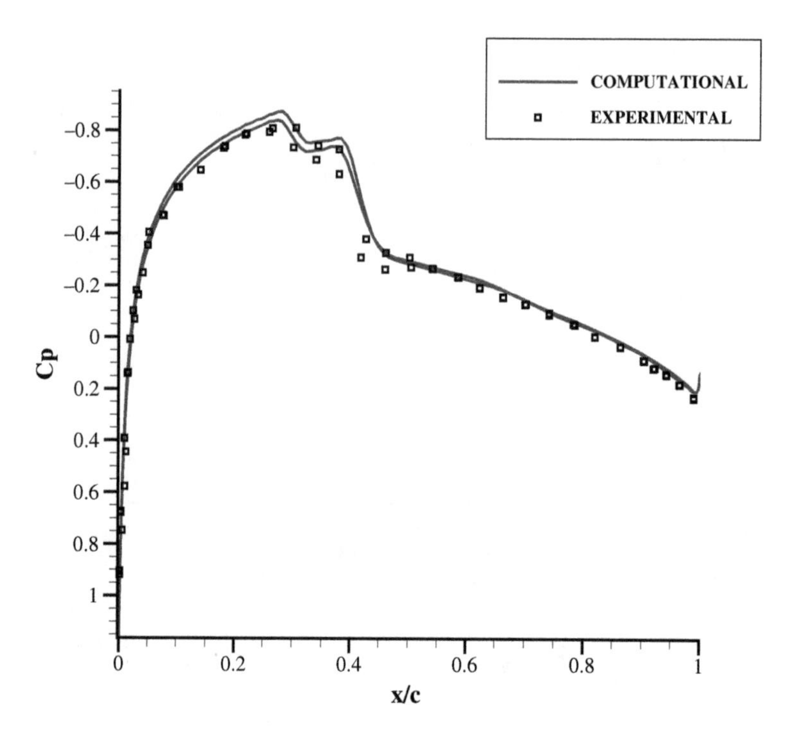

Figure 11.4 Comparison of computed and experimental C_p-distribution for flow around NACA 0012 airfoil with $Re_\infty = 3 \times 10^6$ at $\alpha = -0.14°$ and $M_\infty = 0.779$. Computed data is time-averaged over $t = 10$ to 45

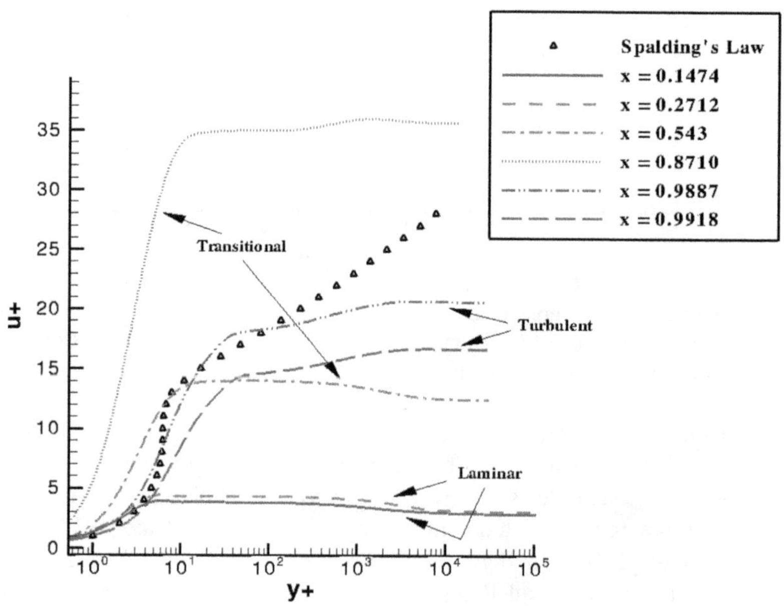

Figure 11.5 Distribution of mean streamwise velocity at various streamwise locations in wall coordinates, for flow around NACA 0012 airfoil with $Re_\infty = 3 \times 10^6$ at $\alpha = -0.14°$ and $M_\infty = 0.779$. (See colour version of this figure in colour plate section)

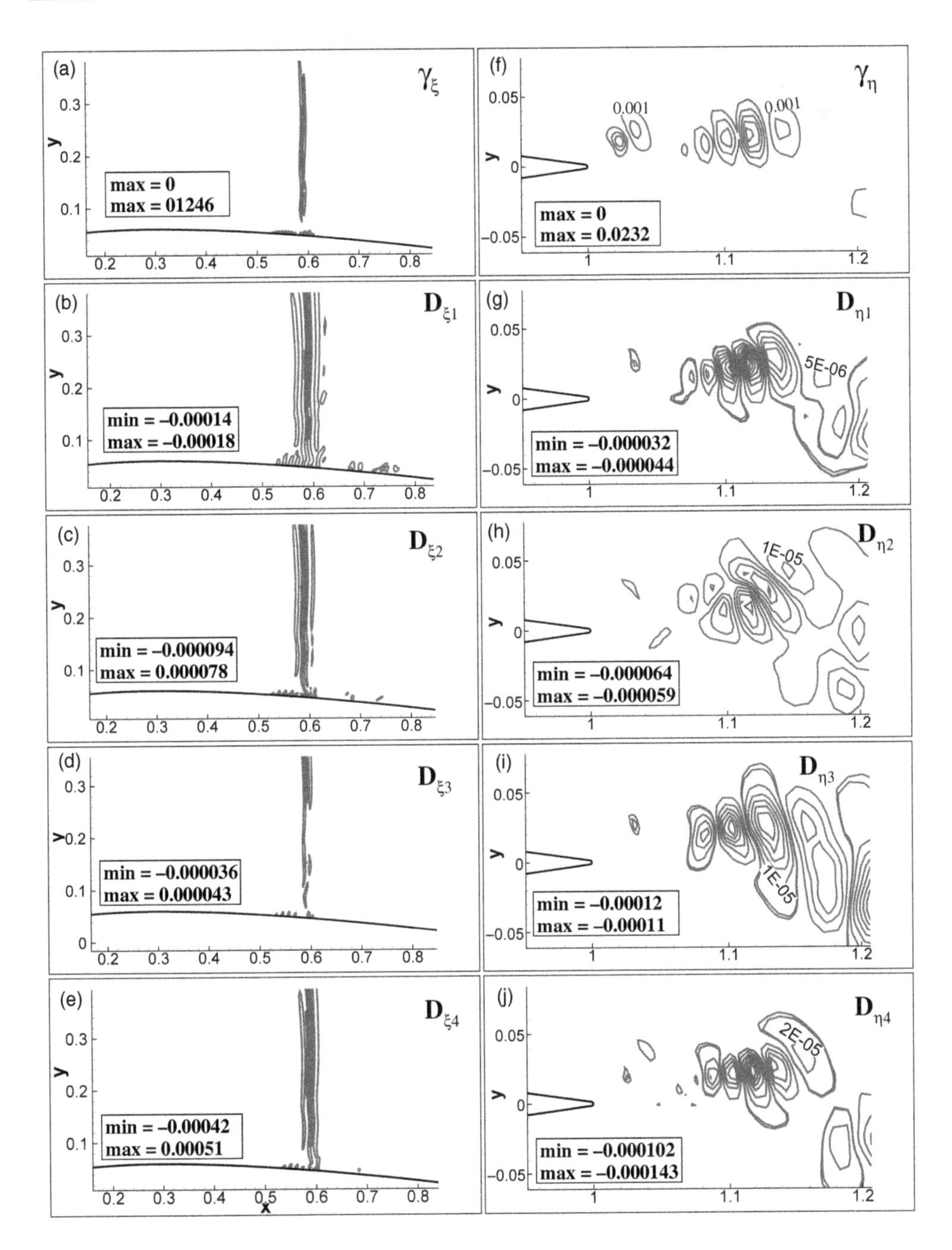

Figure 11.6 Added diffusion terms for the case of flow over NACA 0012 airfoil shown in Figure 11.4: Frames (a) and (f) show contours for values of $\Upsilon_{i,j}$ terms (Equation (11.9)) in ξ- and η-directions, respectively. Frames (b, c, d and e) show contours for values of D_ξ's (Equation (11.7)) added in the numerical solution of Equation (11.2), in ξ-direction. Frames (g, h, i and j) show contours for values of D_η's (Equation (11.8)) added in the numerical solution Equation (11.2) for the JST model, in η-direction

To demonstrate that the selectively added diffusion terms do not affect solution accuracy, apart from controlling numerical instability and dispersion, in Figure 11.6 the magnitudes of added terms proportional to κ_2 and κ_4 are shown for flow past NACA 0012 airfoil case shown in Figure 11.4. In each frame, the maximum and minimum values of the quantities involved are also marked. It is readily evident that the second diffusion term is only added near the shock and only a limited region of the flow field is affected by these added terms.

To further test capabilities of the developed procedure, results for computed flow over NLF SHM-1 airfoil is shown for $M_\infty = 0.62$ and $Re_\infty = 13.6 \times 10^6$ at $\alpha = 0.27°$ in Figure 11.7. These simulations are performed with convective flux derivatives evaluated using both OUCS2 and OUCS3 schemes and results compared in Figure 11.7, with experimental and computed C_p-distribution (Fujino *et al.* 2003). Although, the OUCS2 scheme has fifth order accuracy, as compared to second order accuracy of the OUCS3 scheme, identical results are obtained with both the schemes in capturing flow features correctly. This is due to both the schemes having comparable spectral resolution (Sengupta 2013). Figures 11.8(a), 11.9(a) and 11.10(a) show computed M, p and ρ-contours, respectively, at different time instants for the flow using OUCS2 scheme, while Figures 11.8(b), 11.9(b) and 11.10(b) show the corresponding computed M, p and ρ-contours computed with OUCS3 scheme at the same time instants. One notices identical flow features captured by both the schemes.

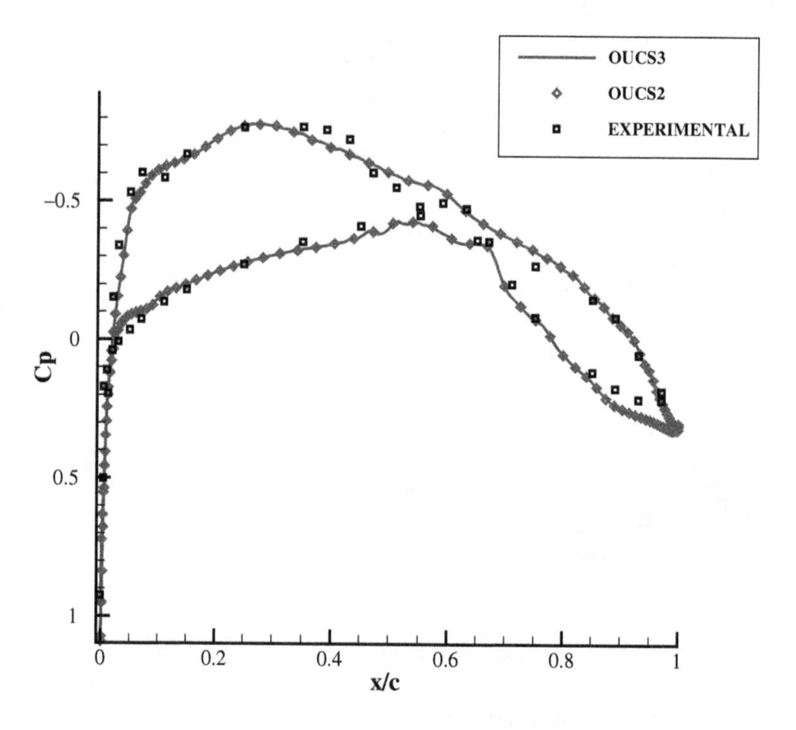

Figure 11.7 Comparison of computed and experimental C_p-distribution for flow around SHM-1 airfoil with $Re_\infty = 13.6 \times 10^6$ at $\alpha = 0.27°$ and $M_\infty = 0.62$. Computed data is time-averaged over $t = 10$ to 20

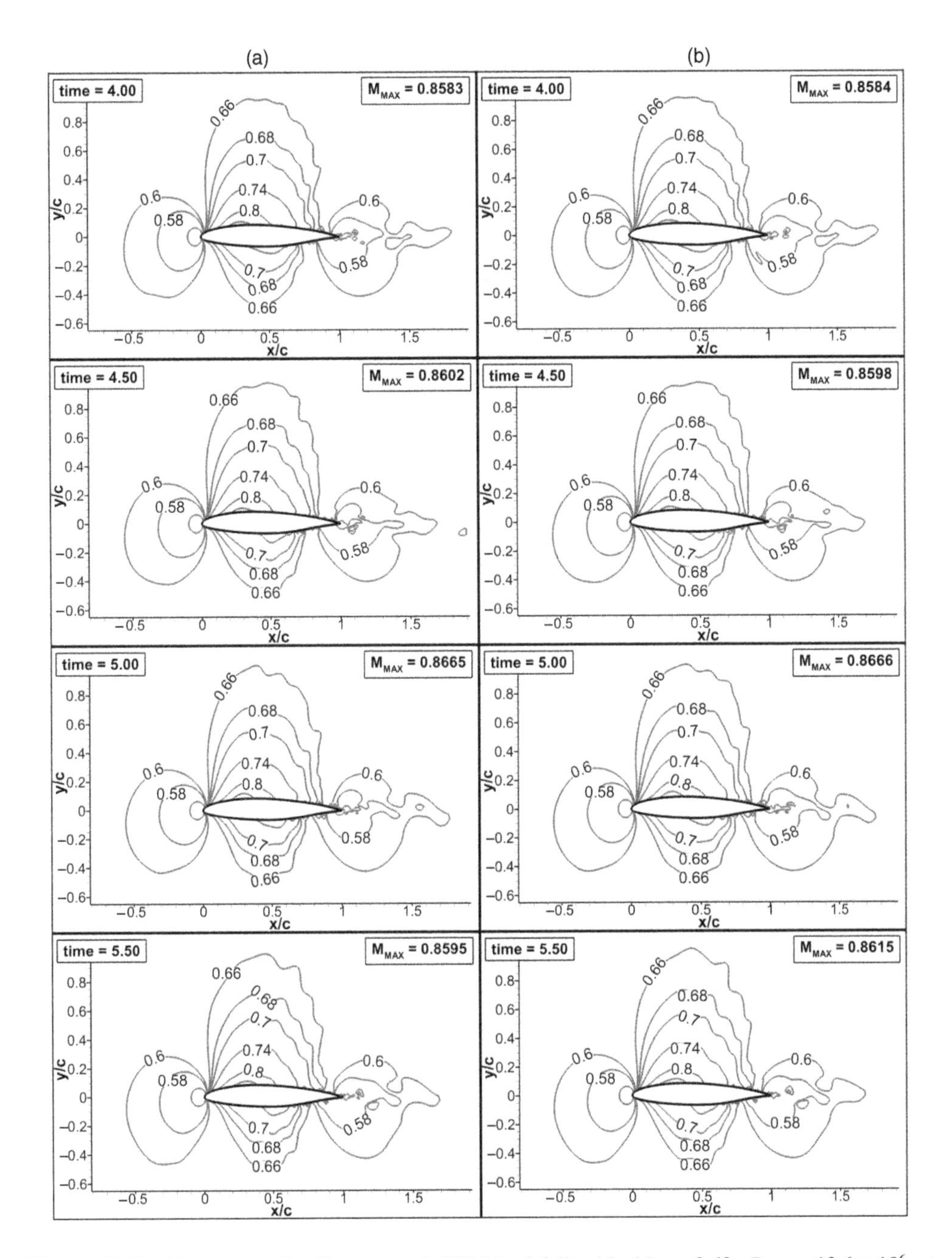

Figure 11.8 M contours for flow around SHM-1 airfoil with $M_\infty = 0.62$, $Re_\infty = 13.6 \times 10^6$ at $\alpha = 0.27°$ computed with scheme used for evaluating convective fluxes: (a) OUCS2 (b) OUCS3

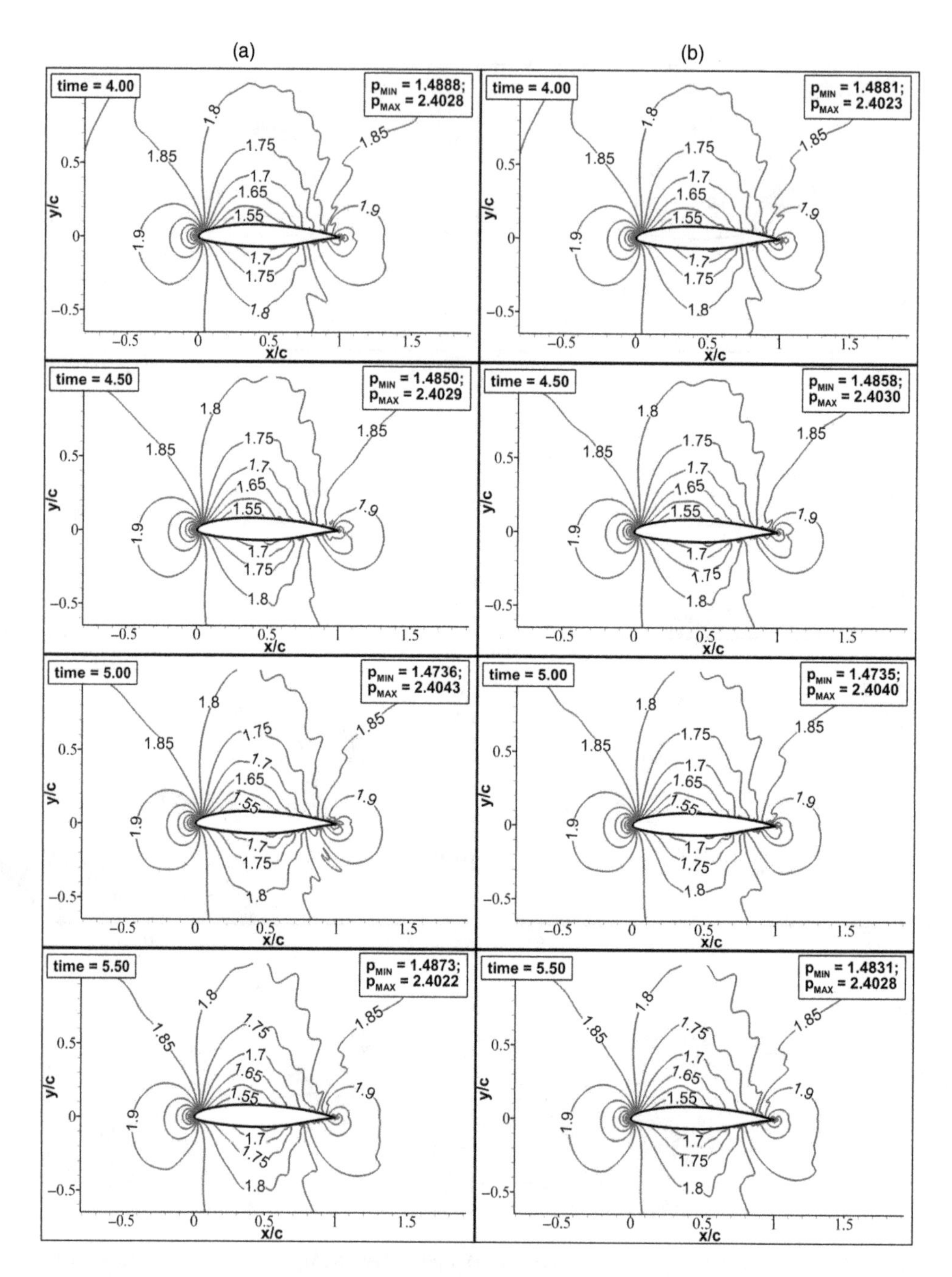

Figure 11.9 ρ contours for flow around SHM-1 airfoil with $M_\infty = 0.62$, $Re_\infty = 13.6 \times 10^6$ at $\alpha = 0.27°$ computed with scheme used for evaluating convective fluxes: (a) OUCS2 (b) OUCS3

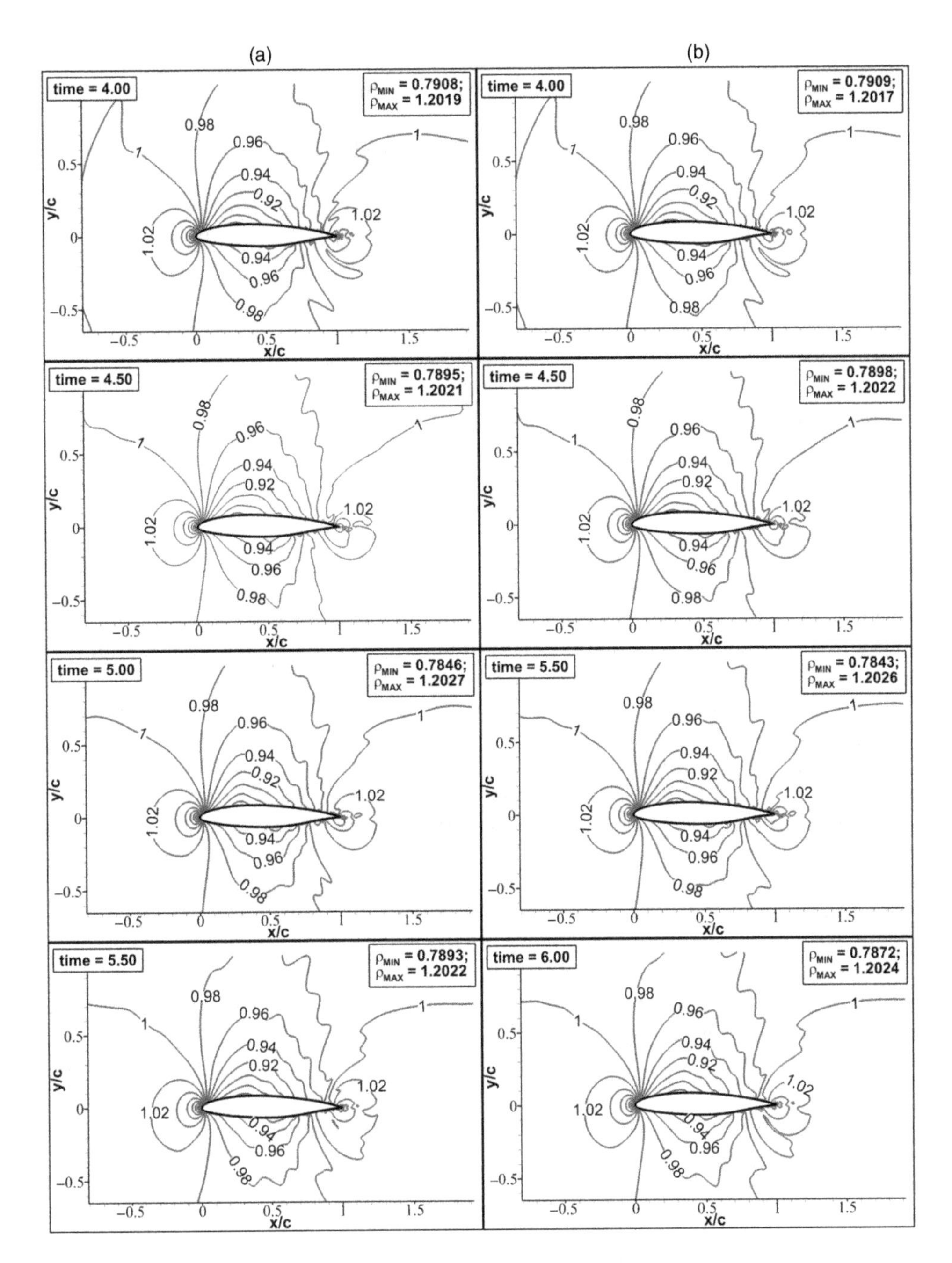

Figure 11.10 p contours for flow around SHM-1 airfoil with $M_\infty = 0.62$, $Re_\infty = 13.6 \times 10^6$ at $\alpha = 0.27°$ computed with scheme used for evaluating convective fluxes: (a) OUCS2 (b) OUCS3

11.4.2 Computing Strong Shock Cases

Having demonstrated the correctness of the adopted procedures with respect to C_p-distribution for both sub- and super-critical cases, without the use of any models for transition and turbulence for compressible flows, in the following cases higher Mach numbers are reported for transonic flows with shock creating entropy.

In Figures 11.11 and 11.12, computed M and p-contours are shown, respectively, at different time instants for the flow around NACA 0012 airfoil for $M_\infty = 0.779$, $Re_\infty = 3 \times 10^6$ and at $\alpha = -0.14°$. The flow encounters a favourable pressure gradient region on the top surface near the leading edge of the airfoil, up to approximately 30% of the airfoil chord (referred to as x_{cr}, a location downstream of the shock formation region). The flow accelerates creating a pocket of supersonic flow over the top surface. Beyond x_{cr}, the flow tends to recover pressure and retards as a result of adverse pressure gradient. This is noted in the sharp rise of C_p in Figure 11.4 near $x = 0.30c$. Pressure contours in Figure 11.12 indicate shock formation, as noted in Figure 11.10 from Mach contours.

11.4.3 Unsteadiness of Compressible Flows

As noted in Figures 11.8 to 11.10, the flow over the airfoil displays time dependent behaviour, even before the shock is formed. With shock formation, unsteadiness increases due to physical dispersion of higher wavenumber components. For a DRP scheme, such convection of high wavenumbers are naturally captured for flow past NACA 0012 airfoil for $M_\infty = 0.779$ in Figure 11.13. From the near-vicinity of the shock, one notices convection of pressure pulses in the downstream direction, associated with sharp pressure discontinuity responsible for wide-spectrum of physical variables.

However, it has to be emphasized that such unsteadiness with high wavenumbers is not numerical. It is well established that discontinuities are sites of spurious upstream propagating waves due to extreme dispersion, known as q-waves and explained in (Sengupta *et al.* 2012a). In Figure 11.13, no such upstream propagating waves are noted, instead unsteady downstream propagating waves noted are attributes of adverse pressure gradient flow. The time-average of C_p-distribution shows very good match with experimental value, indicating the fact that such pressure measurements are not instantaneous, but averaged over a finite time interval. Similar kind of C_p-distribution results are also reported in (Hermes *et al.* 2013), which show unsteadiness in instantaneous values, whereas time-averaged values show a match with experimental values. In the absence of definitive initial conditions, a uniform flow condition is assumed at $t = 0$. Thus the results at early times are not relevant physically. In most cases, it takes about $t \simeq 10$ for the flow to develop into a meaningful viscous flow state. Hence, in the reported cases, time-averages are taken from the time series from $t = 10$ to $t = 45$.

11.4.4 Creation of Rotational Effects

For flow over NACA 0012 airfoil at small angles of attack, two different streams of flow merge at the trailing edge, which are created on the top and bottom surfaces. Such dissimilar flows create an unstable mixing layer, which roll into vortices of different length scales, as shown in Figure 11.14 for entropy contours, with values calculated from

$$s - s_0 = c_v \ln \frac{p}{\rho^\gamma}.$$

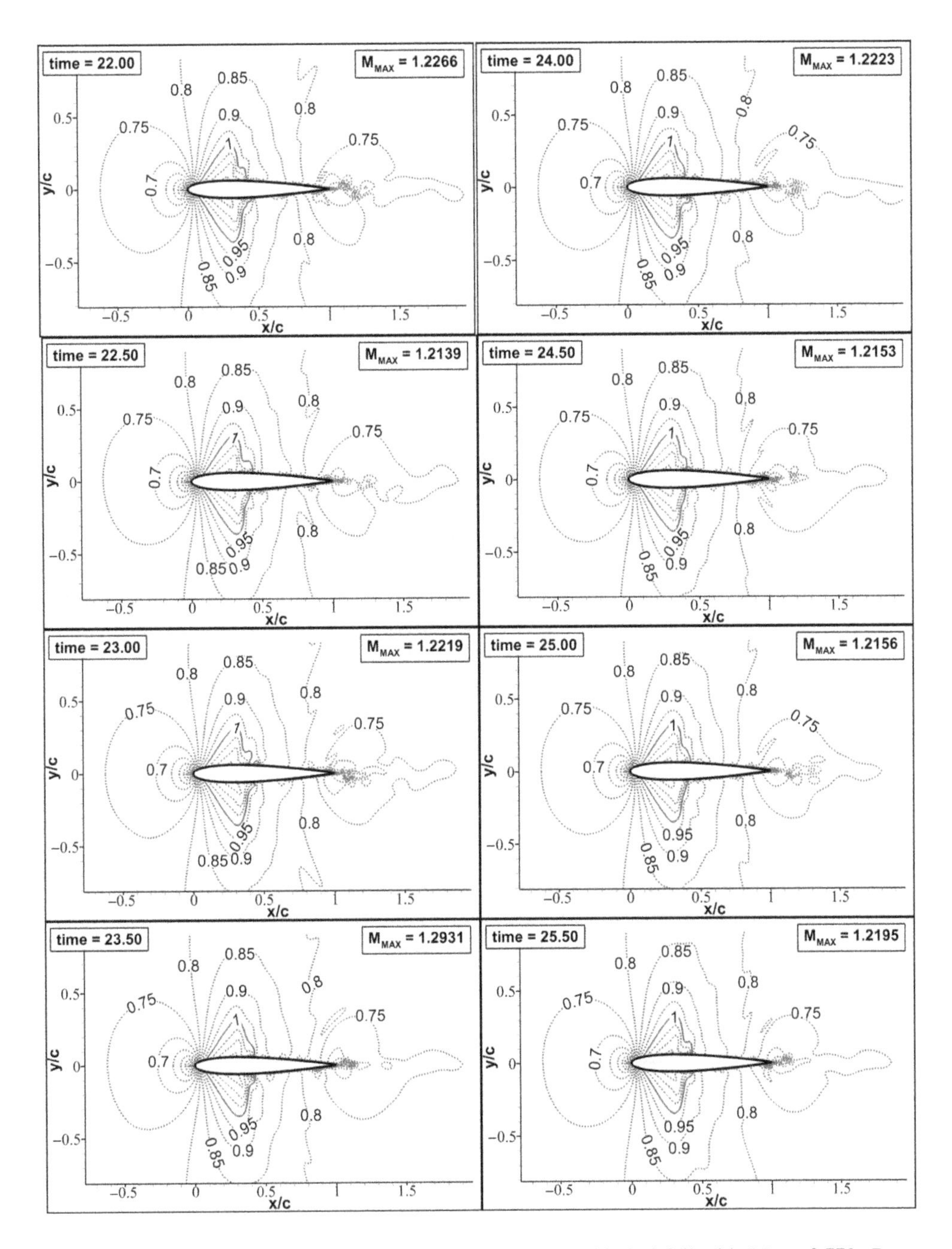

Figure 11.11 Computed M contours for flow around NACA 0012 airfoil with $M_\infty = 0.779$, $Re_\infty = 3 \times 10^6$ at $\alpha = -0.14°$

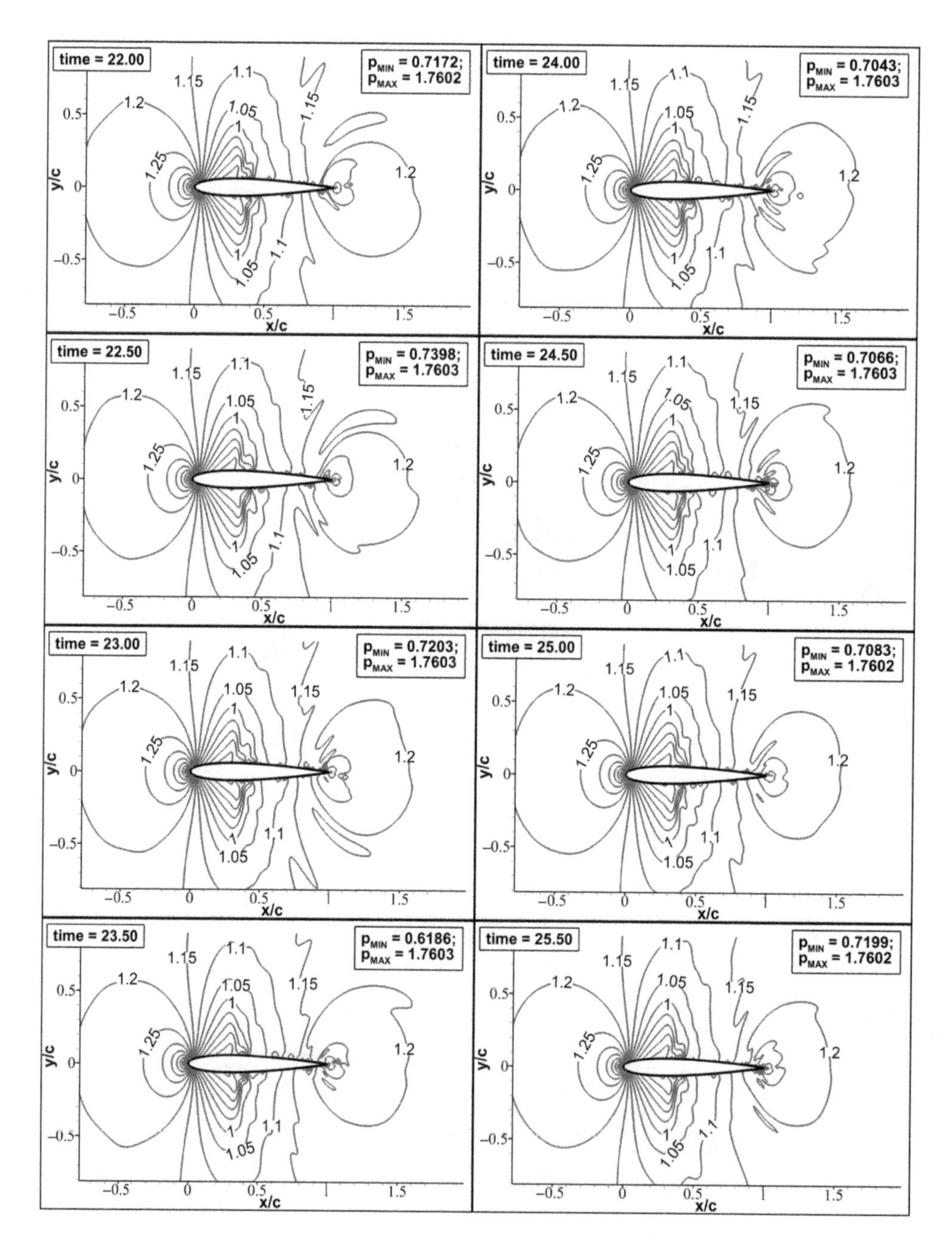

Figure 11.12 Computed p contours for flow around NACA 0012 airfoil with $M_\infty = 0.779$, $Re_\infty = 3 \times 10^6$ at $\alpha = -0.14°$

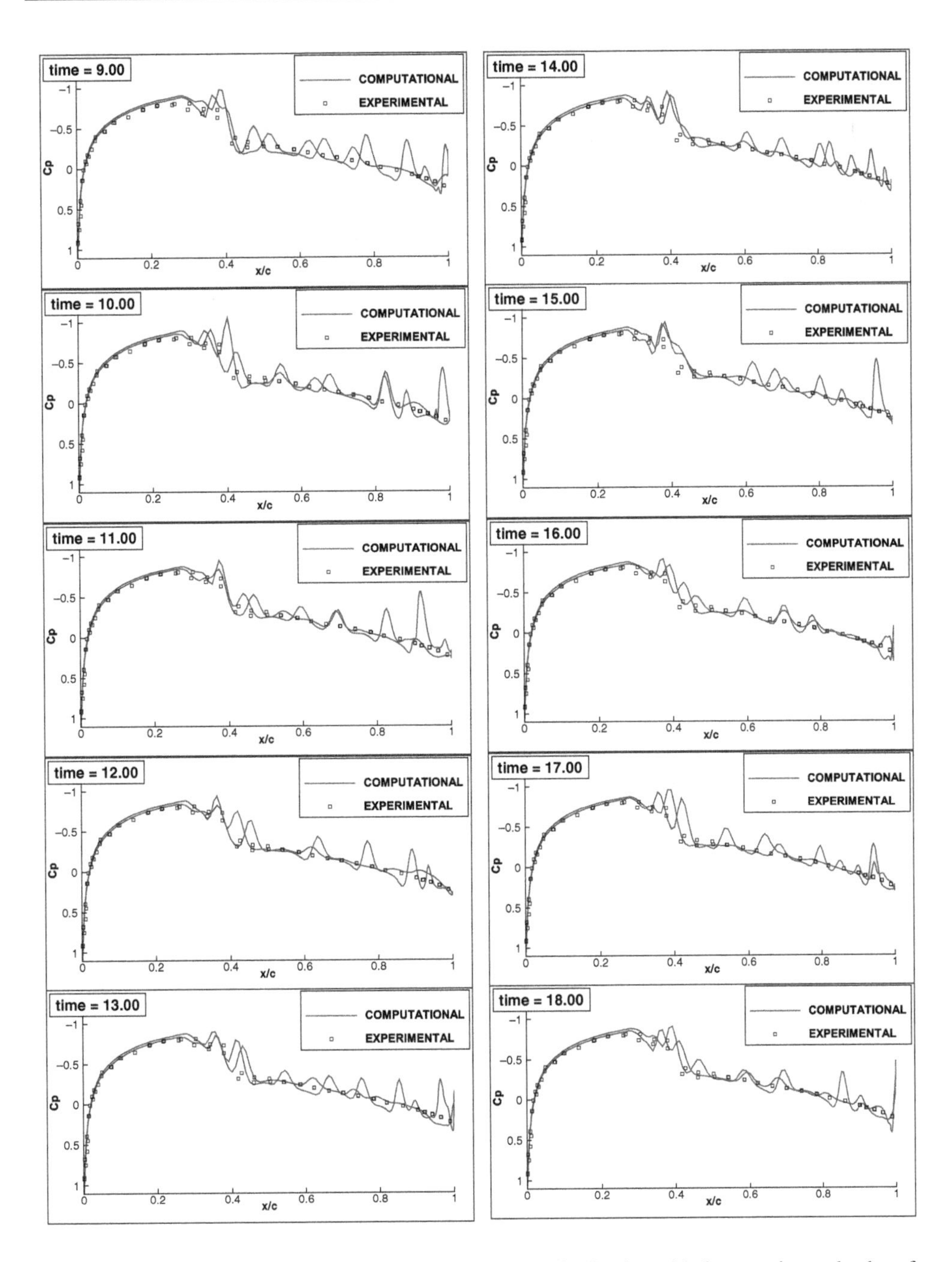

Figure 11.13 Comparison of computed instantaneous C_p-distribution with the experimental values for flow around NACA 0012 airfoil with $M_\infty = 0.779$, $Re_\infty = 3 \times 10^6$ at $\alpha = -0.14°$

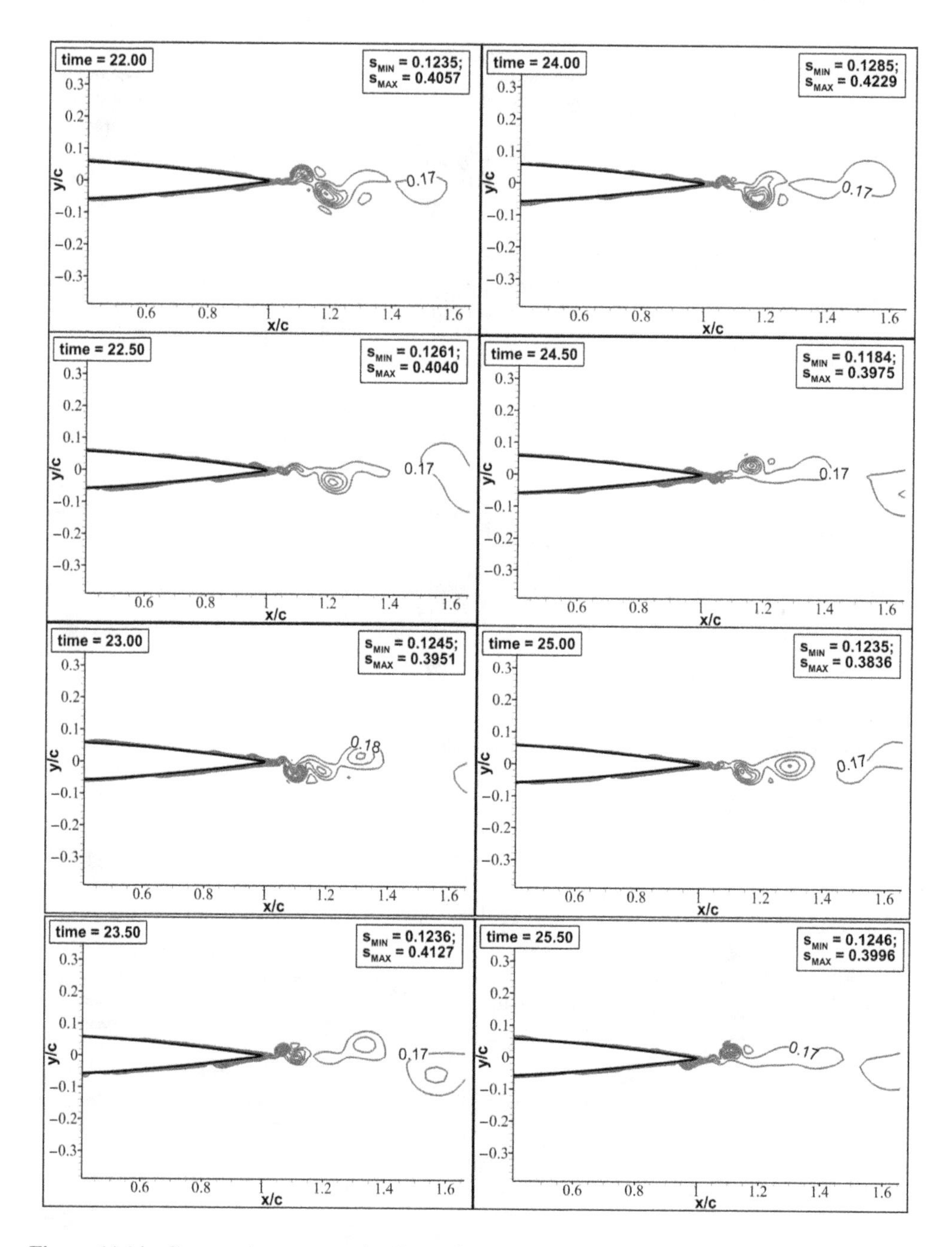

Figure 11.14 Computed s contours for flow around NACA 0012 airfoil with $M_\infty = 0.779$, $Re_\infty = 3 \times 10^6$ at $\alpha = -0.14°$

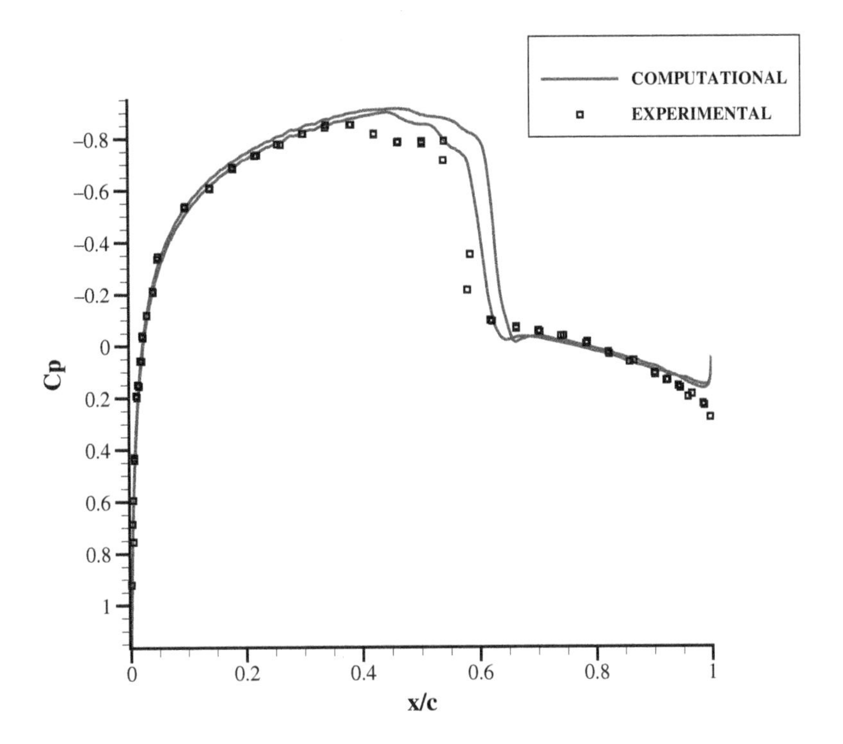

Figure 11.15 Comparison of computed and experimental C_p-distribution for flow around NACA 0012 airfoil with $Re_\infty = 3 \times 10^6$ at $\alpha = -0.14°$ and $M_\infty = 0.82$. Computed data is time-averaged over $t = 10$ to 20

The fact that entropy remains constant for any flow which has uniform total enthalpy, also indicates that such flows are irrotational. Whenever vorticity is created, as near the trailing edge for the mixing layer due to flow instability, the flow creates an entropy gradient across any streamline, which is a consequence of Crocco's theorem given by

$$\vec{\omega} = \frac{T}{u}\frac{dS}{dn}.$$

The relationship between vorticity, total enthalpy and entropy gradient is given by Crocco's theorem (Hirsch 1994) given as

$$T\nabla S + \vec{V} \times \vec{\omega} = \nabla h_0 + \frac{\partial \vec{V}}{\partial t}.$$

11.4.5 Strong Shock and Entropy Gradient

Next, results for $M_\infty = 0.82$ (greater than drag divergence Mach number) are shown for $Re_\infty = 3 \times 10^6$ and $\alpha = -0.14°$ for flow over NACA 0012 airfoil. For this case an orthogonal grid is generated with 2658 points in ξ- and 360 points in η-directions, respectively. Near-wall resolution of the grid is given by $\Delta s_\eta = 8.70 \times 10^{-6}c$. Very strong shocks are observed on upper and lower surfaces of the airfoil for this case in Figure 11.15, with computed

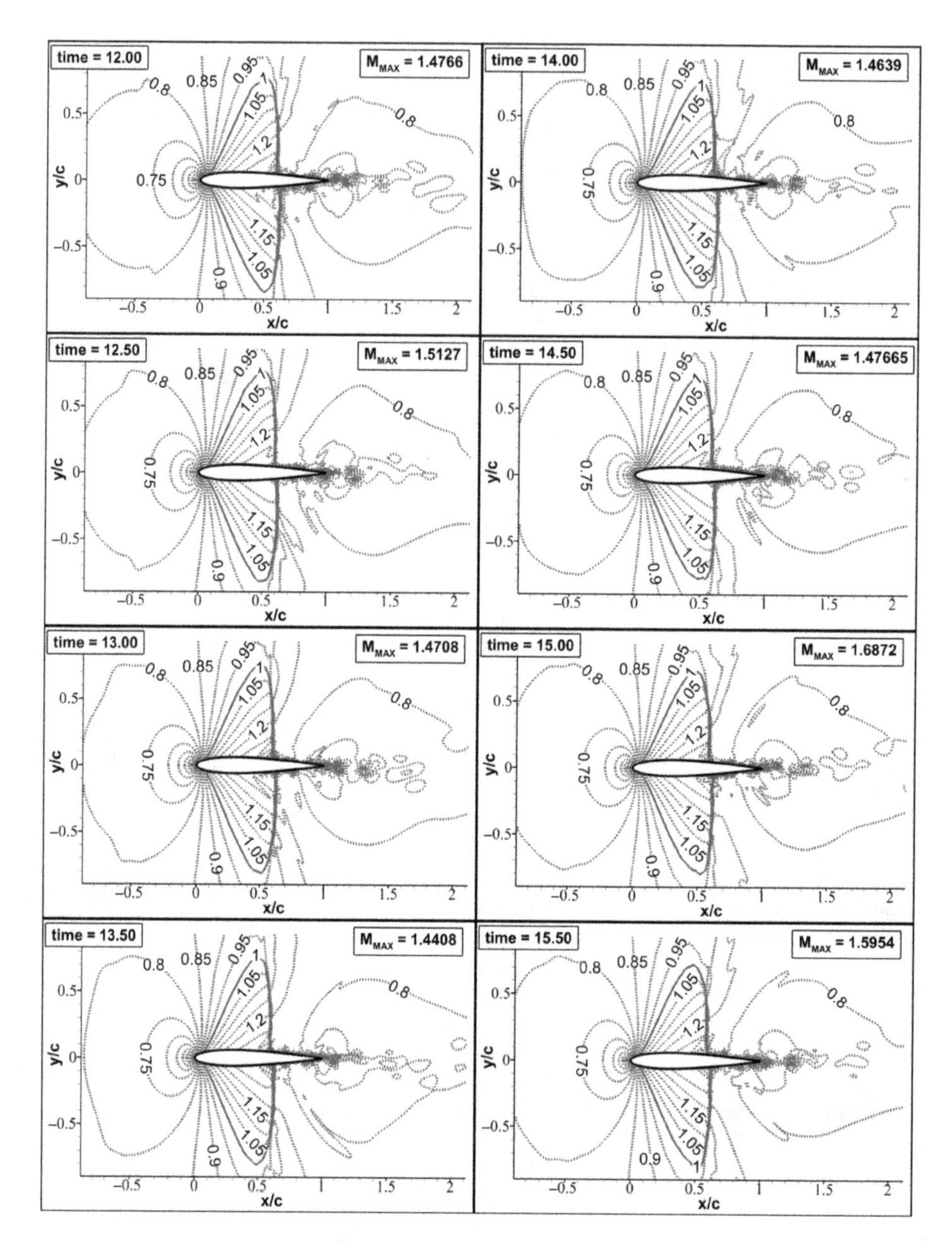

Figure 11.16 Computed M contours for flow around NACA 0012 airfoil with $M_\infty = 0.82$, $Re_\infty = 3 \times 10^6$ at $\alpha = -0.14°$. (See colour version of this figure in colour plate section)

C_p-distribution matching well with the experimental value (Harris 1981). Contours of computed M and p in the vicinity of the airfoil for this case are shown in Figures 11.16 and 11.17, respectively. Since the oncoming flow has higher value of M_∞, the flow accelerates near the leading edge to higher value of M, as noted at $t = 15.0$, when maximum M reaches a high

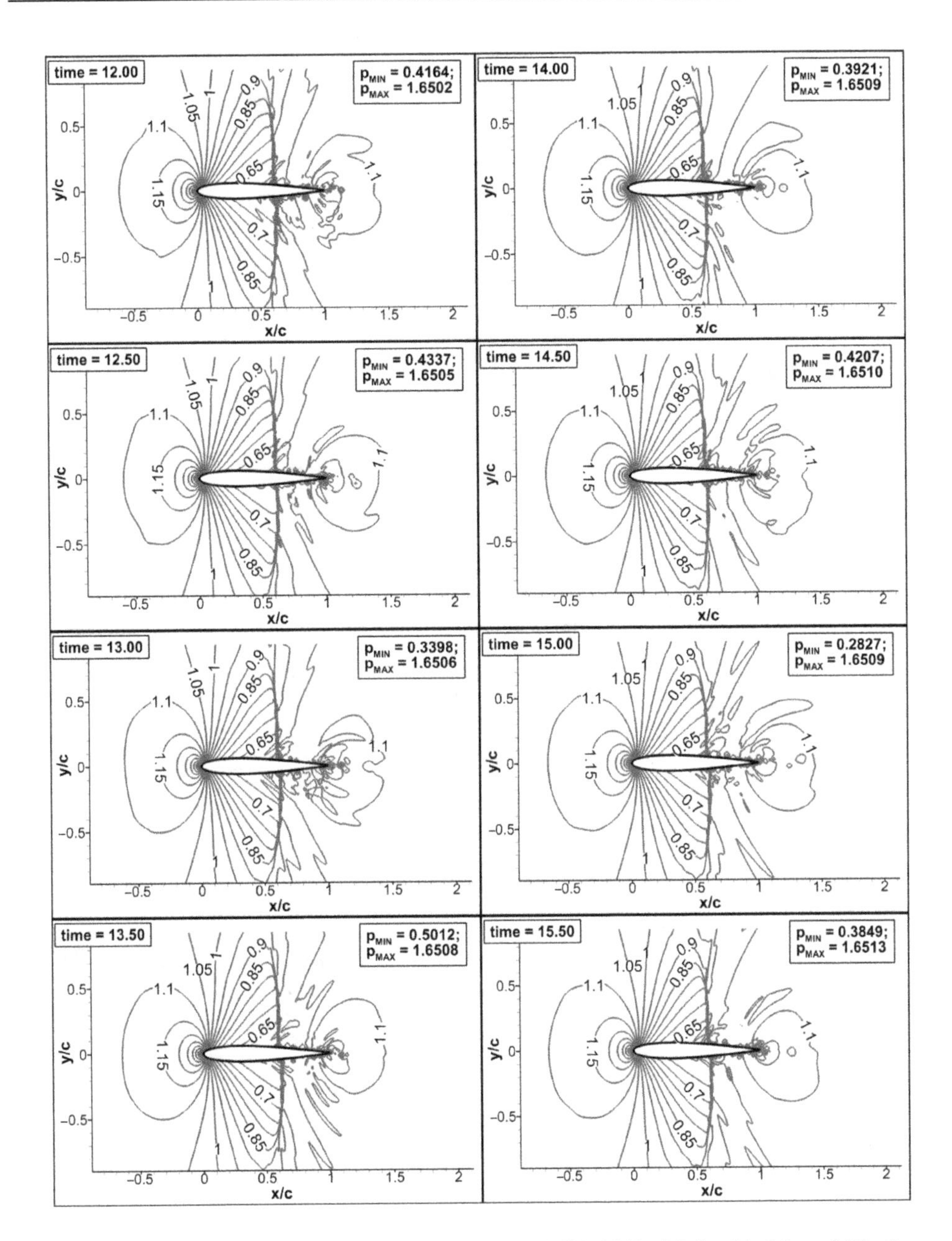

Figure 11.17 Computed p contours for flow around NACA 0012 airfoil with $M_\infty = 0.82$, $Re_\infty = 3 \times 10^6$ at $\alpha = -0.14°$. (See colour version of this figure in colour plate section)

value of 1.6872. The computed ρ-contours at the corresponding times are presented in Figure 11.18. In Figure 11.19(a) and 11.19(b), M and s-contours are plotted, respectively, for the case which display very strong shock. Strength of the shock is evident via creation of additional entropy from the location of the shock. Furthermore, one notices the creation of entropy in the shear layer near trailing edge and in the near-wake.

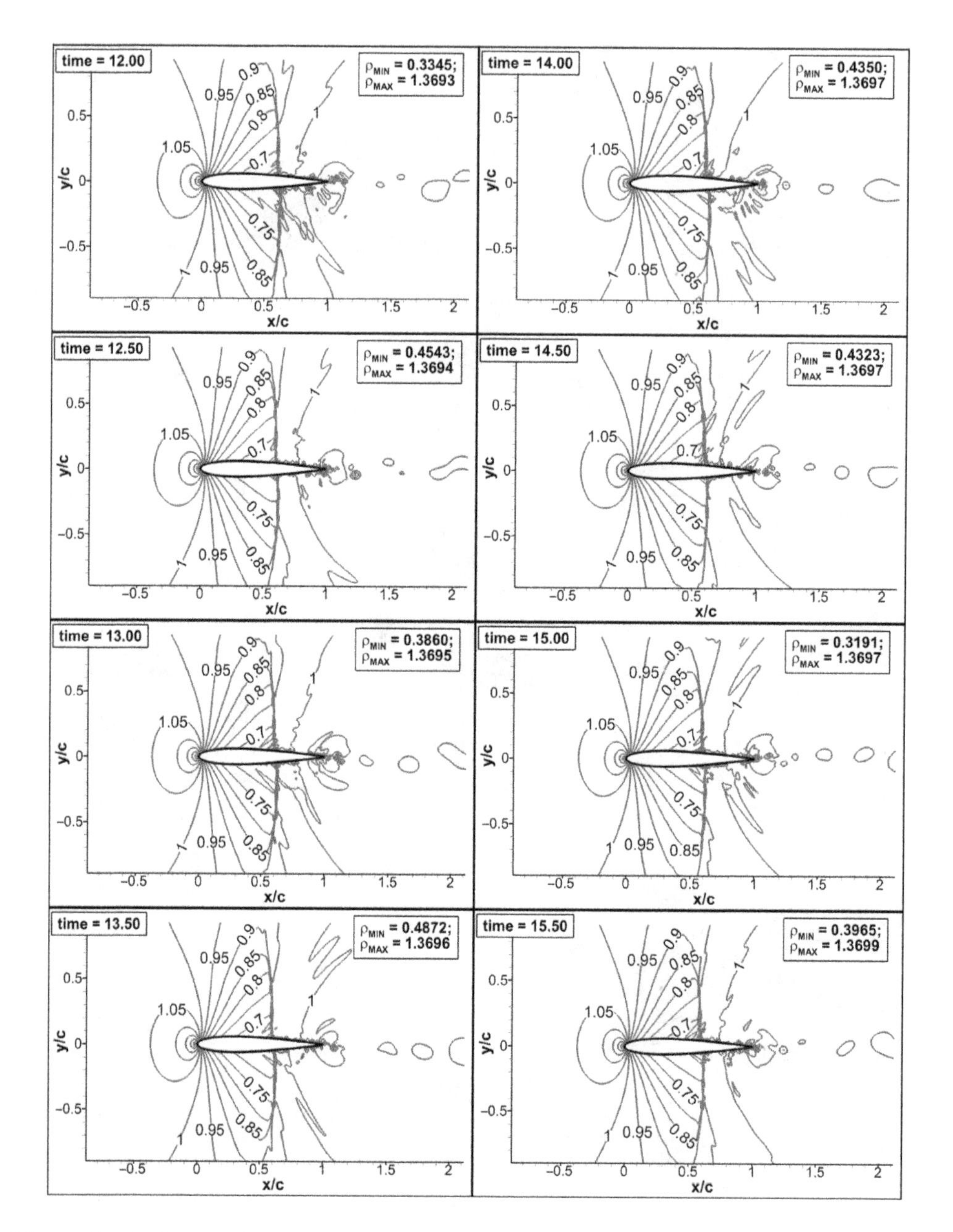

Figure 11.18 Computed ρ contours for flow around NACA 0012 airfoil with $M_\infty = 0.82$, $Re_\infty = 3 \times 10^6$ at $\alpha = -0.14°$. (See colour version of this figure in colour plate section)

11.4.6 Lift and Drag Calculation

The coefficients of lift (C_l) and drag (C_d) are computed by integrating the static pressure and shear stress over the airfoil surface and compared with available experimental values. For flow over NACA 0012 airfoil with $M_\infty = 0.6$ and $Re_\infty = 3 \times 10^6$ at $\alpha = -0.14°$ experimental

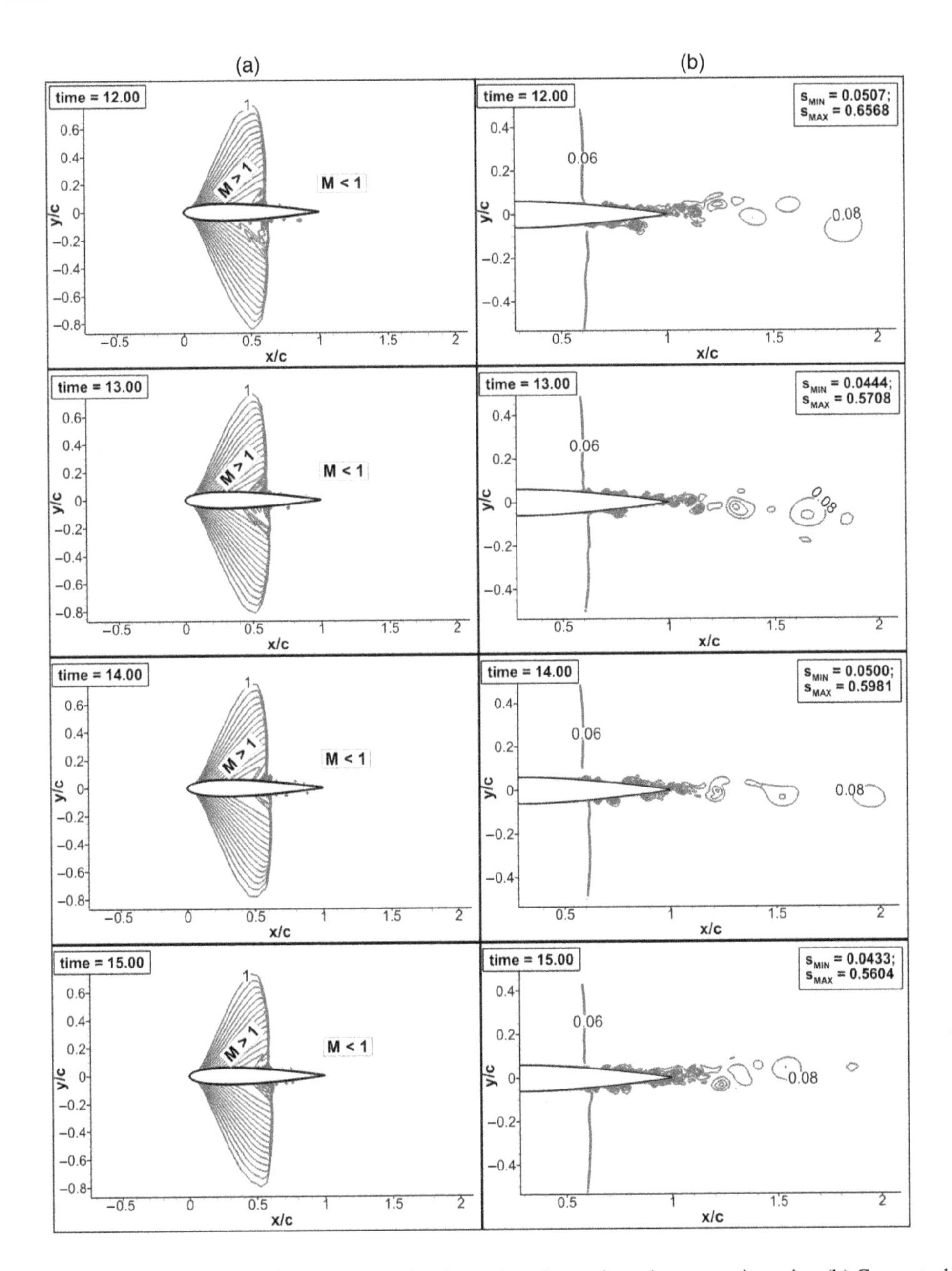

Figure 11.19 (a) Computed M contours showing subsonic, sonic and supersonic region (b) Computed s contours showing entropy generation in the region of the shock for $M_\infty = 0.82$ and $Re_\infty = 3 \times 10^6$ at $\alpha = -0.14°$. (See colour version of this figure in colour plate section)

Table 11.1 Computed and experimental (Harris (1981)) values of C_l and C_d for viscous flow past NACA 0012 airfoil

Sr. No	M_∞	Re_∞	α (in deg)	Computed		Experimental	
				C_l	C_d	C_l	C_d
1	0.6	3×10^6	−0.14	−0.0161	0.0022	−0.016	0.006
2	0.779	3×10^6	−0.14	−0.00989	0.0057	−0.013	0.007
3	0.82	3×10^6	−0.14	−0.0406	0.0238	NA	NA

Source: Courtesy of NASA

values of C_l and C_d are provided in Harris (1981) as −0.016 and 0.006, respectively. The computed values of C_l and C_d for the same case are noted as −0.0161 and 0.0022, respectively, with the data time-averaged in the range $10 \leq t \leq 55$. For $M_\infty = 0.779$, $Re_\infty = 3 \times 10^6$ and $\alpha = -0.14°$ experimental values of C_l and C_d are provided in Harris (1981) as −0.013 and 0.007, respectively, while the computed values are noted as −0.00989 and 0.0057, respectively, with data time-averaged for $10 \leq t \leq 45$. The time-averaged C_l and C_d values obtained by computations for the flow around NACA 0012 airfoil for $M_\infty = 0.82$; $Re_\infty = 3 \times 10^6$ and $\alpha = -0.14°$ are −0.0406 and 0.0238, respectively, for $10 \leq t \leq 25$. For this case, experimental value for C_l is given as −0.008 in Harris (1981). All these computed values are put together in Table 11.1.

Bibliography

Allaneu, Y. and Jameson, A. (2009) Direct numerical simulations of a two-dimensional viscous flow in a shock tube using a kinetic energy preserving scheme. *AIAA Paper*, 2009–3797.

Alshabu, A., Olivier, H. and Klioutchnikov, I. (2006) Investigation of upstream moving pressure waves on a supercritical airfoil. *J. Aero. Sci. Tech.*, **10**, 465–473.

Barnwell, R.W. (1979) A similarity rule for sidewall-boundary-layer effect in two-dimensional wind tunnels. *AIAA Paper*, 79-0108.

Beam, R.M. and Warming, R.F. (1978) An implicit factored scheme for the compressible Navier-Stokes equations. *AIAA J.*, **16**(4), 393–402.

Binion, T.W. (1979) Limitations of available data, *AGARD-AR* **138**, 2.1–2.8.

Chiu, E.K., Wang, Q. and Jameson, A. (2011) A conservative meshless scheme: General order formulation and application to Euler equations. *AIAA Paper*, 2011-651.

Dipankar, A. and Sengupta, T.K. (2006) Symmetrized compact schemes for receptivity study of 2D channel flow. *J. Comp. Phys.*, **215**, 245–273.

Fujino, M., Yoshizaki, Y. and Kawamura, Y. (2003) Natural-laminar-flow airfoil development for a lightweight business jet. J. Aircraft, **40**(4), 609–615.

Garbaruk, A., Shur, M., Strelets, M. and Spalart, P. (2003) Numerical study of wind tunnel wall effects on transonic airfoil flow. *AIAA J.*, **41**, 1046–1054.

Harris, C.D. (1981) Two-dimensional aerodynamic characteristics of the NACA0012 airfoil in the Langley 8-foot transonic pressure tunnel. *NASA TM* **81927**.

Hermes, V., Klioutchnikov, I. and Olivier, H. (2013) Numerical investigation of unsteady wave phenomena for transonic airfoil flow. *J. Aero. Sci. Tech.*, **25**, 224–233.

Hirsch, C. (1994) *Numerical Computation of Internal and External Flows.* **1**, *Fundamentals of Numerical Discretization*, Wiley-Interscience Publication, New York, USA.

Hoffmann, K.A. and Chiang, S.T. (2000) *Computational Fluid Dynamics: vol. II*, Engineering Education System, Wichita, Kansas, USA.

Jameson, A. and Ou, K. (2011) 50 years of transonic aircraft design. *Prog. Aero. Sci.*, **47**(5), 308–318.

Jameson, A., Schmidt, W. and Turkel, E. (1981) Numerical simulation of the Euler equations by finite volume methods using Runge-Kutta time-stepping schemes. *AIAA Paper 1981-1259, AIAA, 14th Fluid and Plasma Dynamics Conf.*, Palo Alto, California.

Liu, T.C.X. and Osher, S. (1994) Weighted essentially non-oscillatory schemes. *J. Comp. Phys.*, **115**, 200–212.

Maybey, D.G. (1989) Physical phenomena associated with unsteady transonic flows, in *Unsteady Transonic Aerodynamics*. **120**, *Progress in Astronautics and Aeronautics, AIAA Series*, Washington, DC, USA.

Maybey, D.G. (1976) Some remarks on the design of transonic tunnels with low levels of flow unsteadiness. *NASA CR* **2722**.

Mercer, J.E., Geller, E.W., Johnson, M.L. and Jameson, A. (1981) Transonic flow calculations for a wing in a wind tunnel. *AIAA J.*, **18**(9), 707–711.

Moulden, T.H. (1984) *Fundamentals of Transonic Flow*. Wiley-Interscience Publication, Canada.

Nair, M.T. and Sengupta, T.K. (1998) Orthogonal grid generation for Navier-Stokes computations. *Int. J. Num. Methods Fluids*, **28**, 215–224.

Otto, H. (1976) Systematical investigation of the influence of wind tunnel turbulence on the results of force measurements. *AGARD CP* **174**.

Pate, S.R. and Schueler, C.J. (1968) Radiated aerodynamic noise effects on boundary layer transition in supersonic and hypersonic wind tunnels, *AIAA J.*, **7**, 450–457.

Panton, R.L. (2005) *Incompressible Flow*. 3rd. edn., Wiley, New York, USA.

Pirozzoli, S., Bernardini, M. and Grasso, F. (2010) Direct numerical simulation of transonic shock/boundary layer interaction under conditions of incipient separation. *J. Fluid Mech.*, **657**, 361–393.

Sengupta, T.K. (2004) *Fundamentals of Computational Fluid Dynamics*. Universities Press, Hyderabad, India.

Sengupta, T.K. (2013) *High Accuracy Computing Methods: Fluid Flows and Wave Phenomena*, Cambridge University Press, USA.

Sengupta, T.K., Bhaumik, S. and Bhumkar, Y. (2012) Direct numerical simulation of two-dimensional wall-bounded turbulent flows from receptivity stage. *Phys. Rev. E*, **85**(2), 026308.

Sengupta, T.K., Bhaumik, S. and Bose, R. (2013) Direct numerical simulation of two-dimensional transitional mixed convection flows: Viscous and inviscid instability mechanisms. *Phys. Fluids*, **25**, 094102.

Sengupta, T.K., Bhumkar, Y., Rajpoot, M. and Saurabh, S. (2012a) Spurious waves in discrete computation of wave phenomena and flow problems. *App. Math. Comp.*, **218**(18).

Sengupta, T.K., Dipankar, A. and Rao, A.K. (2007) A new compact scheme for parallel computing using domain decomposition. *J. Comp. Phys.*, **220**, 654–677.

Sengupta, T.K., Ganeriwal, G. and De, S. (2003) Analysis of central and upwind compact schemes. *J. Comp. Phys.*, **192**, 677–694.

Sengupta, T.K. Ganeriwal, G. and Dipankar, A. (2004) High accuracy compact schemes and Gibbs' phenomena. *J. Sci. Comput.*, **21**(3), 253–268.

Sengupta, T.K., Sircar, S.K. and Dipankar, A. (2006) High accuracy schemes for DNS and acoustics. *J. Sci. Comput.*, **26**, 151–193.

Visbal, M.R. (1990) Dynamic stall of a constant-rate pitching airfoil. *J. Aircraft*, **27**, 400–407.

12

Low Reynolds Number Aerodynamics

12.1 Introduction

Unsteady flows at low to moderate Reynolds numbers (*Re*) are of significant interest due to their importance in engineering devices and in nature; for example, in unmanned aerial vehicle (UAV) and micro air vehicle (MAV) operations; for insect/ bird flights and for motion in aquatic environment. Flapping and hovering flight of bird or insect are fine examples of optimum motion of aerodynamic surfaces which develop necessary thrust for forward motion and sustained lift to keep the creature airborne. In comparison, conventional aircrafts are suboptimal as lift and thrust are created by different subsystems. Furthermore, lift and thrust created in aircraft are explained by steady flow equations (by modelling the high frequency fluctuations associated with turbulence), while for natural fliers one is forced to use unsteady aerodynamics to explain effects of articulating surfaces which simultaneously generate lift and thrust. Comprehensive reviews of natural fliers can be found in Rozhdestvensky and Ryzhov (2003), Ho *et al.* (2003), Lian *et al.* (2003) and Sane (2003), primarily providing the state of art in experiments. Biologists have focused attention on kinematics of motion for birds and insects, as explained in Norberg (1990) and Pennycuick (1997), while fluid dynamicists have used simplified models to explain mechanisms of flights, in the limit of quasi-steady and unsteady operations. Current interest in engineering on bird and insect flights is in view of carrying payload in MAVs. Also from the constraint of limited lift generation by conventional planforms, the weight is also kept very small for the perceived mission requirements. With restriction on size of such devices (often around 15 cm or less) and speeds (of about 10 m/s), operating *Re*'s are in the range of 10^4 to 10^5. Low *Re* flows display massive boundary layer separation, flow unsteadiness and bypass transition. Associated large vortical structures make the study difficult by simplified theories and current theoretical and computational approaches are based on solution of Navier-Stokes equation (NSE). The importance of this subject is understood, when one notes that birds and insects at low *Re* display extraordinary maneuverability, agility, low speed flying capability and even the ability to hover due to

Theoretical and Computational Aerodynamics, First Edition. Tapan K. Sengupta.

Companion Website: www.wiley.com/go/sengupta

their ability to create and manipulate large vortices over lifting surfaces. To mimic such flight modes in engineering devices, first and foremost, one needs to understand low *Re* aerodynamics.

During bird/insect flight, the lifting surface kinematics are complex combinations of translational and rotational motion of lifting surfaces. Low *Re* aerodynamics is highly viscous dominated. Understanding the complex relationship between wing kinematics (input) and system response (output) is of central consideration to design any engineering devices with simultaneously high aerodynamic and propulsive efficiencies as the objective function. In the absence of definitive reproducible experimental results, computational and theoretical approaches assume importance. Even for such studies the visual signatures of wing kinematics recorded for example in Tobalske and Dial (1996) and Nachtigall (1974) are vital. Freymuth (Freymuth 1990) has experimentally investigated the motion of an airfoil in combined harmonic plunging and pitching oscillatory motion with phase difference to generate thrust in still air environment. The role of vortex street acting as a jet stream in the wake is discussed as the mechanism for generating lift and thrust in this reference. The author identified three generic cases during hover realized by a combination of horizontal and pitching oscillations. Zero baseline angle of attack of the airfoil, with 90° phase difference between horizontal and pitch oscillation is termed as the *first hover* or the *water-treading* mode. In the *second hover* or *degenerate figure-of-eight* mode, the baseline angle of attack is 90° and pitching oscillation has a phase lag of 90° with respect to translational oscillation. For the *third mode*, also called the *oblique* or *dragonfly* mode, the lifting surface is at an angle of attack nominally between zero and 90°, while the phase difference between horizontal and pitch oscillations is kept close to 90°. In Sengupta *et al.* (2005), the *second hover* mode case was computed and analyzed.

Steady-state aerodynamics is of limited value and unsteady inviscid flow model has been used in DeLaurier (1993) for flapping wing, with Spedding (1992) providing extensive review of early aerodynamic models of flapping flight. To incorporate unsteady viscous flow features of separation and transition in the presence of large vortices, computational fluid dynamical techniques have been used to study flapping flight, with airfoil executing heaving oscillations in uniform flow. In Ho *et al.* (2003), a commercial software based on finite volume primitive variable formulation was used to solve NSE in Lagrangian-Eulerian framework, with the critical remark that it is essential to satisfy the conservation law, otherwise false mass is created leading to large errors. In Lian *et al.* (2003), 3D incompressible NSE in primitive variables have been solved in strong conservation form. Sun and Tang (Sun and Tang 2002) have reported solving 3D NSE using the artificial compressibility method (Rogers and Kwak 1990). In Sun and Tang (2002), spatial discretization was performed by third-order upwind flux-difference splitting and time integration by second-order Adams-Bashforth technique (which is known to display spurious computational mode with large error for unsteady flows in Sengupta and Dipankar (2004) and Sengupta (2013) providing details).

To study flapping and hover modes of motion, 3D computations around a deforming wing is desirable to understand low *Re* aerodynamics. However, the present-day scenario of high accuracy computing remains identical to that noted by Wang (2000) that

> 'from a practical point of view, while it is possible to resolve two-dimensional flows at Reynolds numbers relevant to insect flight, it remains to be seen whether one can do the same for three-dimensional flows.'

In this context, commercial software based on primitive variable formulation are not able to resolve relevant length and time scales involved in solution of unsteady NSE, while satisfying mass conservation accurately. In computing 2D flows, NSE using (ψ, ω)-formulation has been used in Sengupta *et al.* (2005). This formulation satisfies mass conservation identically and allowed taking larger number of grid points, as the number of unknowns reduces from three to two. In Gustafson and Leben (1991) and Gustafson *et al.* (1992) this formulation was also used to compute hovering flight of an elliptic cylinder. Unfortunately, the governing vorticity transport equation (VTE) written in the moving frame had the angular acceleration term missing, which is essential for hovering flight, with the lifting surface executing pitching oscillations. In Wang (2000), the same formulation was used for heaving oscillation in uniform flow, which does not require this angular acceleration term. The author reported flapping motion shedding vortex and optimal flapping frequency based on time scales associated with shedding of leading and trailing edge vortices identified, using a fourth-order compact finite difference scheme. In Wang (2000), Gustafson and Leben (1991) and Gustafson *et al.* (1992), NSE is solved for an elliptic airfoil and results qualitatively compared with the experiments in Freymuth (1990) performed with an airfoil with rectangular cross section with rounded off edges of significantly lower thickness ratio.

The lower accuracy of primitive variable formulations, encouraged the authors in Sengupta *et al.* (2005) to reformulate the problem of general flapping and hover mode of motion in 2D with actual airfoil. In this study, unsteady aerodynamics of hovering and flapping airfoil was reported with the following improvements: (a) a correct formulation of the problem using ψ and ω as dependent variables; (b) calculating loads and moment by a method to solve the governing pressure Poisson equation (PPE) in a truncated part of the computational domain on a collocated grid; (c) obtaining accurate solution by high accuracy compact difference scheme described in Chapter 8 for the VTE and (d) accelerating computations by using higher order filters (Sengupta (2013)), after each time step of integration. These have been used to solve NSE for flow past flapping and hovering NACA 0014 and 0015 airfoils at typical *Re*'s relevant to the study of unsteady aerodynamics of MAV and insect/bird flight. Orthogonal grids around airfoils were generated following the hyperbolic grid generation method described in Chapter 8. The PPE is solved for the total mechanical energy using the orthogonal collocated grid to calculate loads and moment by solving the PPE in a subdomain with exact boundary conditions on this boundary.

In a combined experimental and computational investigation for flapping wing aerodynamics, in Jones *et al.* (2002) results were presented including wind tunnel investigation on finite aspect ratio wing to directly measure forces, time-accurate LDV measurements and flow visualization for a NACA 0014 airfoil. The airfoil in the experiments was made to execute only plunging oscillation given by

$$y(t) = h\cos(kt). \tag{12.1}$$

A plunge amplitude of $h = 0.4c$ was considered for a range of reduced frequencies, k for different angles of attack. In Sengupta, *et al.* (2005), this geometry and motion were considered to obtain the flow for $Re = 20\,000$ with the airfoil set at zero angle of attack and the reduced frequency was $k = 0.4$. In addition to this, the *second hover* mode case of Freymuth (1990) was computed for oscillating NACA 0015 airfoil in the pitch plane with the rotating centre at the mid-chord for $Re = 27\,000$. In Freymuth (1990), the Reynolds number chosen was less than $Re = 600$. According to the definition of this mode of motion, the mean angle of attack is 90° with the pitching angle oscillation amplitude as 5°. The airfoil executed horizontal oscillation with an amplitude that is equal to the chord of the airfoil.

It is to be noted that the present problem has similarities with the unsteady airfoil aerodynamics often studied for rotary wing devices. Some experimental and numerical contribution in this field are recorded in McCroskey (1982), Carr (1988), Bousman (2000) and Berton *et al.* (2004) and other references contained therein. However, the *Re* ranges are much higher for helicopter rotor blades and flow behaviour is qualitatively different from what is encountered in insect and bird flights.

12.2 Micro-air Vehicle Aerodynamics

Micro-air vehicles can be used for variety of missions and these vehicles are either remotely piloted or flown autonomously. Thus, these come under the general category of unmanned aerial vehicles (UAV's) and a timely review of the same is available in Mueller and DeLaurier (2003). While a high altitude, long-endurance remotely piloted vehicle aerodynamic design has been described by Maughmer and Somers (1989) with reference to the drag polar shown in Figure 10.5, here we describe some of the attributes and requirements of low altitude UAVs. Such vehicles fly with speeds in the range of 20 to 100 kmph, at altitudes ranging between 3 and 300 metres, with typical wing span less than 6 metres and weight less than 25 kgs. Due to miniaturization of the sensors and actuators used in such vehicles, weight restrictions do not seem to pose a serious constraint. However, for the combinations of low speeds and small length scales, relevant *Re* vary in the range from fifteen thousands to half a million. Thus, these vehicles

> 'require efficient low Reynolds number airfoils that are not overly sensitive to wind shear, gusts, and the roughness produced by precipitation'

as reported by Mueller and DeLaurier (2003). At the same time, these authors also note the absence of published studies which address such problems. According to them,

> 'quantitative studies of unsteady aerodynamics directly related to small UAVs, with the exception of flapping-wing vehicles, at low Reynolds numbers have only recently become of interest (e.g. Broeren and Bragg 2001).'

We have already noted that airfoils designed for *Re* greater than 500 000 functions poorly at lower *Re* due to adverse boundary layer properties. Mainly the culprit is the flow separation due to thick boundary layer for low Reynolds numbers. A good survey of low Reynolds number aerofoil is due to Carmichael (1981) which also focused attention in the range $30\,000 \leq Re \leq 500\,000$ for the importance of it for model aircrafts and MAVs. For airfoils thicker than 6% one observes lift curve hysteresis, which is undesirable. Additionally, presence of separation causes significant pressure drag due to flow separation. Such separated shear layers can also suffer transition to turbulence. As noted in Chapter 10, such flows after suffering transition can reattach with the airfoil surface, forming separation bubbles. This bubble formation due reattachment of separated shear layer happens when *Re* is in excess of 50 000. We will see in later part of this chapter that unsteady effects on flow dynamics is severe in the presence of bubbles, as noted experimentally and computationally. Also, at relatively lower angles of attack, the point of separation approaches the leading edge, causing the airfoil to stall.

For the range of $70\,000 \leq Re \leq 200\,000$ mostly relevant to MAV's and small UAV's, flow over the airfoil remains essentially laminar and the problem arises only due to separation. In the presence of low frequency events, accurate computations of such flows are tougher, as one cannot resort to RANS calculations any more. Moreover, none of the turbulence or transitional flow models is useful for such low *Re* flow field. One has to perform high accuracy computations using DRP schemes. This is one of the remaining challenge and opportunity for computations of such low *Re* aerodynamics. We exclude from our considerations the various designs by empirical methods using panel methods coupled to boundary layer developments, as will be readily apparent of its limited use after going through the contents of this chapter.

Another major impediment for the operation of MAVs is their intended use at lower altitudes, where atmospheric gusts are very much present. Very little attention has been devoted to this topic, with the exception of studies reported in Pelletier and Mueller (2000) and Mueller (2000). Some recent computational efforts are reported in Krishnan (2013) and in Bagade *et al.* (2014).

12.3 Governing Equations in Inertial and Noninertial Frames

In the following, the equations are derived for both the inertial and moving frame of references with variables/operators represented with a subscript I indicating the quantities in the inertial frame and a subscript r indicating the corresponding moving frame quantities. NSE written in inertial frame is given by

$$\frac{\partial \vec{V}_I}{\partial t} + \vec{V}_I \cdot \nabla \vec{V}_I = -\frac{1}{\rho}\nabla p + \nu \nabla^2 \vec{V}_I. \tag{12.2}$$

This is written for the moving frame of reference, with its origin translating with a velocity $\vec{V}_{oI}$ and rotating with an angular velocity $\vec{\Omega}$ with respect to the inertial frame given by the following equation

$$\frac{\partial \vec{V}_r}{\partial t} + 2(\vec{\Omega} \times \vec{V}_r) + (\vec{\Omega} \times \vec{\Omega} \times \vec{R}_r) + \frac{\partial \vec{\Omega}}{\partial t} \times \vec{R}_r + \vec{V}_r \cdot \nabla \vec{V}_r = -\frac{1}{\rho}\nabla p + \nu \nabla^2 \vec{V}_r. \tag{12.3}$$

The acceleration terms are obtained in the respective reference frames, with $\vec{R}_r$ representing position vector of a point in the noninertial frame of reference. VTE for 2D flows are obtained in both the frames of reference by taking curl of Equations (12.2) and (12.3) to obtain

$$\frac{\partial \vec{\omega}_I}{\partial t} + (\vec{V}_I \cdot \nabla)\vec{\omega}_I = \nu \nabla^2 \vec{\omega}_I \tag{12.4}$$

$$\frac{\partial \vec{\omega}_r}{\partial t} + (\vec{V}_r \cdot \nabla)\vec{\omega}_r = \nu \nabla^2 \vec{\omega}_r - 2\frac{\partial \vec{\Omega}}{\partial t}, \tag{12.5}$$

where ω_I is the only nonzero component of vorticity in the out-of-plane for the 2D flow and defined by $\omega_I = (\nabla \times \vec{V}_I).\hat{k}$. The velocity vector is related to the stream function by $\vec{V}_I = \nabla \times \vec{\Psi}_I$, where the stream function is a special form of vector potential given by $\vec{\Psi}_I = (0, 0, \psi_I)$. Similar relations exist for vorticity, velocity, and stream function in the moving frame of reference. The stream function is related to the corresponding vorticity fields by the kinematic definition given by the stream function equations (SFEs), $\nabla^2\psi_I = -\omega_I$ and $\nabla^2\psi_r = -\omega_r$.

The vorticity fields in the inertial and moving frame of references are related by $\vec{\omega}_I = \vec{\omega}_r + 2\vec{\Omega}$. The stream functions for the inertial and moving frame of references are related in 2D by

$$\psi_I = \psi_r - \frac{|\vec{\Omega}|}{2} R_r^2 + f(x_r, y_r), \tag{12.6}$$

with f an unknown function that can be defined in terms of the velocity field. For 2D flows stream functions, vorticity and $\vec{\Omega}$ are all in a direction perpendicular to the plane of flow. Hence we will not use vector notation any more, without any confusion. However, one can define a new stream function, ψ_N, by

$$\psi_N = \psi_r - \frac{\Omega}{2}(R_r^2 - R_{oI}^2 + u_{oI} y_I - v_{oI} x_I), \tag{12.7}$$

such that ψ_N satisfies the Poisson equation: $\nabla^2\psi_N = -\omega_I$. The motion of the origin of the moving frame of reference is defined by the following relationships:

$$x_{oI} = (x_{oI})_m + a_x \cos(k_x t) \tag{12.8}$$

$$y_{oI} = (y_{oI})_m + a_y \cos(k_y t) \tag{12.9}$$

and the pitching motion of the airfoil is given by

$$\alpha = \alpha_m + \alpha_a \cos(k_\alpha t + \phi). \tag{12.10}$$

In the above equations, quantities with subscript m signify mean quantities and ϕ is the phase difference between the pitching and heaving/horizontal oscillation executed by the airfoil. Here, different reduced frequencies for the translational and rotational motion of the airfoil are prescribed.

The (ψ, ω)-formulation avoids the problems of pressure-velocity coupling and satisfaction of mass conservation everywhere in the flow field. However, to calculate the load accurately, one needs to solve a PPE instead, which is obtained by taking the divergence of Equations (12.2) and (12.3) and after some simplification yields the following equations:

$$\nabla^2\left(\frac{p}{\rho} + \frac{|\vec{V}|_I^2}{2}\right) = \nabla \cdot (\vec{V}_I \times \vec{\omega}_I) \tag{12.11}$$

and

$$\nabla^2\left(\frac{p}{\rho} + \frac{|\vec{V}|_r^2}{2}\right) = (\vec{V}_r \cdot \nabla^2\vec{V}_r + \omega_r^2) + 2\vec{\Omega} \cdot \vec{\omega}_r. \tag{12.12}$$

The quantity in parenthesis, on the left-hand side are the total pressure (P), which is a measure of mechanical energy of the flow (see Sengupta *et al.* (2003) and Sengupta (2012) for an in-depth analysis of this topic). To calculate loads and moment, it would be preferable to solve Equation (12.11), as it involves fewer terms. Above equations are solved in appropriate nondimensional forms which are obtained by introducing the chord of the airfoil (c) as the length scale and U_o as the velocity scale. The nondimensional VTE solved in Sengupta *et al.* (2005) is given by

$$\frac{\partial \vec{\omega}_r}{\partial t} + (\vec{V}_r \cdot \nabla)\vec{\omega}_r = \frac{1}{Re}\nabla^2 \vec{\omega}_r - 2\frac{\partial \vec{\Omega}}{\partial t}, \tag{12.13}$$

where the Reynolds number is given by $Re = \frac{U_o c}{\nu}$. Vorticities, frequencies and angular rotation rates are nondimensionalized by $\frac{U_o}{c}$. For the flapping motion, the oncoming free-stream speed is chosen as the velocity scale. The velocity field for a given vorticity distribution is calculated by solving two Poisson equations given by

$$\nabla^2 \psi_I = -\omega_I \tag{12.14}$$

$$\nabla^2 \psi_N = -\omega_I. \tag{12.15}$$

Having obtained the velocity and vorticity field, one solves the PPE given by Equation (12.11) to obtain the pressure field. Obtained pressure and vorticity fields are integrated to calculate the pressure and viscous forces acting on the airfoil. Note that the kinematic parameters of the airfoil motion given in Equations (12.8)–(12.10), retain the same form when nondimensionalized. For the pure hover mode of motion, when the airfoil oscillates in pitch and in horizontal direction in the absence of mean motion, the velocity scale is taken as $\frac{k_x}{2\pi}c$.

12.3.1 Pressure Solver

For the orthogonal curvilinear coordinate system the governing PPE given by Equation (12.11) can be rewritten as

$$\frac{\partial}{\partial \xi}\left(\frac{h_2}{h_1}\frac{\partial P_I}{\partial \xi}\right) + \frac{\partial}{\partial \eta}\left(\frac{h_1}{h_2}\frac{\partial P_I}{\partial \eta}\right) = \frac{\partial}{\partial \xi}(h_2 v_I \omega_I) - \frac{\partial}{\partial \eta}(h_1 u_I \omega_I), \tag{12.16}$$

where

$$P_I = \frac{p}{\rho} + \frac{V_I^2}{2},\ u_I = \frac{1}{h_2}\frac{\partial \psi_I}{\partial \eta} \text{ and } v_I = -\frac{1}{h_I}\frac{\partial \psi_I}{\partial \xi}.$$

Equation (12.16) is solved subject to the boundary condition derived from the normal momentum equation as applied on the airfoil and the outer boundary. These are obtained from the normal (η) momentum equation given by

$$\frac{h_1}{h_2}\frac{\partial P_I}{\partial \eta} = -h_1 u_I \omega_I + \frac{1}{Re}\frac{\partial \omega_I}{\partial \xi} - h_1 \frac{\partial v_I}{\partial t}. \tag{12.17}$$

12.3.2 Proof of Equation (12.17)

The normal pressure gradient is obtained by projecting NSE in rotational form in the normal direction as

$$\frac{\partial P_I}{\partial n} = \hat{n} \cdot \left[-\frac{1}{Re} \nabla \times \vec{\omega}_I + \vec{V}_I \times \vec{\omega}_I - \frac{\partial \vec{V}_I}{\partial t} \right], \tag{12.18}$$

where $\hat{n}$ is the unit normal vector. Various quantities on the right-hand side is obtained with $\hat{\rho}$ indicating unit vector in normal direction as

$$\nabla \times \vec{\omega}_I = \frac{1}{h_1 h_2} \begin{bmatrix} h_1 \hat{\xi} & h_2 \hat{\eta} & \hat{\rho} \\ \frac{\partial}{\partial \xi} & \frac{\partial}{\partial \eta} & \frac{\partial}{\partial \rho} \\ 0 & 0 & \omega_I \end{bmatrix}$$

$$= \frac{1}{h_1 h_2} \left[h_1 \hat{\xi} \frac{\partial \omega_I}{\partial \eta} - h_2 \hat{\eta} \frac{\partial \omega_I}{\partial \xi} \right].$$

Thus

$$\nabla \times \vec{\omega}_I = \frac{1}{h_2} \frac{\partial \omega_I}{\partial \eta} \hat{\xi} - \frac{1}{h_1} \frac{\partial \omega_I}{\partial \xi} \hat{\eta}.$$

Similarly one obtains

$$\vec{V} \times \vec{\omega}_I = \begin{bmatrix} \hat{\xi} & \hat{\eta} & \hat{\rho} \\ u_I & v_I & 0 \\ 0 & 0 & \omega_I \end{bmatrix}.$$

Thus

$$\vec{V} \times \vec{\omega}_I = \hat{\xi} (v_I \, \omega_I) - \hat{\eta} (u_I \, \omega_I).$$

Collating all these terms in the right-hand side, one obtains

$$\text{RHS} : \hat{\xi} \left(\frac{1}{h_2} \frac{\partial \omega_I}{\partial \eta} \left[-\frac{1}{Re} \right] + (v_I \omega_I) - \frac{\partial u_I}{\partial t} \right)$$

$$+ \hat{\eta} \left(\frac{1}{Re} \frac{1}{h_1} \frac{\partial \omega_I}{\partial \xi} - (u_I \omega_I) - \frac{\partial v_I}{\partial t} \right).$$

Thus

$$\frac{1}{h_2} \frac{\partial P_I}{\partial \eta} = \frac{1}{Re} \frac{1}{h_1} \frac{\partial \omega_I}{\partial \xi} - (u_I \omega_I) - \frac{\partial v_I}{\partial t}.$$

On the surface, we have $u_I, v_I = 0$ and the above condition simplifies. However, that will not yield a very high accuracy estimate of P_I, as one is forced to take one-sided difference formula of lowest order. To avoid such inaccuracy, one computes $\partial P_I / \partial \eta$ at half node above the surface. Hence the proof.

In Abdallah (1987), the steady state boundary condition was considered in a Cartesian frame with uniform nonstaggered grid system for a driven cavity problem. The formulation for the PPE in Sengupta *et al.* (2005) is a major development which did not require the Neumann boundary condition to be steady at the far-field boundary of the curvilinear orthogonal grid system.

In Abdallah (1987), it has been shown that the existence of solution for PPE requires the satisfaction of a compatibility condition that relates the source terms with the Neumann boundary condition, as a consequence of Green's theorem used for the PPE. This condition is not satisfied automatically in a collocated grid system and solution drifts without convergence. This is explained for the Poisson equation: $\nabla^2 P = \sigma$ in a 2D plane. Existence of the solution to this with Neumann boundary condition requires application of divergence theorem

$$\iint_{Area} \sigma dA = \oint \frac{\partial P}{\partial \eta} dl. \tag{12.19}$$

Satisfaction of this is equivalent to the ensuring the following identities:

$$LHS = RHS = 0, \tag{12.20}$$

where the LHS and RHS are finite difference expression of the left- and right-hand side, respectively, for Equations (12.16) and (12.17) summed over all the nodes in the computing domain. To ensure that this is indeed true, specific stencils are to be chosen for the discretization. For example, Equation (12.16) is discretized as

$$\begin{aligned} &\frac{1}{\Delta\xi^2}\left(\left.\frac{h_2}{h_1}\right|_{(i+1/2,j)} P_{I(i+1,j)} + \left.\frac{h_2}{h_1}\right|_{(i-1/2,j)} P_{I(i-1,j)}\right) - \left(\frac{1}{\Delta\xi^2}\left[\left.\frac{h_2}{h_1}\right|_{(i+1/2,j)} + \left.\frac{h_2}{h_1}\right|_{(i-1/2,j)}\right]\right. \\ &\quad \left. + \frac{1}{\Delta\eta^2}\left[\left.\frac{h_1}{h_2}\right|_{(i,j+1/2)} + \left.\frac{h_1}{h_2}\right|_{(i,j-1/2)}\right]\right) P_{I(i,j)} \\ &\quad + \frac{1}{\Delta\eta^2}\left(\left.\frac{h_1}{h_2}\right|_{(i,j+1/2)} P_{I(i,j+1)} + \left.\frac{h_1}{h_2}\right|_{(i,j-1/2)} P_{I(i,j-1)}\right) \\ &= \frac{(h_2 v\omega)_{(i+1/2,j)} - (h_2 v\omega)_{(i-1/2,j)}}{\Delta\xi} - \frac{(h_1 u\omega)_{(i,j+1/2)} - (h_1 u\omega)_{(i,j-1/2)}}{\Delta\eta}, \end{aligned} \tag{12.21}$$

for $2 \leq i \leq n$ and $2 \leq j \leq m$, where the subscript i denotes constant ξ-lines and the subscript j denotes constant η-lines. For the ease of representation, subscript I is dropped for u, v and ω, which is otherwise understood. The half-node quantities are taken as the arithmetic average of adjacent cell values. The right-hand side quantities of Equation (12.21) are discretized in the following manner:

$$(h_2 v\omega)_{(i+1/2,j)} = \frac{1}{8}\{h_2|_{(i,j)} + h_2|_{(i+1,j)}\}\{v_{(i,j)} + v_{(i+1,j)}\}\{\omega_{(i,j)} + \omega_{(i+1,j)}\}$$

$$(h_2 v\omega)_{(i-1/2,j)} = \frac{1}{8}\{h_2|_{(i,j)} + h_2|_{(i-1,j)}\}\{v_{(i,j)} + v_{(i-1,j)}\}\{\omega_{(i,j)} + \omega_{(i-1,j)}\}$$

$$(h_1 u\omega)_{(i,j+1/2)} = \frac{1}{8}\{h_1|_{(i,j)} + h_1|_{(i,j+1)}\}\{\omega_{(i,j)} + \omega_{(i,j+1)}\}\{u_{(i,j)} + u_{(i,j+1)}\}$$

$$(h_1 u\omega)_{(i,j-1/2)} = \frac{1}{8}\{h_1|_{(i,j)} + h_1|_{(i,j-1)}\}\{\omega_{(i,j)} + \omega_{(i,j-1)}\}\{u_{(i,j)} + u_{(i,j-1)}\}.$$

Neumann boundary condition (12.17) on the airfoil surface is discretized as follows:

$$\frac{P_{I(i,2)} - P_{I(i,1)}}{\Delta\eta} = \frac{1}{Re}\left(\frac{h_2}{h_1}\frac{\partial\omega}{\partial\xi}\right)_{(i,3/2)} - (h_2 u\omega)_{(i,3/2)} - \left[h_2\frac{\partial v}{\partial t}\right]_{(i,3/2)}. \tag{12.22}$$

To make LHS equal to zero, the above equation has been multiplied by $\frac{1}{\Delta\eta}[\frac{h_1}{h_2}]_{(i,3/2)}$ on both sides. Individual terms are discretized as indicated above. The last term of the right-hand side is represented as

$$\left[h_1\frac{\partial v}{\partial t}\right]_{(i,3/2)} = -\frac{\partial}{\partial t}\left(\frac{\partial\psi_I}{\partial\xi}\right)_{(i,3/2)}. \tag{12.23}$$

The right-hand side quantities are obtained as central averages with individual quantities discretized using second-order central difference scheme. Similarly, the Neumann boundary condition at the outer boundary is written as

$$\frac{P_{I(i,m-1)} - P_{I(i,m)}}{\Delta\eta} = \frac{1}{Re}\left(\frac{h_2}{h_1}\frac{\partial\omega}{\partial\xi}\right)_{(i,m-1/2)} - (h_2 u\omega)_{(i,m-1/2)} - \left[h_2\frac{\partial v}{\partial t}\right]_{(i,m-1/2)}. \tag{12.24}$$

This equation is multiplied by $\frac{1}{\Delta\eta}[\frac{h_1}{h_2}]_{(i,m-1/2)}$ to make LHS equal to zero. The discretization procedure is similar to that adopted for the Neumann boundary condition applied on the airfoil surface. Discretized equations are then solved using the conjugate-gradient algorithm of Van Der Vorst (1992). The above formulation of the PPE with the boundary condition of (12.17) is solved following the discretization as indicated above and it can also be solved by taking a smaller domain as compared to the domain used for solving SFEs and the VTE. This is due to the fact that the boundary condition used here are exact up to the accuracy by which the velocity and vorticity fields are numerically calculated.

12.3.3 Distinction between Low and High Reynolds Number Flows

Low Reynolds number flows are prone to steady and/ or unsteady separation, which obviously makes the flow inherently unsteady. In the presence of unsteadiness at lower frequencies, the flow cannot be predicted by RANS equation using turbulence models. Neither will such flows be computed by classical LES with subgrid scale (SGS) models. To understand the flow features at low Reynolds numbers, in Figure 12.1, we show the computed vorticity field for flow past SHM-1-airfoil by solving Equations (12.4) and (12.14) for the 2D flow. The flow is computed as simulated by the monotone implicit large eddy simulation (MILES) technique, which uses the multidimensional filter of Sengupta and Bhumkar (2010). This is a 15% thick airfoil designed for a cruise Reynolds number of 10.3 million. In this figure, the snapshots

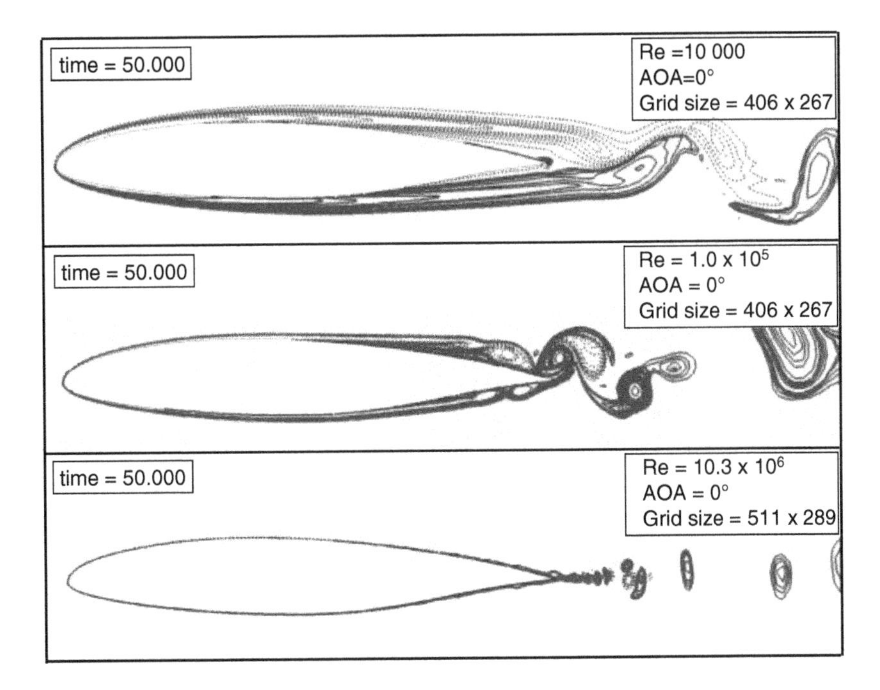

Figure 12.1 Flow past SHM-1 airfoil at the indicated Reynolds numbers, with the section placed at zero angle of attack. (See colour version of this figure in colour plate section)

of vorticity field around the airfoil is plotted at $t = 50$ for the indicated *Re*, following an impulsive start, with the airfoil kept at zero angle of attack. It is seen that at $Re = 10^4$ and 10^5, boundary layers are thicker and exhibit steady separation near maximum thickness location. Such elongated bubbles are prone to Kelvin-Helmholtz instability of the free shear layer at the bubble boundary. As a consequence, the drag experienced by the section is high. It is important to realize that for a commercial aircraft, the wing is used to carry fuel and that dictates the large thickness requirement. However for MAV, there is no need to have such thicker sections. Having said that, one also notes that wind turbine blades are significantly thicker, whose operation would be characterized also at low *Re*. Another aspect of operation in low *Re* aerodynamics is high background free stream disturbances. As a consequence, the flow also becomes turbulent by bypass mechanism of transition. It is noted in Figure 12.1, that at cruise Reynolds number the boundary layers are thin and very small coherent vortices are shed from the trailing edge. Despite the size of these vortices, the vorticity values embedded in the vortices are very large. At the cruise *Re*, bypass transition occurs in the aft portion of the airfoil section, with details given in Bhumkar (2012).

In Figure 12.2, coefficient of pressure for the three *Re* cases shown in Figure 12.1 are displayed and one notes that despite the higher boundary layer thickness at $Re = 10^4$, the flow is essentially steady, which is reflected in the C_p distribution. When the Reynolds number is increased tenfolds, boundary layer thickness comes down significantly as compared to the case of $Re = 10^4$. However, on both the airfoil surfaces near the trailing edge the flow suffers unsteady separation additionally. For the cruise Reynolds number of $Re = 10.3 \times 10^6$, one notices unsteady separation, which one associates with bypass transition from $x/c \simeq 0.60$

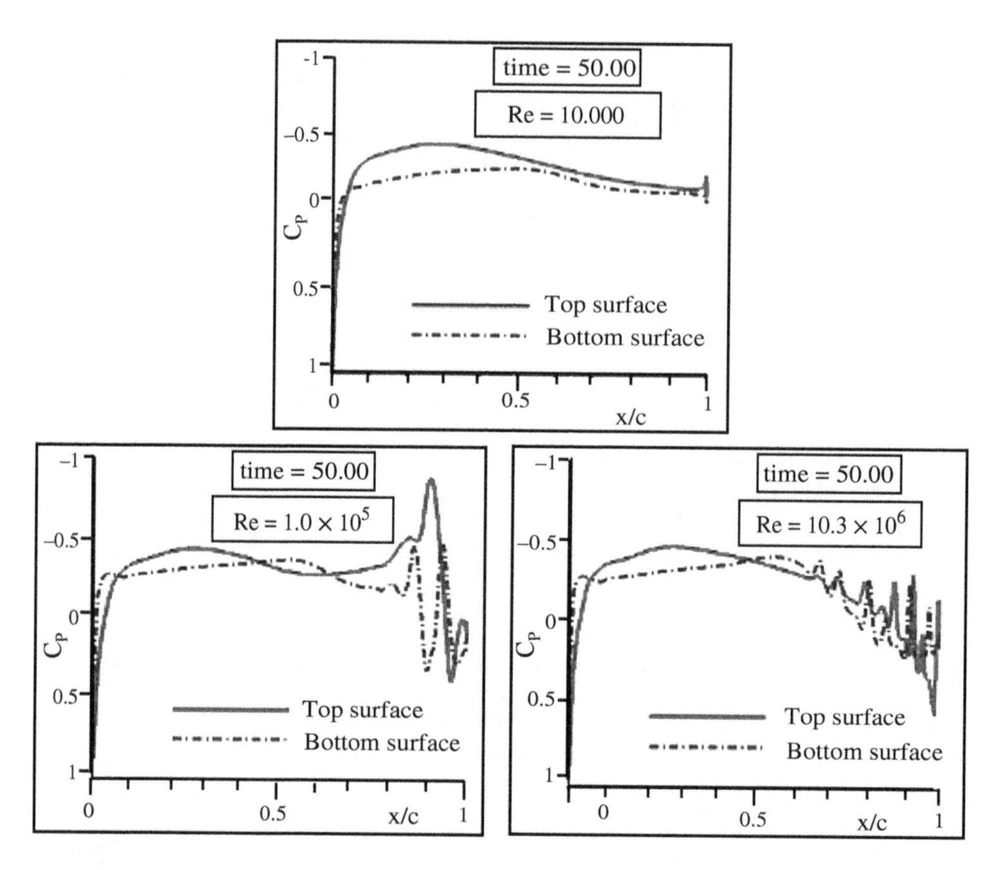

Figure 12.2 Pressure distribution over SHM-1 airfoil at the indicated Reynolds numbers for the cases shown in Figure 12.1

onwards. Thus, the flows at lower Reynolds would be dominated by pressure drag, while at the cruise Reynolds number the drag would be mostly contributed by skin friction. These observations noted here are for zero angle of attack and at higher angles of attack, the flow will be more prone to unsteadiness, more for lower *Re* due to thicker boundary layer and in such cases both the pressure and skin friction drag will increase.

Before we proceed further, it is necessary to validate the numerical process with some low *Re* experimental results with nonuniform or accelerating flows.

12.3.4 Validation Studies of Computations

To substantiate the methods and procedures described, some unsteady flow cases are computed for comparison with experimental results. First, the unsteady accelerated flow past aerofoil cases are discussed.

The impulsive start used for the simulations shown in Figure 12.1 is given by

$$\left.\begin{aligned} U_\infty(t) &= 0 \quad \text{for} \quad t < 0 \\ &= \bar{U}_\infty \quad \text{for} \quad t \geq 0 \end{aligned}\right\}. \tag{12.25}$$

Assumption of impulsive start results in significant simplification, as boundary conditions become time independent and still one can study unsteady effects. One applies no-slip boundary condition for $t \geq 0$. Immediately after an impulsive start, flow field is inviscid and irrotational everywhere and this is used as the initial condition. Imposition of no-slip condition at $t = 0$ creates instantaneously a boundary layer of infinitesimal thickness at the wall. Role of VTE is to diffuse and convect generated wall vorticity. Theoretically, singular acceleration of oncoming flow would excite a wide-band of circular frequencies, which is limited by supplied energy to the flow, measured in terms of Reynolds number. If one views this by perturbation analysis, then it can be shown that fluid dynamical system is not at all receptive to high frequencies, instead the response to high frequency excitation is the highly viscous Stokes' flow. Very high viscous diffusion at early times moderates singular start of the flow. For this reason, inaccurate evaluation of wall vorticity at $t = 0$ is of less concern.

In Figure 12.3, the acceleration used in Morikawa and Grönig (1995), Freymuth (1985) and Sengupta *et al.* (2007a) are shown. In Morikawa and Grönig (1995), a NACA 0015 airfoil is tested in a gravity-fed vertical water tunnel. The final free stream speed is established via application of continuous acceleration shown by the tanh function in Figure 12.3. The case shown by the thick line is the ramp start used in Freymuth (1985).

The ramp-start case of Figure 12.3 is given by

$$\left.\begin{aligned} U_\infty(t) &= \bar{U}_\infty t/\tau \quad \text{for} \quad t \leq \tau \\ &= \bar{U}_\infty \quad \text{for} \quad t > \tau \end{aligned}\right\}, \tag{12.26}$$

where τ is a characteristic acceleration time. This type of accelerated start-up is occasionally studied experimentally with a view to study effects of uniform acceleration, as in Freymuth (1985). The continuous acceleration case in Figure 12.3 was used in Morikawa and Grönig (1995), where free stream speed was varied according to

$$U_\infty(t) = \bar{U}_\infty \tanh\left(\frac{t}{\tau}\right), \tag{12.27}$$

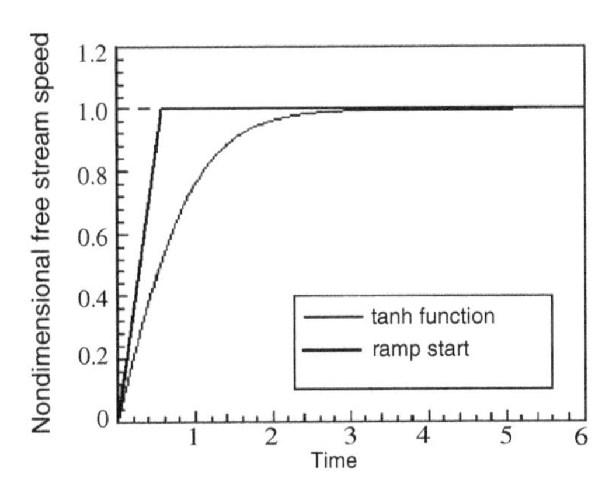

Figure 12.3 Sketch of nondimensional velocity time history for tangent hyperbolic and ramp start cases

where $\bar{U}_{\infty}$ is the final steady state free stream speed and τ is the characteristic acceleration time. Here, acceleration is uniform during early stages, followed by a duration when it is nonuniform and finally beyond 3τ, it decays to zero smoothly.

In Morikawa and Grönig (1995), the aerofoil is set at 30° angles of attack for $Re = 35\,000$, with the oncoming free stream speed varied as $\bar{U}_{\infty} = U_{\infty} \tanh(t/\tau)$. The Reynolds number is calculated based on the chord of the airfoil and the final free stream speed. In the computation, the flow start-up is implemented through the Neumann boundary condition applied at the outflow for Equation (8.44), for the stream function.

In Figure 12.4, the computed streamline contours are compared with the flow visualization picture of Morikawa and Grönig (1995). The final steady mean flow velocity is 64 cm/s and τ is the characteristic acceleration time given by 50 ms, which is equal to 0.60 in the nondimensional unit used in the formulation. The mean flow nonuniform acceleration is for a time up to 100 *ms*, beyond which the mean flow remains steady, but the unsteady effects persist much longer. The experimental results shown in Figure 12.4(a), corresponds to a nondimensional time of $t = 2.903$ and the computed flow field shown in Figure 12.4(b) with the initial start up according to Equation (12.27) matches in essential details of the experimental visualization, in terms of size, shape and orientations of the primary, secondary and tertiary

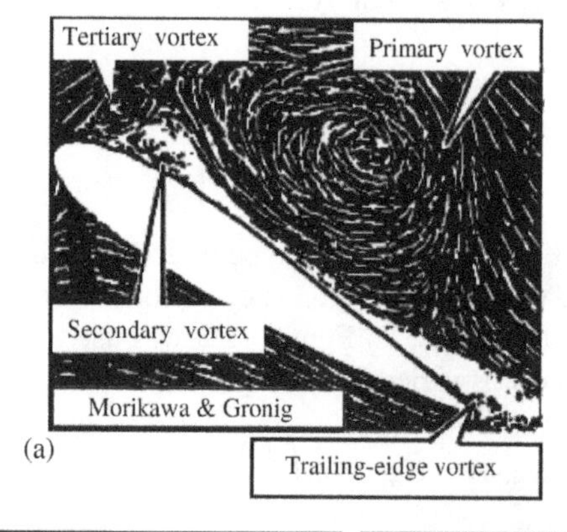

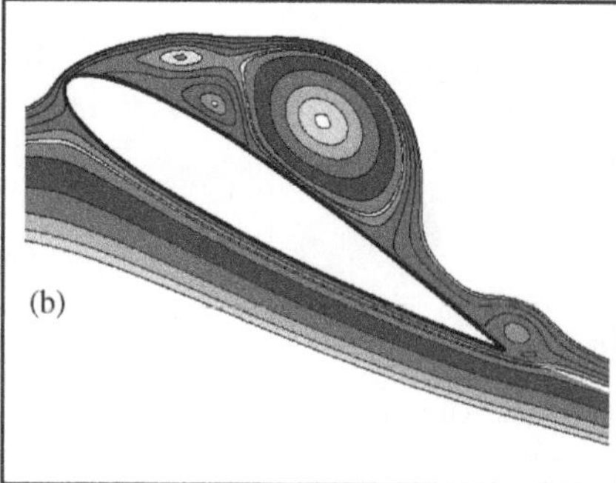

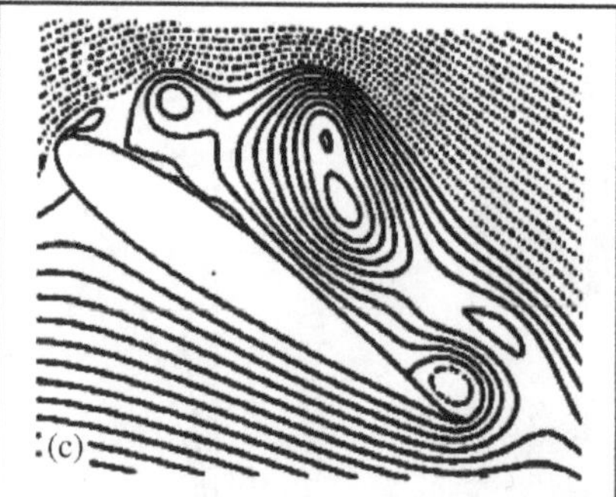

Figure 12.4 Comparison of (a) experimental visualization of Morikawa and Gröenig (1995) with the computations reported in Sengupta, Lim, Sajjan, Ganesh and Soria (2007a). Reproduced with permission from Cambridge University Press for the cases of (b) tangent hyperbolic start-up given by Equation (12.27) and (c) ramp start given by Equation (12.26). All the results are shown at $t = 2.903$. (See colour version of this figure in colour plate section)

vortices. In comparison, computation done by the same method is shown in Figure 12.4(c), which correspond to the acceleration given by Equation (12.26). It is clearly evident that the tangent hyperbolic start case provides the best match with the experimental results.

Another case is considered here for validation studies reported in Sengupta *et al.* (2007a), for NACA 0015 airfoil held at 30° angle of attack, for which the flow is accelerated uniformly from rest. The Reynolds number based on airfoil chord of 80 mm and the steady post-acceleration flow velocity of 100 mm/s, is 7968. Here the flow is accelerated from zero to the final flow velocity in 1, 2 and 3 s duration by a constant amount for the cases with the acceleration rates 100 mm/s^2, 50 mm/s^2 and 33.33 mm/s^2, respectively, as shown in Figure 12.5. Locations of discrete symbols indicate the instants at which PIV measurements were taken. In the experiments, cross-correlation digital PIV results were measured and the data were ensemble averaged.

In Figure 12.6, ensemble averaged velocity field are shown for the indicated acceleration rates for the indicated times. It is evident that the formation of separation bubbles and their evolution depends upon the length of time the acceleration is applied and not upon the strength of the acceleration. Thus, at the same nondimensional times the separation effects are accentuated for the higher acceleration case, as can be seen at $U_\infty = 80$ mm/s. Although one has problems gathering data very close to the wall, one still can identify the separation bubbles.

In Figure 12.7, the ensemble averaged vorticity distributions are shown for the same three acceleration cases. Here also one can notice the effect of sustenance of acceleration in the flow, rather than the strength of the applied acceleration. Thus, from Figures 12.6 and 12.7, one notices that the maximum effect of acceleration is noted for the case of lowest acceleration rate of 33.33 mm/s^2. Corresponding computational results have been presented in Sengupta *et al.* (2007a). Here just one such representative case is shown.

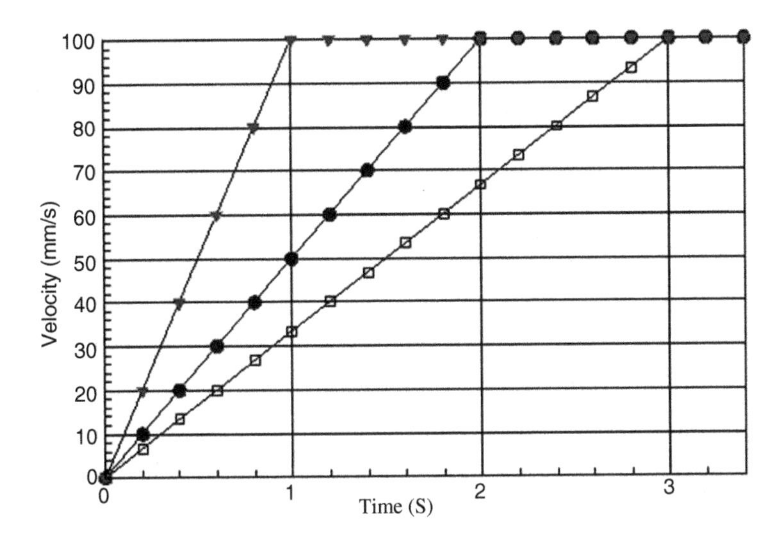

Figure 12.5 Oncoming free stream speed as a function of time obtained by imposed piston velocity for initial acceleration rates of 33.33 mm/s^4 (open rectangle), 50 mm/s^2 (filled circle) and 100 mm/s^2 (filled triangle) followed by constant velocity of 100 mm/s

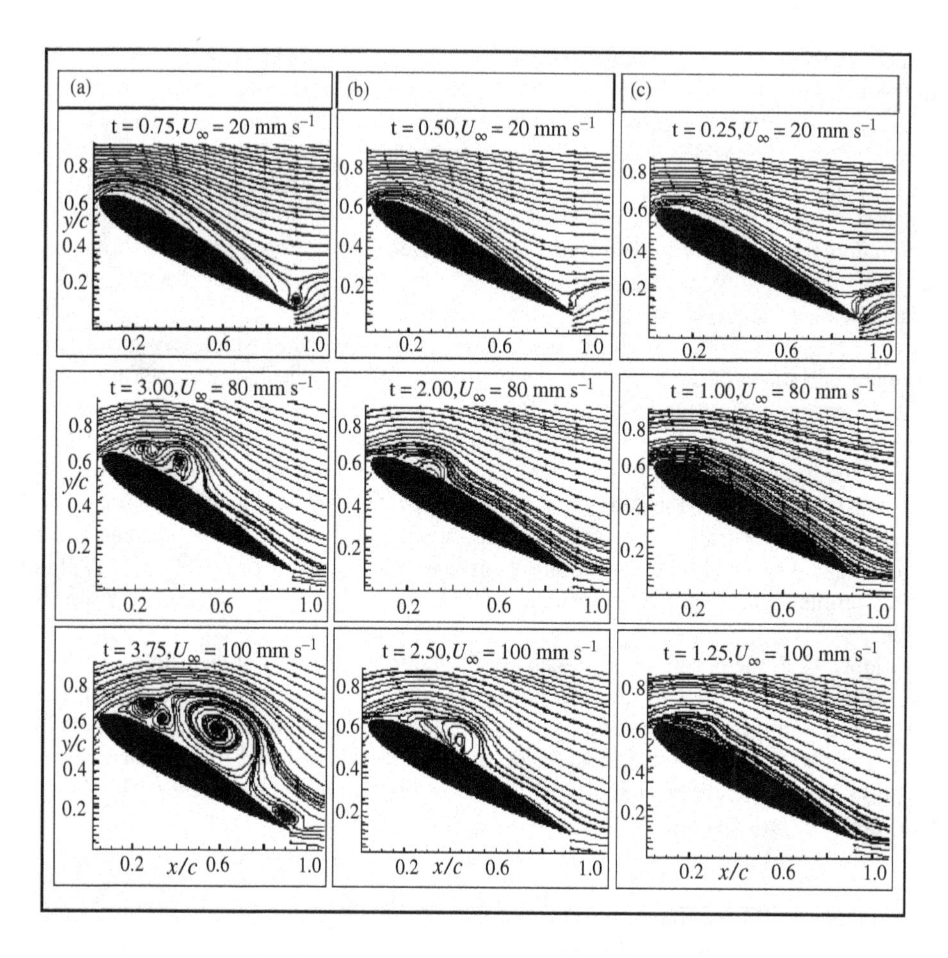

Figure 12.6 Ensemble-averaged velocity field for the three acceleration cases at the indicated times for the acceleration rates shown in Figure 12.5

We are idealizing that the ramp start case corresponds to a uniform acceleration created by a piston motion in Sengupta *et al.* (2007a). Due to structural and mechanical problems in the design of a plexiglass piston tank filled with water, the acceleration is never truly constant. This is demonstrated in Figure 12.8, where the case is considered for which the piston is accelerated for the first $3s$ and this is followed by constant deceleration of same magnitude for the next $3s$, so that the flow velocity is zero beyond $t = 6s$. In Figure 12.8(a), the piston velocity is shown as a function of time for the case where the magnitude of acceleration and deceleration was 33.33 mm/s^2. While the piston velocity is shown in Figure 12.8(a), actual flow velocity at the centreline of the tunnel was also measured and the time history of this is presented in Figure 12.8(b). One can see that despite the linear profile of the piston velocity schedule, the fluid velocity show the presence of kinks for the water tunnel. When one operates the tunnel with air, such uneven acceleration is not noted. This particular case was computed and compared with experimental results in Sengupta *et al.* (2007a).

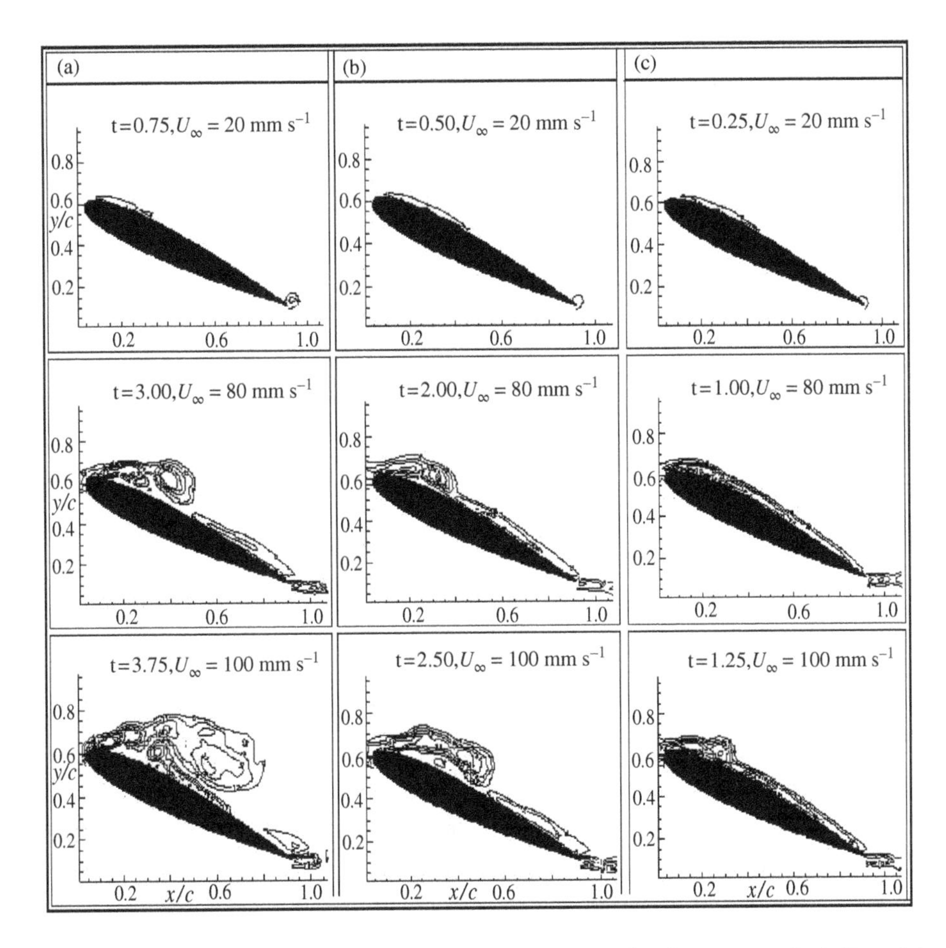

Figure 12.7 Ensemble-averaged velocity field for the three acceleration cases at the indicated times for the acceleration rates shown in Figure 12.5

Experimentally obtained instantaneous integral curves generated from the instantaneous velocity field are compared with the computed streamline contours obtained up to $t = 3.75$ in Figure 12.9. The experimental results are displayed on the left column, while the computed cases are shown in the middle and right columns for the velocity time-history of Figures 12.8(a) and (b), respectively. Once again the computed results match the details shown in the experimental result at $t = 2.75$. However, at later times match is poorer, yet the result at $t = 3.75$ show better qualitative match with the experimental results for the actual velocity history given in Figure 12.8(b), excepting the unsteady separation seen near the trailing edge. Although the airfoil model had an aspect ratio of 2.475 and the 3D effects may not be negligible, the efficacy of computing flows is demonstrated yet again for unsteady flows.

12.4 Flow Past an AG24 Airfoil at Low Reynolds Numbers

In discussing the flow past SHM-1 airfoil in Figures 12.1 and 12.2, it was noted that the section is thick and as a consequence, it experiences more of pressure drag, almost mimicking a bluff

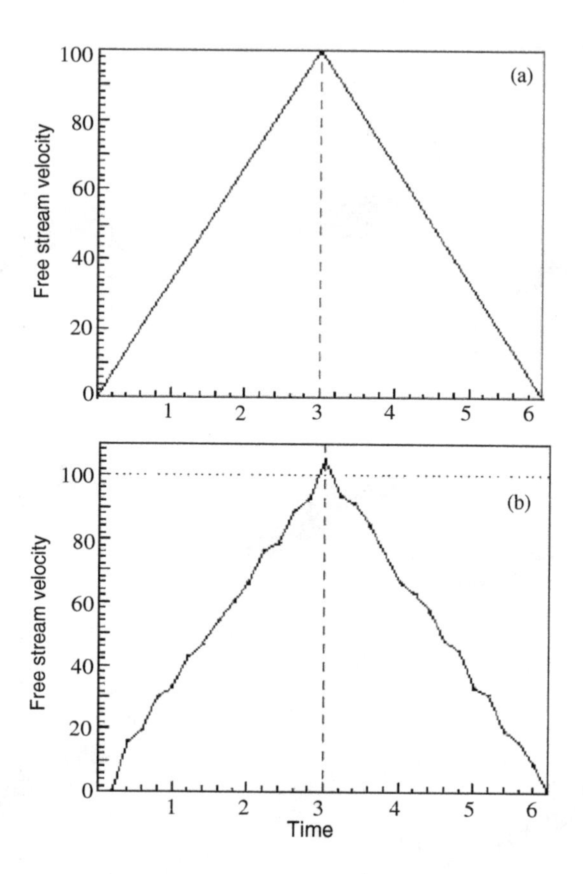

Figure 12.8 A special case of acceleration and deceleration in a close circuit water tunnel is shown via the time history of (a) the speed of piston and (b) tunnel centreline velocity are depicted

body. If the sole purpose of a design is to consider low *Re* aerodynamics, then it makes more sense to consider thinner section. One such section AG24, has been introduced for its use for radio controlled aero-models. This airfoil has a maximum thickness of 8.4% located at 26% chord location. This also has a nonnegligible maximum camber of 2.2% at 0.47c. The detailed coordinates of the section is given for the upper and lower surface in Table 12.1 (Williamson *et al.* (2012)).

The design of AG24 airfoil is based on classical viscous-inviscid interaction between potential flow and boundary layer solutions. Despite the fact that the boundary layer assumption is violated for thicker airfoil, one can consider the design process for thinner section to be guided similarly. The airfoil has a leading edge radius of 0.6% and its zero lift angle is reported to be −2.5°, while the stall angle is 6°. The small value of stall angle is due to lower thickness. Of particular interest is the maximum L/D ratio of the section to be moderately high. However, such high value of L/D occurs at 5°, which is very close to the stall angle. Actual experimental drag polar of this airfoil is provided later in the chapter, where the same will be computed using presented methodology.

The inviscid irrotational flow past AG24 airfoil is computed by panel method and result shown in Figure 12.10 for zero angle of attack. As the zero lift angle is negative, the pressure

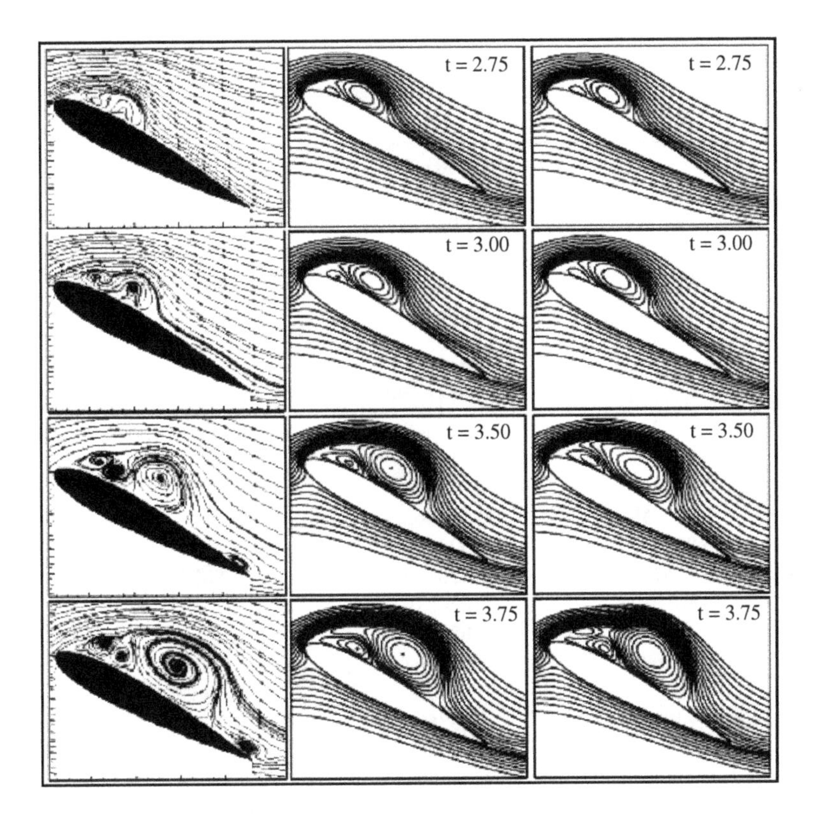

Figure 12.9 Flow field of accelerating flow over NACA 0015 airfoil for acceleration value of 33.33 mm/s^2 for $Re = 7968$ and $\alpha = 30°$. Comparison of experimental results (left) with computed streamline contours calculated with the data of Figure 12.8(a) (middle) and with the data of Figure 12.8(b) for the indicated times

distribution shown in this figure shows the presence of extensive range of adverse pressure gradient over both the surfaces, following an acceleration near the leading edge on both surfaces for zero angle of attack. The adverse pressure gradient is not too steep to cause flow separation for this inviscid pressure distribution. As inviscid results are a good description of the flow at higher *Re*, this pressure distribution would be compared with flow field at lower *Re*. Relatively, upper surface experiences larger adverse pressure gradient, whose effect can be tracked from the solution of NSE. In Figure 12.11, C_p distribution over AG24 airfoil is shown at $\alpha = 5°$ for which L/D attains a maximum value for *Re* lower than 100 000. One can readily see that the lift experienced by the airfoil is significantly increased, as compared to that shown in Figure 12.10. Additionally, the upper surface experiences a stronger acceleration on a small stretch near the leading edge, which is followed by higher severe adverse pressure gradient. This airfoil section experiences stall at 6° angle of attack, due to its thinner section. In Figure 12.12, the C_p distribution for this angle of attack is shown, which is qualitatively same as that for $\alpha = 5°$. However, the acceleration near the leading edge and subsequent flow deceleration are both severe, which will be more so on the upper surface. It is for this reason that the flow will show higher tendency to separate and suffer transition for $\alpha = 6°$, which

Table 12.1 AG-24 airfoil data

	Upper surface		Lower surface	
	x/c	y/c	x/c	y/c
1.	1.000000	0.000312	0.000001	−0.000230
2.	0.994048	0.001043	0.000293	−0.001719
3.	0.982038	0.002630	0.001184	−0.003235
4.	0.968488	0.004486	0.002714	−0.004738
5.	0.954647	0.006421	0.004994	−0.006301
6.	0.940769	0.008378	0.008364	−0.008034
7.	0.926890	0.010328	0.013488	−0.010053
8.	0.913006	0.012258	0.021193	−0.012377
9.	0.899118	0.014165	0.031412	−0.014726
10.	0.885230	0.016048	0.043212	−0.016796
11.	0.871339	0.017908	0.055804	−0.018519
12.	0.857451	0.019744	0.068780	−0.019918
13.	0.843559	0.021562	0.081987	−0.021042
14.	0.829670	0.023357	0.095351	−0.021938
15.	0.815779	0.025130	0.108824	−0.022639
16.	0.801883	0.026879	0.122377	−0.023175
17.	0.787989	0.028602	0.135989	−0.023565
18.	0.774091	0.030297	0.149652	−0.023827
19.	0.760188	0.031964	0.163351	−0.023977
20.	0.746285	0.033600	0.177085	−0.024026
21.	0.732376	0.035204	0.190844	−0.023986
22.	0.718461	0.036776	0.204622	−0.023867
23.	0.704547	0.038313	0.218421	−0.023675
24.	0.690631	0.039813	0.232237	−0.023420
25.	0.676709	0.041277	0.246064	−0.023106
26.	0.662788	0.042702	0.259909	−0.022740
27.	0.648868	0.044087	0.273764	−0.022326
28.	0.634943	0.045432	0.287631	−0.021871
29.	0.621017	0.046736	0.301514	−0.021381
30.	0.607093	0.047997	0.315408	−0.020856
31.	0.593166	0.049214	0.329313	−0.020301
32.	0.579236	0.050389	0.343234	−0.019718
33.	0.565310	0.051517	0.357166	−0.019111
34.	0.551385	0.052600	0.371111	−0.018482
35.	0.537456	0.053635	0.385067	−0.017835
36.	0.523528	0.054622	0.399034	−0.017173
37.	0.509606	0.055558	0.413004	−0.016498
38.	0.495682	0.056443	0.426962	−0.015813
39.	0.481756	0.057276	0.440907	−0.015121
40.	0.467835	0.058053	0.454842	−0.014424
41.	0.453918	0.058774	0.468769	−0.013723
42.	0.440000	0.059437	0.482685	−0.013021
43.	0.426087	0.060040	0.496593	−0.012319
44.	0.412180	0.060579	0.510494	−0.011617
45.	0.398274	0.061053	0.524394	−0.010919

Table 12.1 (*Continued*)

	Upper surface		Lower surface	
	x/c	y/c	x/c	y/c
46.	0.384372	0.061458	0.538286	−0.010224
47.	0.370481	0.061792	0.552174	−0.009535
48.	0.356594	0.062050	0.566059	−0.008852
49.	0.342711	0.062231	0.579942	−0.008177
50.	0.328841	0.062328	0.593817	−0.007511
51.	0.314984	0.062338	0.607689	−0.006857
52.	0.301131	0.062254	0.621556	−0.006214
53.	0.287296	0.062074	0.635419	−0.005585
54.	0.273479	0.061789	0.649276	−0.004971
55.	0.259672	0.061394	0.663131	−0.004372
56.	0.245889	0.060883	0.676986	−0.003788
57.	0.232132	0.060248	0.690839	−0.003221
58.	0.218395	0.059480	0.704693	−0.002670
59.	0.204696	0.058572	0.718548	−0.002136
60.	0.191034	0.057514	0.732407	−0.001619
61.	0.177412	0.056295	0.746265	−0.001122
62.	0.163854	0.054904	0.760127	−0.000645
63.	0.150354	0.053325	0.773993	−0.000194
64.	0.136934	0.051545	0.787859	0.000230
65.	0.123613	0.049542	0.801731	0.000622
66.	0.110400	0.047296	0.815611	0.000975
67.	0.097342	0.044784	0.829491	0.001286
68.	0.084457	0.041971	0.843374	0.001550
69.	0.071816	0.038832	0.857263	0.001759
70.	0.059466	0.035323	0.871153	0.001909
71	0.047533	0.031413	0.885046	0.001991
72.	0.036200	0.027083	0.898943	0.001997
73.	0.025838	0.022393	0.912839	0.001920
74.	0.017129	0.017634	0.926737	0.001758
75.	0.010712	0.013341	0.940634	0.001501
76.	0.006482	0.009857	0.954532	0.001149
77.	0.003763	0.007107	0.968401	0.000707
78.	0.001997	0.004865	0.982019	0.000185
79.	0.000855	0.002963	0.994039	−0.000364
80.	0.000198	0.001287	1.000000	−0.000659

lead to stalling of the section at lower *Re*. As the panel method is based on large *Re* for which viscous and inviscid interaction is weak, hence it is preferable to solve NSE for lower *Re*, for which the viscous diffusion is much stronger and analyzing viscous and inviscid part of the flow separately is questionable.

The following detailed results of flow past AG24 airfoil obtained by solving NSE are due to Krishnan (2013) obtained using the parallel compact scheme described in Sengupta *et al.* (2007) and Sengupta (2013). In Krishnan (2013), simulations are reported with and without considering the effects of free stream turbulence (FST) and are compared with available

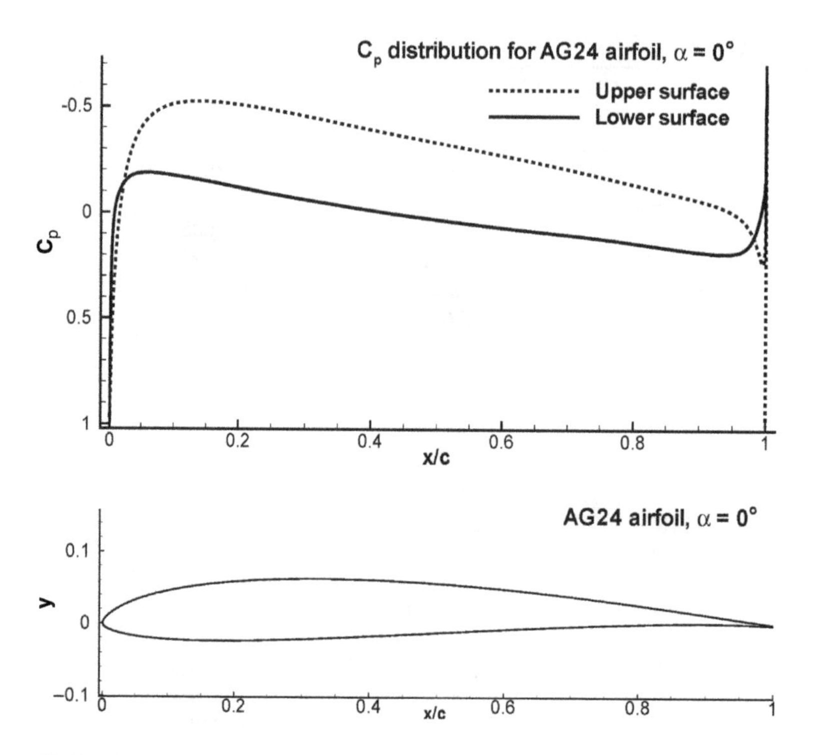

Figure 12.10 Computed pressure distribution over AG24 airfoil by panel method for $\alpha = 0°$

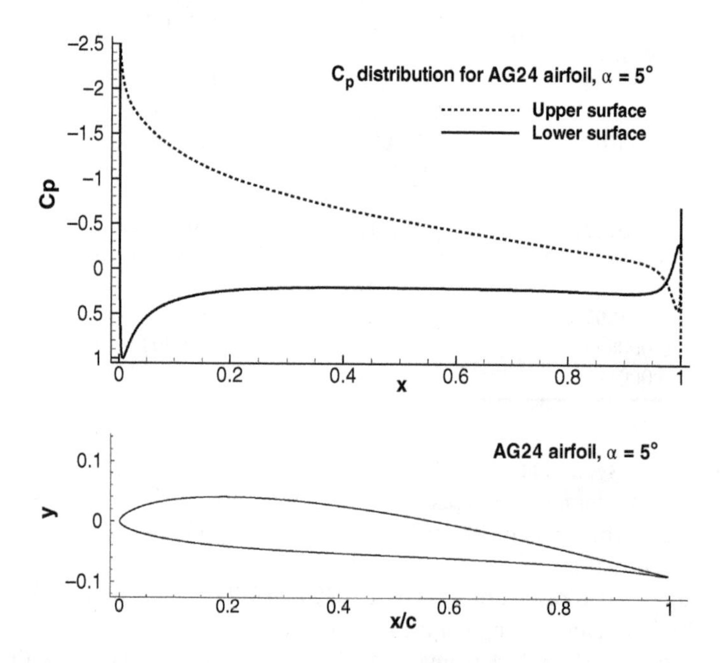

Figure 12.11 Computed pressure distribution over AG24 airfoil by panel method for $\alpha = 5°$

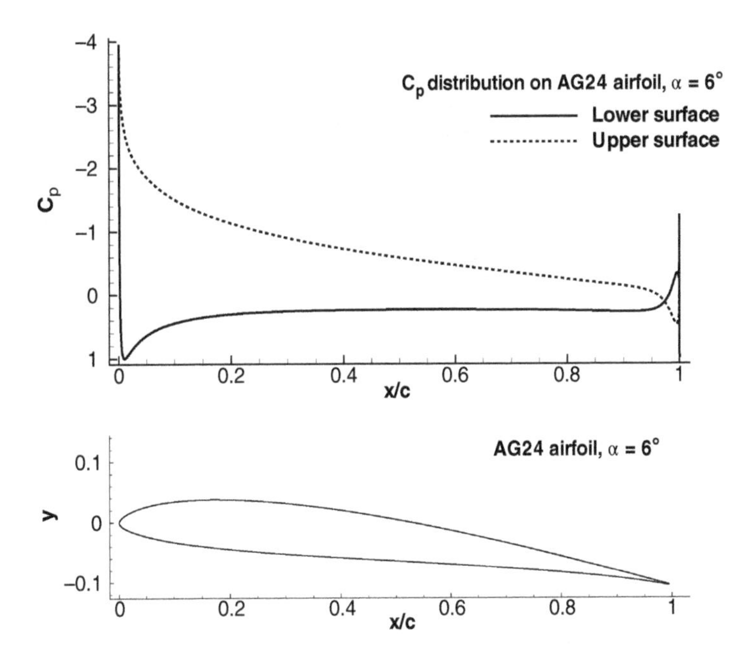

Figure 12.12 Computed pressure distribution over AG24 airfoil by panel method for $\alpha = 6°$

experimental results. Chord based *Re*'s considered are 60 000 and 400 000 and only the cases without FST are shown here.

Time evolution of the flow is shown in Figure 12.13, by plotting the streamline and vorticity contours for $Re = 60\,000$ and zero angle of attack case. Both these contour plots show that the shear layer is thin and which is compatible with the C_p distribution shown in Figure 12.10 indicating the presence of favourable pressure gradient over a longer stretch. The flow does change appreciably over the airfoil surface with time, except that a separation bubble is noted near the trailing edge on the lower surface at later times.

Direct simulation of flow past AG24 airfoil is reported for the range of angles of attack from $-6°$ to $6°$ at chord-based Reynolds numbers of 60 000 and 400 000. Obtained results are compared with experimentally available results for the same in Williamson *et al.* (2012). As noted earlier, low *Re* aerodynamics is characterized by unsteady separation bubbles and intense vortex shedding from the trailing edge and can be seen from computational results. In Figure 12.14, plotted stream function and vorticity contours show effects of angle of attack on the upper and lower surfaces of the airfoil by the results shown for the nondimensional time of $t = 50$. Near the negative stall angle at $\alpha = -6°$ one notices the formation of unsteady separation bubbles on the lower surface, whose intensity decreases for $\alpha = -4°$, but the vortices are noted right near the leading edge on the lower surface. For all negative angles of attack including $\alpha = 0$, the flow over the top surface remains attached. For positive angles of attack, one progressively starts noticing unsteady separation bubbles on top surface, whose intensity increase with α. However, the top surface vortices at any α is always smaller than the vortices on the lower surface, for the corresponding negative α, due to the camber.

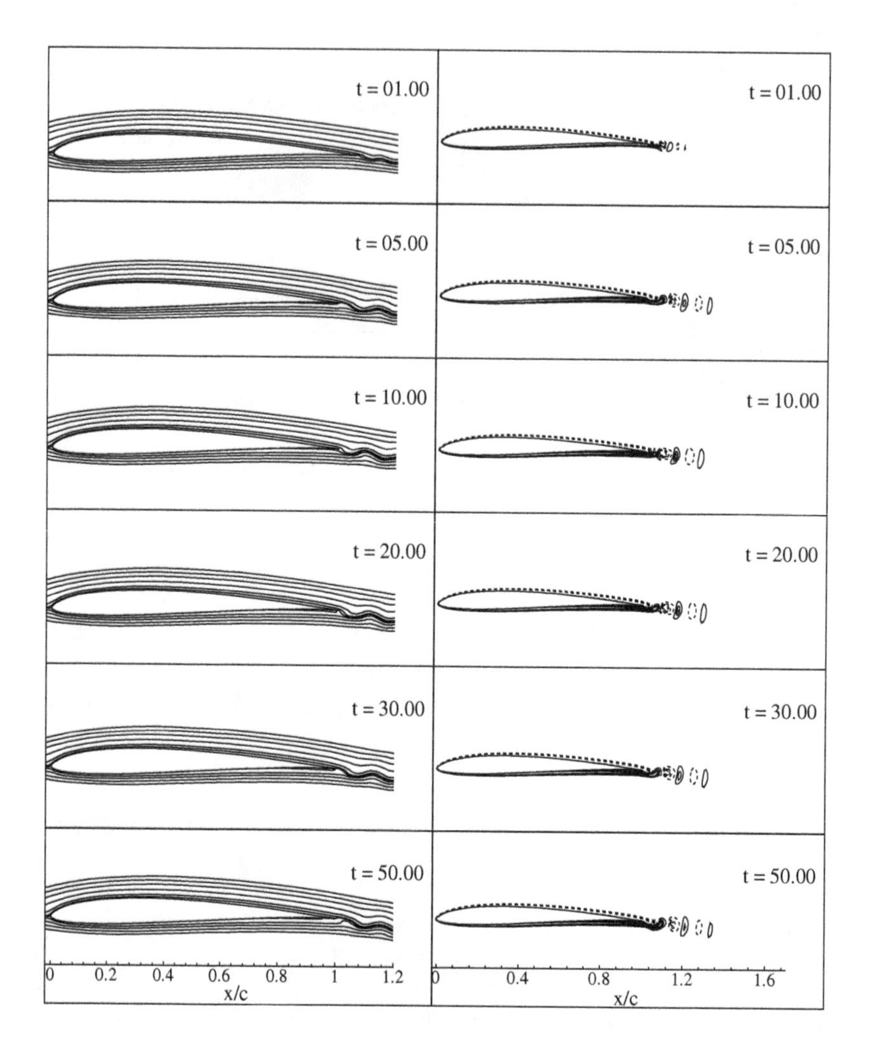

Figure 12.13 Computed stream function and vorticity contours for AG24 airfoil for $Re = 60\,000$ obtained by solving NSE for zero angle of attack, at the indicated time instants

The following have been reported in Krishnan (2013), which is noted only without showing any figures. Among the positive values of α considered, notable bubble formation is first seen for an angle of 3° at the time $t = 3.7$ which extends from $0.65c$ to the trailing edge. For $\alpha = 4°$, the onset time come down to $t = 2.7$ and the bubble extends to a greater distance, from $0.45c$ to the trailing edge. The time comes down further to $t = 1.9$ for $\alpha = 5°$, in case of which the bubble formation begins from $0.35c$. With further increase in angle of attack to 6°, bubble formation starts as early as at $t = 1.5$ and covers the entire chord length from $0.2c$ downstream.

In a similar pattern, as α is decreased, though extremely small bubble formation can be seen for $\alpha = -1°$, clear separation bubble formation can only be noted from $\alpha = -2°$, for which it happens at $t = 5.4$, at the trailing edge and with time the onset extends till $0.65c$. As α is brought down to $-3°$, the onset time comes down to $t = 4.1$ and extent increases

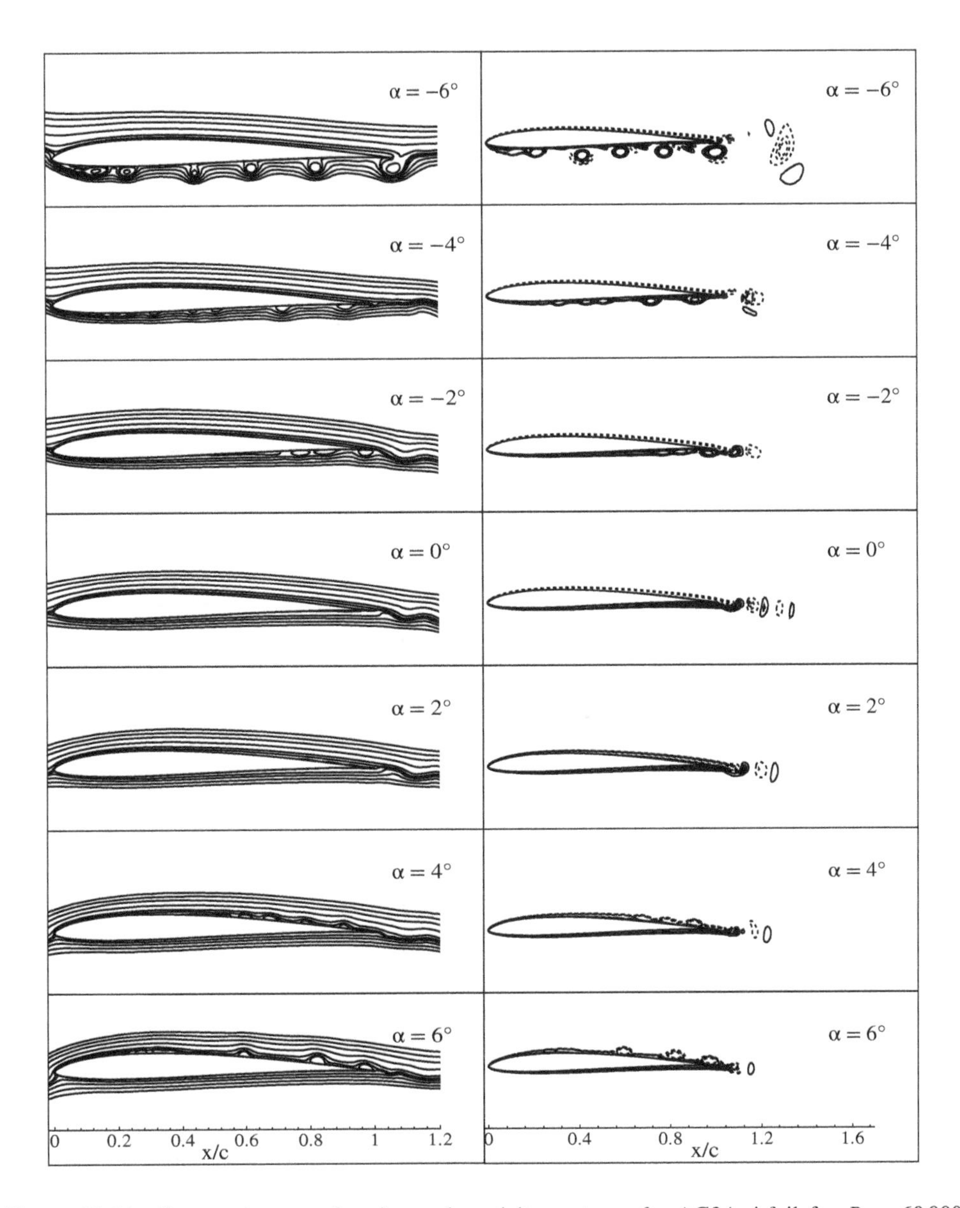

Figure 12.14 Computed stream function and vorticity contours for AG24 airfoil for $Re = 60\,000$ obtained by solving NSE for the indicated angles of attack, at $t = 50$

to $0.45c$. With further decrease in $\alpha = -4°$, notable separation bubbles are generated from $t = 1.8$ onwards and their domain increases from $0.15c$, which further increases to $0.75c$ for $\alpha = -5°$. It is seen that for the highest negative angle considered ($-6°$), separation occurs right at the leading edge itself. Very clear presence of these laminar separation bubbles can be observed from $t = 0.2$ onwards, which are found to be generated at less than $0.05c$. However, it can be observed that there is no complete flow separation even at this high adverse pressure

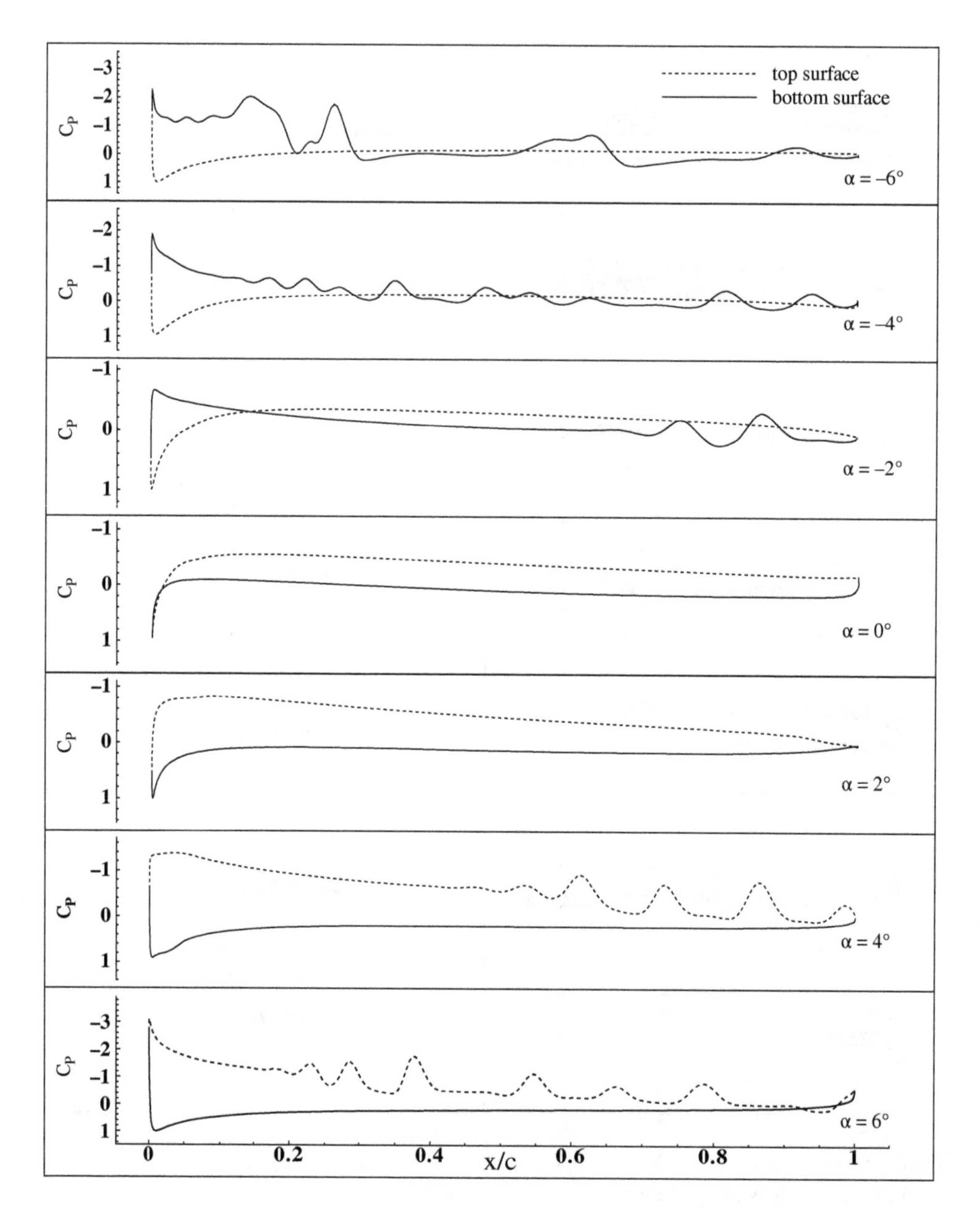

Figure 12.15 Time-averaged C_p plots of AG24 airfoil at mentioned angles of attack and Reynolds number of 60 000 without the effects of FST till time, $t = 50$

gradient created on the lower surface. The separation bubbles formed on the airfoil surface convect downstream and finally from trailing edge very clear vortex shedding can be seen.

Figure 12.15 shows the time-averaged C_p plots for $Re = 60\,000$ at the considered α values for $0 \leq t \leq 50$. Observed fluctuations on top and bottom surfaces are indicative of the existing adverse pressure gradient and also the intensity of unsteady separation. It is seen that the

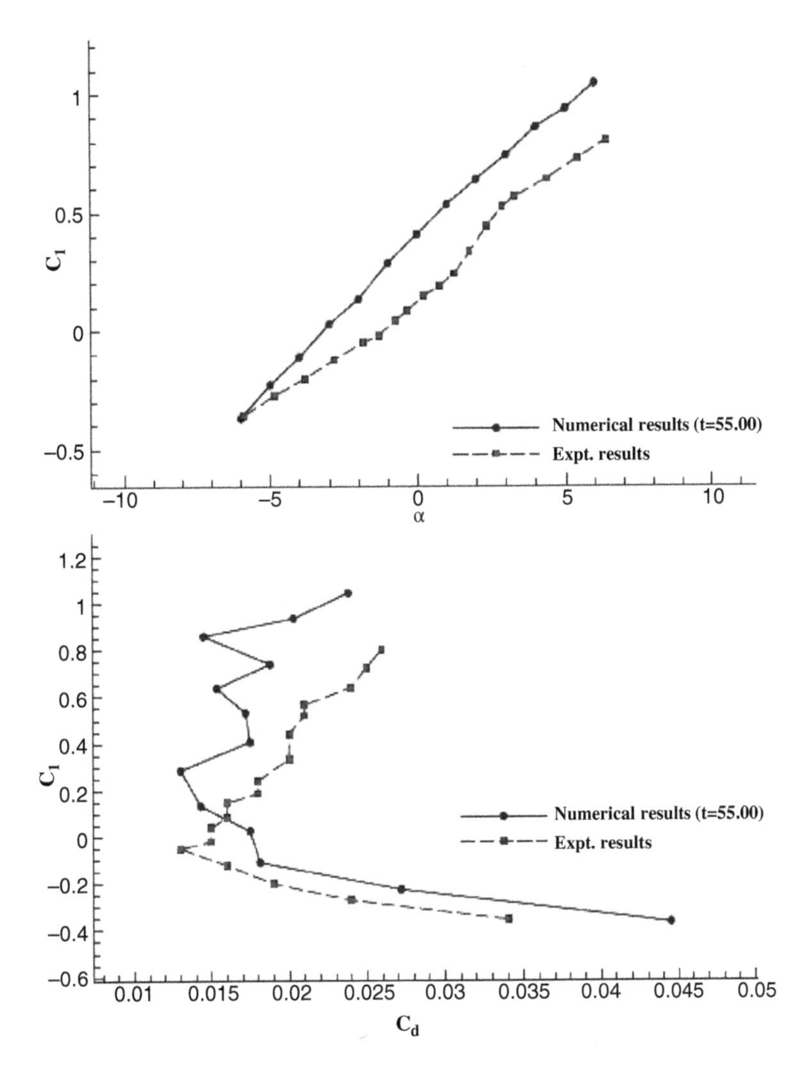

Figure 12.16 Comparison of computed lift curve (top) and drag polar (bottom) obtained using time-averaged pressure values till time, t = 50 for AG24 airfoil with the experimental results, in the considered range of angle of attack, without considering the effects of FST and at a Reynolds number of 60 000

mean pressure on top surface is more positive, than the bottom surface for higher positive values of α. Since AG24 airfoil is a low-cambered airfoil, it can directly be inferred that the corresponding values of C_l are also higher for these α's. In a similar way, it can be noted that there is a greater negative C_l for higher negative values of α.

Figure 12.16 shows the lift curve and drag polar for $Re = 60\,000$ using time-averaged lift and drag coefficients for the considered α for $0 \leq t \leq 50$. The numerical results obtained here are compared with the experimental data in Williamson *et al.* (2012). It can be seen that the obtained numerical results are in good agreement with the experimental results. However, it

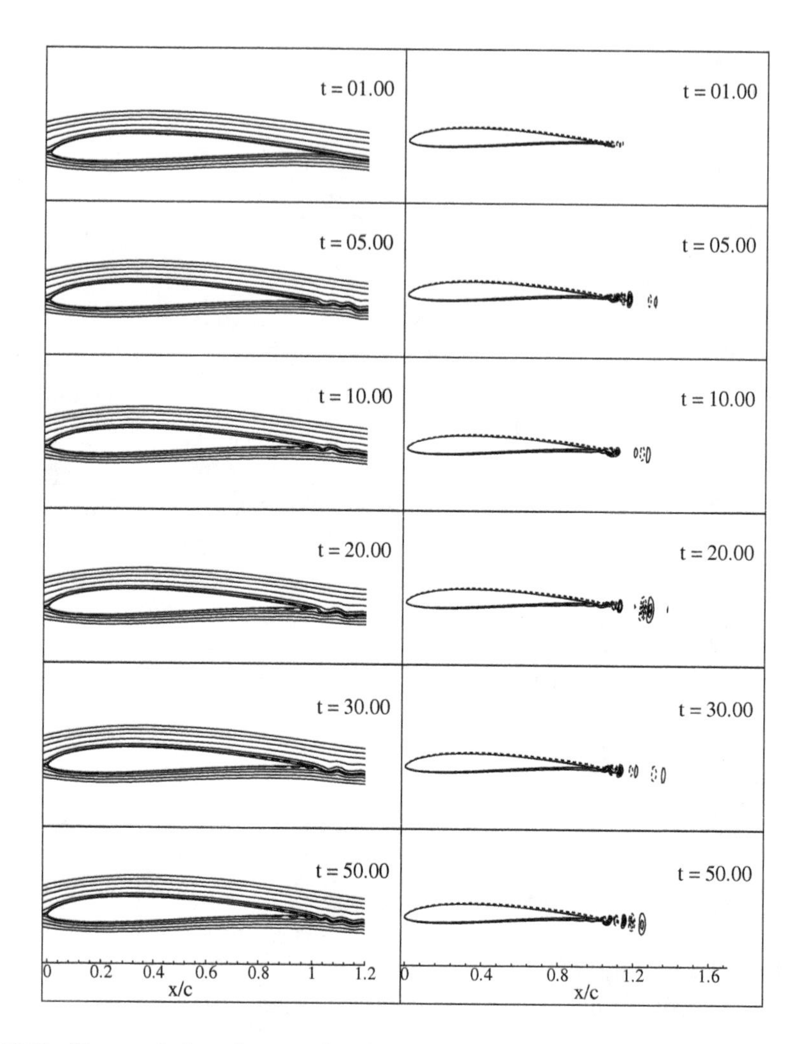

Figure 12.17 Time evolution of stream function (left) and vorticity (right) contours at 0° angle of attack of AG24 airfoil at $Re = 400\,000$

is seen that the numerically obtained C_d is over-predicted in the computational results, for higher negative values of α and under-predicted for higher positive values of α.

Next, consider a flow with higher energy indicated by higher $Re = 400\,000$, for which Figure 12.17 shows the time evolution of the flow at $\alpha = 0°$. Separation bubble size is very much reduced and a delay is noted in the time of appearance. The domain of existence of these separation bubbles are also reduced as compared to the case of Reynolds number of 60 000. Although the figures are not shown here, for $\alpha = 6°$ separation bubble formation is noted from $t = 1.9$ onwards, visible only beyond mid-chord. For negative α distinct separation bubbles can be seen, though a similar pattern of reduced bubble size, delayed bubble formation and reduced domain of existence is noted here. For $\alpha = -6°$, the bubble formation is found to begin at $t = 0.3$ and the bubbles exist on the entire lower surface of the airfoil beginning from $0.05c$. Size of these bubbles are remarkably smaller as compared to $Re = 60\,000$ case.

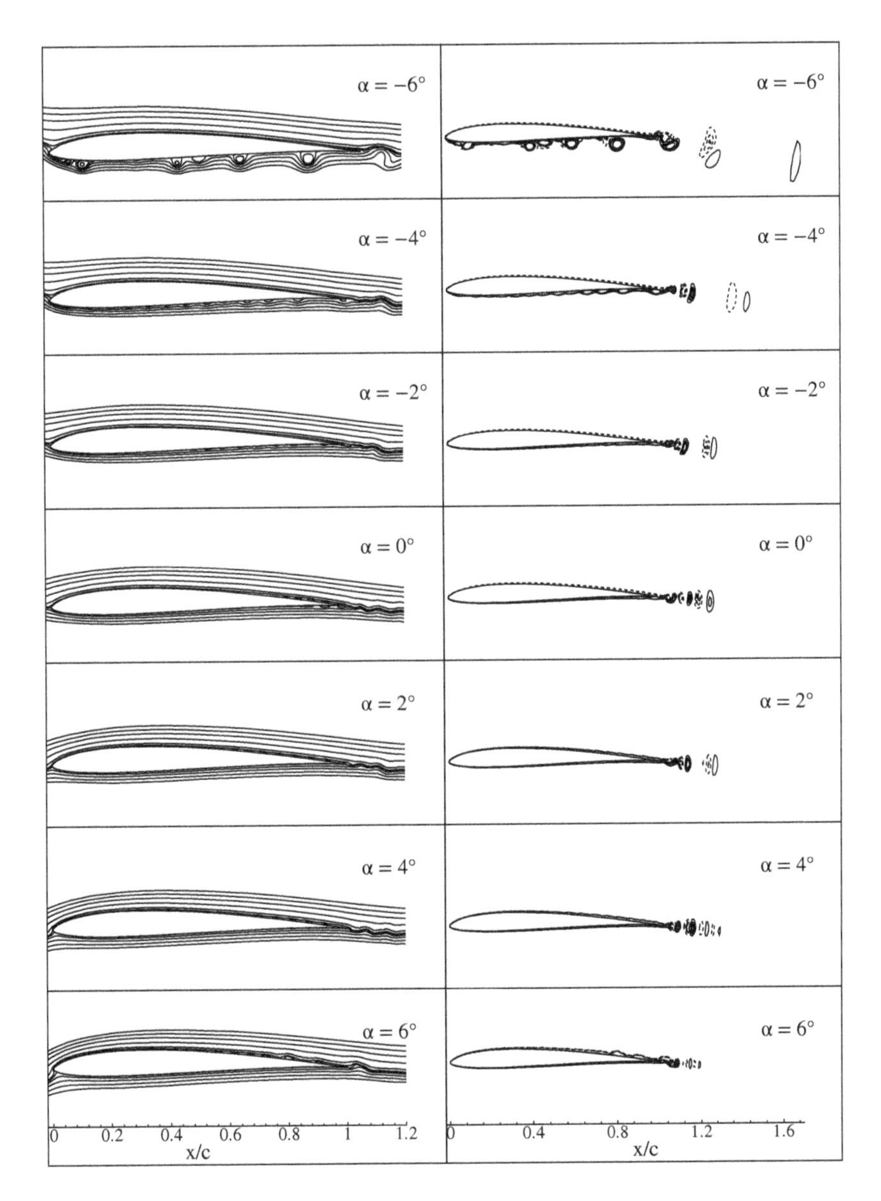

Figure 12.18 Stream function (left) and vorticity (right) contours of AG24 airfoil at indicated angles of attack for $Re = 400\,000$ at $t = 50$

Figure 12.18 show instantaneous streamline and vorticity contours at $t = 50$ for the indicated angles of attack for the case of $Re = 400\,000$. It is seen that major vortices are shed along the mean line with a narrower wake. The unsteady separation bubble in this case are significantly smaller in dimension, more for the positive α, than for the negative α. The shedding pattern is such that the shed vortices alternate in sign.

Figure 12.19 shows the time-averaged C_p plots for $Re = 400\,000$ at the indicated α values for $0 \leq t \leq 50$, which explains corresponding separation phenomenon and lift coefficient; the

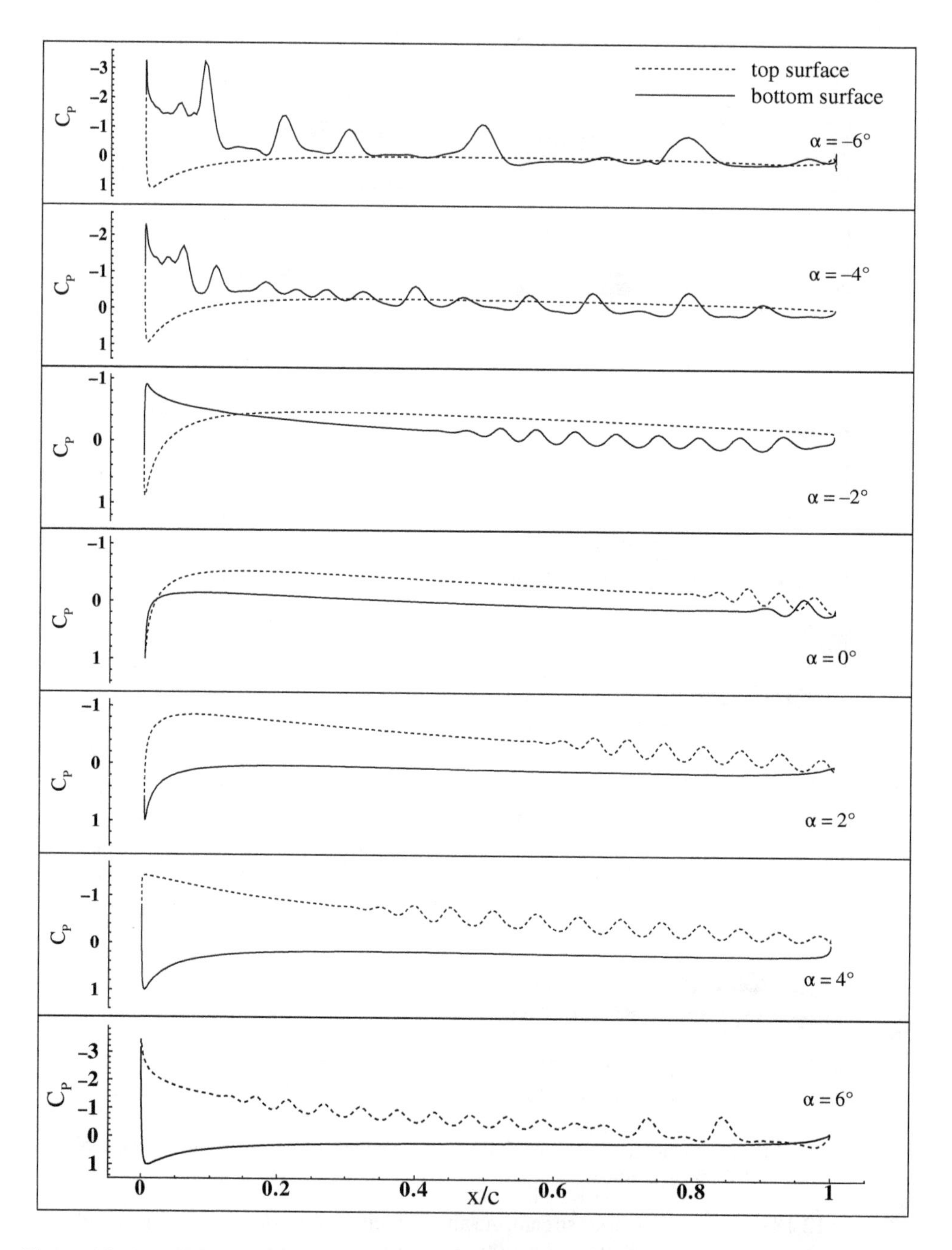

Figure 12.19 Time-averaged C_p distribution for AG24 airfoil at indicated α values for $Re = 400\,000$ for data till $t = 50$

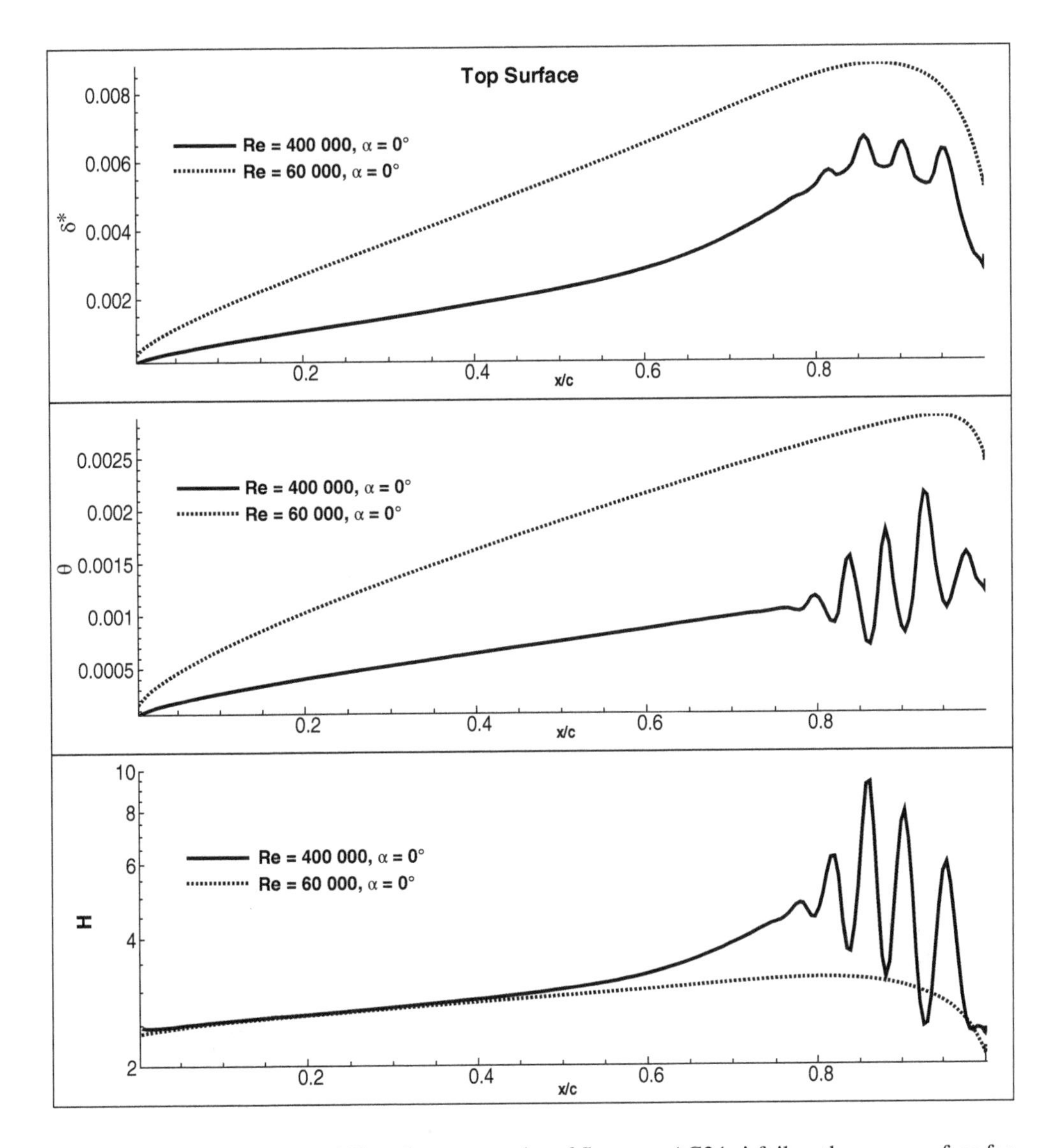

Figure 12.20 Comparison of Shear layer properties of flow past AG24 airfoil on the upper surface for $Re = 60\,000$ and $Re = 400\,000$

fluctuations in the plot corresponding to the zones of separation and reattachment. For $\alpha = -6°$, it is noted that increasing Re, increases the value of averaged lift, but with higher pitch up tendency of the airfoil. Also, the unsteady separation initiates right from the leading edge of the bottom surface. With increasing α, unsteady component and average lift progressive decreases. Similar events are noted on the upper surface with smaller unsteady separation bubbles forming, but forming over lower streamwise extent. However, the readers will be advised to not confuse the effects of unsteady separation with the size of recirculating bubble.

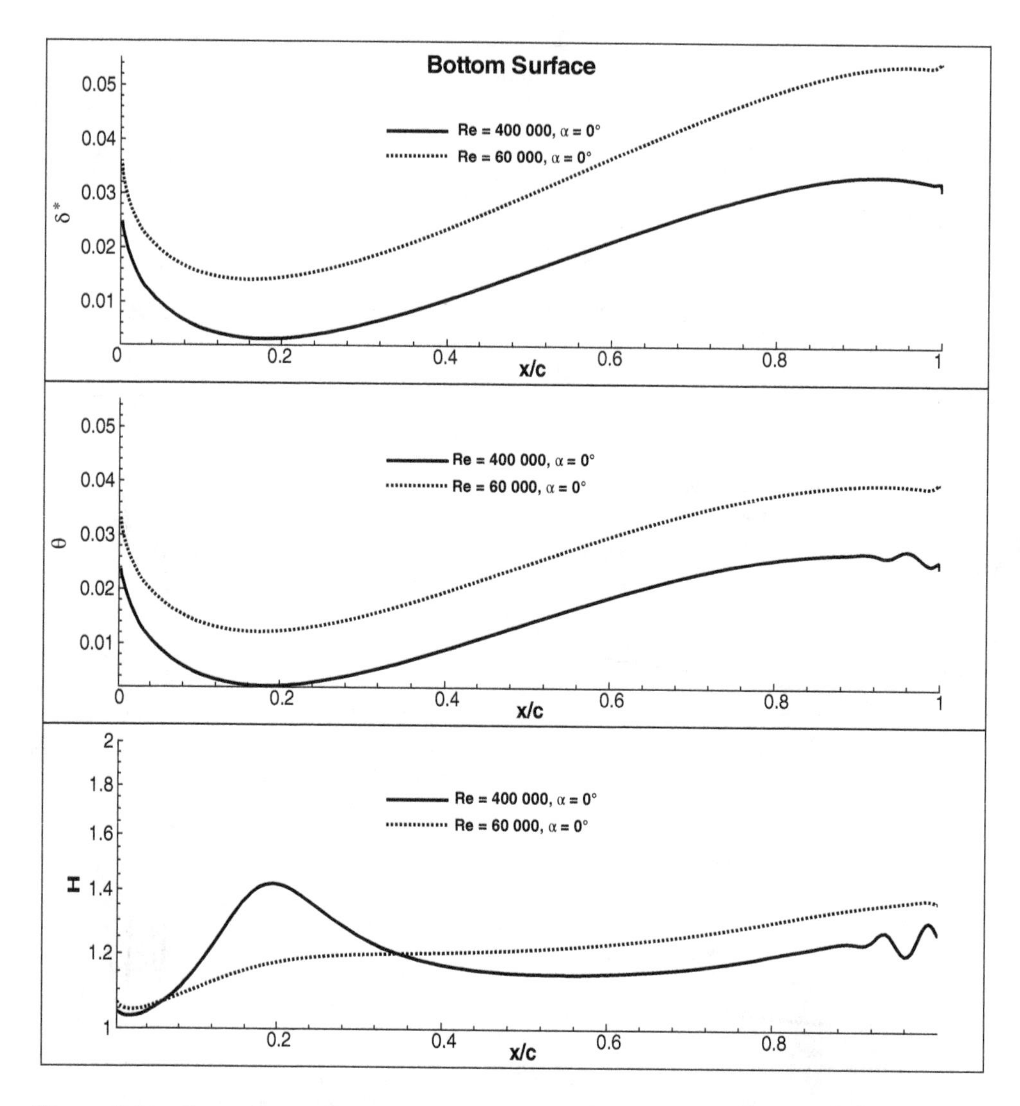

Figure 12.21 Comparison of Shear layer properties of flow past AG24 airfoil on the lower surface for $Re = 60\,000$ and $Re = 400\,000$

In the following, we compare the boundary layer forming over the AG24 airfoil for $Re = 60\,000$ and $400\,000$ in terms of displacement and momentum thicknesses (δ^* and θ) and their ratio providing the shape factor (H). In Figures 12.20 and 12.21, we display these properties for the upper and lower surfaces of AG24 airfoil for these two Reynolds number when the airfoil is kept at $\alpha = 0$. The δ^* and θ on the top and bottom surfaces show correspondence with the C_p data shown in Figures 12.15 and 12.19. As we have also noted earlier from inviscid pressure distribution that for $\alpha = 0$, both the surfaces of the airfoil experiences favourable pressure gradient from $x = 0.05c$ on the upper surface and $x = 0.15c$ on the lower surface. However for the lower Re, δ^* is much higher, as compared to the higher Re case on both

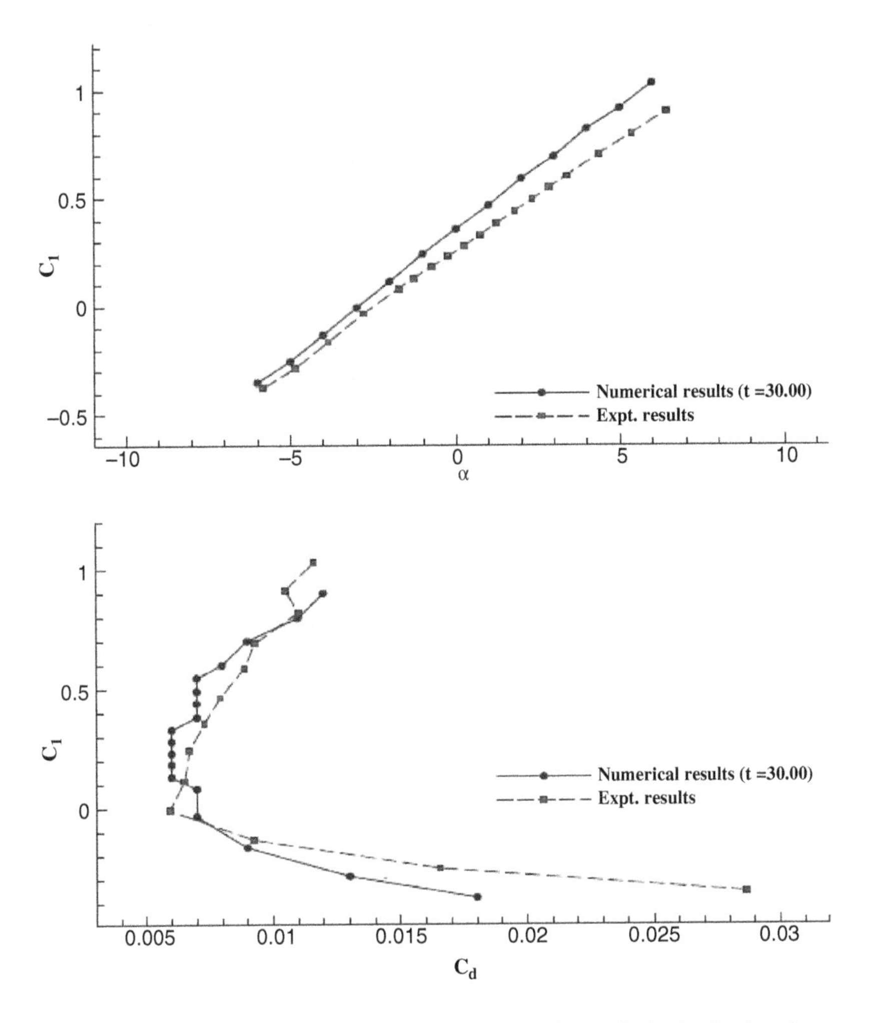

Figure 12.22 Comparison of lift curve (top) and drag polar (bottom) obtained using time-averaged pressure values till $t = 50$ for AG24 airfoil for $Re = 400\,000$ with the experimental results, when no FST has been used

the surfaces. The δ^* on the bottom surface is significantly higher, as compared to that on the upper surface, despite the fact that both the surfaces experience favourable pressure gradient. This is a special feature of low *Re* aerodynamics as compared to what is noted for higher Reynolds numbers. Such a feature is also noticeable when one looks at the shape factor (H). We know that for zero pressure gradient boundary layer, the laminar flow has a shape factor close to 2.6, which decreases to half its value following flow transition. Here, for these low *Re* flows the airfoil exhibit varying pressure gradient and as a consequence, the value of H is significantly different for upper and lower surfaces. The upper surface value remains close to the value for Blasius boundary layer, while the lower surface value is closer to corresponding turbulent flows. The oscillations noted in all the properties including the pressure gradient is indicative of unsteady separation.

The flow at higher Re has greater energy to overcome the adverse pressure gradient and hence laminar separation bubble is greatly reduced. The formation of separation bubble is delayed and the bubble size is also reduced, by ensuring quicker transition to turbulence and thus helping quicker reattachment. In Figure 12.22, the lift curve and drag polar are shown for AG24 airfoil for $Re = 400\,000$ using time-averaged lift and drag coefficients for the considered angles of attack for $0 \leq t \leq 50$. Numerical results are compared with experimental results in Williamson *et al.* (2012). It is notable that a better match between these can be observed compared to $Re = 60\,000$. Here also, it can be seen that the numerically obtained C_d is slightly over-predicted in the computations, for negative values of α. Aerodynamic performance parameter, i.e. the L/D ratio, shows a reasonable increase for $Re = 400\,000$, owing to drastic decrease in C_d, though there is also a marginal loss of lift noted in C_l. This decrease in lift can be associated with the loss in nonlinear lift due to reduced separation bubble formation and vortex interaction. This results in a reduction in induced drag. The reduction in drag can also be attributed to a reduction in effective profile, thus reducing profile or pressure drag.

The difference between numerically obtained value and the experimentally available results can be attributed to the effects of FST, which is present inherently in every experimental setup. The decrease in this difference with increase in Re can be correlated to the inverse dependence of FST on Re or speed. As such, an accurate modelling of FST proves to be indispensable in solving the NSE and subsequent flow predictions at low Re. The effects of FST on low Re is beyond the scope of this book and not dealt with here.

Bibliography

Abdallah, S. (1987) Numerical solutions for the pressure Poisson equation with Neumann boundary conditions using a nonstaggered grid-I. *J. Comput. Phys.*, **70**, 182–192.

Bagade, P.M., Krishnan, S.B. and Sengupta, T.K. (2014) DNS of low Reynolds number aerodynamics in the presence of free stream turbulence. (Private communication).

Berton, E., Maresca, C. and Favier, D. (2004) A new experimental method for determining local airloads on rotor blades in forward flight. *Exp. Fluids*, **37**(3), 455–457.

Bousman, W.G. (2000) Airfoil dynamic stall and rotorcraft maneuverability. *NASA TM* **2000-209601**.

Bhumkar, Y.G. (2012) *High Performance Computing of Bypass Transition*, Ph.D. thesis, Aerospace Engineering Dept., IIT Kanpur.

Broeren, A.P. and Bragg, M.B. (2001) Unsteady stalling characteristics of thin airfoils at low Reynolds number. In *Fixed and Flapping Wing Aerodynamics for Micro-Air Vehicle Applications* (Ed. TJ Mueller), **195** 191–213 Reston, VA, USA.

Carmichael, B.H. (1981) Low Reynolds number airfoil survey. **1** *NACA CR* **165803**

Carr, L.W. (1988) Progress in analysis and prediction of dynamic stall. *J. Aircraft*, **25**(1), 6–17.

DeLaurier, J.D. (1993) An aerodynamic model for flapping wing flight. *Aeronaut. J.*, **97**, 125–130.

Freymuth, P. (1990) Thrust generation by an airfoil in hover mode. *Expts. Fluids*, **9**, 17–24.

Gustafson, K. and Leben, R. (1991) Computation of dragonfly aerodynamics. *Comput. Phys. Commun.*, *65*, 121–132.

Gustafson, K., Leben, R. and McArthur, J. (1992) Lift and thrust generation by an airfoil in hover modes, *Comput. Fluid Dyn. J.*, **1**(1), 47–57.

Ho, S., Nassef, H., Pornsinsirirak, N., Tai, Y.-C. and Ho, C.M. (2003) Unsteady aerodynamics and flow control for flapping wing flyers. *Prog. Aerospace Sci.* **39**, 635–681.

Jones, K.D., Castro, B.M., Mahmoud, O., Pollard, S.J., Platzer, M.F., Neef, M.F., Gonet, K. and Hummel, D. (2002) A collaborative numerical and experimental investigation of flapping-wing propulsion. *AIAA Paper* 2002-0706.

Krishnan, S.B. (2013) *Low Speed Aerodynamics and Effects of Free Stream Turbulence*, M.Tech. thesis, Aerospace Engineering Dept., IIT Kanpur.

Lian, Y., Shyy, W., Viieru, D. and Zhang, B. (2003) Membrane wing aerodynamics for micro air vehicles. *Prog. Aerosp. Sci.*, **39**, 425–465.

McCroskey, W.J. (1982) Unsteady airfoils. *Ann. Rev. Fluid Mech.*, **24**, 285–311.

Morikawa, K. and Grönig, H. (1995) Formation and structure of vortex systems around a translating and oscillating airfoil. *Z. Flugwiss. Weltraumforsch.*, **19**, 391–396.

Mueller, T.J. (2000) Aerodynamic measurements at low Reynolds number for fixed wing micro-air vehicles. In *Development and Operation of UAVs for military and Civil Applications*. Quebec: Canada Commun. Group Inc. 302.

Nachtigall, W. (1974) *Insects in Flight*. Mc Graw-Hill, New York, USA.

Norberg, U.M. (1990) *Vertebrate Flight: Mechanics, Physiology, Morphology, Ecology and Evolution*. Springer Verlag, New York, USA.

Pelletier, A. and Mueller, T.J. (2000) Low Reynolds number aerodynamics of low-aspect-ratio, thin/ flat/ cambered plate wings. *J. Aircraft*, **37**, 825–832.

Pennycuick, C.J. (1997) Actual and 'optimum' flight speeds: field data reassessed. *J. Exp. Biol.* **200**, 2355–2361.

Rogers, S.E. and Kwak, D. (1990) Upwind differencing scheme for the time-accurate incompressible Navier-Stokes equation. *AIAA J.* **28**, 253–262.

Rozhdestvensky, K.V. and Ryzhov, V.A. (2003) Aerohydrodynamics of flapping-wing propulsors. *Prog. Aerospace Sci.* **39**, 585–633.

Sane, S.P. (2003) The aerodynamics of insect flight. *J. Exp. Biol.* **206**, 4191–4208.

Sengupta, T.K. (2012) *Instabilities of Flows and Transition to Turbulence*. CRC Press, USA.

Sengupta, T.K. (2013) *High Accuracy Computing Methods: Fluid Flows and Wave Phenomena* Cambridge University Press, USA.

Sengupta, T.K. and Bhumkar, Y.G. (2010) New explicit two-dimensional higher order filters. *Comput. Fluids*, **39**, 1848–1863.

Sengupta, T.K., De, S. and Sarkar, S. (2003) Vortex-induced instability of incompressible wall-bounded shear layer. *J. Fluid Mech.*, **493**, 277–286.

Sengupta, T.K. and Dipankar, A. (2004) A comparative study of time advancement methods for solving Navier-Stokes equations. *J. Sci. Comp.* **21**(2), 225–250.

Sengupta, T.K., Dipankar, A. and Rao, A.K. (2007) A new compact scheme for parallel computing using domain decomposition. *J. Comput. Physics*, **220**, 654–677.

Sengupta, T.K., Lim, T.T., Sajjan, S.V., Ganesh, S. and Soria, J. (2007a) Accelerated flow past a symmetric aerofoil: Experiments and computations. *J. Fluid Mech.*, **591**, 255–288.

Sengupta, T.K., Vikas, V. and Johri, A. (2005) An improved method for calculating flow past flapping and hovering airfoils. *Theo. and Comput. Fluid Dyn.* **19**(6), 417–440.

Spedding, G.R. (1992) The aerodynamics of flight. *Adv. Comp. Env. Physiol.*, **11**, 51–111.

Sun, M. and Tang, J. (2002) Lift and power requirements of hovering flight in Drosophila virilis. *J. Exp. Biol.*, **205**, 2413–2427.

Tobalske, B.W. and Dial, K.P. (1996) Flight kinematics of black-billed magpies and pigeons over a wide range of speed. *J. Exp. Biol.*, **199**, 263–280.

Van Der Vorst, H. (1992) A Bi-CGSTAB: A fast and smoothly converging variant of Bi-CG for the solution of nonsymmetric linear systems. *SIAM J. Sci. Stat. Comput.* **13**(2), 631–644.

Wang, Z.J. (2000) Vortex shedding and frequency selection in flapping flight. *J. Fluid Mech.*, **410**, 323–341.

Williamson, G.A., McGranahan, B.D., Broughton, B.A., Deters, R.W., Brandt, J.B. and Selig, M.S. (2012) *Summary of Low Speed Airfoil Data*, **5**, 301–302, Dept. of Aerospace Engineering, University of Illinois at Urbana-Champaign.

13

High Lift Devices and Flow Control

13.1 Introduction

For low speed level flight of an aircraft in the longitudinal plane, the resultant force acting on the aerodynamic surfaces usually consists of two components: lift and drag. Despite such demarcation, it is to be remembered that these two components are dependent. This is very well demonstrated by the relation between drag and lift given by: $D = c_1 q + c_2 L^2/q$, where q is the dynamic pressure, $1/2\rho U_\infty^2$ and c_1, c_2 depend on the aircraft geometry. The first term is the parasitic drag expression contributed by viscous effects in terms of skin friction and pressure distribution. The second term has been reasoned in Chapter 1 as lift dependent drag, which has been explained in greater detail in Chapter 4, as the induced drag due to finite aspect ratio of the wing.

For level flight at nonaccelerating speeds, the lift equals the weight of the aircraft and one obtains minimum drag with respect to speed (U_∞) from

$$\frac{dD}{dV} = \rho c_1 U_\infty - \frac{2c_2 W^2}{(\rho U_\infty^3)} = 0. \tag{13.1}$$

This gives the optimum dynamic pressure as, $q_o = W\sqrt{2c_2/c_1}$, which gives identical contribution from parasitic and induced drag. That is, the minimum drag speed corresponds to when lift dependent drag constitute 50% of the minimum drag, i.e. for L/D maximum, we have maximum aerodynamic efficiency. However, for an aircraft cruising at higher speeds than this optimum speed, the induced drag is approximately about 40% of the total drag.

In Table 13.1, profile drag breakdown for a complete transport jet aircraft is provided considering the flow over the lifting surfaces to be either fully turbulent or fully laminar. This table tells us that it is possible to reduce drag on the complete aircraft by drag reduction on

Theoretical and Computational Aerodynamics, First Edition. Tapan K. Sengupta.

Companion Website: www.wiley.com/go/sengupta

Table 13.1 Profile drag break-up of a transport jet aircraft (Bushnell and Heffner (1990))

	Fully turbulent flow condition for all surfaces		Fully laminar flow on lifting surfaces only	
	Drag count	Percentage	Drag count	Percentage
Wing	0.0060	31.8%	0.0020	15.3%
Fuselage	0.0092	48.7%	0.0092	70.2%
Empennage	0.0027	14.3%	0.0009	6.9%
Nacelle/misc.	0.0010	5.2%	0.0010	7.6%
Total profile drag	0.0189		0.0131	

Source: Reproduced with permission from the American Institute of Aeronautics and Astronautics, Inc.

other components also, apart from reducing drag over the lifting surfaces, e.g. reducing drag of the fuselage and nacelle and this is also discussed here.

If the flow is fully turbulent then the fuselage accounts for 48.7% of total drag. If the flow is constrained to be fully laminar on the lifting surfaces only, then the fuselage contribution becomes even higher to 70.2% of the total drag. This makes a stronger case for investigating ways and means for reducing fuselage and nacelle drags. For example, Bushnell and Hefner (1990) report that an axisymmetric body at zero angle of attack for a Reynolds number of 40.86 million, has a drag coefficient of 0.0260 when the point of transition is located at $x/c = 0.322$. When the transition location point advances to $x/c = 0.05$, the drag coefficient increases to 0.0374 – an increase of about 44%.

Although from aerodynamic efficiency point of view, the desirable design goal would be to maximize L/D, yet flying at such optimum speed derived using Equation (13.1) would not be practical. This is simply due to the fact that the objective function is a combination of multiple requirements of logistics, economics etc. For example, the time taken to travel from one point to other has to be also optimized and which would require flying at a speed higher than that is given by minimizing drag, as in Equation (13.1).

At lower speeds during landing and take-off, situation is completely different, when the aircraft produces more lift than during cruise and hence the induced drag contributes more to the drag and it could be as high as 90% of the total. A good review of induced drag and its prediction is given in Kroo (2001). Higher lift is essential during landing and take-off due to constraints of speed limit during landing and take-off, which in turn is dictated by runway length. Thus, high lift requirement for aircraft design is one of necessity and not an option.

13.1.1 High Lift Configuration

For an aircraft flying at a given altitude and at constant speed, higher lift can be obtained by a combination of the following ways:

(a) It can be increased by increasing the wing area, as the lift can be expressed in terms of the lift coefficient as, $L = 1/2\rho_\infty U_\infty^2 SC_L$. Thus, for fixed ρ_∞ and U_∞, lift can be augmented by increasing S, if there are no means to change C_L. In defining the lift in this manner, one usually takes the wing planform as the reference area. There are structural and aerodynamical considerations by which one is limited to increasing wing area. The structural reason is related to increased weight versus the added lift obtained. Also, increased structural weight can lead to increased bending moments. To mitigate the same, one would require thicker sections near the

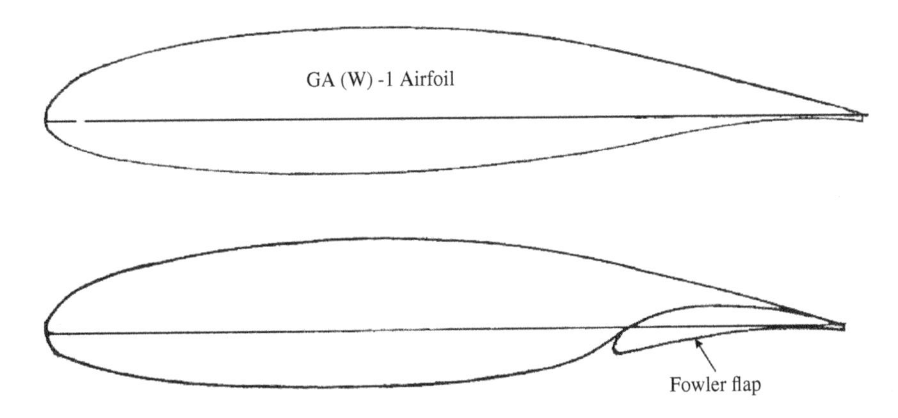

Figure 13.1 Single element GA(W)-1 airfoil on top and the same airfoil with a nested Fowler flap shown in the bottom

wing root adding further structural weight. The aerodynamic reason is related to increasing the planform area and thereby increasing lift, yet avoiding the possibility of flow separation. We will shortly see how this is attained in actual aircraft wing, by introducing high lift devices, which also increases area, as in the Fowler flap, apart from changing the effective airfoil section. A sketch depicting the geometry of the Fowler flap are shown in Figure 13.1, as fitted to GA(W)-1 airfoil section, taken from Wentz and Seetharam (1974). Fowler flaps increase wing area during landing. The experimental results presented in Wentz and Seetharam (1974) provide a typical example of Fowler flap deployed with a chord 91.4 cm, while the nested flap airfoil chord is given as 61.0 cm, thereby indicating a 50% increase in chord of the wing covered by the flap.

(b) The lift experienced by the wing can be increased by a choice of the basic airfoil section, which can mean an increase in equivalent lift via the improved combination of thickness and camber distributions. This essentially boils down to designing the basic single element airfoil, so that the resultant pressure distribution creates as high a lift as possible, without the adverse effects of sectional pitching moment coefficient. Such desirable pressure distribution must be obtained by controlling the developing boundary layer over the suction surface. Also, such an acceleration should not lead the flow to become supercritical so that shock waves form over the upper surface, increasing drag in the form of wave drag. All the early airfoil designs took into consideration avoiding flow separation for high lift flight conditions. Post-Second World War development in aeronautics witnessed spectacular increase in flight speed due to advent of jet engines. This necessitated development of shock-free airfoil for transonic flow regime. It should be noted that transonic flow regime is most aerodynamically efficient, provided one can control wave drag. This was achieved by the design of supercritical airfoils. One of the pioneering effort in this direction was by R. T. Whitcomb and his team at NASA Langley Research Center. This entails creating a roof-top pressure distribution over the upper surface of the airfoil. Such pressure distribution also displays lack of adverse pressure gradient which is a strong precursor of transition to turbulence. The common element of lack of adverse pressure gradient is thus common to supercritical airfoil and NLF airfoil. One such airfoil was designed for the purpose of general aviation starting from a supercritical airfoil and is

Table 13.2 GA(W)-1 airfoil coordinates

Upper surface		Lower surface	
x/c	z/c	x/c	z/c
0.00000	0.00000	0.00000	0.00000
.00200	.01300	.00200	−.00930
.00500	.02040	.00500	−.01380
.01250	.03070	.01250	−.02050
.02500	.04170	.02500	−.02690
.03750	.04965	.03750	−.03190
.05000	.05589	.05000	−.03580
.07500	.06551	.07500	−.04210
.10000	.07300	.10000	−.04700
.12500	.07900	.12500	−.05100
.15000	.08400	.15000	−.05430
.17500	.08840	.17500	−.05700
.20000	.09200	.20000	−.05930
.25000	.09770	.25000	−.06270
.30000	.10160	.30000	−.06450
.35000	.10400	.35000	−.06520
.40000	.10491	.40000	−.06490
.45000	.10445	.45000	−.06350
.50000	.10258	.50000	−.06100
.55000	.09910	.55000	−.05700
.57500	.09668	.57500	−.05400
.60000	.09371	.60000	−.05080
.62500	.09006	.62500	−0.4690
.65000	.08599	.65000	−.04280
.67500	.08136	.67500	−.38400
.70000	.07634	.70000	−.03400
.72500	.07092	.72500	−.02940
.75000	.06513	.75000	−.02490
.77500	.05907	.77500	−.02040
.80000	.05286	.80000	−.01600
.82500	.04646	.82500	−.01200
.85000	.03988	.85000	−.00860
.87500	.03315	.87500	−.00580
.90000	.02639	.90000	−.00360
.92500	.01961	.92500	−.00250
.95000	.01287	.95000	−.00260
.97500	.00609	.97500	−.00400
1.00000	−.00070	1.00000	−.00800

shown in Figure 13.1, known as the GA(W)-1 airfoil. The letter W in the parenthesis is in honour of important contributions made by Whitcomb on different aspects of compressible aerodynamics.

The coordinates of the GA(W)-1 airfoil is provided in Table 13.2, which represents a 17% thickness to chord ratio section, which has a significant camber near the trailing edge with a cruise design lift coefficient of 0.4, with a wide drag-bucket. This latter feature of optimizing the airfoil over a wide range of C_l is the special design practice of all modern day airfoils, distinct from the six-digit series airfoils designed in NACA. It is also important to note that the airfoil has a cusped trailing edge, reminiscent of Jukowski airfoil. Such parallel trailing edge is specifically designed to reduce the adverse pressure gradient in this region to delay flow separation. However, we also point out some critical issues with such cusped aerofoil.

Readers' attention is drawn to the review paper by Smith (1975), which contains discussion on high lift aerodynamics. In this reference, using panel methods the author tries to explain the operation of multi-element airfoils and is explained later. Adverse pressure gradient cannot be completely ruled out due to the requirement of the flow upon acceleration near the leading edge of the airfoil must relax back to the free stream condition upon arrival at the trailing edge. One looks for an optimum pressure recovery in the rear half of the airfoil and one such distribution is due to Stratford (1959) for turbulent flow pressure distribution, which always remains on the verge of separation.

(c) Lift can also be maximized by active control of the flow over the airfoil by application of suction, blowing or other forms of control, which in essence injects higher momentum fluid close to the airfoil surface to avoid separation and transition. In the following, we will mostly discuss passive control of flow, as well as active control by plasma actuation, a topic of current research interest yet to be tested on a prototype aircraft.

13.2 Passive Devices: Multi-Element Airfoils with Slats and Flaps

In Figure 13.2, we show explanatory sketches of multi-element airfoils. On the top frame, a schematic shows a four-element airfoil consisting of (i) main airfoil, (ii) a slat ahead of the

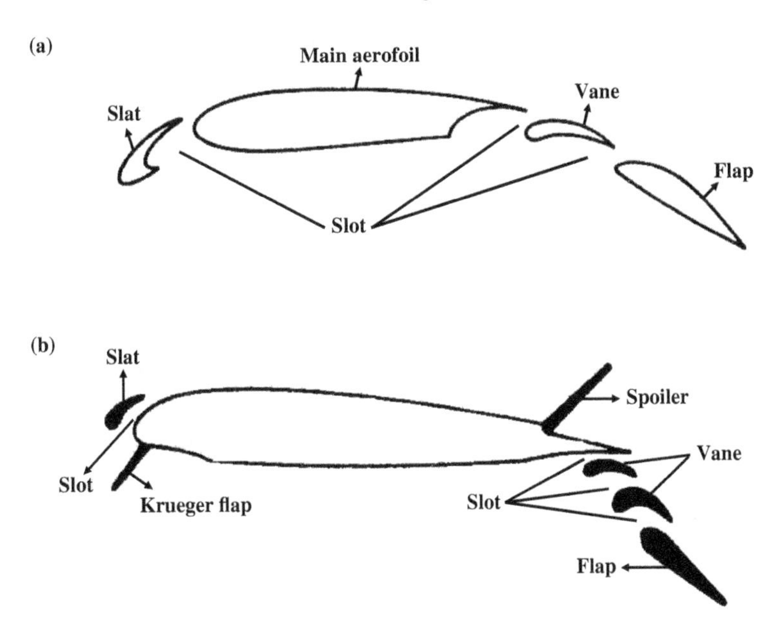

Figure 13.2 (a) Schematic of a typical four-element airfoil and (b) an operational multi-element airfoil used in an aircraft utilizing all possible slats, Krueger flap, multiple vanes and a flap

main airfoil; (iii) a vane which is just aft of the main airfoil and which is followed by the flap, all shown in a hypothetically deployed condition. In undeployed condition, all the elements are expected to be nested together to form an integral unit, which will appear as a single element airfoil. The sketch only shows a cross-section and how these elements are knitted together is one of mechanical description involving detailed aerodynamical analysis and optimal functioning of each unit. We avoid detailed description of operational deployment aspect of multi-element lifting surfaces, to focus upon the basic aerodynamic principles. However, multi-element aerofoil requires a more detailed understanding of aerodynamical features of individual elements, as well as their functioning in conjunction. In the deployed condition, air is allowed to flow between successive element through the indicated slots. The flow over the top and bottom surfaces therefore get closely coupled and conventional approaches of viscous and inviscid flow analysis becomes difficult, as we will indicate here for a Fowler flap. Current state of art in development of high lift devices depend on qualitative understanding of flow, which is supplemented by extensive testing. Thus, a primary emphasis here is on following some experimental results in detail.

In Figure 13.2(b), a more complex operational multi-element system is shown, which is typical of some commercial aircrafts. Additionally, one notices here the presence of the Krueger flap, which though it may appear to be similar to slats or slots, is deployed differently. This device is hinged at its leading edge and rotates in a manner by which it moves forward, remaining under the wing. The slat on the contrary extends forward, remaining above the upper surface. The Krueger flap is an appendage and the main wing upper surface is not changed. This is usually used between the fuselage and the nearest engine to it. This multi-element airfoil in Figure 13.2(b), is also characterized by the presence of two vanes, which creates three slots near the trailing edge, apart from another slot near the leading edge created by the slat. While flaps sit in the main airfoil in a cove, the slat snug-fits the leading edge of the main airfoil without causing significant discontinuity, in nested condition.

To understand the aerodynamics of multi-element aerofoil in the following, we first describe a two-element airfoil, in which the GA(W)-1 airfoil is fitted with a $0.29c$ Fowler flap and the results are from Wentz and Seetharam (1974) tested for *Re* in the range of 2.2 million to 2.9 million based on the chord (c). Such an optimized flap system is shown to generate a $C_{l,max} = 3.8$ for a $0.30c$ Fowler flap. The geometric arrangement is shown in Figure 13.3, showing the nested flap and the deployed flap at an angle of $\delta_f = 10°$. The coordinates of the flap are given in Table 13.3. As mentioned earlier, when the flap is deployed it leads to increase in wing area, as well as a change in camber. One of the feature of this flap system is that the upper surface does not change all the way up to $0.96c$, where it creates a step, followed by which the flap forms the last 4% of the airfoil upper surface in nested condition. It is readily evident that on the lower surface, the junction of the flap with the main airfoil creates a discontinuity. To avoid the problem of uncertainties related to transition point for different physical and geometric conditions, the point of transition was fixed at $x/c = 0.05$ on both upper and lower surfaces by using 2.5 mm wide strips of 80% carborundum grit for the results presented in Wentz and Seetharam (1974).

13.2.1 Optimization of Flap Placement and Settings

The flap settings have to be optimized for different segment of flight envelope. For example, during take-off one must have sufficient $C_{l,max}$ at an angle of attack which is achieved by

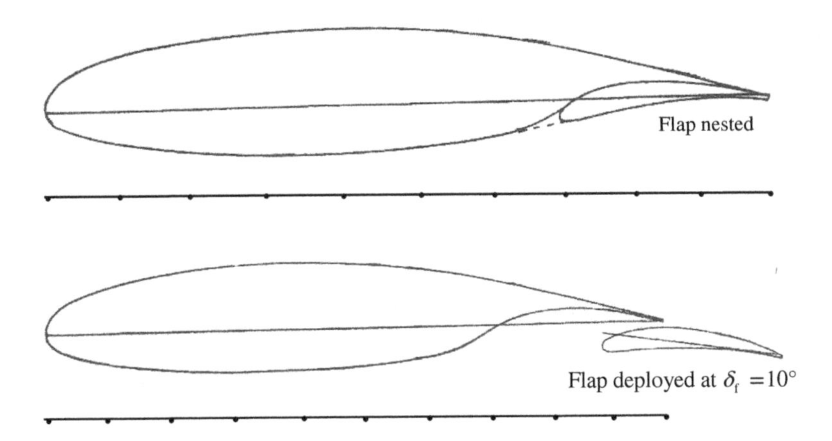

Figure 13.3 GA(W)-1 airfoil with nested Fowler flap (on top) and the same flap deployed at $\delta_f = 10°$ shown in the bottom

rotation of the complete aircraft about the main gear, without fuselage coming in contact with the runway. As noted in Wentz and Seetharam (1974),

> 'it is desirable to have maximum airplane lift-drag ratio occur at climb lift coefficient or higher in order to avoid the difficulties associated with flight operations in the ''region of reversed command'' or ''back side of power curve''.'

The region of reverse command means flight in which a higher air speed requires lower power setting and vice versa, for constant altitude steady flight. At flight speeds below the speed for maximum endurance requires higher power settings with a decrease in air speed. As the need to increase power to reduce speed is counter-intuitive, the regime of flight speeds between the speed for minimum required power and stalling speed is termed the region of reversed command.

While one is talking of airplane L/D ratio, which is reflected in the airplane drag polar, it is understood that the basic airfoil must have maximum C_l/C_d. To achieve higher L/D during climb, it is imperative that the drag contribution from fuselage and nacelle must be minimum and this is achieved when the angle of incidence for fuselage and nacelle is kept as close to zero as possible. During aircraft touchdown, one requires shortest runway length and hence have low approach speed, which dictates achieving $C_{L,max}$. However during landing, it is necessary to have higher drag also for deceleration. This is achieved by deploying spoilers, simultaneously with the flaps deflected for very high lift.

The flap deflections are in terms of rotations from the nested position about a pivot point which moves with the flap. Thus the flap position is defined in terms of two positional coordinates and the flap rotation about the pivot point. These are listed in Table 13.4 and the corresponding location for the pivot point is shown in Figure 13.4, by crosses for different flap rotation angle. These locations are the results of optimization process, which maximizes the lift generated upon deployment of the flap and various constraints of operation of the deployment mechanism and the overall operation of the aircraft. Also from Table 13.4, one notices that from the nested position ($\delta_f = 0°$) the pivot shifts from $x/c = 0.730$ to $x/c = 0.880$ for a flap

Table 13.3 29% c Fowler flap configuration

Upper surface		Lower surface	
X_f/c	Z_f/c	X_f/c	Z_f/c
0.00000	−.02350	0.00000	−.02350
.00030	−.02000	.00100	−.02700
.00200	−.01790	.00200	−.02880
.00400	−.01550	.00400	−.03000
.00800	−.01130	.00800	−.03100
.01200	−.00780	.01200	−.03040
.01800	−.00330	.02000	−.02880
.02300	.00000	.03000	−.02700
.02800	.00230	.05000	−.02350
.03800	.00700	.07000	−.01980
.04800	.01100	.09000	−.01600
.05800	.01410	.11000	−.01300
.06800	.01680	.13000	−.01000
.07800	.01900	.15000	−.00770
.08800	.02070	.17000	−.00580
.09800	.02180	.19000	−.00360
.10800	.02230	.21000	−.00270
.11800	.02280	.23000	−.00280
.12800	.02300	.25000	−.00350
.13800	.02340	.27000	−.00500
.14800	.02280	.29000	−.00800
.15800	.02230		
.16800	.02190		
.19000	.01980		
.21000	.01680		
.23000	.01380		
.25000	.00980		
.27000	.00590		
.29000	−.00070		

Table 13.4 Flap pivot point locations – computer design setting

δ_f	x/c	z/c
0° (Nested)	.730	−.40
10°	.880	−.061
15°	.900	−.055
20°	.920	−.049
25°	.930	−.046
30°	.940	−.043
40°	.950	−.040

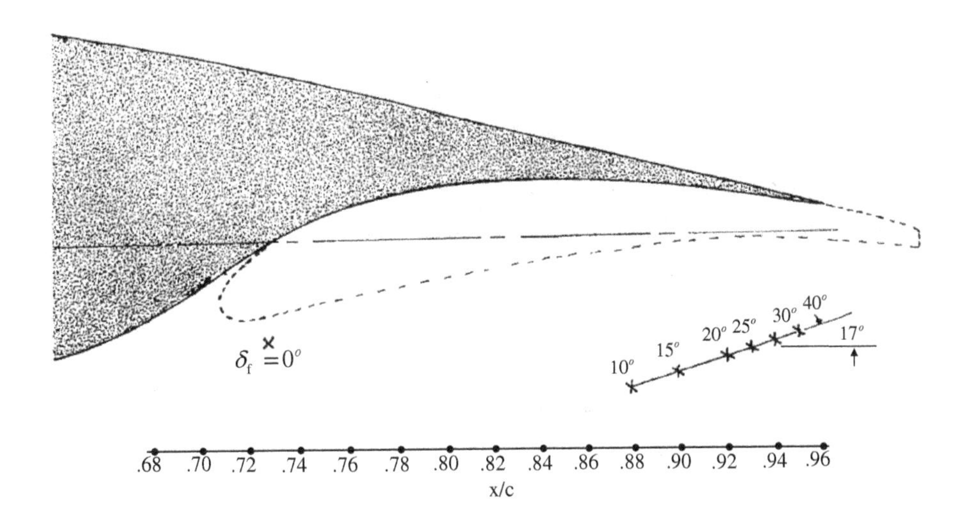

Figure 13.4 Flap pivot point locations for the Fowler flap attached to the GA(W)-1 airfoil

deflection of $\delta_f = 10°$. Thereafter the pivot location moves along a line which is inclined at 17° to the chord line.

13.2.2 Aerodynamic Data of GA(W)-1 Airfoil Fitted with Fowler Flap

In Figures 13.5 and 13.6, C_l and C_m variations with airfoil α are shown for the indicated δ_f for the indicated *Re* and *M* for the experiments in Wentz and Seetharam (1974). These figures also indicate the case with the flap nested. As compared to the nested flap case, the flap deflection causes the lift curve to shift laterally with $C_{l,max}$ progressively increasing with δ_f up to 40°. All the flap deflected cases show earlier stall at around 15° mean airfoil angle of attack, whereas for the nested case the stall angle is in excess of 20°. Also, for flapped airfoil cases, stall becomes progressively sharper with higher δ_f. For the basic thicker airfoil section, the lift curve slope does not show linear variation of C_l with α. The C_m for the nested flap case shows significant value for $C_l = 0$ case, which is a consequence of large camber near the trailing edge.

The coefficient of drag for the GA(W)-1 airfoil section is shown in Figure 13.7, for different δ_f as drag polar data. It is noted from the figure that progressively the rate of increase of C_d with C_l decreases as δ_f increases. It is noted that the increase of $C_{l,max}$ with δ_f is associated with increase in C_d. As observed earlier, such simultaneous increase in C_l and C_d should be desirable during landing.

It is noted in Wentz and Seetharam (1974) that the nested-flap and the $\delta_f = 40°$ cases did not show good match with theoretical procedure of solving the flow field by using viscous-inviscid interaction procedure given in Stevens *et al.* (1971) developed specifically for multi-element aerofoil. According to the authors,

> 'discrepancy between theory and experiment for the flap nested case is disturbing, in that the progressive loss of lift prior to stall indicates premature boundary layer thickening, possibly with separation.'

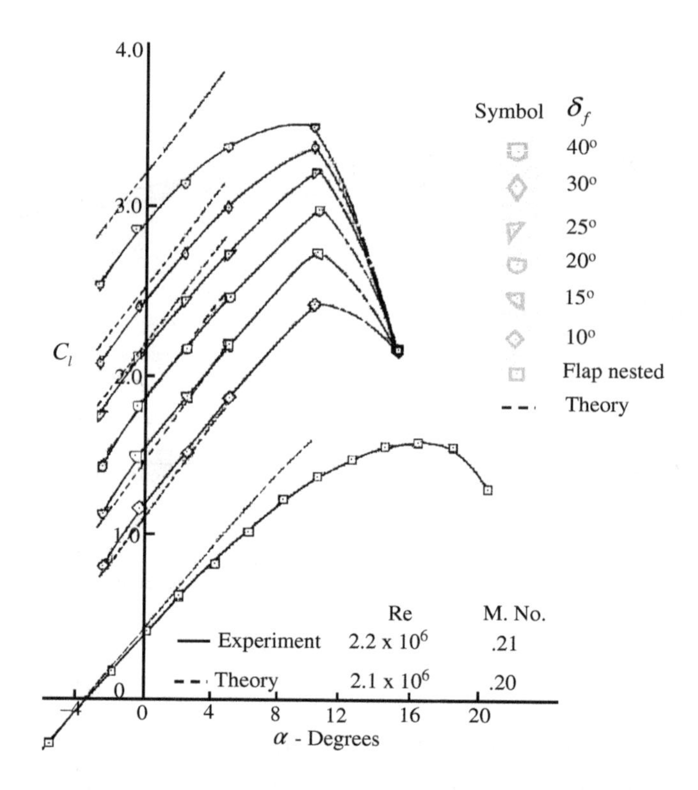

Figure 13.5 Lift coefficient curve for GA(W)-1 airfoil fitted with Fowler flap plotted as function of angle of attack of the airfoil for different angles of flap deflection as compared to the nested-flap case, for the indicated Reynolds and Mach numbers

The measured pressure distribution in Figure 13.8, clearly indicates the location of point of separation moving aftward with decrease in angle of attack on the upper surface. It is interesting to note that this point of separation on the upper surface is located at the flat roof-top pressure distribution with the express intention of preventing adverse pressure gradient and thereby also preventing flow transition for NLF application and shock formation for supercritical airfoil consideration. In this particular model tested in Wentz and Seetharam (1974), the trailing edge has finite thickness and the lower surface pressure distribution shows nonsmooth variation at the junction of the main airfoil and the Fowler flap, as indicated in the figure. This localized pressure variation on the lower surface is communicated to the upper surface causing flow separation, as shown in Figure 13.8.

The above discussion indicates the direct connection between the upper and lower surface, if the rear stagnation point is not fixed by having a sharp leading edge which isolates the flow over the two surfaces allowing the boundary layer over both the surfaces to develop independently in any viscous-inviscid interaction computing procedure, as in Stevens *et al.* (1971). The boundary value nature of the problem associated with Navier-Stokes equation in such procedures are evident only while solving the inviscid flow part by solving the Laplace's equation by panel methods described in Chapter 5. However, the presented experimental results clearly indicates the inadequacy of decoupling viscous and inviscid flows and supports

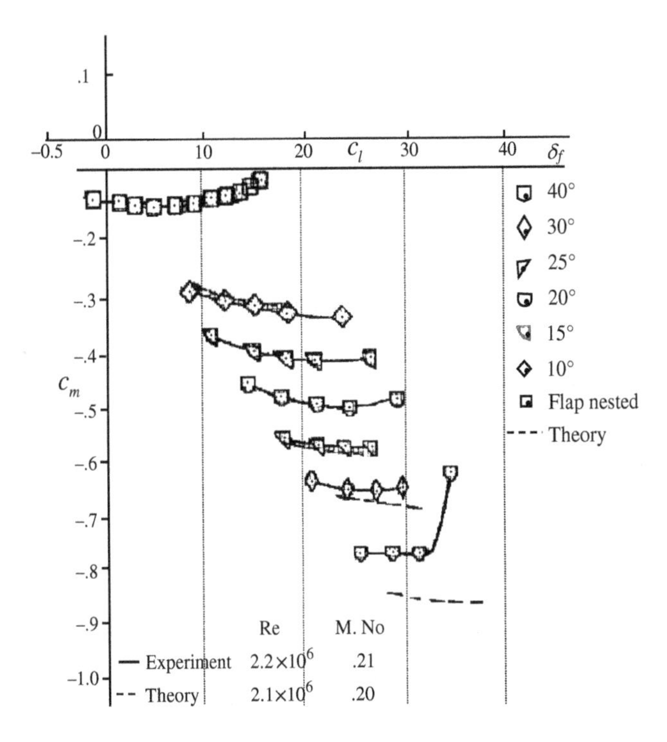

Figure 13.6 Pitching moment coefficient curve for GA(W)-1 airfoil fitted with Fowler flap plotted as function of angle of attack of the airfoil for different angles of flap deflection compared to the nested-flap case, for the indicated Reynolds and Mach numbers

the need to solve full Navier-Stokes equation. Physically also, one notices the importance of designing the rear part of the airfoil section, as it has global effects.

This is particularly important about the large camber provided near the trailing edge. In Figure 13.9, we show the performance of the GA(W)-1 airfoil fitted with the Fowler flap, when the rear bottom surface is reconstructed to remove this camber, as well as, the discontinuity at the juncture of Fowler flap and the main airfoil. This recontoured airfoil acts like a single element airfoil, whose aerodynamic properties are compared with the basic GA(W)-1 airfoil. From the lift curve plot versus α, one notices degradation of lift generation for all angles of attack, which is severely noted in the value of $C_{l,max}$, although the stall angle remains the same. The drag polar and the pitching moment coefficient also shows significant degradation.

The deflection of the Fowler flap is also defined in terms of slot gap and overlap, as shown in the top of Figure 13.10. Below this, the $C_{l,max}$ contours are shown with respect to slot gap and overlap expressed in percentage of chord, for the two flap deflection angles, $\delta_f = 35°$ and 40°, for seeking out optimum values. It is noted that a maximum of $C_{l,max} = 3.8$ is achieved for the higher flap angle case, which is marked in the frame.

13.2.3 Physical Explanation of Multi-element Aerofoil Operation

It is noted in Figure 13.2 that the deployment of slat, vane and flaps creates various slots, which alters the flow globally. In Nakayama *et al.* (1990), it has been shown that an optimum

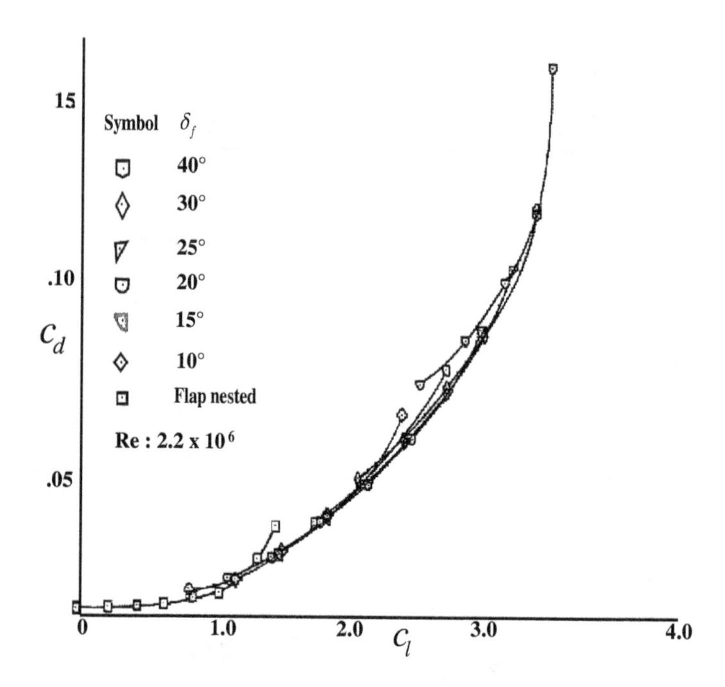

Figure 13.7 Drag coefficient curve for GA(W)-1 airfoil fitted with Fowler flap plotted as function of angle of attack of the airfoil for different angles of flap deflection compared to the nested-flap case, for the indicated Reynolds and Mach numbers

performance of multi-element airfoil depends upon the boundary layers emanating from different elements to flow confluently without interacting with each other. This behaviour was attributed by earlier researchers to fluids at higher pressure below the airfoil to create a flow towards the suction side of the airfoil. However, it has also been noted in Houghton and Carpenter (1993) that both the side have the same total pressure, i.e. the energy is same on upper and lower side which cannot be the reason for the flow through the slots. Instead, flow through the slot is created by vortical effects caused by the localized vortices created at all the elements. For example, the slat which is ahead of the main airfoil is the site of a clockwise vortex which is placed more towards the upper surface. Such a vortex creates an induced velocity via Biot-Savart interaction to weaken the suction peak and hence reduces the causation of flow separation tendency on the upper surface. At the same time, Biot-Savart interaction by the slat-vortex further lowers the local velocity on the lower surface. Thus while the upper surface velocity distribution causes a loss of lift, the same is more than compensated on the lower surface. One can similarly explain the increase in aerodynamic efficiency caused by vanes and flaps behind the main airfoil. This is how the action of multi-element airfoil experiencing its higher lift was explained by Smith (1975) in terms of potential flow model. However, it is important to note that the action of localized vortex is to delay separation on critical surfaces. Hence study of multi-element airfoil is to be followed using viscous flow models and most preferably by solving Navier-Stokes equation. In the following, we therefore cite another example of action of localized vortex created by specially designed passive vortex generators.

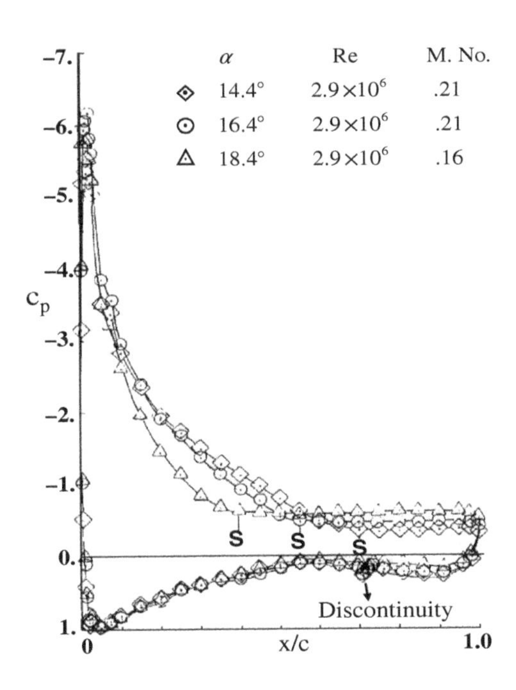

Figure 13.8 Measured pressure distribution for GA(W)-1 airfoil [in *Source:* Wentz, Jr and Seetharam (1974). Courtesy of NASA] with the Fowler flap nested. Results are shown for three angles of attack for the indicated Reynolds and Mach number. S indicates the location of point of separation noted from visualization of tufts fitted on airfoil surface

13.2.4 Vortex Generator

These devices are mostly used for use on sweptback wings. However, we first show its applications in controlling sectional properties. In Figure 13.11, sectional properties of NLF(1)-0414F airfoil is shown for $Re = 6$ million, for which results were obtained in NASA Langley Low Turbulence Pressure Tunnel and reported in Murri *et al.* (1987). Here the effects of positioning a roughness element at $0.075c$ have been studied on lift, drag and pitching moment coefficients by comparing the case without the roughness element case. It is noted that the profile drag changes significantly, while lift and pitching moment coefficients do not change appreciably. However, the figure for C_l shows an interesting feature for the lift curve slope at and around $\alpha = 4°$, which reduces due to flow separation in the upper surface pressure recovery region. Sometimes this effect is purposely added as a design feature to control pitching moment, as in (Fujino *et al.* 2003) for SHM-1 airfoil. However if it is considered undesirable, then it can be eliminated by using boundary layer re-energizer or vortex generator placed on the suction surface.

It has been shown by Schubauer and Spangenberg (1958) that this separation in pressure recovery region can be controlled by employing simple plow and vortex generator, whose function is to scour the boundary layer by creating alternate strip of higher and lower velocity streaks. The higher velocity streaks have the tendency to entrain surrounding low momentum fluids. Such forced mixing causes the pressure recovery region to spread over twice the streamwise distance causing the concave pressure recovery milder, precluding turbulent

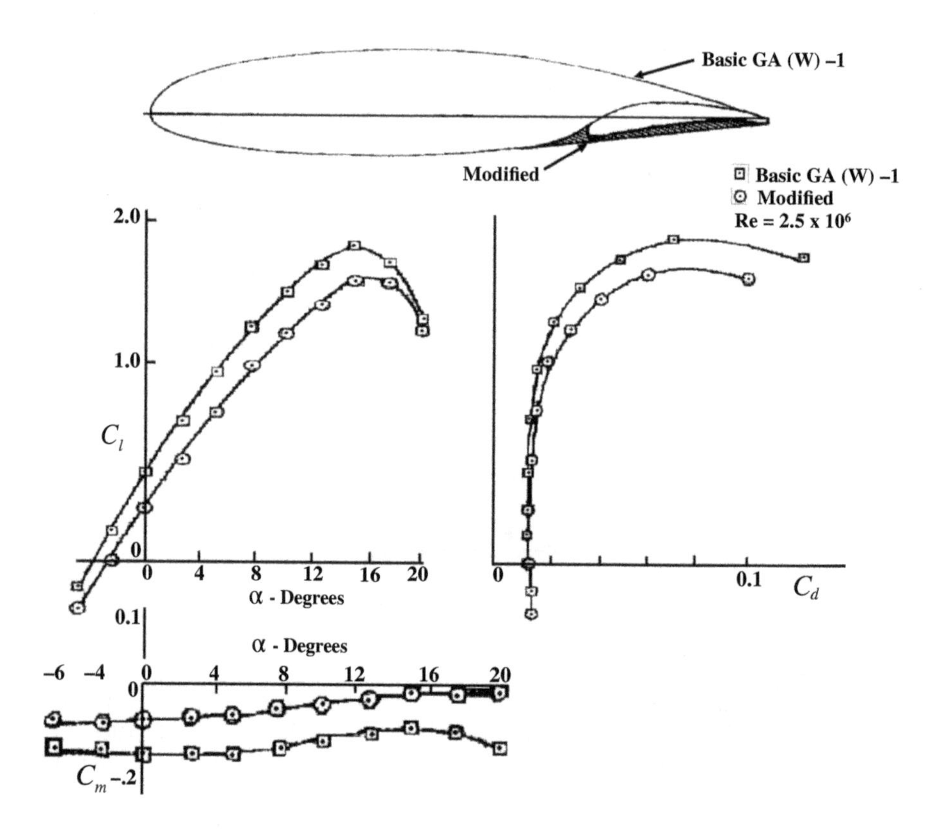

Figure 13.9 Effects of aft camber of GA(W)-1 airfoil shown in *Source:* Wentz, Jr and Seetharam (1974). Courtesy of NASA with the Fowler flap nested and bottom surface of the airfoil is reconstituted. Results are shown for $Re = 2.5 \times 10^6$ and the comparison of results with basic GA(W)-1 airfoil and the modified section clearly indicates the major role played by aft camber

separation. In Murri *et al.* (1987), results have been obtained using a vortex generator whose results are shown in Figure 13.12, with and without the vortex generator for $Re = 3 \times 10^6$. The vortex generators are low aspect ratio small wing-like protrusions positioned near $x = 0.6c$ on the suction surface, set at high angles of attack, with respect to local flow direction. The vortex generators are 0.2 inches high, spaced 1.6 inches apart. The use of vortex generators completely removes the reduction of lift curve slope, while $C_{l,max}$ remains more or less the same. Because of removal of turbulent flow separation, drag performance also improves above $\alpha = 4°$, while drag increases for lower angles of attack.

In another effort reported in Wentz and Seetharam (1974), the effect of vortex generator has been recorded to increase $C_{l,max}$, unlike the case shown in Figure 13.12 where addition of vortex generators did not alter $C_{l,max}$. In this case, the flap-nested condition indicate significant boundary layer thickening resulting in flow separation for $Re \leq 3 \times 10^6$. Using vortex generators in the form of half-delta wings protruding on the upper surface causes the desired increase in $C_{l,max}$, as shown in Figure 13.13. In this case also, the vortex generators were located at $x = 0.6c$, with semi-span of 7.62 mm and root chord of 2.54 cm. Such a geometry creates a leading edge sweep angle of 73.3°, and if one considers the flow to be 2D,

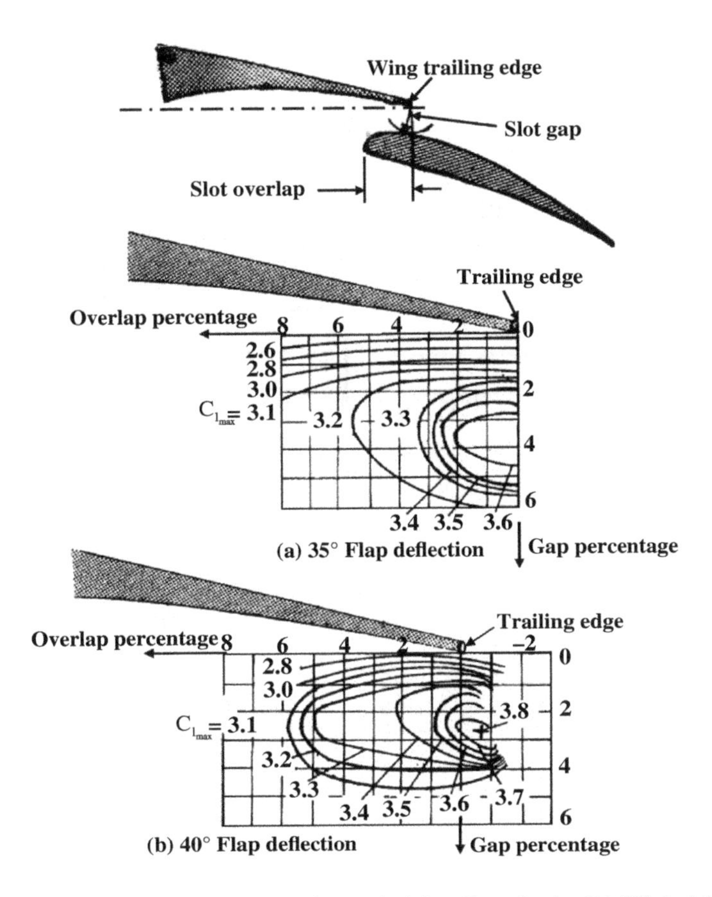

Figure 13.10 Optimization of slot geometry for maximizing $C_{l,max}$ for the GA(W)-1 airfoil in *Source:* Wentz, Jr and Seetharam (1974). Courtesy of NASA with a 0.30*c* Fowler flap shown for two large angles of flap deflection

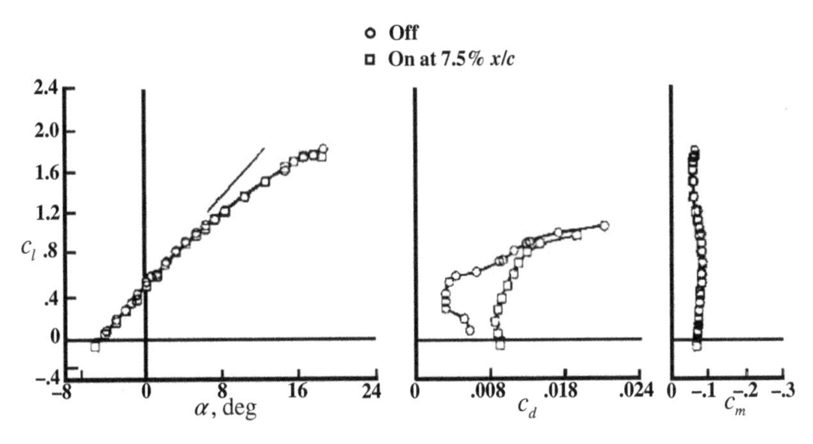

Figure 13.11 Effects of fixed roughness at $x = 0.075c$ for NLF(1)-0414F airfoil shown via the section properties for $Re = 6 \times 10^6$. Note the change in lift curve slope at 4° angle of attack

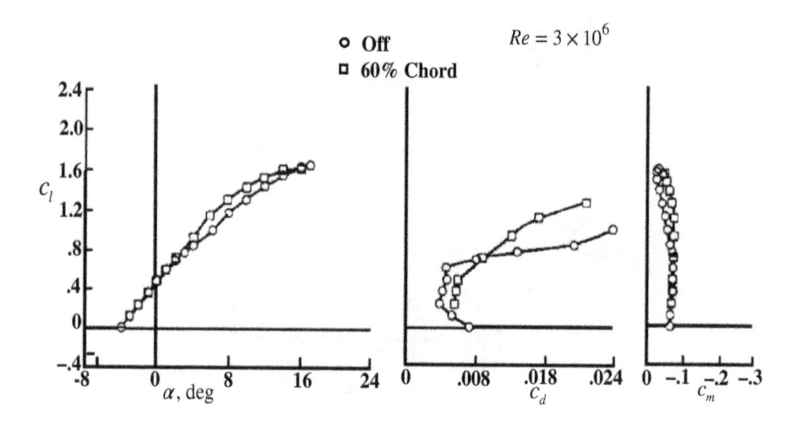

Figure 13.12 Effects of vortex generator used in the turbulent recovery region for NLF(1)-0414F airfoil for $Re = 3 \times 10^6$. Note the recovered lift curve slope above $\alpha = 4°$

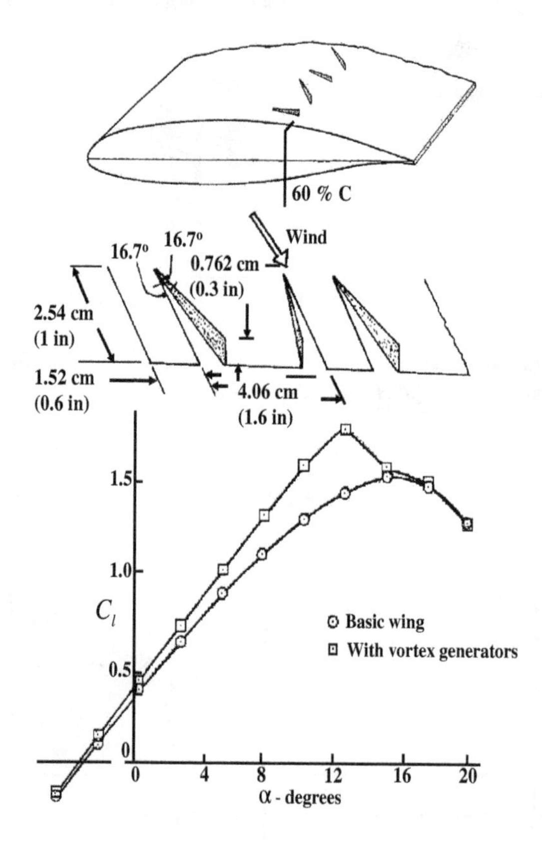

Figure 13.13 Effects of vortex generator used in increasing $C_{l,max}$ for the GA(W)-1 airfoil with nested Fowler flap. Note the drop in C_l above $\alpha = 13°$ and identical stall behaviour for the nested-flap airfoil, with and without vortex generator

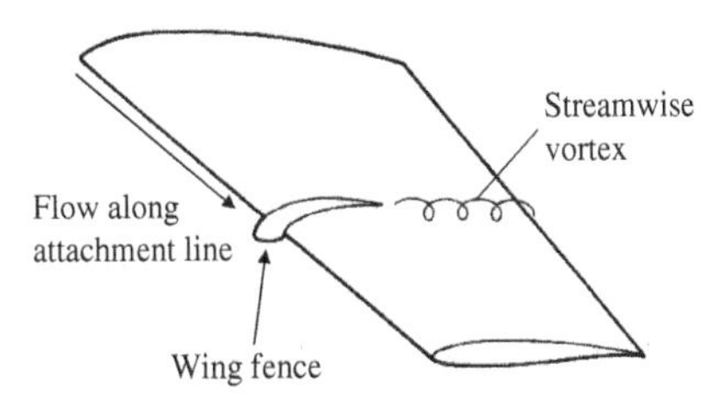

Figure 13.14 Wing fence attached to the leading edge of a swept wing, which prevents cross flow along the leading edge and issues a streamwise vortex over the main wing

as shown in the vortex generator details, then the half-delta wings experience a 16.7° angle of attack. The sizing of the half-delta is done on the basis of turbulent boundary layer at $x = 0.6c$ and the above estimate of angle of attack for these vortex generator is significantly below the angle of vortex bursting (see Chapter 6 for the phenomenon of vortex bursting) and the leading edge vortex of this half-delta structures create strong and coherent vortical flow energizing the local shear layer and delaying the tendency of flow separation. It is fairly obvious that the $C_{l,max}$ in the experiment could be increased because the flow is 2D and the half-delta wings are effective for the estimated angle of attack.

The same principle can be used to prevent three-dimensionality of the flow over a swept wing by preventing crossflow by the leading edge vortex shielding the successive sections from each other. Such prevention of cross-flow avoids tip-stall from occurring and vortex generator is commonly found in many aircrafts. Another similar device is the wing-fence which is usually fitted to the leading edge of a swept wing, as shown in Figure 13.14. Primarily, the protruding wing fence physically prevents a flow in the attachment line plane along the leading edge of the swept wing. This avoids the so-called leading edge contamination (see Sengupta 2012 for details). At the same time, the fence also issues a streamwise vortex which prevents crossflow over the main wing. Avoidance of such three-dimensionality also reduces drag due to reduced tendency of cross flow instability.

13.2.5 Induced Drag and Its Alleviation

In writing the expression for drag polar at low speed flight and obtaining the optimum speed for low drag, we noted the ideal case where induced drag contributes to half of the total drag. For an actual aircraft in operation, we saw that the percentage contribution of induced drag comes down due to many other unaccounted form of the drag in the simplified drag polar expression. These could arise due to compressibility and various interference effects. However, still the lift induced drag constitute a significant proportion of total drag. In discussing finite wing theory in Chapter 4, we found the elegant treatment due to Lanchester and Prandtl in estimating this component of drag experienced by the finite lifting wing in an inviscid model, where the lift was replaced by an ensemble of elementary horseshoe vortices. It has been shown that this component of drag is essentially due to finite span of the wing, with aspect ratio determining this aspect. Thus, it appears that increasing the aspect ratio might provide immediate solution to reducing this component of drag. However, increase in aspect ratio also brings in attendant increase in skin friction or parasitic drag. Additionally, the added tip portion would increase root bending moment and hence would require strengthening the wing structural design, which will lead to increase in weight. The latter aspect becomes more important for powered aircraft due to the high prevalent dynamic pressure. For gliders due to lower flight speed, increase in aspect ratio is readily practised.

Thus, wing-tip devices have been designed to improve aerodynamic efficiency (L/D ratio) for fixed wing aircraft. In Figure 13.15, we show two representative wing-tip devices. Although these might work somewhat differently for different geometries shown in the figure, the essential goal in all these is to weaken tip vortices, which are present always for finite aspect ratio wings. Formation of the tip vortices is a steady drain of energy and it is legitimate to reduce induced drag. From previous discussions, it is apparent that the tip vortices will be strongest when the aircraft is experiencing maximum lift. Thus, the tip vortices becomes strongest during landing and take-off conditions and the long endurance of tip vortices is a concern of flight safety for other aircrafts landing and taking-off.

Historically, the importance and alleviation of the induced drag was also suggested by Lanchester who suggested the use of wing end-plates to control tip vortices. Although a patent was received more than a century ago, its first practical use was by Dr S. F. Hoerner who suggested a drooped wing-tip facing rearwards to cause a weaker tip vortices flowing away from the upper surface. Such drooped wing-tips, also goes by the name of Hoerner tips and have been used in gliders and light aircrafts. However, it was again R. T. Whitcomb's work in NASA during 1970s, which gave a strong impetus to wing-tip device research in the form of winglets, as described in Bargsten and Gibson (2011). In Figure 13.15, Whitcomb's winglet is shown.

In its original form following the work of Whitcomb, winglets are near-vertical extension of the main wing canted upwards smoothly in one of the versions seen in Boeing 737-800 aircraft. It is variously noted that winglets increases operational empty weight between 100 to 236 kg for Boeing 737 (depending upon the model), while it increases range of about 241 km or payload increase from 1996 kgs to 2303 kgs. In the blended winglet used in Boeing 737, the device is attached to the wing as a smooth extension of six foot length and is stated to provide decrease fuel consumption by about 4–6%. Avoiding sharp corners at the junction of wing-tip and the winglet, leads to lesser interference.

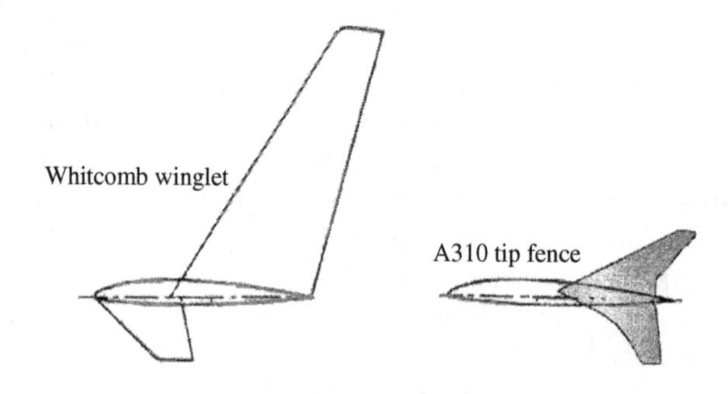

Figure 13.15 Various forms of wing-tip devices used as winglets

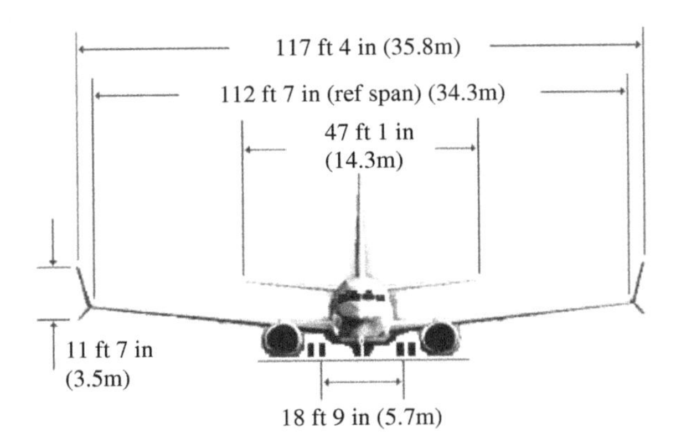

Figure 13.16 Split Scimitar winglet intended for Boeing 737-800 aircraft. Note the new winglet design

In Figure 13.15, also shown is the wing-tip fence, another variant of winglet seen in many airbus designs, which is similar in concept to that proposed by Whitcomb and this incorporates surfaces extending both above and below the wing-tip. Both these extensions are shorter in dimension to produce the same aerodynamic effects. Another form of wing-tip device is the use of raked wing-tip used in many aircrafts, including Boeing 787. Here, the wing-tip has a higher degree of sweep than the main wing sweep angle. This provided added fuel efficiency, climb performance and shorter take-off runway length. Intuitively it is thought that the purpose of a raked wing is to effectively increase the aspect ratio, leading to induced drag reduction.

On the new Boeing 737 MAX aircraft, a new hybrid design of airfoil is being incorporated, which is a combination of blended winglet, wing-tip fence and a raked wing-tip shown in Figure 13.16. The manufacturer claims that such incorporation delivers an additional 1.5% improvement in fuel economy.

13.2.6 Theoretical Analysis of Induced Drag

Induced drag is inter-alia dependent on viscous action which creates wake vortices, whose spanwise gradient determines the strength of tip-vortices. Despite this observation, early investigations assuming the existence of vortices replacing the lifting wing estimated this as inviscid form of drag for a finite wing. This is the lost kinetic energy in the trailing vortical wake. One of the concepts of reducing this component of drag involves understanding the flow over nonplanar wing. Munk (Munk 1921) established a number of results showing nonplanar wing to have significantly lower drag as compared to planar wing with the same span and lift. Mangler (Mangler 1938) considered end-plate configuration, which was generalized in Lundry and Lissaman (1968), for calculating induced drag. In Lowson (1990), the winglet was represented by polynomials, identifying elliptic functions to provide lower induced drag. A good discussion on this topic is given in Hicken and Zingg (2010), who studied induced drag minimization by numerical solution of Euler equation. Apart from studying winglets, they also analyzed the box wing, split wing-tip configuration and concluded that for the same spanwise and vertical bound constraints, an optimized split-tip geometry is little better than optimized box-wing.

13.2.7 Fuselage Drag Reduction

Vijgen and co-authors (Vijgen *et al.* 1986) have presented an estimate of decrease in total aircraft drag when laminar boundary layer is realized over the fuselage. Flow accelerates over the forward portion of the fuselage of a typical transport aircraft, from the nose to the beginning of the almost constant cross-section cylindrical cabin section- except the region in the vicinity of the wind-shield, as discussed in Georgy-Falvy (1965). Further reduction of drag would necessitate turbulent drag reduction over the fuselage and reduced wing-fuselage interaction problems. Dodbele and co-authors (Dodbele *et al.* 1987) reported a significant run of laminar flow over business-jet and commuter-type fuselages obtained by aerodynamic shaping of the fuselage. For Mach numbers of 0.6 and upwards, they reported about as much as 30–40% of the fuselage length to retain a stable laminar boundary layer. A 30% stretch of laminar flow over fuselage of a business jet translates into 7% overall drag reduction. Primarily, the body shape determining the axial and circumferential pressure gradients, dictates streamwise and cross flow instabilities. Receptivity to ambient noise and FST can furthermore enhance the growth rate of disturbances. Additionally, manufacturing excrescences on the fuselage surface can induce bypass transition that also can further strengthen by entrainment of acoustic noise and FST.

Nonaxisymmetric geometries have been tested in flight tests and in wind tunnels. For example, Vijgen and Holmes (1987) report the NASA/Cessna T-303 NLF fuselage flight experiments which optimized fuselage design during cruise and climb conditions. Fujino (Fujino and Kawamura 2003) reported wave drag characteristics of an over-the-wing nacelle for a business-jet configuration in an effort to reduce aircraft drag. This reference also contains wind tunnel results for the nonaxisymmetric fuselage and optimal location of the nacelle to reduce total aircraft drag.

In the T-303 flight tests, the forebody was smoothed with body filler and subsequently sanded the rivet-lines and lap-joints to delay transition; otherwise these become the sources of premature transition. Furthermore, the proximal propellers and engine also have significant effects on transition over small transport aircrafts e.g. the propeller streamtube effect, propeller rotation direction have significant effects on flow transition. The effect of rotation direction can be understood from the following. If the propeller is rotating in a counter-clockwise direction (as viewed from the front), then one would notice a local acceleration and deceleration near the propeller plane on the port and starboard side, respectively, due to superposition of the propeller-induced flow field over the basic flow field over the nose. In Vijgen and Holmes (1987) transition location was noted by sublimating naphthalene technique for flow over the fuselage in the presence of the propeller. Transition was noted to occur about 12 inches ahead of the propeller plane. The authors also reported experimental results using hot-film arrays for zero angle of attack and zero side slip angle. These are shown for a Reynolds number of $Re = 1.7 \times 10^6$, for the indicated speed and altitude for the fuselage top and starboard sides. A reduction in Tu was noted downstream of the propeller plane, due to a favourable pressure gradient immediately behind the propeller plane. It was also noted that when the Reynolds number was reduced to almost half the value, the laminar run on both the top and side increased by a large distance.

13.2.8 *Instability of Flow over Nacelle*

We have noted in Table 13.1, that the nacelle accounts for only about 5–8% of total profile drag, whether the flow is fully turbulent or fully laminar over the lifting surfaces only. This flow field was studied in a joint NASA-GE flight experiments and reported in Hastings *et al.* (1986, 1987), Obara *et al.* (1986). The dual objectives of these experiments were to study the extent of a full scale nacelle in actual flight and to find the effects of acoustic disturbance on the location of transition over the nacelle. A full-scale, flow-through NLF nacelle was evaluated by mounting it below the wing of the experimental NASA OV-1 aircraft. One controlled noise source was located in an underwing pod outboard of the nacelle and a second one located in the nacelle centre-body to study the effects of noise on transition on nacelle.

It is an attractive proposition to have a NLF nacelle for large turbofan-powered aircraft. The nacelles are more or less like fuselage, but have a shorter characteristic length with lower characteristic Reynolds numbers. Because of flow through the nacelle involving an interaction of an internal flow with an external flow and the fact that nacelles are nonlifting, there is less restriction on its pressure distribution from the point of view of lift and moment. The comparison of a conventional nacelle with a NLF nacelle, in terms of geometry and pressure distribution are given in Sengupta (2012).

13.3 Flow Control by Plasma Actuation: High Lift Device and Drag Reduction

In recent times, attention has been focused on flow control by plasma actuation due to its various advantages. The plasma device is usually concealed as one of the electrodes which is embedded inside a dielectric medium, while the second electrode is placed asymmetrically and exposed to the ambient condition. Actuation of this device by a high voltage AC source creates a body force, which induces a wall jet in the direction of the flow and thereby alleviates the tendency of the flow to separate due to adverse pressure gradient experienced by a lifting surface at high angles of attack. Such a secondary flow can also induce Coanda effect and thereby prevent flow transition. This effect, based on a single-dielectric barrier discharge (SDBD) mechanism, can be operational in atmospheric conditions for aerodynamic applications as noted in Corke *et al.* (2010). According to these authors, although this was considered initially to be useful only at low speeds, plasma actuators are now finding uses at

> 'high subsonic, transonic and supersonic Mach numbers, owing largely to more optimized actuator designs that were developed through better understanding and modelling of the actuator physics.'

In the plasma actuator consisting of two electrodes, one uncoated and exposed to the air and the other encapsulated by the dielectric material, upon application of high enough AC voltage causes the air over the covered electrode to weakly ionize creating less than 1 ppm weakly ionized gas, whose emission strength is so low, as to require visualizing it in darkness. This ionized air produces the body force which acts on the neutrally charged air. Although many of the early models of plasma considered this as the conservative body force represented in terms of electrostatic potential, it is worth remembering that to control separation and transition, the body force will have to be more general to introduce rotational effects associated with viscous flows. According to Corke *et al.* (2010) the plasma discharges are in an equilibrium, i.e. the total density of negatively and positively charged particles in any region of the plasma is approximately equal. The discharges require the presence of free electrons for initiation of the process and this can be purposely added to accelerate the process. Also, the plasma is only created when the applied electric field is above the breakdown electric field, which is a function of the driving frequency. Any plasma model must therefore incorporate such dependence on frequency of the applied electric field. The sustained electric field needed to maintain the plasma discharges is always lower than the breakdown electric field.

A current flows between the electrodes due to the plasma conductivity. A normal glow discharge is said to be created when the current density in the boundary region is independent of current flowing through the circuit as reported in Roth (1995). For constant current, the current density in the normal glow discharge is proportional to the square of the static pressure. With increasing pressure at constant current, the cross-sectional area of plasma decreases and vice versa. In Versailles *et al.* (2010), importance of pressure and temperature on the performance of plasma actuators and this study is important as it tells one, if this could be used in aircraft operation at extreme weather conditions, if one is interested in using it to control separation. One can use it as a high lift device during landing and take-off, thereby reducing weight and mechanical complications associated with traditional high lift devices. In such a scenario, the operation is near the ground level for which the temperature and pressure would be nominal ground level values. However, if the goal of plasma actuation is to reduce drag during cruise configuration, then the results from Versailles *et al.* (2010) would be of importance. At cruise altitude, pressure and temperature are significantly lower as compared to ground level values and drag reduction would be feasible, provided the plasma actuation works at such reduced ambient pressure and temperature.

In the following, we discuss the scenario where plasma actuation is utilized for delay of bypass transition for flow over an airfoil. This requires solving the Navier-Stokes equation with an appropriate model for plasma and the contents here are as reported in Sengupta *et al.* (2012).

13.3.1 Control of Bypass Transitional Flow Past an Aerofoil by Plasma Actuation

The control of flow past an aerofoil would entail simultaneous increase of lift and reduction of drag. Continual effort persists to reduce lifting surface drag by keeping the flow laminar as far downstream as possible. This is due to the fact that turbulence leads to drag increase by almost an order of magnitude for streamlined bodies. In general for optimizing commercial aircraft performance, an aerofoil can be designed for many performance optimization goals like range, endurance etc. However, here attention is confined solely on transition delay during cruise, so that the drag is reduced and one obtains better fuel economy.

Fluid flow transition can take place via multiple routes. A recent and up-to-date account of the subject is treated in Sengupta (2012). A classical route of flow transition is attributed to evolution of viscous tuned Tollmien-Schlichting (TS) waves. While there are instances where this mechanism is not present and researchers have looked for alternative routes, it was Morkovin (1969) who coined the term bypass transition to indicate such routes. A detailed account of bypass transition can be found in Sengupta and Poinsot (2010), Bhumkar (2012) and Sengupta (2012) and many other references contained therein.

As noted above, progresses have been made in magneto- and electro-hydrodynamic flow controls where body forces arising as Lorentz forces help mitigate conditions which lead to transition and flow separation over lifting surfaces. Plasma actuators hold promise due to lesser complications of construction and operation, having no moving parts, high bandwidth quick response, while adding lesser weight without creating noise and vibration.

Plasma actuators have been employed and their operating principles have been studied for a wide range of flows. First such published effort for transition delay is due to Velkoff and Ketcham (1968) and a more recent overview of plasma flow control is given in Corke and Post (2005). Plasma control has been utilized for lift augmentation of aerofoils by Corke and his co-authors (Corke *et al.* 2002, Patel *et al.* 2007); turbine blade separation control (Huang *et al.* 2003, List *et al.*2003); bluff body flow control (Asghar *et al.* 2006, Thomas *et al.* 2006); separation delay (Hultgren and Ashpis 2003, Riherd and Roy 2011); control of vortical flows (Visbal and Gaitonde 2006); high lift devices (Corke *et al.* 2004, Post and Corke 2004) and noise control (Huang *et al.* 2010).

Use of plasma actuator for turbulent flow drag reduction has been reported in Wilkinson (2003). Efforts at controlling flow transition caused by TS waves have been experimentally investigated in Grundmann and Tropea (2007, 2008). Computational efforts utilizing unsteady Navier-Stokes equation with appropriate plasma models have been reported in Cristofolini *et al.* (2011), Gaitonde *et al.* (2005), Kotsonis *et al.* (2011), Rizzetta and Visbal (2011), Zhang *et al.* (2011) and Sengupta *et al.* (2012). In these efforts, models are used for the electrodynamic field of the plasma.

One of the first electrodynamic field models was developed by Massines *et al.* (1998), who developed a 1D model based on simultaneous solutions of continuity equations for charged and excited particles to obtain spatio-temporal distribution for the plasma. We note that fluid transport processes have characteristic velocity scales of the order of 10–300 ms^{-1}, while the velocity scales of the charged species, e.g. the electron velocities are of the order of 10^5 to 10^6 ms^{-1} (Orlov 2006). Such scale separation allows one to treat the plasma dynamics independently of the fluid dynamics, by treating the former to be instantaneous and independent of the latter. However, effects of the plasma dynamics are embedded in the fluid flow equation via the body force terms. The models existing in literature (Suzen *et al.* 2005, Orlov 2006 and Lemire and Vo 2011) provide the plasma generated body force. In Palmeiro and Lavoie (2011), comparative analysis of SDBD models was performed and the authors noted that the hybrid model of Lemire and Vo (2011) predicted the trend of the body force better, as compared to other models. It was noted as one model which exhibited the correct nonlinear velocity scaling with voltage. In this reference, the authors have reported experimental measurements of maximum induced velocity in Table 2 of the reference for different plasma set-up configurations in quiescent air condition. These measurements show that the plasma induced velocity near the electrode (also termed as the ionic wind) is in the range of 1.5 to 2 m/s for the supply voltage of 10 kV with 3 kHz frequency. Such induced velocity near the wing surface would be adequate to mitigate separation and delay transition.

Plasma actuators consist of two electrodes separated by insulation (dielectric) covering the lower electrode completely, while the upper one is exposed to the flow, as shown in Figure 13.17. In Figure 13.18, the computational domain and various governing equations solved for the electrodynamic field are also shown.

In Kotsonis (2011), Hanson *et al.* (2010, 2010a) and Seraudie *et al.* (2011), experimental results pertaining to plasma control of external flows are reported. In Seraudie *et al.* (2011),

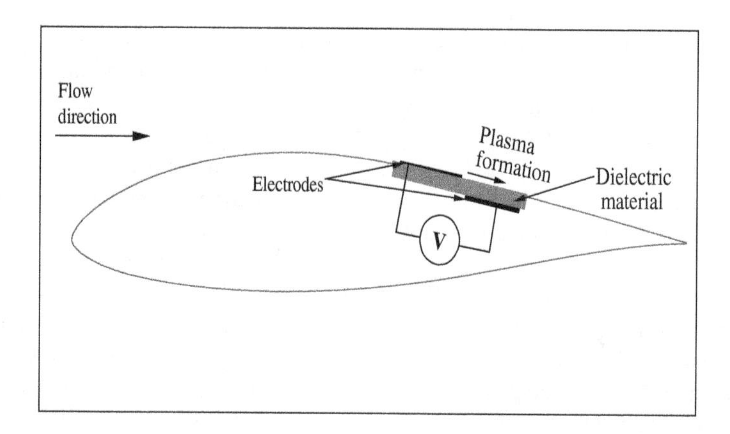

Figure 13.17 Schematic of a plasma actuator mounted on top surface of DU96-W-180 aerofoil

a DBD plasma actuator was placed near the leading edge of a modified ONERA-D wing section. According to the authors, the boundary layer was stabilized and was additionally verified by the empirical e^N method using linear stability theory. Incidentally the so-called TS wave frequencies for a mean flow of 7 m/s, 10 m/s and 12 m/s were noted to have frequency close to 400 Hz, 600 Hz and 700 Hz, respectively. Increase in flow velocity caused a decrease in turbulence intensity in the tunnel. However, how increase in flow velocity augments the unstable TS wave frequency is not apparent. The authors also note that increase of mean velocity makes the flow less amenable to control by plasma actuation. In contrast, Hanson *et al.* (2010, 2010a) created experimentally the transition process by a transient growth and not by TS waves, for a Blasius boundary layer. In these experiments, bypass transition was created via the placement of cylindrical roughness elements and the resultant flow was controlled by the plasma actuator. In Kotsonis *et al.* (2011), body force models are provided as a function of voltage and compared with experiments. The authors noted that applied 8 kV (peak to peak) voltage to the electrodes was the onset value for creating an effective plasma. It was also noted that increasing the peak to peak voltage above 16 kV did not alter the plasma anymore. We note here that the required voltage across the electrodes is a strong function of gap between the electrodes. For example, in Suzen *et al.* (2005), a voltage source of 5 kV with 5 kHz frequency was used to study flow separation control in a low pressure turbine cascade when the gap between the electrode was 0.127 mm.

13.4 Governing Equations for Plasma

In case of plasma actuator on the aerofoil, typically a high voltage source is connected to two different electrodes which are separated by a dielectric material, as shown in Figure 13.17. One electrode is kept on the surface of the aerofoil, while the other electrode is embedded in the surface below the dielectric material. Due to a high voltage difference between two electrodes, air in the vicinity of the electrodes gets weakly ionized. Presence of electric field, exerts a body-force on the flow-field through momentum transfer which can be modelled and incorporated in the Navier-Stokes equation.

In a fluid dynamic system excited by plasma actuation, three different time scales exist (Orlov 2006). The charge distribution occurring in the plasma has the shortest time scale

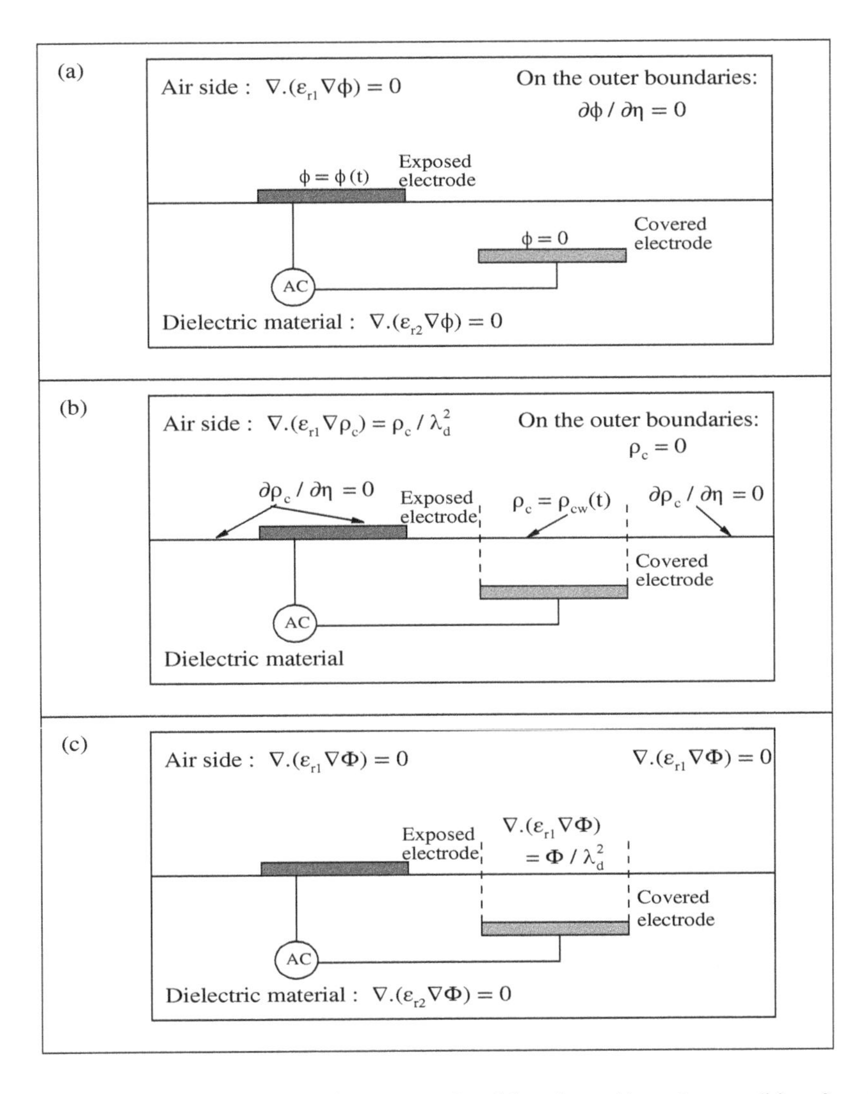

Figure 13.18 Schematic diagram showing computational domains and boundary conditions for solving (a) Equation (13.6) for the electric field; (b) Equation (13.8) for the charge density given by the models described in the text and (c) Equation (13.9) for the electrostatic potential given by the model in Orlov (2006)

($\sim 10^{-8}$ sec). The next time scale is present at the level of actuation time required for the plasma actuator which is ($\sim 10^{-4}$ sec). Finally, the third time scale ($\sim 10^{-2}$ sec) is associated with different fluid flow structures. As the first two time scales are very small compared to the actual flow dynamics time scale, one can assume that the charge formation and distribution are almost instantaneous and plasma formation process can be approximated as a quasi-steady process (Orlov 2006). This assumption forms a basis for adopting equations of electrostatics for describing discharges. A set of Maxwell's equations is used as governing equations to model plasma dynamics. The electro-hydrodynamic force can be expressed as

Suzen *et al.* (2005)

$$\vec{f}_B = \rho_c \vec{E} + \vec{J} \times \vec{B}. \tag{13.2}$$

As plasma formation and charge redistribution processes are assumed to be instantaneous, the following electrostatics equations are adopted. Electric field ($\vec{E}$) can be derived from the gradient of a scalar potential (Φ) as in Suzen *et al.* (2005),

$$\vec{E} = -\nabla\Phi. \tag{13.3}$$

Using Maxwell's equation for charge density (ρ_c) one can write

$$\nabla.(\epsilon \vec{E}) = \rho_c. \tag{13.4}$$

If ϵ_r is a relative permittivity of the medium and ϵ_0 is a permittivity of free-space then the permittivity of the medium is expressed as $\epsilon = \epsilon_r \epsilon_0$. Using Equations (13.3) and (13.4) we obtain

$$\nabla.(\epsilon \nabla\Phi) = -\rho_c. \tag{13.5}$$

Next, we discuss three important algebraic models developed by different researchers to model body force experienced due to plasma actuation following Equation (13.2). We first discuss Suzen *et al.* (2005)'s model which solves two different equations for potential due to the external field and potential due to the net charge density in the plasma to model the body force using Equation (13.2). Next is Orlov (2006)'s model and has higher complexity. It assumes the air and dielectric material as a parallel network of resistors and capacitors to calculate the charge build-up on the surface of the dielectric. A spatio-temporal lumped-element circuit model has been proposed to obtain body force using Equation (13.2). Finally, we discuss Lemire and Vo (2011)'s model which is a combination of the above two approaches and is superior, as shown in Palmeiro and Lavoie (2011). We have used Lemire and Vo (2011)'s model for the present computations.

13.4.1 Suzen et al.'s Model

In the model proposed by Suzen *et al.* (2005), potential Φ in Equation (13.3) is composed of the potential due to external electric field (ϕ) and another part which is contributed by the net charge density in the plasma (φ). Characteristic length for electrostatic shielding in a plasma is denoted by the Debye thickness (λ_d). In Suzen *et al.* (2005)'s model, it is assumed that λ_d is small and the charge on the wall is not large, which results in two separate equations for the two potentials given by

$$\nabla \cdot (\epsilon \nabla\phi) = 0 \tag{13.6}$$

$$\nabla \cdot (\epsilon_r \nabla\varphi) = -\frac{\rho_c}{\epsilon_o}. \tag{13.7}$$

Using the net charge density at any point within the plasma (Suzen *et al.* 2005) given by $\varphi = (-\rho_c \lambda_d^2)/\epsilon_o$ in the above equation, the equation for charge density is obtained as

$$\nabla \cdot (\epsilon_r \nabla \rho_c) = \frac{\rho_c}{\lambda_d^2}. \tag{13.8}$$

In Figure 13.18(a), a schematic of the computational domain along with the boundary conditions for the solution of Equation (13.6) are shown. The Neumann boundary condition is specified at the outer boundaries of the domain while on the electrodes, a time dependent Dirichlet boundary condition is specified. The schematic of a computational domain along with the boundary conditions used in solving Equation (13.8) are shown in Figure 13.18(b). The boundary condition for ρ_c is given by $\rho_{c,w}$ whose form is based on an assumption and from empirical data. In Suzen *et al.*'s model, downstream of the exposed electrode and on the embedded electrode, the charge density is prescribed as $\rho_{c,w}(x,t) = \rho_c^{max} G(x) f(t)$, so that it is synchronized with applied voltage $f(t)$. Here, ρ_c^{max} is the maximum value of charge density and $G(x)$ is a half Gaussian distribution function chosen to resemble plasma distribution over electrode surface in Suzen *et al.*'s model. Figure 7 in Suzen *et al.* (2005) describes the procedure to obtain Debye length λ_d and maximum charge density on the wall ρ_c^{max}. A comparison between the experimental and the numerical results show that $\lambda_d = 0.001$ m and $\rho_c^{max} = 0.0008$ C/m^3 are appropriate choices. Body force can be computed from Equation (13.2), after obtaining a solution for Equations (13.6) and (13.8) and used subsequently in the Navier-Stokes equations for studying the discharge effects on the flow.

13.4.2 Orlov's Model

In Orlov's model, the governing equation for the potential Φ is derived again using Equation (13.5). Assuming steady state, isothermal plasma gas formation and ignoring the diffusion process, the number density of charged particles (n) in the plasma gas is obtained using the following relation

$$n = n_o \exp\left(\frac{e\Phi}{kT}\right).$$

Here the number density of charged particles in the plasma and number density of molecules which are separated into ions and electrons are given by n_i and n_e, respectively. In the above equation, e is the charge of an electron with k is the Boltzmann constant and T is the gas temperature.

If T_i and T_e denote temperatures of ion and electron species and n_i and n_e are the number densities of ions and electrons, respectively, then the net charge density (ρ_c) is related to the electrostatic potential Φ by $\rho_c = e(n_i - n_e) \simeq -en_o \left(\frac{e\Phi}{kT_i} + \frac{e\Phi}{kT_e}\right)$. Substituting this in Equation (13.5) we obtain

$$\nabla \cdot (\epsilon_r \nabla \Phi) = \frac{\Phi}{\lambda_d^2}, \tag{13.9}$$

where the Debey length is defined approximately as 0.00017 m and the charge density is of the order of 10^{16} particles/m^3 for plasma generated in ambient pressure (Roth 1995). One can define λ_d as

$$\lambda_d = \left[\frac{e^2 n_o}{\epsilon_o}\left(\frac{1}{kT_i} + \frac{1}{kT_e}\right)\right]^{-\frac{1}{2}}.$$

Equation (13.9) is solved in the domain, as shown in Figure 13.18(c). Potential on the electrodes is given by the applied AC voltage: $\Phi|_{electrodes} = \pm\Phi_o$, while on the outer boundaries, the electric potential Φ is set to 0. Electric field $\vec{E}$ given in Equation (13.3) can be obtained from the solution of Equation (13.9). Body force per unit volume can be determined from Equation (13.2) with magnetic field neglected ($\vec{B} \equiv 0$) using the relation $\rho_c = -\frac{\epsilon_o}{\lambda_d^2}\Phi$, as

$$\vec{f}_B = -\left(\frac{\epsilon_o}{\lambda_d^2}\right)\Phi\vec{E}. \tag{13.10}$$

The results obtained from the electrostatic model do not account for the charge built up on the surface of the dielectric or change in the volume of plasma as a function of the input voltage. Thus, the results obtained with this model contradicts all the experimental observations (Orlov 2006). Therefore, a spatial and spatial-temporal lumped-element circuit models were proposed by Orlov (2006) to improve upon the existing electrostatic models, but was not implemented by him.

13.4.3 Spatio-temporal Lumped-element Circuit Model

In this model, air and dielectric material are represented as a parallel network of resistors and capacitors to calculate the charge build-up on the surface of the dielectric, as shown in Figure 13.19(a). This model has been developed with an aim to represent the ionization process and predict the body force without any experimental calibration (Orlov 2006). Each parallel network consists of an air capacitor, a dielectric capacitor and a resistive element, as shown in Figure 13.19(b). The Zener diodes in the circuit of Figure 13.19(b), set the voltage threshold levels for the ionization to occur. Each circuit element consists of two Zener diodes to represent the ionization process, with one of the diodes active during a part of the applied AC voltage cycle. Resistors in the circuit, model the plasma formation and its discharge characteristics.

Consider ϵ_a as the relative permittivity of air, l_n as the representative distance over the dielectric surface and A_n as a cross-sectional area of the capacitor, which is the product of spanwise size of the actuator (z_n) and the height of the capacitive element (h_n). For the nth subcircuit, if one denotes ρ_a as the effective resistivity of air, then the value of the air capacitor C_{na} and plasma resistance R_n in the nth subcircuit is given by

$$C_{na} = \frac{\epsilon_o \epsilon_a A_n}{l_n} \tag{13.11}$$

$$R_n = \frac{\rho_a l_n}{A_n}. \tag{13.12}$$

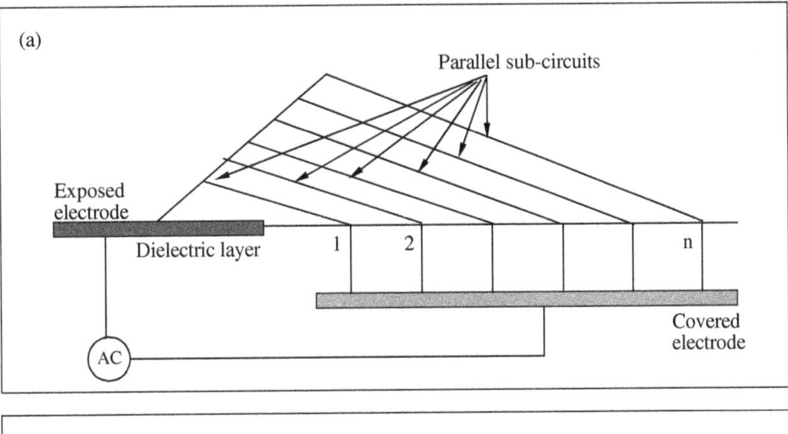

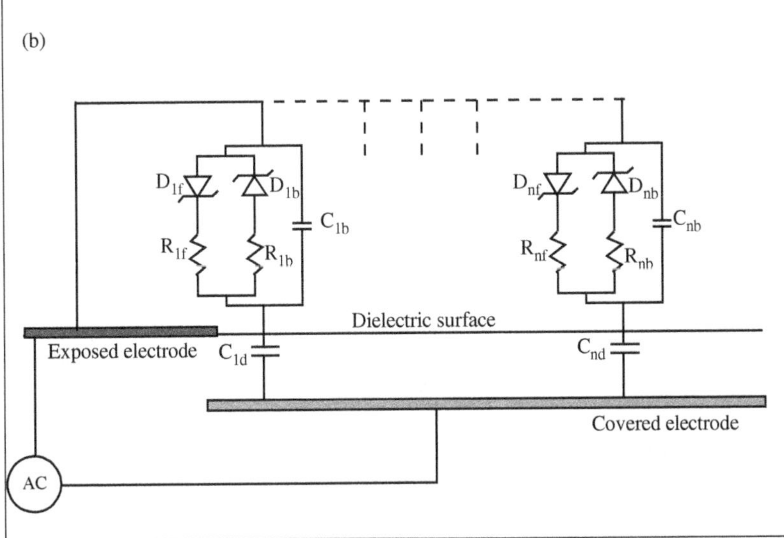

Figure 13.19 Schematic diagram showing (a) plasma formation region divided into a number of parallel subcircuits and (b) subcircuit model of a plasma actuator corresponding to the spatio-temporal model of *Source:* Orlov (2006)

Similarly the value of the dielectric capacitor for the subcircuit is given by

$$C_{nd} = \frac{\epsilon_o \epsilon_d A_d}{l_d}, \tag{13.13}$$

where ϵ_d is the relative permittivity of the dielectric medium, l_d is the thickness of the dielectric material and A_d is the cross-sectional area, which is equal to the product of (z_n) and the width (d_n) of the dielectric capacitor. We observe that all the circuit elements are functions of their distances from the exposed electrode. Based on the actuator geometry and the properties of the materials used, one can evaluate the numerical values of the circuit elements. We can also infer from Equations (13.11) and (13.13) that the farthest subcircuits have lower air capacitance and larger air resistance. This is in accordance with the physical nature of the problem, where plasma formation is restricted in the vicinity of the two electrodes, due to rapidly decreasing electric field strength away from the electrode. Thus, the effects of plasma are also going to be very localized, allowing a stricter control of the flow.

The voltage on the surface of the dielectric for the nth subcircuit is determined from

$$\frac{dV_n(t)}{dt} = \frac{dV_{app}(t)}{dt}\left(\frac{C_{na}}{C_{na}+C_{nd}}\right) + k_n\frac{I_{pn}(t)}{C_{na}+C_{nd}}. \tag{13.14}$$

Here, the applied AC voltage is denoted as V_{app}; k_n depends on the state of the Zener diode (with $k_n = 1$, when plasma forms, otherwise $k_n = 0$); I_{pn} is the current carried through the plasma resistor. This is an initial value problem, which can be solved numerically using standard solvers. In the following, a detailed description of the algorithm to calculate the applied body force is given.

13.4.4 Algorithm for Calculating Body Force

The following algorithm describes the solution procedure adopted for the body force calculations.

- Determine the circuit parameters and solve the subcircuit equations for the applied voltage on the dielectric surface, above the embedded electrode. This gives the value of the potential field (Φ) on the surface.
- Use the above values of potential as the Dirichlet boundary condition to solve Equation (13.9) with the Debye length specified. The Debye length is chosen based on experimental results and usually is of the order of 0.0001 m. The time dependent potential (Φ) as the boundary condition, incorporates the space-time variation of the plasma.
- Calculate the electric field ($\vec{E}$) as described earlier and compute the body force using Equations (13.10).

The body force thus obtained is incorporated into the Navier-Stokes equation for simulating the effects of plasma on the flow.

13.4.5 Lemire and Vo's Model

A hybrid model (Lemire and Vo 2011) combining the features of Suzen *et al.* (2005) and Orlov (2006)'s models was proposed by Lemire and Vo (2011), which gives a spatial body force distribution resembling that is obtained by more elaborate models used in Gaitonde and co-authors (Gaitonde *et al.* 2006 and Roy *et al.* 2006). Here, Equations (13.6), (13.8) are adopted in solving for Φ and ρ_c, during each time step. The boundary conditions over the dielectric layer above the covered electrode $V_n(t)$ and the current $I_{pn}(t)$ are obtained from the spatial-temporal circuit model for the nth element of the parallel circuit model used by Orlov. Voltage distribution on the dielectric surface over the covered electrode ($\Phi = V_n(t)$) thus obtained is used as one of the boundary conditions in solving Equation (13.6). This reproduces the physical nature of the output of the plasma actuator, which is noted to arrange charges in such a way as to annul the applied external electric field (Enloe *et al.* 2003), as much as possible. This also allows us to define the extent of the plasma created over the dielectric surface. This extent is known to depend on the breakdown voltage gradient of air (Lemire and Vo 2011), as modelled by the Zener diode. In the Orlov's model, $V_n(t)$ and $I_{pn}(t)$ are obtained from the amplitude and frequency of the input voltage source.

In Suzen *et al.*'s model, the charge density distribution on the dielectric surface over the covered electrode ($\rho_{cw}(t)$) is specified empirically. In Lemire and Vo's model, it is computed from the current distribution ($I_{pn}(t)$) obtained by the Orlov's model, providing it on the surface by

$$\rho_{cw}(t) = \frac{I_{pn}(t) \times \Delta t}{\text{Volume}}, \tag{13.15}$$

where the denominator represents the volume of the nth electric subcircuit and Δt corresponds to the time step increment used in computing $V_n(t)$ and $I_{pn}(t)$ by solving Equation (13.14). Thereafter, the body force is obtained from $\vec{f}_B = -\rho_c \nabla \Phi$.

13.5 Governing Fluid Dynamic Equations

We obtain numerical solution of Navier-Stokes equation using (ψ, ω)-formulation. This ensures solenoidality of velocity and vorticity field simultaneously and is more accurate, as compared to other formulations. Additionally, an orthogonal grid is used for the incompressible flow. The resultant accurate numerical solution follows the physical nature of the flow. The governing Navier-Stokes equation is decoupled into kinematics and kinetics of the flow in the (ψ, ω)-formulation and expressed in an orthogonal grid by the following equations

$$h_1 h_2 \frac{\partial \omega}{\partial t} + h_2 u \frac{\partial \omega}{\partial \xi} + h_1 v \frac{\partial \omega}{\partial \eta} = \frac{1}{Re}\left[\frac{\partial}{\partial \xi}\left(\frac{h_2}{h_1}\frac{\partial \omega}{\partial \xi}\right) + \frac{\partial}{\partial \eta}\left(\frac{h_1}{h_2}\frac{\partial \omega}{\partial \eta}\right)\right] + \frac{\partial (h_2 F_2)}{\partial \xi} - \frac{\partial (h_1 F_1)}{\partial \eta} \tag{13.16}$$

$$\frac{\partial}{\partial \xi}\left[\frac{h_2}{h_1}\frac{\partial \psi}{\partial \xi}\right] + \frac{\partial}{\partial \eta}\left[\frac{h_1}{h_2}\frac{\partial \psi}{\partial \eta}\right] = -h_1 h_2 \omega, \tag{13.17}$$

where h_1 and h_2 are the scale factors used in mapping physical (x, y)-plane to the computational (ξ, η)-plane (Sengupta 2004), where ξ coordinate is in azimuthal direction and η is normal to it. The scale factors are given by, $h_1 = \sqrt{x_\xi^2 + y_\xi^2}$ and $h_2 = \sqrt{x_\eta^2 + y_\eta^2}$. Plasma actuation causes the body force $\vec{f}_B$ with components given as (F_1, F_2).

For the presented computation, an impulsive start has been used which requires prescribing the irrotational flow as the initial solution. The following boundary conditions are used in solving the governing equations. On the aerofoil surface, no-slip conditions given below are used

$$\psi = \text{constant}; \quad \frac{\partial \psi}{\partial \eta} = 0. \tag{13.18}$$

These conditions also help fix the wall vorticity, which is required as the boundary condition for the vorticity transport equation, Equation (13.16). From the stream function equation, the wall vorticity is calculated as

$$\omega|_{body} = -\frac{1}{h_2^2}\frac{\partial^2\psi}{\partial\eta^2}\bigg|_{body}. \tag{13.19}$$

In O-grid topology, one introduces a cut starting from the leading edge of the aerofoil to the outer boundary. Periodic boundary conditions apply at the cut, which are introduced to make the computational domain simply-connected. For Equation (13.17), Sommerfeld boundary condition is used on the η-component of the velocity field at the outer boundary. Resultant value of stream function is used to calculate the vorticity value at the outer boundary.

The high accuracy compact scheme SOUCS3 (Dipankar and Sengupta 2006) has been used to discretize the convection terms in Equation (13.16). This scheme has higher spectral resolution as compared to conventional explicit schemes. For time advancement of the solution, an optimized DRP OCRK3 scheme is used, which is developed in Sengupta *et al.* (2011). For discretizing Equation (13.17) and the diffusion terms of the vorticity transport equation, a second order central differencing scheme is used.

13.6 Results and Discussions

Here, a case of flow past DU96-W-180 aerofoil used for wind turbine applications is considered (Ghosh 2012). Uniform flow past this 18% thick airfoil at zero angle of attack (AOA) is considered such that the flow Reynolds number is 4.0×10^5 based on chord (c). An orthogonal grid has been numerically generated with 4972 points in the azimuthal direction and 945 points in the wall normal direction. The first point in the wall normal direction is located at a distance of $10^{-5}c$ from the surface of the aerofoil. A time step of 1.5×10^{-5} is used while solving Equation (13.16).

In the absence of plasma actuation, this flow has a tendency to form unsteady, small convecting vortices from $x/c = 0.54$ onwards. This is the bypass transition event in the context of enhanced drag experienced by a streamlined body. Having located the onset point of bypass transition, a plasma actuator is located on the surface of the aerofoil with the essential purpose of delaying transition. The plasma actuator is set to work with a potential difference of 5 kV between the electrodes and with a frequency of 5 kHz. Length of the electrodes in the streamwise direction is $0.005c$, while the thickness of the electrodes is around $7 \times 10^{-5}c$. Thickness of the dielectric material in between the electrodes is of the order of $10^{-4}c$. One notes that the actuation frequency is far above the unstable frequencies of TS waves for the varying pressure gradient flow. The effect of actuation at such large frequencies is primarily to cause a steady streaming, which alters the stresses created by the imposed disturbance field, in the opposite way by which Reynolds stresses alters the mean of a turbulent flow. It is also noted that the present simulation is for the time accurate flow field, as opposed to various simulation results presented for time-averaged flows in Cristofolini *et al.* (2011) and Gaitonde *et al.* (2005). Thus, in the present study we show the time accurate flow evolution, with and without the plasma actuation.

Due to plasma actuation, the flow experiences a body force which is evaluated using Equation (13.2). In this model, the electrostatic equations are solved at each time step and the corresponding body force evaluated is incorporated for solving the Navier-Stokes equation. Variations of body force per unit volume very near to the exposed electrode at different instants are shown in Figure 13.20, only for the top surface. In this figure, the maximum and the minimum components of the body force are indicated in all the frames. A positive value of F_1 indicates a force component in the downstream direction. Similarly, a negative value of F_2 indicates the force component to be towards the aerofoil. One can observe the localized nature of the body force distribution, with the maximum noted in the gap between the electrodes and which progressively diminishes to a negligible value away from the electrodes. At $t = 0.10$, one observes a large force vector in the streamwise direction near the tip of the electrode. The overall body force distribution is in the streamwise direction, which delays flow transition caused by adverse pressure gradient near the trailing edge. Such body force distribution is also seen at $t = 1.0$ and 4.5. At $t = 3.0$, the force vectors are directed towards the surface causing reduction of boundary layer thickness in adverse pressure gradient region, which is otherwise seen to cause wall normal vortical eruptions during bypass transition. Thus the plasma actuation assists the flow on the aerofoil to remain laminar by suppressing events during bypass transition.

The effects of plasma actuation on top surface of the aerofoil at different streamwise locations and at different times is shown next. In Figure 13.21(a), the variation of streamwise velocity component at $t = 0.10$, with and without the plasma actuation for different streamwise locations is shown. One can observe that with plasma actuation, the flow has fuller streamwise velocity profiles in the boundary layer region as compared to no actuation case. Maximum value of streamwise component of the velocity also decreases as one moves from $x = 0.40c$ to $x = 0.95c$, due to adverse pressure gradient. However at this early time, one does not observe any presence of small scale unsteady separation bubbles in the actuated and nonactuated cases, whose presence would have been evident from localized reverse flow velocity profile. In Figure 13.21(b) at $t = 1.0$, for the case of no plasma actuation, one notices the growth of boundary layer in the wall-normal direction, which is highest at $x = 0.95c$ for the compared velocity profiles. This is due to strong adverse pressure gradient near the trailing edge, with the flow showing higher tendency to form separation bubbles. In contrast, for the case with plasma actuation one does not note any large growth of the boundary layer. In Figure 13.21(c) for the flow at $t = 3.0$, velocity profiles corresponding to no-actuation case show flow separation after $x = 0.54c$ onwards, while for the case with plasma actuation, flow retardation caused by adverse pressure gradient is not seen, with the velocity profiles significantly fuller – an indication of favourable pressure gradient caused by plasma actuation. This observation is also seen at the later time of $t = 4.50$ in Figure 13.21(d), implying continual stabilizing tendency of plasma actuation.

In Figure 13.22, variation of vorticity at different instants on the top surface of the aerofoil, with and without plasma actuation is shown. For the no-actuation case, one notices large wall vorticity gradients due to bypass transition causing unsteady separations. However, at the early time of $t = 0.10$, one cannot observe presence of such large vorticity gradients for both the cases. This also happens to be the case at $t = 1.0$, where large gradients are not noted for either cases. Instead for the case of plasma actuation, one notices low wavenumber variation of wall vorticity which does not trigger any separation. Presence of large vorticity disturbances for the no-actuation case are observed at $t = 3.0$ and 4.50 in Figure 13.22. For the case with plasma actuation, such unsteady separations have been suppressed at the later instances.

Comparison of velocity profiles at different stations on aerofoil surface shows that the case with plasma actuation does not show any localized flow reversal, as a signature of bypass transition, while the nonactuated case shows velocity profiles with flow reversal beyond

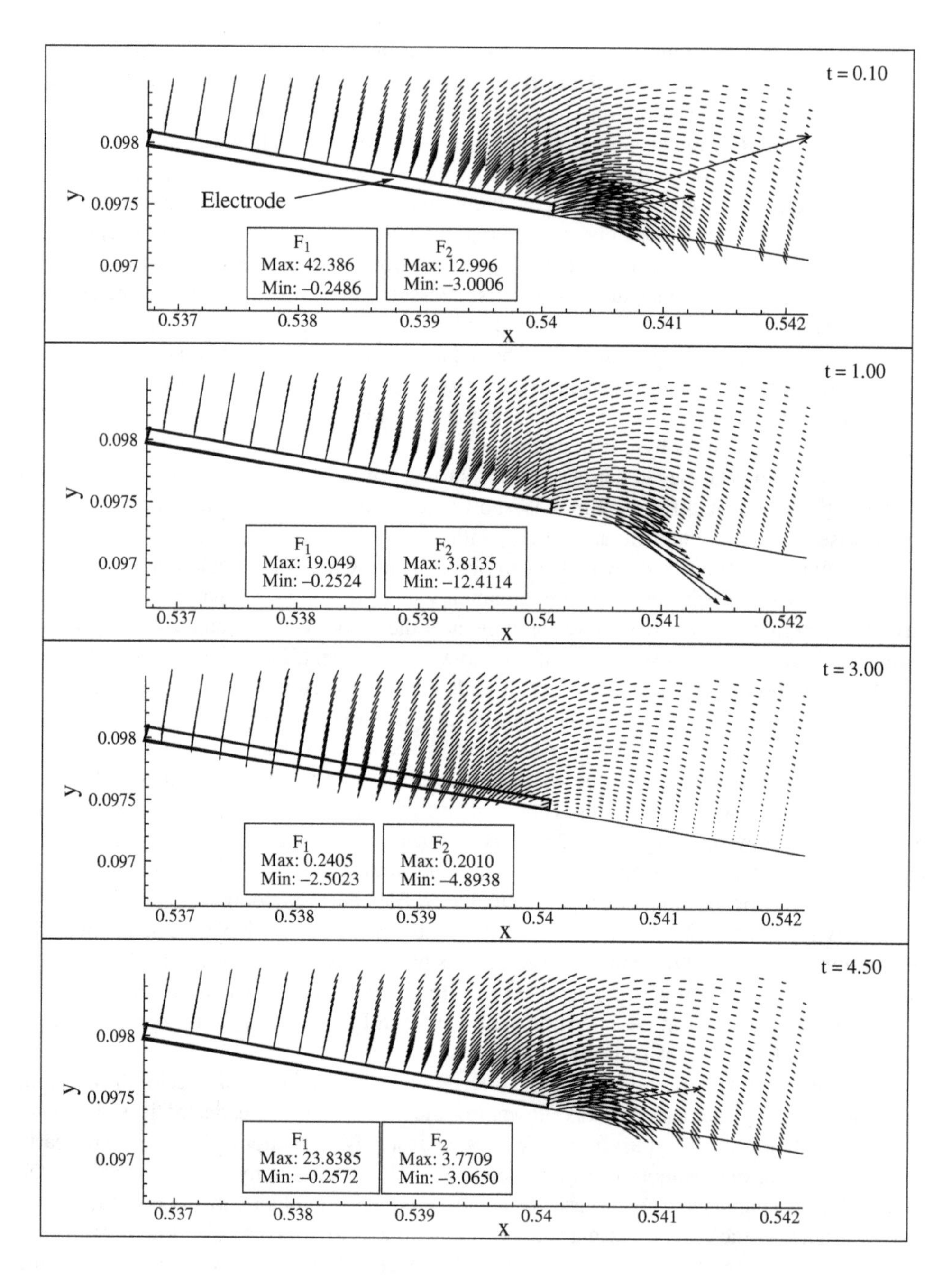

Figure 13.20 Instantaneous variation of body force vectors (F_1, F_2) near the electrode on the surface of the aerofoil. Note the maximum and minimum values of (F_1, F_2) provided at corresponding time

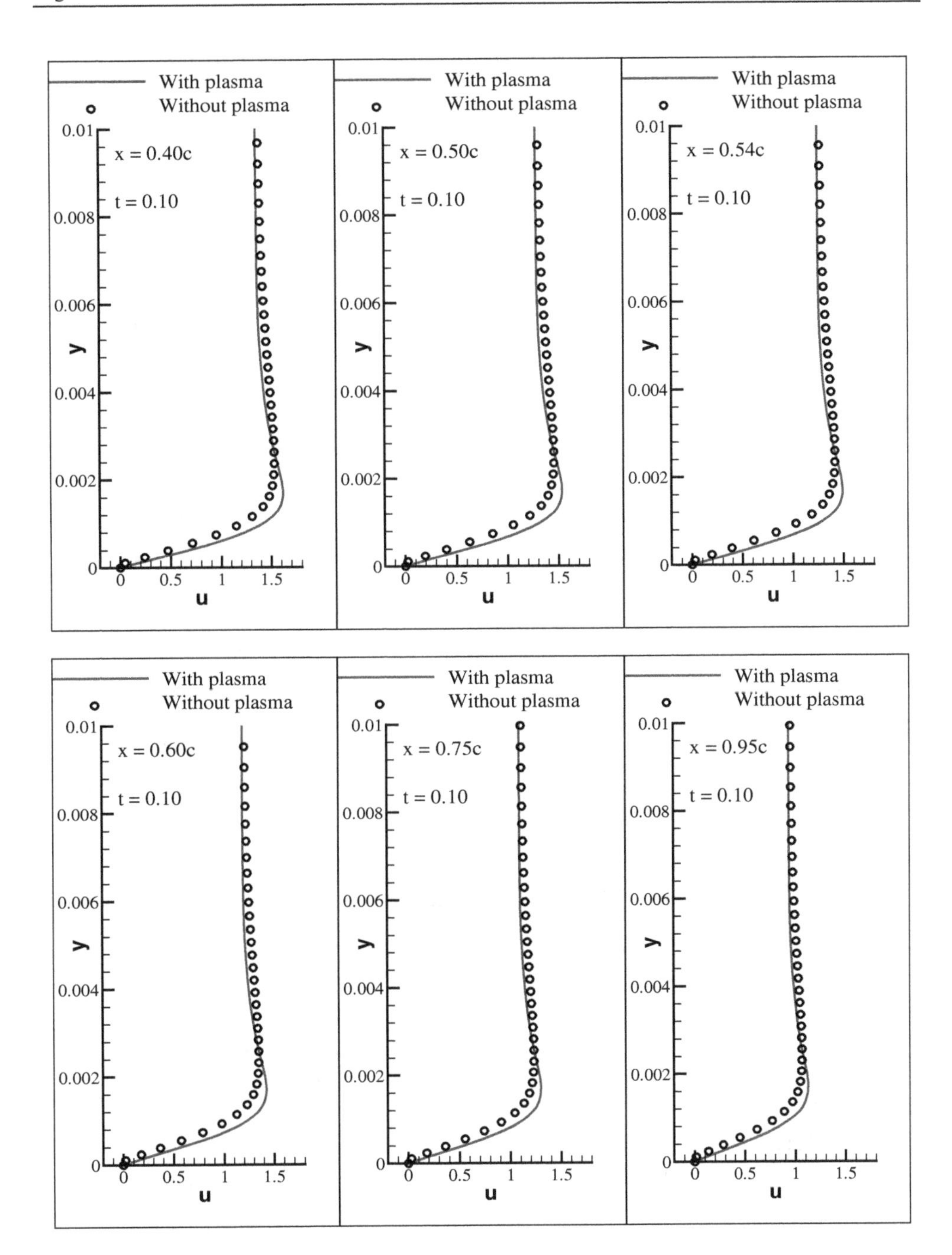

Figure 13.21(a) Variation of azimuthal component of the velocity with and without plasma actuation at different locations on top surface of the aerofoil at $t = 0.10$

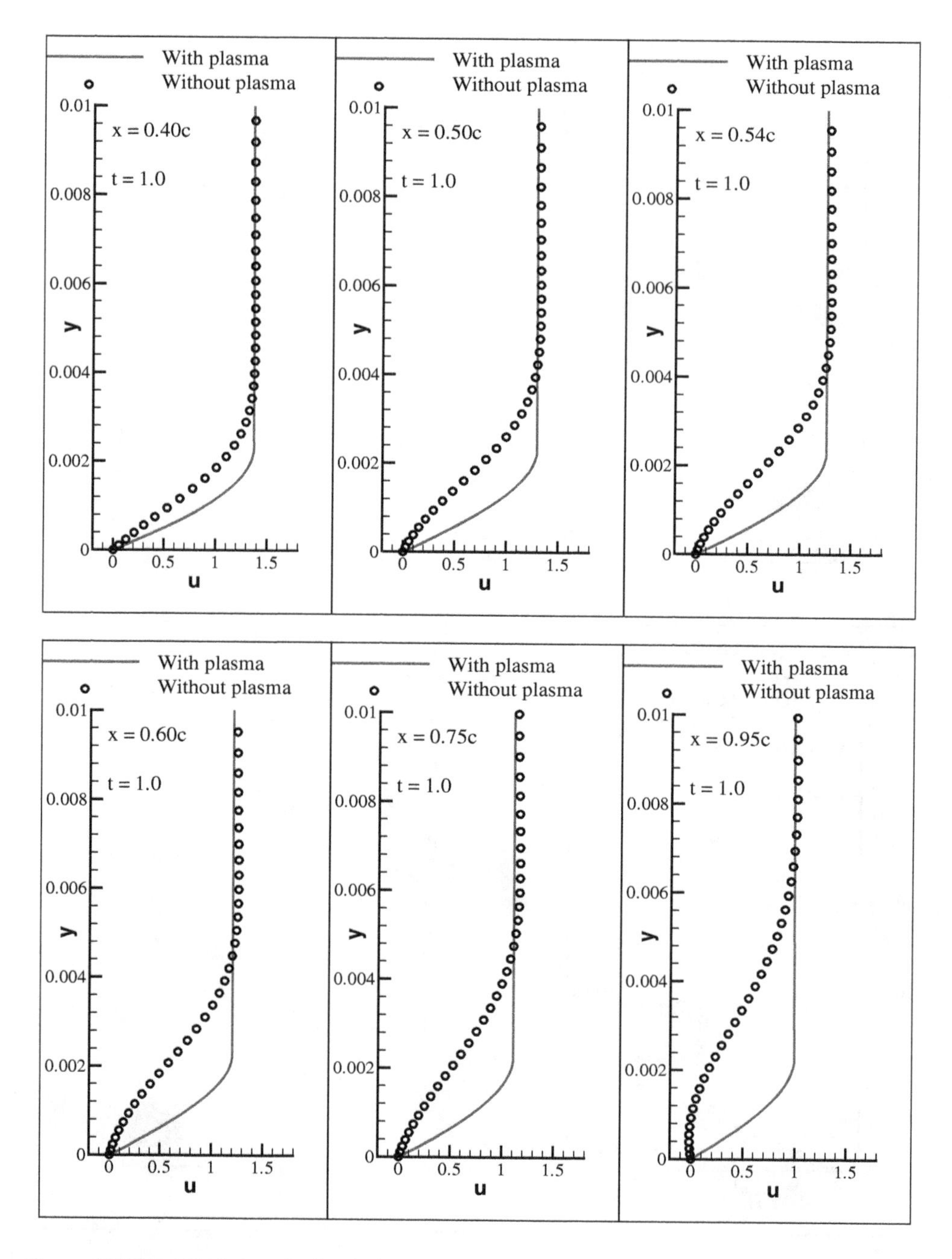

Figure 13.21(b) Variation of azimuthal component of the velocity with and without plasma actuation at different locations on top surface of the aerofoil at t = 1.00

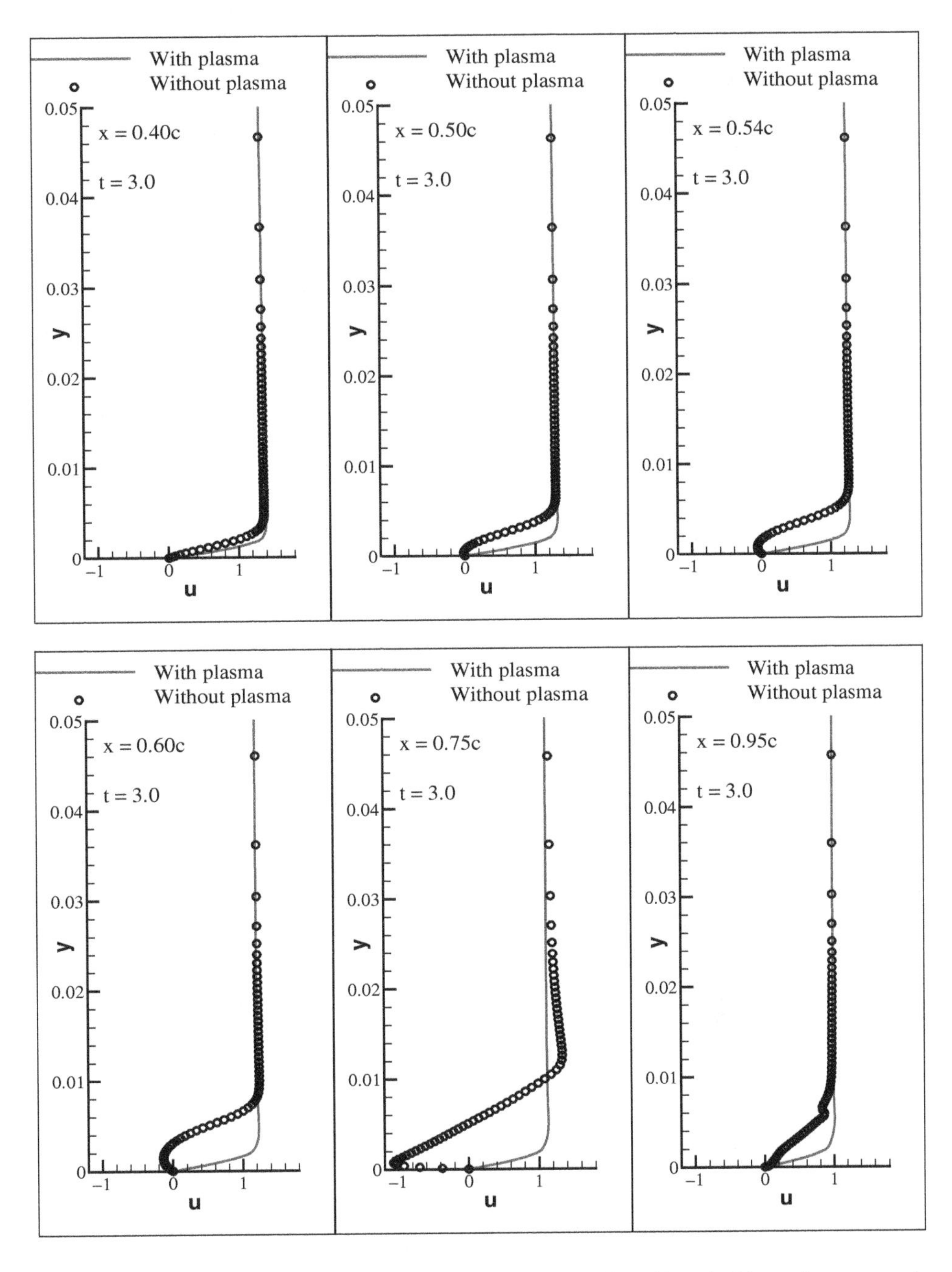

Figure 13.21(c) Variation of azimuthal component of the velocity with and without plasma actuation at different locations on top surface of the aerofoil at $t = 3.00$

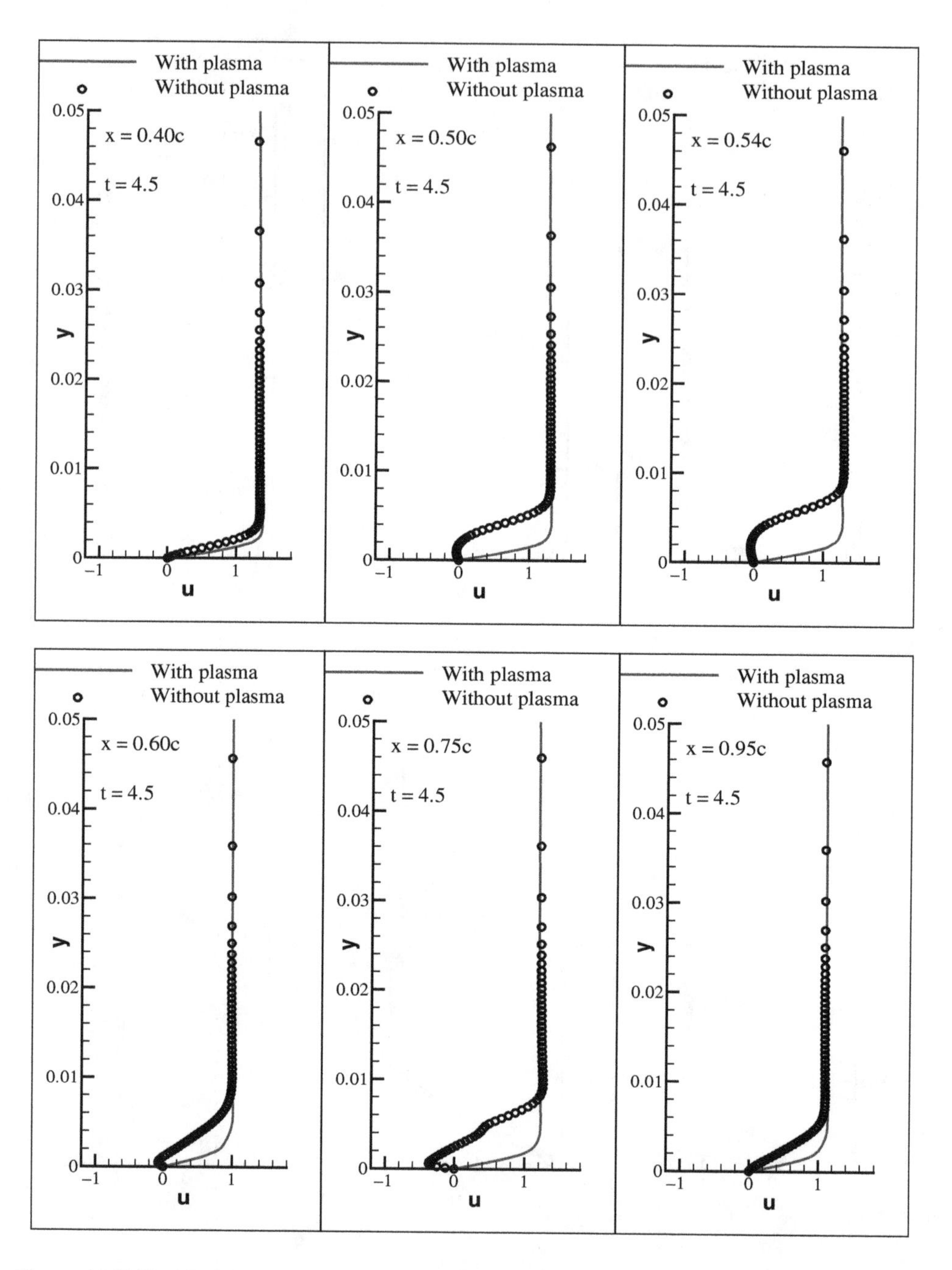

Figure 13.21(d) Variation of azimuthal component of the velocity with and without plasma actuation at different locations on top surface of the aerofoil at $t = 4.50$

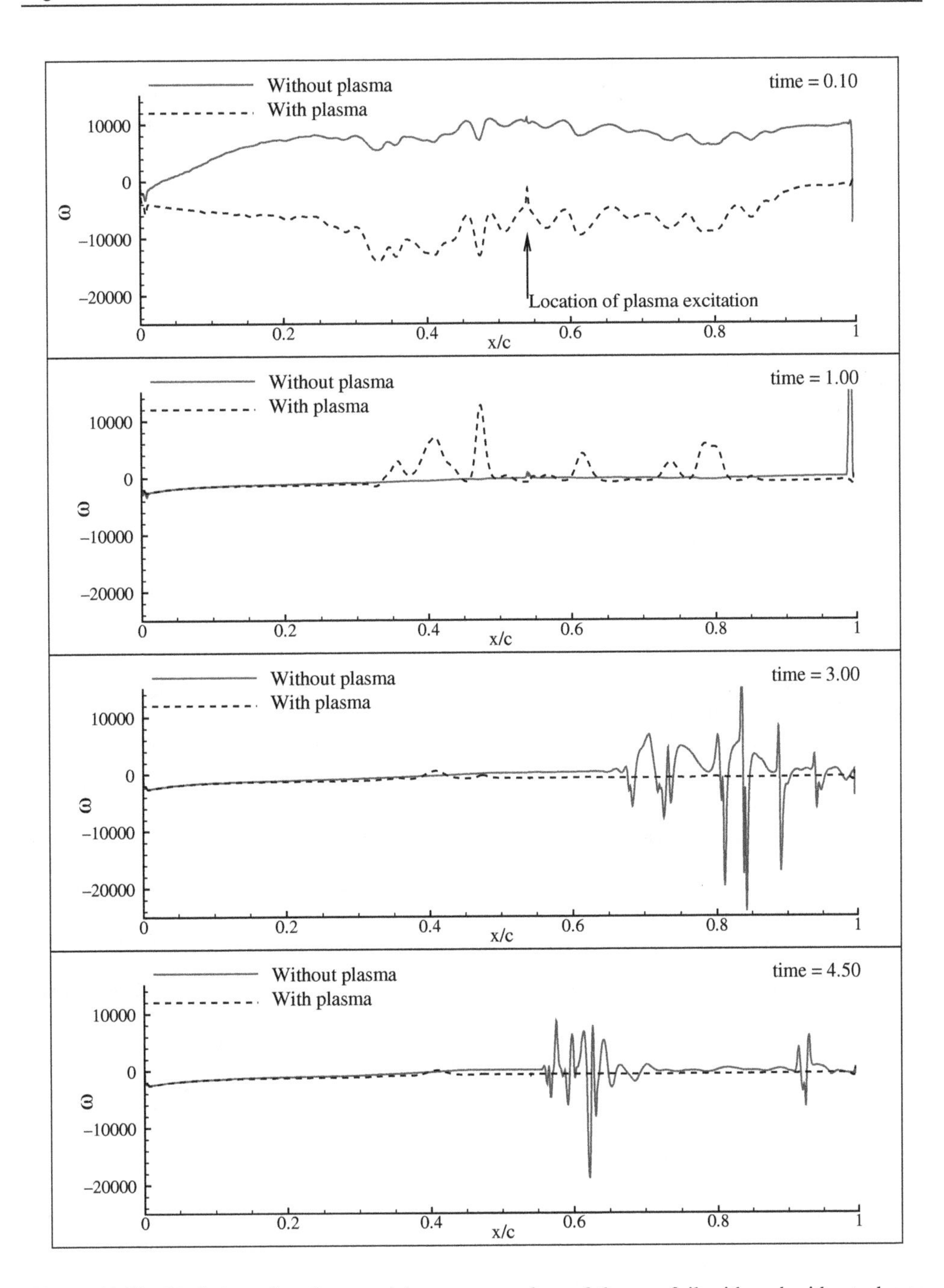

Figure 13.22 Variation of surface vorticity on top surface of the aerofoil with and without plasma actuation at different instants

$x = 0.54c$. Without plasma actuation, formation of steep vorticity gradient which leads to unsteady flow separation beyond mid-chord of the aerofoil is noted, while such disturbances are absent for the case with plasma actuation for 5 kV applied voltage between the electrodes at 5 kHz frequency.

Bibliography

Asghar, A., Juniper, E.J. and Corke, T.C. (2006) On the use of Reynolds number as the scaling parameter for the performance of plasma actuator in a weakly compressible flow. *AIAA Paper* 2006–21.

Bargsten, C.J. and Gibson M.T. (2011) NASA innovation in aeronautics: Select technologies that have shaped modern aviation. *NASA TM* **2011–216987**.

Bhumkar, Y.G. (2012) *High Performance Computing of Bypass Transition*. Ph.D. thesis, Dept. of Aerospace Engineering, IIT Kanpur.

Bushnell, D.M. and Hefner, J.N. (1990) *Viscous Drag Reduction in Boundary Layers. AIAA Prog. in Aeronaut. and Astronaut.*, **123**, USA.

Corke, T.C., Enloe, C.L. and Wilkinson, S.P. (2010) Dielectric barrier discharge plasma actuators for flow control. *Ann. Rev. Fluid Mech.*, **42**, 505–529.

Corke, T.C., He, C. and Patel, M. (2004) Plasma flaps and slats: an application of weakly-ionized plasma actuators. *AIAA Paper* 2004–2127.

Corke, T.C., Jumper, E.J., Post, M.L., Orlov, D.M. and McLaughlin, T.E. (2002) Application of weakly-ionized plasmas as wing flow-control devices. *AIAA Paper* 2002–0350.

Corke, T.C. and Post, M.L. (2005) Overview of plasma flow control: concepts, optimization and applications. *AIAA Paper* 2005–563.

Cristofolini, A., Neretti, G., Roveda, F. and Boghi, C.A. (2011) Experimental and numerical investigation on a DBD actuator for airflow control. *AIAA Paper* 2011–3912.

Dipankar, A. and Sengupta, T.K. (2006) Symmetrized compact scheme for receptivity study of 2D transitional channel flow. *J. Comput. Phys.*, **215**, 245–273.

Dodbele, S.S., Van Dam, C.P., Vijjgen, P.M.H.W. and Holmes, B.J. (1987) Shaping of airplane fuselages for minimum drag. *J. Aircraft,* **24(5)**, 298–304.

Enloe, C.L., McLaughlin, T.E., VanDyken, R.D., Kachner, K.D., Jumper, E.J. and Corke, T.C. (2003) Mechanisms and responses of a single dielectric barrier plasma. *AIAA Paper* 2003–1021.

Fujino, M. and Kawamura, Y. (2003) Wave drag characteristics of an over-the-wing nacelle business-jet configuration. *J. Aircraft,* **40(6)**, 1177–1184.

Fujino, M. Yoshizaki, Y. and Kawamura, Y. (2003) Natural-laminar-flow airfoil development for a lightweight business jet. *J. Aircraft,* **40(4)**, 609–615.

Gaitonde, D.V., Visbal, M.R. and Roy, S. (2005) Control of flow past a wing section with plasma-based body forces. *AIAA Paper* 2005-5302.

Gaitonde, D.V., Visbal, M.R. and Roy, S. (2006) A coupled approach for plasma-based flow control simulations of wing sections. *AIAA Paper* 2006–1205.

Georgy-Falvy, D. (1965) Effect of pressurization on airplane fuselage drag. *J. Aircraft,* **2(6)**, 531–537.

Ghosh, S. (2012) *Plasma actuation for active control of wind turbine power*. M.Sc. thesis, Ecole Polytechnique de Montreal, Canada.

Grundmann, S. and Tropea, C. (2007) Experimental transition delay using glow-discharge plasma actuators. *Experiments Fluids*, **42**, 653–657.

Grundmann, S. and Tropea, C. (2008) Active cancellation of artificially introduced Tollmien-Schlichting waves using plasma actuators. *Experiments Fluids*, **44**, 795–806.

Hanson, R.E., Lavoie, P. and Naguib, A.M. (2010) Effect of plasma actuator excitation for controlling bypass transition in boundary layers. *AIAA Paper* 2010–1091.

Hanson, R.E., Lavoie, P., Naguib, A.M. and Morrison, J.F. (2010) Transient growth instability cancellation by a plasma actuator array. *Experiments Fluids*, **49**, 1339–1348.

Hastings, E.C., Faust, G.K., Mungur, P., Obara, C.J., Dodbele, S.S., Schoenster, J.A. and Jones, M.G. (1987) Status report on a natural laminar-flow nacelle flight experiment. *In Research in Natural Laminar Flow and Laminar-Flow Control. NASA CP* **2487**.

Hastings, E.C., Schoenster, J.A., Obara, C.J. and Dodbele, S.S. (1986) Flight research on NLF nacelles: A progress report. *AIAA Paper* 86-1629.

Hicken, J.E. and Zingg, D.W. (2010) Induced-drag minimization of nonplanar geometries based on the Euler equations. *AIAA J.*, **48**(11), 2564–2575.

Houghton, E.L. and Carpenter, P.W. (1993) *Aerodynamics for Engineering Students*. Fourth Edn, Edward Arnold Publishers Ltd, UK.

Huang, J., Corke, T.C. and Thomas, F.O. (2003) Plasma actuators for separation control of low pressure turbine blades. *AIAA Paper* 2003–1027.

Huang, X., Zhang, X. and Li, Y. (2010) Broadband flow-induced sound control using plasma actuators. *J. Sound and Vibration*, **329**, 2477–2489.

Hultgren, L.S. and Ashpis, D.E. (2003) Demonstration of separation delay with glow-discharge plasma actuators. *AIAA Paper* 2003–1025.

Kotsonis, M., Giepman, R. and Veldhuis, L. (2011) Numerical study on control of Tollmien-Schlichting waves using plasma actuators. *AIAA Paper* 2011–3175.

Kroo, I. (2001) Drag due to lift: Concepts for prediction and reduction. *Ann. Rev. Fluid Mech.*, **33**, 587–617.

Lemire, S. and Vo, H.D. (2011) Reduction of fan and compressor wake defect using plasma actuation for tonal noise reduction. *J. Turbomachinery*, **133(1)**, 011017.

List, J., Byerley, A.R., McLaughlin, T.E. and VanDyken. R.D. (2003) Using a plasma actuator to control laminar separation on a linear cascade turbine blade. *AIAA Paper* 2003–1026.

Lowson, M.V. (1990) Minimum Induced drag for wings with spanwise camber. *J. Aircraft*, **27**(7), 627–631.

Lundry, J.L. and Lissaman, P.B.S. (1968) A numerical solution for the minimum induced drag of non-planar wings. J. Aircraft, **5**(1), 17–21.

Mangler, W. (1938) The lift distribution of wings with end plates. *NACA Tech. Rept.* **856**.

Munk, M. (1921) The minimum induced Drag of Aerofoils. *NACA Tech. Rept.* **121**.

Massines, F., Rabehi, A., Decomps, P., Ben Gadri, R., Segur, P. and Mayoux, C. (1998) Experimental and theoretical study of a glow discharge at atmospheric pressure controlled by dielectric barrier. *J. Applied Physics*, **83**, 41–52.

Morkovin, M.V. (1969) On the many faces of transition. In *Viscous Drag Reduction* (Wells CS, editor). Plenum Press, New York, USA.

Murri, D.G., McGhee, R.J., Jordan, F.L., Davis, P.J. and Viken, J.K. (1987) Wind tunnel results of the low speed NLF(1)-0414F airfoil. *NASA CP* **2487**(3), 673–695.

Nakayama, A, Kreplin, H.-P. and Morgan, H.L. (1990) Experimental investigation of flow field about a multi-element airfoil. *AIAA J.*, **26**, 14–21.

Obara, C.J., Hastings, E.C., Schoenster, J.A., Parrott, T.L. and Holmes, B.J. (1986) Natural laminar flow flight experiments on a turbine engine nacelle fairing. *AIAA Paper* 86–9756.

Orlov, D.M. *Modelling and simulation of single dielectric barrier discharge plasma actuators*. Ph.D. thesis, University of Notre Dame, 2006.

Palmeiro, D. and Lavoie, P. (2011) Comparative analysis on single dielectric barrier discharge plasma actuator models. *7th Int. Symp. on Turbulence and Shear Flow Phenomena*, Ottawa, Canada.

Patel, M.P., Sowle, Z.H., Corke, T.C. and He, C. (2007) Autonomous sensing and control of wing stall using a smart plasma slat. *J. Aircraft*, **44**, 516–527.

Post, M.L. and Corke, T.C. (2004) Separation control using plasma actuators- stationary and oscillatory airfoils. *AIAA Paper* 2004–0841.

Riherd, M and Roy, S. (2011) Numerical investigation of serpentine plasma actuators for separation control at low Reynolds number. *AIAA Paper* 2011–3990.

Rizzetta, D.P. and Visbal, M.R. (2011) Exploration of plasma-based control for low-Reynolds number airfoil/gust interaction. *AIAA Paper* 2011–3252.

Roth, J. (1995) *Industrial Plasma Engineering: Volume 1: Principles*. CRC Press, USA.

Roy, S., Singh, K.P., Haribalan, K., Gaitonde, D.V. and Visbal, M.R. (2006) Effective discharge dynamics for plasma actuators. *AIAA Paper* 2006–374.

Schubauer, G.B. and Spangenberg, W.G. (1958) Forced mixing in boundary layers. *National Bureau Standards Rept.*, **6107**.

Sengupta, T.K. (2004) *Fundamentals of Computational Fluid Dynamics*. Universities Press, Hyderabad, India.

Sengupta, T.K. (2012) *Instabilities of Flows and Transition to Turbulence*. CRC Press, USA.

Sengupta, T.K. and Poinsot, T. (2010) *Instabilities of Flows: With and Without Heat Transfer and Chemical Reaction*. Springer-Wien, New York, USA.

Sengupta, T.K., Rajpoot, M.K. and Bhumkar, Y.G. (2011) Space-time discretizing optimal DRP schemes for flow and wave propagation problems. *Comput. Fluids*, **47**, 144–154.

Sengupta, T.K, Suman, V.K. and Bhumkar, Y.G. (2012) Control of bypass transitional flow past an aerofoil by plasma actuation. *Int. J. Emerging Multidisc. Fluid Sci.*, **3**(2), 117–134

Seraudie, S., Vermeersch, O. and Arnal, D. (2011) DBD plasma actuator effect on a 2D model laminar boundary layer. Transition delay under ionic wind effect. *AIAA Paper* 2011–3515.

Smith, A.M.O. (1975) High lift aerodynamics. *J. Aircraft*, **12**, 501–530.

Stevens, W.A., Goradia, S.H. and Braden, J.A. (1971) Mathematical model for two-dimensional multi-component airfoils in viscous flow. NASA CR **1843**.

Stratford, B.S. (1959) The prediction of separation of the turbulent boundary. *J. Fluid Mech.*, **5**(1), 1–16.

Suzen, Y., Huang, P.G., Jacob, J.D. and Ashpis, D.E. (2005) Numerical simulation of plasma based flow control applications. *AIAA Paper* 2005–4633.

Thomas, F.O., Kozlov, A. and Corke, T.C. (2006) Plasma actuators for bluff body flow control. *AIAA Paper* 2006–2845.

Velkoff, H. and Ketcham, J. (1968) Effect of an electrostatic field on boundary-layer transition. *AIAA*, **6**(7), 1381–1383.

Versailles, P., Gingras-Gosseline, V. and Vo, H.D. (2010) Importance of pressure and temperature on the performance of plasma actuators. *AIAA J.*, **48**(4), 859–862.

Vijgen, P.M.H.W., Dodbele, S.S., Holmes, B.J. and Van Dam, C.P. (1986) Effects of compressibility on design of subsonic natural laminar flow fuselages. *AIAA* 4th *Applied Aerodynamics Conf.*, 86–1825 CP.

Vijgen, P.M.H.W. and Holmes, B.J. (1987) Experimental and numerical analyses of laminar boundary-layer flow stability over an aircraft fuselage forebody. In *Research in Natural Laminar Flow and Laminar-Flow Control. NASA CP* **2487**(3), 861–886.

Visbal, M.R. and Gaitonde, D.V. (2006) Control of vortical flows using simulated plasma actuators. *AIAA Paper* 2006–505.

Wentz, Jr. W.H. and Seetharam, H.C. (1974) Development of a fowler flap system for a high performance general aviation airfoil. NASA CR **2443**.

Wilkinson, S.P. (2003) Investigation of an oscillating surface plasma for turbulent drag reduction. *AIAA Paper* 2003–1023.

Zhang, X., Luo, X. and Chen, P. (2011) Airflow control using plasma actuation and Coanda effect. *AIAA Paper* 2011–3516.

Index

Theoretical and Computational Aerodynamics, First Edition. Tapan K. Sengupta.

Companion Website: www.wiley.com/go/sengupta

G

H

I

J

K

L

T

U

V

W

Y

Z

www.ingramcontent.com/pod-product-compliance
Lightning Source LLC
Chambersburg PA
CBHW080758300125
20834CB00043B/27